Dynamics and Control of Aerospace Systems

Dynamics and Control of Aerospace Systems

Guest Editors

Ti Chen
Dongdong Li
Junjie Kang
Shidong Xu
Shuo Zhang

Basel • Beijing • Wuhan • Barcelona • Belgrade • Novi Sad • Cluj • Manchester

Guest Editors

Ti Chen
College of Aerospace
Engineering
Nanjing University of
Aeronautics and Astronautics
Nanjing
China

Dongdong Li
College of Aerospace
Engineering
Nanjing University of
Aeronautics and Astronautics
Nanjing
China

Junjie Kang
Department of Earth and
Space Science & Engineering
York University
Toronto
Canada

Shidong Xu
College of Aerospace
Engineering
Nanjing University of
Aeronautics and Astronautics
Nanjing
China

Shuo Zhang
Department of Earth and
Space Science & Engineering
York University
Toronto
Canada

Editorial Office
MDPI AG
Grosspeteranlage 5
4052 Basel, Switzerland

This is a reprint of the Special Issue, published open access by the journal *Actuators* (ISSN 2076-0825), freely accessible at: https://www.mdpi.com/journal/actuators/special_issues/L0F7493HT3.

For citation purposes, cite each article independently as indicated on the article page online and as indicated below:

Lastname, A.A.; Lastname, B.B. Article Title. *Journal Name* **Year**, *Volume Number*, Page Range.

ISBN 978-3-7258-3381-8 (Hbk)
ISBN 978-3-7258-3382-5 (PDF)
https://doi.org/10.3390/books978-3-7258-3382-5

© 2025 by the authors. Articles in this book are Open Access and distributed under the Creative Commons Attribution (CC BY) license. The book as a whole is distributed by MDPI under the terms and conditions of the Creative Commons Attribution-NonCommercial-NoDerivs (CC BY-NC-ND) license (https://creativecommons.org/licenses/by-nc-nd/4.0/).

Contents

Preface vii

Ti Chen, Yue Cao, Mingyan Xie, Shihao Ni, Enchang Zhai and Zhengtao Wei
Distributed Passivity-Based Control for Multiple Space Manipulators Holding Flexible Beams
Reprinted from: *Actuators* **2025**, *14*, 20, https://doi.org/10.3390/act14010020 1

Yingjie Chen, Yankai Wang, Ti Chen, Zhengtao Wei and Javad Tayebi
Test Mass Capture Control for Drag-Free Satellite Based on State-Dependent Riccati
Equation Method
Reprinted from: *Actuators* **2024**, *13*, 434, https://doi.org/10.3390/act13110434 24

Alessandro A. Quarta
Thrust Model and Trajectory Design of an Interplanetary CubeSat with a Hybrid
Propulsion System
Reprinted from: *Actuators* **2024**, *13*, 384, https://doi.org/10.3390/act13100384 43

Xiaopeng Gong, Wanchun Chen and Zhongyuan Chen
Cooperative Integrated Guidance and Control for Active Target Protection in
Three-Player Conflict
Reprinted from: *Actuators* **2024**, *13*, 245, https://doi.org/10.3390/act13070245 60

Zhen Jiang, Xi Wang, Jiashuai Liu, Nannan Gu and Wei Liu
Intelligent Reduced-Dimensional Scheme of Model Predictive Control for Aero-Engines
Reprinted from: *Actuators* **2024**, *13*, 140, https://doi.org/10.3390/act13040140 82

Chunyang Kong, Dangjun Zhao and Buge Liang
Vibration Suppression of a Flexible Beam Structure Coupled with Liquid Sloshing via ADP
Control Based on FBG Strain Measurement
Reprinted from: *Actuators* **2023**, *12*, 471, https://doi.org/10.3390/act12120471 101

**Marco Bassetto, Giovanni Mengali, Karim Abu Salem, Giuseppe Palaia
and Alessandro A. Quarta**
A Sliding Mode Control-Based Guidance Law for a Two-Dimensional Orbit Transfer with
Bounded Disturbances
Reprinted from: *Actuators* **2023**, *12*, 444, https://doi.org/10.3390/act12120444 119

Minghao Chen, Qibo Mao, Lihua Peng and Qi Li
Active Vibration Control Using Loudspeaker-Based Inertial Actuator with Integrated
Piezoelectric Sensor
Reprinted from: *Actuators* **2023**, *12*, 390, https://doi.org/10.3390/act12100390 138

Xiaoxiang Ji, Jiao Ren, Jianghong Li and Yafeng Wu
A Linear Iterative Controller for Software Defined Control Systems of Aero-Engines Based
on LMI
Reprinted from: *Actuators* **2023**, *12*, 259, https://doi.org/10.3390/act12070259 161

Yu Bai, Tian Yan, Wenxing Fu, Tong Li and Junhua Huang
Robust Adaptive Composite Learning Integrated Guidance and Control for Skid-to-Turn
Interceptors Subjected to Multiple Uncertainties and Constraints
Reprinted from: *Actuators* **2023**, *12*, 243, https://doi.org/10.3390/act12060243 182

Quanling Zhang, Ningze Tang, Xing Fu, Hao Peng, Cuimei Bo and Cunsong Wang
A Multi-Scale Attention Mechanism Based Domain Adversarial Neural Network Strategy for Bearing Fault Diagnosis
Reprinted from: *Actuators* **2023**, *12*, 188, https://doi.org/10.3390/act12050188 . . . . **204**

Xiaoxiang Ji, Jianghong Li, Jiao Ren and Yafeng Wu
A Decentralized LQR Output Feedback Control for Aero-Engines
Reprinted from: *Actuators* **2023**, *12*, 164, https://doi.org/10.3390/act12040164 . . . . **224**

Chenlu Feng, Weidong Chen, Minqiang Shao and Shihao Ni
Trajectory Tracking and Adaptive Fuzzy Vibration Control of Multilink Space Manipulators with Experimental Validation
Reprinted from: *Actuators* **2023**, *12*, 138, https://doi.org/10.3390/act12040138 . . . . **248**

Wenfeng Zhang, Zhendong Liu, Xiong Liu, Yili Jin, Qixiao Wang and Rong Hong
Model-Based Systems Engineering Approach for the First-Stage Separation System of Launch Vehicle
Reprinted from: *Actuators* **2022**, *11*, 366, https://doi.org/10.3390/act11120366 . . . . **274**

Preface

This Special Issue, entitled Dynamics and Control of Aerospace Systems, presents a curated selection of innovative research articles that address critical challenges and recent advancements in aerospace engineering. Covering a broad spectrum of topics, this reprint explores distributed control for space manipulators, state-dependent control for drag-free satellites, hybrid propulsion systems for CubeSats, vibration suppression in flexible structures, and intelligent control strategies for aero-engines. These studies provide robust theoretical frameworks and experimental validations to enhance the performance, reliability, and safety of aerospace systems.

The motivation for this Special Issue arises from the rapid evolution of aerospace technologies and the growing demand for efficient, reliable, and adaptive control mechanisms in space exploration, satellite systems, and aviation. By integrating advanced control methods, innovative materials, and state-of-the-art sensing technologies, the contributions included in this reprint offer practical solutions and valuable insights into overcoming real-world challenges in aerospace dynamics and control.

Targeted at researchers, engineers, and practitioners in aerospace engineering and related disciplines, this reprint provides a comprehensive overview of the latest developments and future trends in the field. The authors, recognized experts in their respective domains, have made substantial contributions to advancing aerospace technologies, offering both theoretical and practical perspectives.

We extend our heartfelt gratitude to all of the authors for their exceptional contributions, the reviewers for their meticulous evaluations, and the editorial team of Actuators for their unwavering support and guidance throughout the publication process. Their collective efforts have ensured the high quality and impact of this Special Issue.

Ti Chen, Dongdong Li, Junjie Kang, Shidong Xu, and Shuo Zhang
Guest Editors

Article

Distributed Passivity-Based Control for Multiple Space Manipulators Holding Flexible Beams

Ti Chen, Yue Cao, Mingyan Xie Shihao Ni, Enchang Zhai and Zhengtao Wei *

State Key Laboratory of Mechanics and Control for Aerospace Structures, Nanjing University of Aero-Nautics and Astronautics, No. 29 Yudao Street, Nanjing 210016, China; chenti@nuaa.edu.cn (T.C.); cao.yue@nuaa.edu.cn (Y.C.); xmingy@nuaa.edu.cn (M.X.); nshnuaa@nuaa.edu.cn (S.N.); zhainc0312@nuaa.edu.cn (E.Z.)
* Correspondence: weizhengtao@nuaa.edu.cn

Abstract: This paper proposes a distributed passivity-based control scheme for the consensus and vibration suppression of multiple space manipulators holding flexible beams. A space manipulator holding a flexible beam is essentially a rigid–flexible underactuated system. The bending deformation of the flexible beam is discretized by employing the assumed modes method. Based on Lagrange's equations of the second kind, the dynamics model of each manipulator holding a flexible beam is established. By connecting such underactuated systems with the auxiliary Euler–Lagrange systems, a distributed passivity-based controller is designed under undirected communication graphs. To suppress flexible vibration effectively, a distributed controller with the feedback of the velocity of deflection at the free end of the flexible beam is proposed to achieve the manipulator synchronization and vibration suppression simultaneously. The stability of the proposed controller is analyzed with LaSalle's invariance principle. Numerical simulations and experiments are conducted to show the effectiveness of the designed controllers.

Keywords: underactuated systems; passivity-based control; vibration suppression; distributed control; rigid–flexible

1. Introduction

Space manipulators are playing an increasingly significant role in complex space operations. The space manipulators can independently complete on-orbit servicing operations such as target capture, space observation, extravehicular equipment installation, etc. In addition, the space manipulators can also assist astronauts in completing extravehicular operations, reducing the workload of the astronauts. The application of space manipulators greatly improves the efficiency and safety of space missions. Therefore, space manipulators have attracted widespread attention from researchers in all fields.

The coordinated control of multiple networked manipulators is of great significance for space tasks such as load transportation and human–machine collaboration. The manipulator dynamics model established by the Lagrangian method is essentially an Euler–Lagrangian (EL) system. Therefore, the consensus of multiple manipulators can be solved by using the distributed control algorithms of multiple EL systems. In order to achieve the consensus of multi-agent systems, distributed control is usually considered one of the important methods since it does not need global information. Distributed control coordinates the behavior of the group through the information exchange between intelligent agents [1–3]. The construction of the communication topology is a crucial step for distributed control of multi-agent systems. A common approach to describe the communication topology is the graph theory [4]. Communication graphs usually include directed and undirected graphs. In directed graphs, information exchange between agents is unidirectional, and communication ability is limited [5]. To improve the flexibility of the communication topology, Meng et al. [6] introduced switchable directed graphs into the distributed control of multi-agent systems. Different from the directed graph, undirected graphs allow the path between any nodes being undirected, which results in higher fault tolerance [7,8]. Liu et al. modeled the communication network of multiple EL systems as an undirected graph [9]. Furthermore, for the consensus missions of multi-agent systems, there may or may not exist a leader. Hence, the consensus behavior can be divided into leader–follower and leaderless patterns [10,11]. Wang et al. proposed an adaptive controller based on distributed observers to solve the leaderless consensus problem of heterogeneous networked Euler–Lagrange systems [12]. In certain space missions, such as transportation, it is necessary for each agent to follow the leader to reach the designated position and orientation. The leader–follower pattern is more suitable for such space missions. Song et al. designed an optimized leader–follower consensus controller by combining sliding mode control and reinforcement learning for multiple manipulators [13]. Cong et al. addressed the issue of synchronization control for multiple nonlinear manipulation systems by designing a bounded distributed cooperative controller with a virtual leader [14].

In current space missions, manipulators will probably operate some flexible modules. For a manipulator grasping a flexible appendage, such as solar panels, it is necessary to suppress the elastic vibration of the flexible appendage during the movement of the manipulator. Essentially, flexible appendages have infinite degrees of freedom. Also, there are usually no extra actuators on the flexible appendages. Hence, the system made of the manipulator with the flexible appendage is essentially an underactuated EL system. This poses a significant challenge for the controller design. The main control methods for underactuated systems include backstepping [15,16], feedback linearization [17], Passivity-Based Control (PBC), and so on. For example, Yan et al. proposed an adaptive controller based on the backstepping method to address the control problem of a planar underactuated manipulator with two flexible joints [18]. Du and Li addressed the consensus for multiple underactuated EL systems by designing a distributed control law based on the backstepping method using the cascaded normal form [19]. Cheikh et al. designed a control method that combines sliding mode control and partial feedback linearization to address the issue of underactuated system control caused by joint actuator failures in the manipulator [20]. Matous et al. simplified the dynamic behavior of underactuated underwater vehicles using the output feedback linearization method to allow many control strategies to be employed [21]. For highly coupled nonlinear systems, it is difficult to convert the system into a standard cascade normal form [22,23], so the backstepping and feedback linearization methods may not be applicable to highly coupled nonlinear systems. Ortega et al. [24,25] proposed the PBC theory based on the passive characteristics of the system. The PBC does not require altering the system's nonlinearity and structure, thereby decreasing the difficulty of the controller design [26]. Chen et al. addressed the issue of attitude synchronization during the maneuvering of a group of underactuated flexible spacecrafts. Considering the scenario without system damping, they designed a distributed passivity-based controller that relied only on attitude angle feedback [27]. Introducing vibration information of the flexible beam into the control system is an effective way for vibration suppression in such underactuated systems. Mansour et al. addressed the problem of vibration suppression and stability control of a single-joint flexible manipulator. A vibration suppression method based on deformation feedback to adjust torque was designed [28]. Wang et al. proposed a distributed cooperative controller for the vibration problems of multiple underactuated flexible spacecraft. The controller utilizes proportional and differential feedback, as well as interactive feedback between adjacent control units, to suppress vibrations while achieving attitude consensus among multiple spacecraft [29].

As mentioned above, few studies have thoroughly investigated the issue of vibration suppression of flexible payloads during the consensus of space manipulators [30–32], particularly by integrating the feedback of the Velocity of Deflection at the Free End of the Beam (VDFEB) into the passivity-based controller. This paper aims to present distributed passivity-based controllers for networked space manipulators holding flexible appendages. Compared with the studies on distributed control of fully actuated and rigid manipulators in [4,7,12,14], this paper proposed distributed passivity-based controllers for a team of rigid–flexible underactuated systems to realize the consensus manipulation and vibration suppression of the flexible appendages. Moreover, different from the distributed passivity-based controller in [27], the VDFEB is introduced into passivity-based controllers as the feedback signal.

The rest of the paper is organized as follows. Problem formulation is given in Section 2. The distributed passivity-based controllers with or without the feedback of the VDFEB are designed in Section 3. Sections 4 and 5 present the results of numerical simulations and experiments, respectively. Conclusions are drawn in Section 6.

2. Problem Formulation

2.1. Research Objective

This paper will investigate the control problem of a group of fixed-base four degrees-of-freedom manipulators. Each manipulator holds a flexible beam without any actuators. The control purpose is to realize the coordinated motion of multiple manipulators and vibration suppression of the flexible beams. The space mission scenario is illustrated in Figure 1.

Figure 1. Mission concept of cooperative operation of flexible beams using multiple space manipulators.

2.2. Graph Theory

Consider a multi-agent system composed of multiple manipulators. The information exchange among the multiple intelligent agents can be described as a communication topology graph. The graph consists of a set of nodes and edges, denoted as $G = (V, E)$. Nodes represent agents, and edges represent the paths of information flow between adjacent agents. In graph $G = (V, E)$, $V = \{1, 2 \ldots N\}$ represents the set of all nodes, and $E \subseteq V \times V$ represents the set of all edges connecting the nodes. The edge (i, j) represents the information flow from node i to node j, and node i is a neighbor of node j. A graph G is undirected if there is an edge (j, i) for any edge (i, j); otherwise, the graph is directed. The set of neighbors of node i is denoted by $N_i = \{j \in V | (i, j) \in E\}$. The path from node i_l to node i_k is a series of ordered edges (i_l, i_{l+1}), where $l = 1, \ldots, k-1$. If there exists at least one path between any two nodes, the graph G is connected. The adjacency matrix of graph G is defined as $A = \{a_{i,j}\} \in \mathbb{R}^{N \times N}$, where $a_{i,j} \geq 1$ if $(j,i) \in E$; otherwise, $a_{i,j} = 0$. The in-degree of the ith node is denoted as d_i, which is defined as the sum of the elements in the ith row of the adjacency matrix A, i.e., $d_i = \sum_{j=1}^{N} a_{ij}$. The matrix D is then defined as $D = \text{diag}\{d_i\}$, where D is a diagonal matrix with d_i as its diagonal elements. The Laplacian matrix L of the graph G is then defined as $L = D - A$.

Assume that the desired position of the manipulator is determined by a virtual leader, denoted as node 0. Node i is said to be reachable from the leader if there exists a path from the leader to node i. If there exists a path from the leader to every node, the leader is said to be globally reachable. If node i can receive information from the leader without having to go through other nodes, then b_i is set to 1; otherwise, $b_i = 0$. The matrix H is defined as $H = L + B$, where $B = \text{diag}\{b_i\}$.

Lemma 1. *In graph G, if there is at least one follower that can receive information from the leader and the communication among followers is a connected undirected graph, the leader is globally reachable* [33] *and the matrix H is positive definite* [34].

2.3. System Dynamics

The schematic diagram of a 4-DOF manipulator grasping a flexible beam in a three-dimensional space is shown in Figure 2. The beam is slender such that it is considered as an Euler–Bernoulli beam.

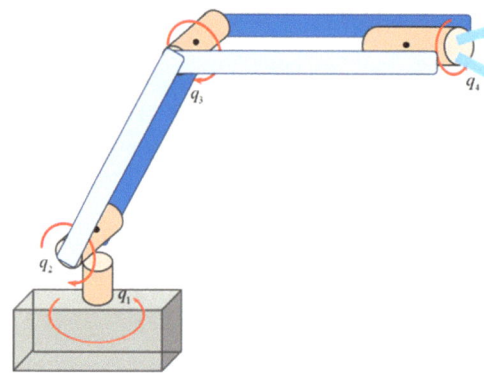

Figure 2. A manipulator with 4 joints grasping a flexible beam.

As illustrated in Figure 3a, the cross-section of the beam is parallel to the ozy plane. To facilitate understanding, the three-dimensional floating frame is projected onto the oxy plane along the height direction of the flexible beam. $w(x,t)$ denotes the bending deformation of the flexible beam, which occurs in the plane oxy, as shown in Figure 3b. $w(x,t)$ is discretized by using the assumed mode method and shown as follows

$$w(x,t) = \sum_{i=1}^{\infty} W_i(x) p_i(t) \quad (1)$$

where $W_i(x)$ is the ith order mode shape and $p_i(t)$ is the corresponding modal coordinate, respectively. Without loss of generality, only the first bending mode of the cantilever beam is considered in this study. Therefore, the first-order mode shape function is

$$W_1(x) = \cosh s_1 x - \cos s_1 x + v_1(\sinh s_1 x - \sin s_1 x) \quad (2)$$

where
$$v_1 = -\frac{\sinh s_1 l - \sin s_1 l}{\cosh s_1 l + \cos s_1 l} \tag{3}$$

and s_1 is the first-order wave number. Then, $s_1 l_b = 1.8571$, and l_b is the length of the flexible beam. To facilitate subsequent writing, the $p_1(t)$ is uniformly replaced with $\eta(t)$.

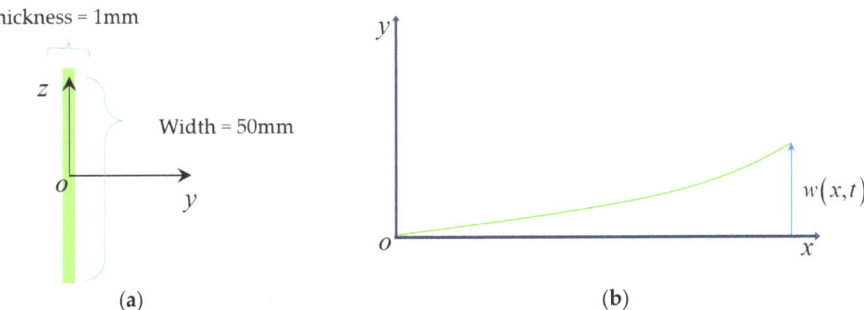

Figure 3. The bending deformation of the flexible beam in the floating frame; (**a**) the cross-section of the beam; (**b**) the deformation plane of the beam.

To give a short expression, (*t*) after a variable will be uniformly omitted in the following context. Hence, the dynamic model of a space manipulator grasping a flexible beam system is given based on the Lagrange equation of the second kind as follows:

$$M(q)\ddot{q} + C(q, \dot{q})\dot{q} + Kq = u \tag{4}$$

where $u = \begin{bmatrix} u_a & 0 \end{bmatrix}^T$ represents the generalized driving torque vector, and $q \in \mathbb{R}^n$ represents the generalized coordinates of the system. u_a is the input torque of the manipulator's joints. The generalized coordinates of the system can be written as $q = \begin{bmatrix} q_a^T & \eta \end{bmatrix}^T$, where $q_a = [q_1\ q_2\ q_3\ q_4]^T \in \mathbb{R}^4$ and $\eta \in \mathbb{R}$. q_1, q_2, q_3 and q_4 represent the generalized coordinates of the manipulator's joints. η represents the modal coordinate of the first bending mode of the flexible beam. $M = \begin{bmatrix} m_{11} & m_{12} \\ m_{21} & m_{22} \end{bmatrix}$ is the mass inertia matrix of the system. $C = \begin{bmatrix} c_{11} & c_{12} \\ c_{21} & c_{22} \end{bmatrix}$ includes the centrifugal and Coriolis matrix of the system. $K = \begin{bmatrix} 0 & 0 \\ 0 & K \end{bmatrix}$ is the stiffness matrix. It is clear that only the manipulator joint angles are controlled directly. The dynamic model can also be written in the following form of partitioned matrix:

$$\begin{bmatrix} m_{11} & m_{12} \\ m_{21} & m_{22} \end{bmatrix} \begin{bmatrix} \ddot{q}_a \\ \ddot{\eta} \end{bmatrix} + \begin{bmatrix} c_{11} & c_{12} \\ c_{21} & c_{22} \end{bmatrix} \begin{bmatrix} \dot{q}_a \\ \dot{\eta} \end{bmatrix} + \begin{bmatrix} 0 & 0 \\ 0 & K \end{bmatrix} \begin{bmatrix} q_a \\ \eta \end{bmatrix} = \begin{bmatrix} u_a \\ 0 \end{bmatrix} \tag{5}$$

Then, the dynamic equations of the networked manipulator holding flexible beam systems can be rewritten according to Equation (4) as

$$M_s(q)\ddot{q}_s + C_s(q, \dot{q})\dot{q}_s + K_s q_s = u_s \tag{6}$$

where $M_s = \text{diag}\{M_i\} \in \mathbb{R}^{5N \times 5N}$, $C_s = \text{diag}\{C_i\} \in \mathbb{R}^{5N \times 5N}$, $K_s = \text{diag}\{K_i\} \in \mathbb{R}^{5N \times 5N}$, $q_s = [q_{a1}^T\ \eta_1 \ldots q_{aN}^T\ \eta_N]^T$, $u_s = [u_1^T \ldots u_N^T]^T$ for $i = 1, \ldots, N$. Note that N represents the number of the manipulator of concern. The subscript i is added to show that the variables are corresponding to the ith agent.

2.4. System Energy Analysis

The definition of passive is briefly introduced in this subsection firstly. Consider the system governed by the following equation

$$\Sigma: \begin{cases} \dot{x} = f(x,u) \\ y = h(x,u) \end{cases} \tag{7}$$

where a causal dynamic operator $\Sigma: u \to y$ [35] is defined. According to [36,37], Equation (7) is passive if there is a continuously differentiable semidefinite function E such that

$$\dot{E} \leq u^T y, \ \forall (x,u) \tag{8}$$

where E is called the storage function. Moreover, it is said to be lossless if $\dot{E}_s = u^T y$.

For the system in Equation (6) with output $y_s = \dot{q}_{as}$, where $q_{as} = [q_{a1}^T \ldots q_{aN}^T]^T$, the storage function is defined as

$$E_s = \frac{1}{2} \dot{q}_s^T M_s(q) \dot{q}_s + \frac{1}{2} q_s^T K_s q_s \tag{9}$$

where the first and second terms in E_s are the kinetic energy and potential energy for the networked manipulators carrying a flexible beam, respectively. The derivative of E_s can be expressed as

$$\begin{aligned} \dot{E}_s &= \dot{q}_s^T M_s \ddot{q}_s + \frac{1}{2} \dot{q}_s^T \dot{M}_s \dot{q}_s + q_s^T K_s \dot{q}_s \\ &= \frac{1}{2} \dot{q}_s^T \left(2u_s - 2C_s \dot{q}_s - 2K_s q_s + \dot{M}_s \dot{q}_s \right) + q_s^T K_s \dot{q}_s \end{aligned} \tag{10}$$

The following equation can be obtained for the EL system:

$$\dot{q}_s^T (\dot{M}_s - 2C_s) \dot{q}_s = 0 \tag{11}$$

Hence, Equation (10) can be rewritten as

$$\dot{E}_s = \dot{q}_s^T u_s \tag{12}$$

Since $u = \begin{bmatrix} u_a & 0 \end{bmatrix}^T$, $\dot{E}_s$ can be reformulated as

$$\dot{E}_s = \dot{q}_{as}^T u_{as} \tag{13}$$

where $u_{as} = [u_{a1}^T \ldots u_{aN}^T]^T$. Hence, the system in Equation (9) is lossless.

3. Controller Design

In order to achieve the leader–follower consensus of multiple manipulators holding flexible beam systems, the focus of the passivity-based controller is to construct an auxiliary controller dynamic equation with isomorphic structure and form a closed-loop system with the system dynamic equation. In order to further suppress flexible vibrations, the feedback of the VDFEB is added to the controller as an integral form. This enables the manipulator to suppress vibration actively without adding an actuator.

3.1. Control Objective

The controller design is intended to drive N manipulators that hold flexible beams to track a static virtual leader while effectively suppressing flexible vibration under the following Assumption 1. The desired generalized coordinate vector is represented by $q_d = [q_{ad}^T \ 0]^T$, where q_{ad} is the desired value of q_{ai}. The leader–follower synchronization and vibration suppression problems are said to be achieved if $\lim_{t\to\infty}\|q_i\| = q_d$ and $\lim_{t\to\infty}\|\dot{q}_i\| = 0$ hold for $i = 1, \ldots, N$.

Assumption 1. *The communication topology is undirected and connected. At least one follower can receive information from the leader.*

If Assumption 1 is satisfied, it can be clearly concluded from the description in Section 2.1 that the matrix H is positive definite.

3.2. Passivity-Based Control Method

The objective of PBC methods is to render the closed-loop system passive. Essentially, as illustrated in Figure 4, a PBC is realized by interconnecting another auxiliary Euler–Lagrange system with the controlled plant. The resulting controller can be referred to as an Euler–Lagrange controller. This approach allows for the incorporation of additional energy and damping into the closed-loop system.

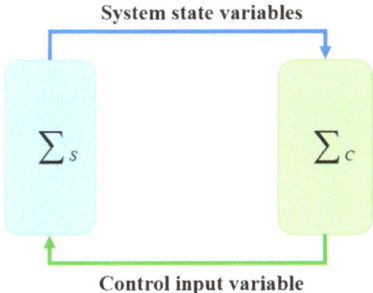

Figure 4. Information exchange mechanisms between the controller and the control plant.

3.3. Design of Distributed Passivity-Based Controller

The purpose of this subsection is to utilize PBC methods to design a distributed controller for achieving synchronization of multiple manipulators gripping a flexible beam. Before proceeding, a necessary assumption is stated as follows.

Assumption 2. *Only the joint angles of the manipulators are measurable.*

In synchronization missions, it is crucial to ensure real-time communication and complete the task in the shortest possible time with minimal information. Therefore, studying the control problem based on joint angle feedback only is highly necessary. According to the requirements of Assumption 2, only q_{as} will be used as feedback variables in the controller design in this subsection.

By introducing the virtual generalized coordinates corresponding to the generalized joint angle coordinates of the manipulator [24], the dynamic equation of the controller can be formulated as follows:

$$M_c \ddot{\theta} + C_c \dot{\theta} + K_c(\theta - q_s) + (H \otimes I_4)(\theta - q_d) = 0 \tag{14}$$

where $M_c = \text{diag}\{M_{ci}\} \in \mathbb{R}^{4N \times 4N}$, $C_c = \text{diag}\{C_{ci}\} \in \mathbb{R}^{4N \times 4N}$ and $K_c = \text{diag}\{K_{ci}\} \in \mathbb{R}^{4N \times 4N}$ are constant positive definite matrices. $M_{ci} \in \mathbb{R}^{4 \times 4}$, $C_{ci} \in \mathbb{R}^{4 \times 4}$ and $K_{ci} \in \mathbb{R}^{4 \times 4}$ are also constant positive definite matrices for $i = 1, \ldots, N$. $\theta \in \mathbb{R}^{4N}$ is a generalized coordinate vector of the controller dynamics. Thus, the controller dynamics corresponding to the ith manipulator are as follows:

$$M_{Ci}\ddot{\theta}_i + C_{Ci}\dot{\theta}_i + K_{Ci}(\theta_i - q_i) + \sum_{j=1}^{N} a_{ij}(\theta_i - \theta_j) + b_i(\theta_i - q_{ad}) = 0 \tag{15}$$

It is evident that, since the controller dynamics correspond to the ith manipulator depends on its own information and that of its neighbors, the controller is distributed. The total energy associated with the controller dynamics can be formulated as

$$E_c = E_{cT}(\dot{\boldsymbol{\theta}}) + E_{cK}(\boldsymbol{\theta}, \boldsymbol{q}_s) \tag{16}$$

where

$$E_{cT}(\dot{\boldsymbol{\theta}}) = \frac{1}{2}\dot{\boldsymbol{\theta}}^T M_c \dot{\boldsymbol{\theta}} \tag{17}$$

and

$$E_{cK}(\boldsymbol{\theta}, \boldsymbol{q}) = \frac{1}{2}(\boldsymbol{\theta} - \boldsymbol{q}_{as})^T K_c (\boldsymbol{\theta} - \boldsymbol{q}_{as}) + \frac{1}{2}(\boldsymbol{\theta} - \boldsymbol{q}_{ad})^T (H \otimes I_4)(\boldsymbol{\theta} - \boldsymbol{q}_{ad}) \tag{18}$$

where K_c is a positive definite matrix to be defined in the following controller.

According to Assumption 2, since the joint angles of the manipulator are measurable, the controller can be designed as follows:

$$u_{as} = K_c(\boldsymbol{\theta} - \boldsymbol{q}_{as}) \tag{19}$$

For the leader–follower consensus under an undirected communication graph, only the controller's generalized coordinates are exchanged among the followers by using the distributed auxiliary Euler–Lagrange dynamics.

Theorem 1. *For the networked manipulators gasping a flexible beam governed by Equation (6), the controller in Equation (19) based on the controller dynamics in Equation (14) can drive the group of manipulators to track the leader and suppress the flexible vibration asymptotically.*

Proof. The following total energy of the system is selected as the Lyapunov function candidate.

$$V_1 = E_s + E_c \tag{20}$$

where E_S and E_C are defined by Equations (9) and (16), respectively. The time derivative of V_1 is

$$\begin{aligned}
\dot{V}_1 &= \dot{\boldsymbol{q}}_{as}^T u_{as} + \dot{E}_c \\
&= \dot{\boldsymbol{q}}_{as}^T K_c(\boldsymbol{\theta} - \boldsymbol{q}_{as}) - \dot{\boldsymbol{\theta}}^T C_c \dot{\boldsymbol{\theta}} - \dot{\boldsymbol{\theta}}^T K_c(\boldsymbol{\theta} - \boldsymbol{q}_{as}) - \dot{\boldsymbol{\theta}}^T (H \otimes I_4)(\boldsymbol{\theta} - \boldsymbol{q}_{ad}) \\
&\quad + (\dot{\boldsymbol{\theta}} - \dot{\boldsymbol{q}}_{as})^T K_c(\boldsymbol{\theta} - \boldsymbol{q}_{as}) + \dot{\boldsymbol{\theta}}^T (H \otimes I_4)(\boldsymbol{\theta} - \boldsymbol{q}_{ad}) \\
&= -\dot{\boldsymbol{\theta}}^T C_c \dot{\boldsymbol{\theta}} \\
&\leq 0
\end{aligned} \tag{21}$$

According to LaSalle's invariance principle, the system will asymptotically converge to the largest invariant set $B = \{\boldsymbol{q}_{as}, \dot{\boldsymbol{q}}_{as}, \boldsymbol{\eta}_s, \dot{\boldsymbol{\eta}}_s, \boldsymbol{\theta}, \dot{\boldsymbol{\theta}} | \dot{V}_1 = 0\}$. Given that C_c is positive definite, the condition $\dot{V}_1 = 0$ implies that $\dot{\boldsymbol{\theta}} = 0$. Under the assumption that $\dot{\boldsymbol{\theta}} \equiv 0$, $\boldsymbol{\theta}$ remains constant; therefore, the controller dynamics described in Equation (14) can be reformulated as

$$K_c(\boldsymbol{\theta} - \boldsymbol{q}_{as}) + (H \otimes I_4)(\boldsymbol{\theta} - \boldsymbol{q}_d) = 0 \tag{22}$$

It can thus be inferred that $\boldsymbol{q}_{as}$ remains invariant. Given that both K_c and $(H \otimes I_4)$ are constant positive definite matrices, it follows that $\dot{\boldsymbol{q}}_{as} = \ddot{\boldsymbol{q}}_{as} = 0$. Consequently, the dynamic equation of the rigid–flexible coupled manipulator, as expressed in Equation (5), can be reformulated as

$$\begin{cases} m_{12}\ddot{\eta}_i + c_{12}\dot{\eta}_i = \text{constant} \\ m_{22}\ddot{\eta}_i + c_{22}\dot{\eta}_i + K_i\eta_i = 0 \end{cases} (i = 1, 2, \cdots N) \tag{23}$$

From the dynamic Equation (5) of the manipulator, there exists at least one row of $[m_{12}\ c_{12}]$ independent of $[m_{22}\ c_{22}]$. Therefore, the aforementioned Equation (23) has a unique solution when $\eta_i \equiv 0$ and $m_{12}\ddot{\eta}_i + c_{12}\dot{\eta}_i = 0$, which holds when $\theta_i = q_{ai} = q_{ad}$. Hence, the unique solution to Equation (23) is $\eta_i \equiv 0$. Consequently, when $\dot{V}_1 = 0$, it follows that $\eta_s \equiv 0$ and $\theta = q_{as} = q_{ad}$. Therefore, according to LaSalle's invariance principle, the controller can achieve distributed consensus control and vibration suppression. Therefore, this system is asymptotically stable. □

Remark 1. *The closed-loop system previously mentioned can be reformulated as follows*

$$\begin{cases} M_s(q)\ddot{q}_s + C_s(q,\dot{q})\dot{q}_s + K_s q_s = \begin{bmatrix} u_{as} \\ 0 \end{bmatrix} \\ M_c\ddot{\theta} + C_c\dot{\theta} + u_{as} + (H \otimes I_4)(\theta - q_d) = 0 \end{cases} \tag{24}$$

Then, the energy function of this closed-loop system V_c can be formed as

$$V_c = \frac{1}{2}\dot{q}_s^T M_s(q)\dot{q}_s + \frac{1}{2}q_s^T K_s q_s + \frac{1}{2}\dot{\theta}^T M_c \dot{\theta} + \frac{1}{2}(\theta - q_{ad})^T (H \otimes I_4)(\theta - q_{ad}) \tag{25}$$

The time derivative of V_c is

$$\begin{aligned} \dot{V}_c &= \dot{q}_{as}^T K_c(\theta - q_{as}) - \dot{\theta}^T C_c \dot{\theta} - \dot{\theta}^T K_c(\theta - q_{as}) \\ &= \left(\dot{q}_{as} - \dot{\theta}\right)^T u_{as} - \dot{\theta}^T C_c \dot{\theta} \\ &\leq \left(\dot{q}_{as} - \dot{\theta}\right)^T u_{as} \end{aligned} \tag{26}$$

where $\dot{q}_{as} - \dot{\theta}$ and u_{as} are the new output and input of this system, respectively. Hence, the closed-loop system considered is passive.

3.4. Distributed PBC with the Feedback of the VDFEB

This section will systematically discuss the controller design with feedback of the VDFEB. As mentioned in Section 2.3, only the first-order bending mode of the flexible beam is considered. Since $W_1(x)$ depends only on x, $W_1(l)$ is a constant. Define the rate of change in the deflection at the end of the flexible beam as

$$g(t) = \dot{w}(l,t) = A\dot{\eta} \tag{27}$$

where $A = W_1(l)$.

Assumption 3. *The angles and angle velocities of manipulator joints, and the velocities of the deflection at the end of the flexible beam, are all measurable.*

According to the above assumption, $\dot{q}_a$, $g(t)$ and q_a can be used to design the following controller. Inspired by [28], the new controller can be reformulated as follows:

$$u_{ai} = K_{ci}\begin{bmatrix} \theta_{i,1} - q_{ai,1} \\ \theta_{i,2} - q_{ai,2} \\ \theta_{i,3} - q_{ai,3} \\ \theta_{i,4} - q_{ai,4} \end{bmatrix} + K_{ci}\begin{bmatrix} \cos^2 q_{ai,4}g(t)\tanh\left(-k\dot{q}_{ai,1}\right)\int_0^t \cos^2 q_{ai,4}\tanh\left(-k\dot{q}_{ai,1}\right)g(\tau)d\tau \\ 0 \\ \sin^2 q_{ai,4}g(t)\tanh\left(-k\dot{q}_{ai,3}\right)\int_0^t \sin^2 q_{ai,4}\tanh\left(-k\dot{q}_{ai,3}\right)g(\tau)d\tau \\ 0 \end{bmatrix} \quad (28)$$

In this context, $q_{ai,1}\cdots q_{ai,4}$ represent the first to fourth joint angles of the ith manipulator, and $\theta_{i,1}\cdots\theta_{i,4}$ are the four elements of $\theta_{i,1}$. The first term in Equation (28) is identical to that in Equation (19), while the second term represents the fine-tuning of the torque of the first and third joints of the manipulator to suppress the vibration of the flexible beam. Such torques applied to the first and third joints to suppress vibrations depend on the angle of the fourth joint. It should be noted that such a control torque for vibration suppression is specifically designed based on the unique configuration of the space manipulators in this paper. The value of k is chosen as a large positive constant. This controller is designed to further suppress the vibration of flexible beams. According to the design of this controller, the vibration information of each beam is only feedback to its own controller, and there is no need to exchange the system states between followers.

Theorem 2. *For the networked manipulators gasping a flexible beam governed by Equation (6), with the controller dynamics in Equation (14), the controller in Equation (28) can drive the group of manipulators to track the leader and suppress the flexible vibration asymptotically.*

Proof. Define the following vectors:

$$\psi_i = \begin{bmatrix} \int_0^t \cos^2 q_{ai,4}\tanh\left(-k\dot{q}_{ai,1}\right)\left(-\dot{q}_{ai,1}\right)g(\tau)d\tau \\ 0 \\ \int_0^t \sin^2 q_{ai,4}\tanh\left(-k\dot{q}_{ai,3}\right)\left(-\dot{q}_{ai,3}\right)g(\tau)d\tau \\ 0 \end{bmatrix} \quad (29)$$

and

$$\delta_i = \begin{bmatrix} \cos^2 q_{ai,4}g(t)\tanh\left(-k\dot{q}_{ai,1}\right)\int_0^t \cos^2 q_{ai,4}\tanh\left(-k\dot{q}_{ai,1}\right)g(\tau)d\tau \\ 0 \\ \sin^2 q_{ai,4}g(t)\tanh\left(-k\dot{q}_{ai,3}\right)\int_0^t \sin^2 q_{ai,4}\tanh\left(-k\dot{q}_{ai,3}\right)g(\tau)d\tau \\ 0 \end{bmatrix} \quad (30)$$

Consider the following Lyapunov function:

$$V_2 = E_s + E_c + \frac{1}{2}\psi_s^T K_c \psi_s \quad (31)$$

where $\psi_s = \left[\psi_1^T, \ldots, \psi_N^T\right]^T$.

The time derivative of V_2 is

$$\begin{aligned}\dot{V}_2 &= \dot{q}_{as}^T u_{as} + \dot{E}_c + \psi_s^T \dot{\psi}_s \\ &= \dot{q}_{as}^T K_c(\theta - q_{as}) + \dot{q}_{as}^T \delta_s - \dot{\theta}^T C\dot{\theta} - \dot{\theta}^T K_c(\theta - q_{as}) - \dot{\theta}^T (H \otimes I_4)(\theta - q_{ad}) \\ &\quad + \left(\dot{\theta} - \dot{q}_{as}\right)^T K_c(\theta - q_{as}) + \dot{\theta}^T (H \otimes I_4)(\theta - q_{ad}) + \psi_s^T K_c \dot{\psi}_s \end{aligned} \quad (32)$$

where $\delta_s = \left[\delta_1^T, \ldots, \delta_N^T\right]^T$. One can obtain

$$\begin{aligned}
\dot{q}_{as}{}^T \delta_s &= \sum_{i=1}^N \dot{q}_{ai}{}^T \delta_i \\
&= \sum_{i=1}^N \begin{aligned}&\dot{q}_{ai,1}\cos^2 q_{ai,4} g(t)\tanh\left(-k\dot{q}_{ai,1}\right)\int_0^t \cos^2 q_{ai,4}\tanh\left(-k\dot{q}_{ai,1}\right) g(\tau)d\tau \\ &+\dot{q}_{ai,3}\sin^2 q_{ai,4} g(t)\tanh\left(-k\dot{q}_{ai,3}\right)\int_0^t \sin^2 q_{ai,4}\tanh\left(-k\dot{q}_{ai,3}\right)\left(-\dot{q}_{ai,3}\right)g(\tau)d\tau\end{aligned}
\end{aligned} \quad (33)$$

and

$$\begin{aligned}
\boldsymbol{\psi}_s{}^T \dot{\boldsymbol{\psi}}_s &= \sum_{i=1}^N \boldsymbol{\psi}_i{}^T \dot{\boldsymbol{\psi}}_i \\
&= \sum_{i=1}^N \begin{aligned}&\left(-\dot{q}_{ai,1}\right)\cos^2 q_{ai,4}\tanh\left(-k\dot{q}_{ai,1}\right) g(t)\int_0^t \cos^2 q_{ai,4}\tanh\left(-k\dot{q}_{ai,1}\right)\left(-\dot{q}_{ai,1}\right) g(\tau)d\tau \\ &+\left(-\dot{q}_{ai,3}\right)\sin^2 q_{ai,4}\tanh\left(-k\dot{q}_{ai,3}\right) g(t)\int_0^t \sin^2 q_{ai,4}\tanh\left(-k\dot{q}_{ai,3}\right)\left(-\dot{q}_{ai,3}\right) g(\tau)d\tau\end{aligned}
\end{aligned} \quad (34)$$

Based on Equations (33) and (34), it can be derived that

$$\dot{q}_{as}{}^T K_c \delta_s = -\boldsymbol{\psi}_s{}^T K_c \dot{\boldsymbol{\psi}}_s \quad (35)$$

Utilizing Equation (35), Equation (32) can be reformulated as

$$\begin{aligned}
\dot{V}_2 &= \dot{q}_{as}{}^T u_{as} + \dot{E}_c + \boldsymbol{\psi}_s{}^T \dot{\boldsymbol{\psi}}_s \\
&= \dot{q}_{as}{}^T K_c(\boldsymbol{\theta}-q_{as}) + \dot{q}_{as}{}^T \delta_{as} - \dot{\boldsymbol{\theta}}^T C\dot{\boldsymbol{\theta}} - \dot{\boldsymbol{\theta}}^T K_c(\boldsymbol{\theta}-q_{as}) - \dot{\boldsymbol{\theta}}^T (H\otimes I_4)(\boldsymbol{\theta}-q_{ad}) \\
&\quad + \left(\dot{\boldsymbol{\theta}}-\dot{q}_{as}\right)^T K_c(\boldsymbol{\theta}-q_{as}) + \dot{\boldsymbol{\theta}}^T(H\otimes I_4)(\boldsymbol{\theta}-q_{ad}) + \boldsymbol{\psi}_s{}^T\dot{\boldsymbol{\psi}}_s \\
&= -\dot{\boldsymbol{\theta}}^T C_c \dot{\boldsymbol{\theta}} \\
&\leq 0
\end{aligned} \quad (36)$$

Similarly to the Proof of Theorem 1, it can be concluded that $\dot{\boldsymbol{\theta}} \equiv 0$ implies $\boldsymbol{\eta}_s \equiv 0$ and $\boldsymbol{\theta} = q_{as} = q_{ad}$. From LaSalle's invariance principle, $\boldsymbol{\theta} \to q_{ad}$, $q_{as} \to q_{ad}$, $\dot{q}_{as} \to \boldsymbol{0}$ and $\dot{\boldsymbol{\eta}}_s \to \boldsymbol{0}$ as time goes to infinity. □

Remark 2. *Similar to Remark 1, the closed-loop system governed by Equation (28) is passive.*

Remark 3. *As shown in the proof of Theorems 1 and 2, the closed-loop system will converge to the largest invariant set $B = \left\{q_{as}, \dot{q}_{as}, \eta_s, \dot{\eta}_s, \boldsymbol{\theta}, \dot{\boldsymbol{\theta}} \mid \dot{\boldsymbol{\theta}} \equiv 0\right\}$, which corresponds to the desired control state. Hence, the proposed control method also works for other underactuated EL systems satisfying such a property that the underactuated variable will be equal to the desired value once the actuated variable is kept constant.*

4. Numerical Simulations

With the distributed system governed by the dynamics equation in Equation (4) as the control platform, some simulations will be introduced to validate the effectiveness of the control scheme proposed in Section 3.4. In this section, the distributed system consists of four space manipulators grasping flexible beams. The undirected communication graph is presented in Figure 5 and it shows that only manipulator numbered 1 can obtain the desired position information from the virtual leader. The parameters of the manipulator and the flexible beam are shown in Table 1. I^1 to I^4 represent the inertia matrices of the various parts of the manipulator. m_i and l_i represent the mass and size of each structure of the manipulator [38], respectively. l_b, EI, ρ and A represent the length of the flexible beam, flexural rigidity, density, and cross-sectional area, respectively.

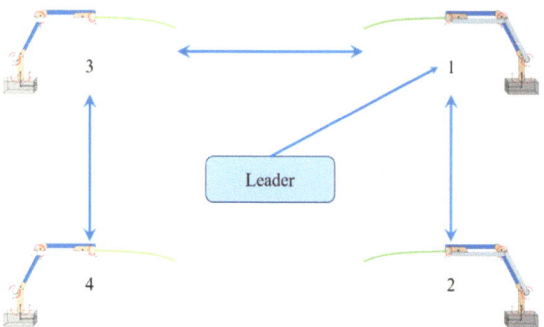

Figure 5. The undirected communication graph for the distributed system.

Table 1. System parameters.

Parameter	Value	Unit
I^1	diag$\{1.5, 1.5, 10\} \times 10^{-3}$	kg·m^2
I^2	diag$\{0.2, 10, 10\} \times 10^{-3}$	kg·m^2
I^3	diag$\{2, 2, 0.3\} \times 10^{-3}$	kg·m^2
I^4	diag$\{1, 1, 0.6\} \times 10^{-3}$	kg·m^2
$m_i (i = 1, 2, 3, 4)$	0.79, 0.46, 0.27, 0.26	kg
$L_i (i = 1, 2, 3)$	0.14, 0.35, 0.4	m
l_b	1	m
EI	2.917	N·m^2
ρ	2700	kg/m^3
A	5×10^{-5}	m^2

To show the effectiveness of the feedback of the VDFEB in the proposed controller, the PBC in Equation (19) together with the proposed controller in Equation (28) are compared with each other. The initial angle vectors of the space manipulators are set as $q_{a1}^* = \begin{bmatrix} -\frac{\pi}{2} & -\frac{\pi}{8} & \frac{\pi}{8} & 0 \end{bmatrix}^T$, $q_{a2}^* = \begin{bmatrix} \frac{\pi}{2} & -\frac{\pi}{8} & \frac{\pi}{8} & 0 \end{bmatrix}^T$, $q_{a3}^* = \begin{bmatrix} \frac{\pi}{2} & -\frac{\pi}{8} & \frac{\pi}{8} & \frac{\pi}{8} \end{bmatrix}^T$ and $q_{a4}^* = \begin{bmatrix} -\frac{\pi}{2} & -\frac{\pi}{8} & \frac{\pi}{8} & \frac{\pi}{8} \end{bmatrix}^T$, respectively. There is no deformation in the flexible beam at the initial time, i.e., both the values the η_i and $\dot{\eta}_i$ for $i = 1 \ldots N$ are set as 0 initially. The desired joint angle vector is $q_{ad} = \begin{bmatrix} 0 & 0 & 0 & 0 \end{bmatrix}^T$. The parameters in controller dynamics in Equation (15) are chosen as $M_{ci} = I_4$, $C_{ci} = 50 I_4$ and $K_{ci} = 100 I_4$. Note that M_{ci}, C_{ci} and K_{ci} remain consistent in the simulations under these two control laws. Furthermore, k is chosen as 1×10^4 in the designed controller in Equation (28).

Figures 6 and 7 display the responses of the joint angles of the manipulators under the control laws in Equations (19) and (28), respectively. As we can see, all the joint angles of the manipulators converge to the desired positions at about 300 s. It is clear that both controllers can effectively drive multiple manipulators to achieve the consensus. According to Equations (19) and (28), the control input will continually drive the joint angle q_{as} to approach the virtual generalized coordinate θ. The time histories of the tracking errors of the joint angles under two control laws are shown in Figures 8 and 9. It can be found that the tracking errors of the joint angles are all within 0.001 rad and converge to zero ultimately. Furthermore, in order to compare the differences between the tracking performance of the joint angles under these two control schemes, the following function is employed.

$$A_{i,j}(t) = \int_0^t |e_{i,j}(\tau)| d\tau \tag{37}$$

where $e_{i,j} = \theta_{i,j} - q_{ai,j}$ represents the tracking error between the *j*th joint angle of the *i*th manipulator and the virtual generalized coordinate. Essentially, Equation (37) represents the area enclosed by the absolute value of the function $e_{i,j}(\tau)$ and the time axis. According to Equation (37), the calculation results in Figures 8 and 9 are presented in Figure 10. As shown by the red bar charts, the areas corresponding to the time responses of the joint angles under the PBC with the feedback of the VDFEB are smaller than those under the controller in the absence of the feedback of the VDFEB. Note that there exists little difference for the fourth joint angles between the areas under these two controllers. The reason for such little difference can be found in Equation (28). As we can see, the feedback of the VDFEB is not employed in the control signal of the fourth joint. Therefore, it can be concluded that the PBC with the feedback of the VDFEB presents greater tracking performance.

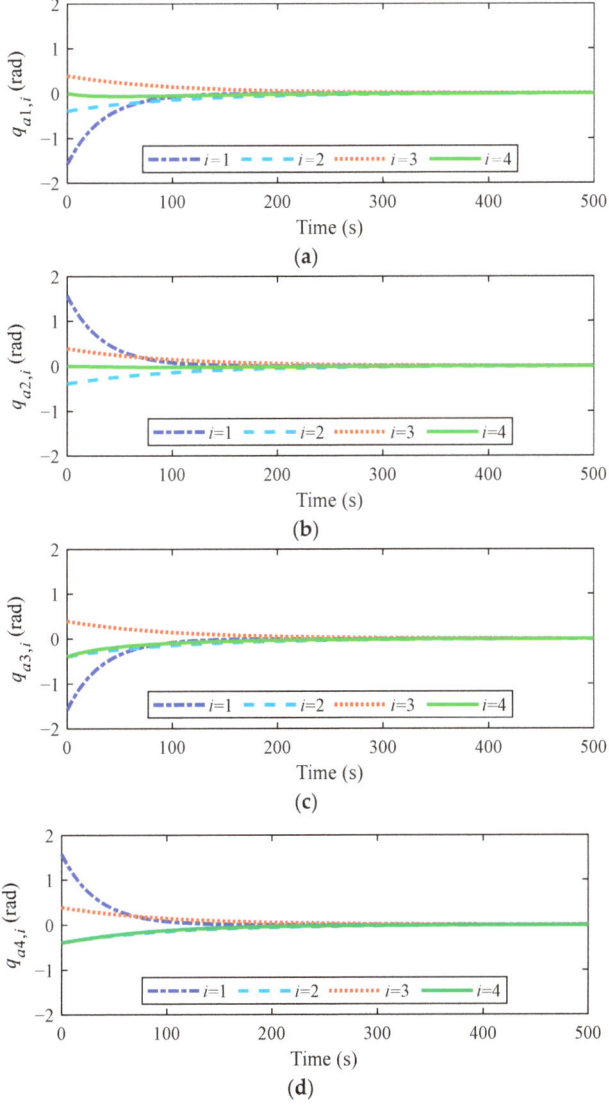

Figure 6. The simulation results of the manipulators under PBC. (**a**) Joint angles of manipulator 1, (**b**) Joint angles of manipulator 2. (**c**) Joint angles of manipulator 3. (**d**) Joint angles of manipulator 4.

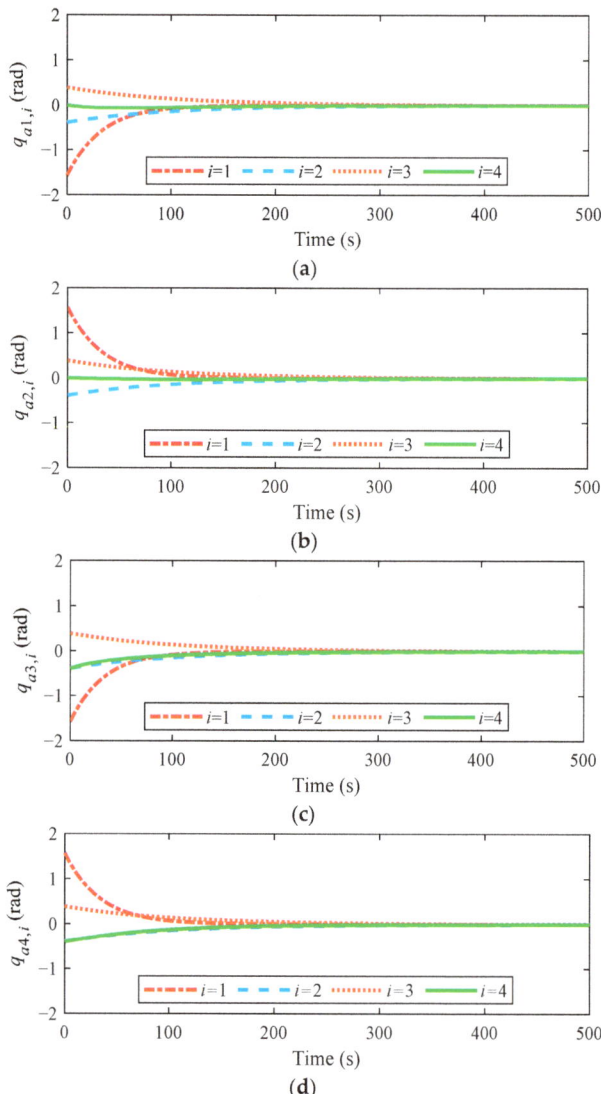

Figure 7. The simulation results of manipulators under PBC with the feedback of the VDFEB. (**a**) Joint angles of manipulator 1. (**b**) Joint angles of manipulator 2. (**c**) Joint angles of manipulator 3. (**d**) Joint angles of manipulator 4.

To further analyze the effect of the feedback of the VDFEB in the proposed controller [39], the comparison of the first bending mode of the flexible beam under the control laws in Equations (19) and (28) is conducted. Figure 11 shows the comparison results, and it can be seen intuitively that the value of the modal coordinates under the control law of Equation (28) is smaller than that of Equation (19). Similarly, to highlight the effectiveness of vibration suppression under the feedback of the VDFEB, the performance function of vibration suppression is defined as follows.

$$B_i(t) = \int_0^t |\eta_i(\tau)| d\tau \tag{38}$$

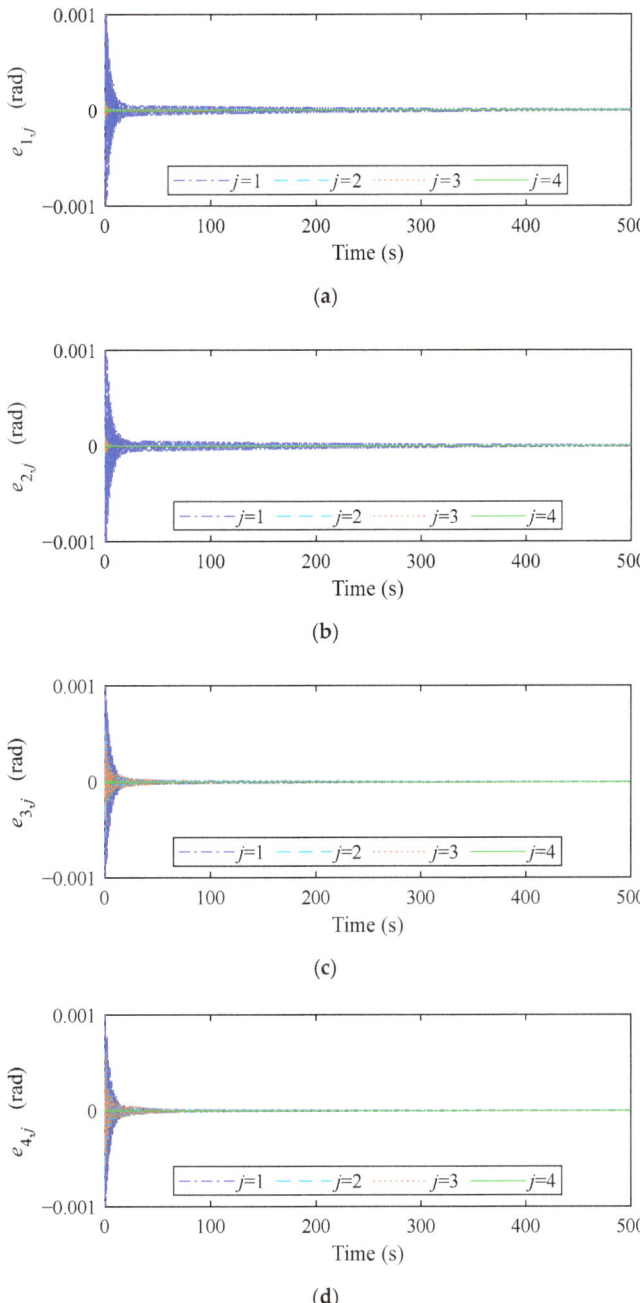

Figure 8. Tracking error of the joint angles of manipulators under PBC. (**a**) Tracking error of the joint angles of manipulator 1. (**b**) Tracking error of the joint angles of manipulator 2. (**c**) Tracking error of the joint angles of manipulator 3. (**d**) Tracking error of the joint angles of manipulator 4.

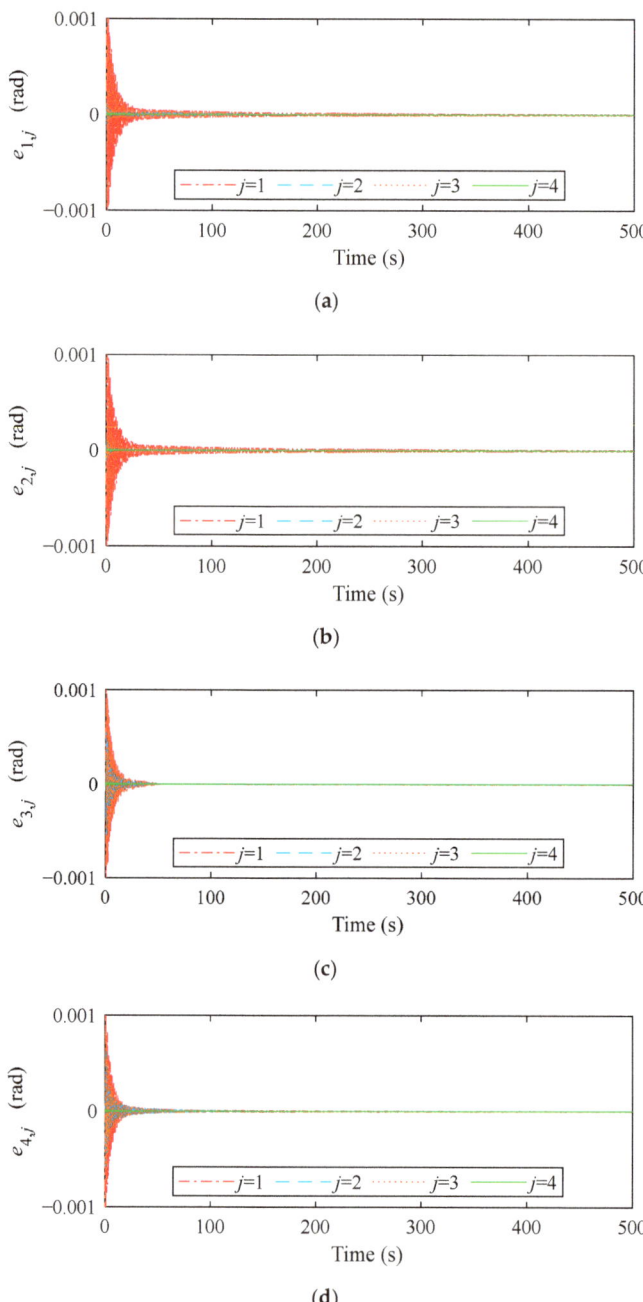

Figure 9. Tracking error of the joint angles of manipulators under PBC with the feedback of the VDFEB. (**a**) Tracking error of the joint angles of manipulator 1. (**b**) Tracking error of the joint angles of manipulator 2. (**c**) Tracking error of the joint angles of manipulator 3. (**d**) Tracking error of the joint angles of manipulator 4.

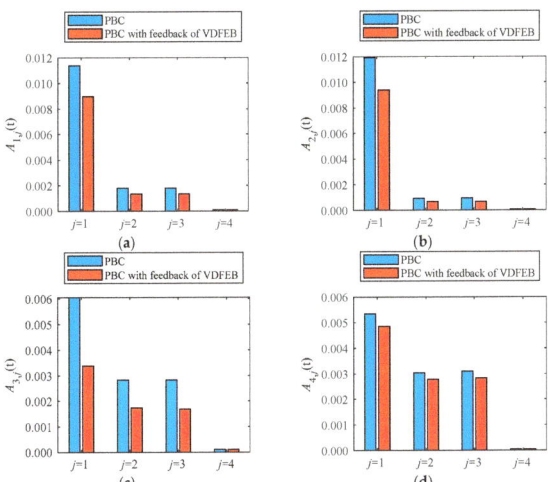

Figure 10. The tracking performance under PBC and PBC with the feedback of the VDFEB. (**a**) The comparison for manipulator 1. (**b**) The comparison for manipulator 2. (**c**) The comparison for manipulator 3. (**d**) The comparison for manipulator 4.

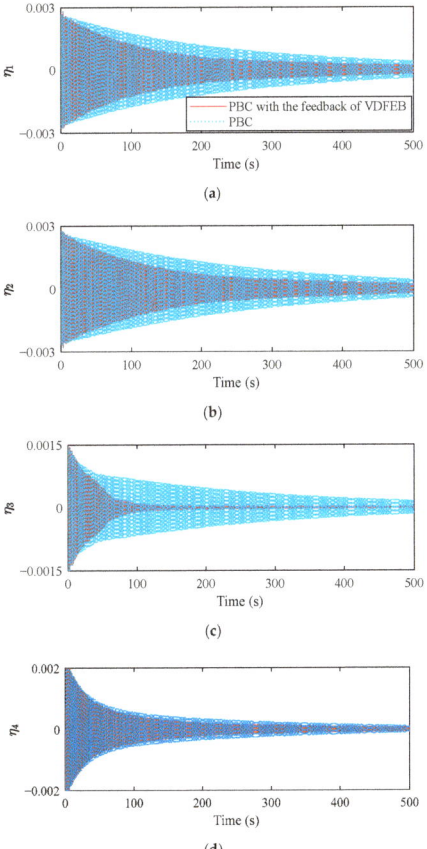

Figure 11. The modal coordinates of the flexible beams. (**a**) Modal coordinates of beam 1. (**b**) Modal coordinates of beam 2. (**c**) Modal coordinates of beam 3. (**d**) Modal coordinates of beam 4.

The calculation results based on Equation (38) are presented in Figure 12 and the vertical axis represents the area enclosed by the absolute value of the modal coordinate and the time-axis, which implies that the degree of vibration suppression can be quantified reasonably. As shown in Figure 12, it is clear that the control method based on the feedback of the VDFEB has an apparent effect on suppressing the vibration of the flexible beam. Hence, it is adopted that the distributed system under the designed controller with the feedback of the VDFEB can achieve consensus with a better performance of the vibration suppression in comparison with that under PBC.

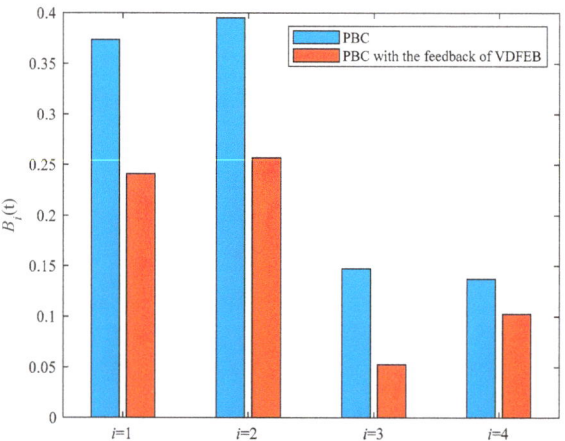

Figure 12. The comparison of the degree of vibration suppression under PBC and PBC with the feedback of the VDFEB.

5. Experimental Verification

In this section, two manipulators from Quanser Consulting Inc. in Toronto, Ontario, Canada are used to verify the effectiveness of the proposed controller of PBC with the feedback of the VDFEB. As shown in Figure 13, the experiment system consists of two manipulators, two flexible beams, and a workstation. The sensing interface can provide data on the joint angle and the velocity of the joint angle. The Intel RealSense D415 RGB-D cameras from Intel Corporation in Santa Clara, California, US with a framerate of 30 Hz are employed to obtain information on the position of the beam ends with the aid of the April tag fixed at the free end of the flexible beam. All the above information is fed back to the workstation by the USB cables. Then, the deflection and the deflection velocity at the end of the flexible beam can be calculated by the workstation in real time. Based on Equations (19) and (28), the workstation can calculate the PBC laws with and without the feedback of the VDFEB, respectively, and drive the manipulators with the QUARC 2023 software from Quanser Consulting Inc. in Toronto, Ontario, Canada. Note that only the joints of manipulators can be controlled directly by the joint motor and there is no extra actuator used to suppress the vibration. That is, the system is underactuated.

To show the effectiveness of the feedback of the VDFEB in the proposed controller, the PBC in Equation (19) together with the proposed controller in Equation (28) are compared with each other. A rectangular cross-section beam, as shown in Figure 3a, was selected, and it was assumed that the deformation of the beam occurs only in the oxy plane. In experimental tests, the gravitational force was controlled to be perpendicular to the deformation plane of the beam to avoid additional deformation caused by gravity. Therefore, the initial angle vectors of the space manipulators are set as $q_{a1}^* = \begin{bmatrix} -\frac{\pi}{4} & 0 & 0 & 0 \end{bmatrix}^T$,

$q_{a2}^* = \begin{bmatrix} \frac{\pi}{4} & 0 & 0 & 0 \end{bmatrix}^T$, respectively. There is no deformation in the flexible beam at the initial time. The desired joint angle vector is $q_{ad} = \begin{bmatrix} 0 & 0 & 0 & 0 \end{bmatrix}^T$. The parameters in controller dynamics in Equation (15) are chosen as the same as that in Section 4. Figure 14 records the state of the manipulators 0 s, 1 s, 5 s, and 50 s under PBC with and without the feedback of the VDFEB. The left column of Figure 14 pictures represents the PBC with the feedback of the VDFEB, while the right column represents the PBC without the feedback of the VDFEB. It can be seen that under the two control laws, the time required for the two manipulators to reach the desired position is roughly the same.

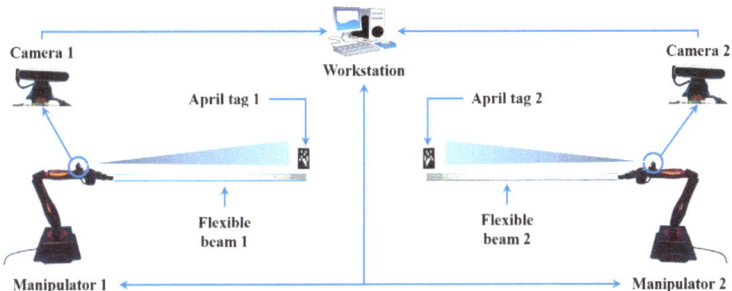

Figure 13. The system of the QARM manipulators holding flexible beams from Quanser, Inc.

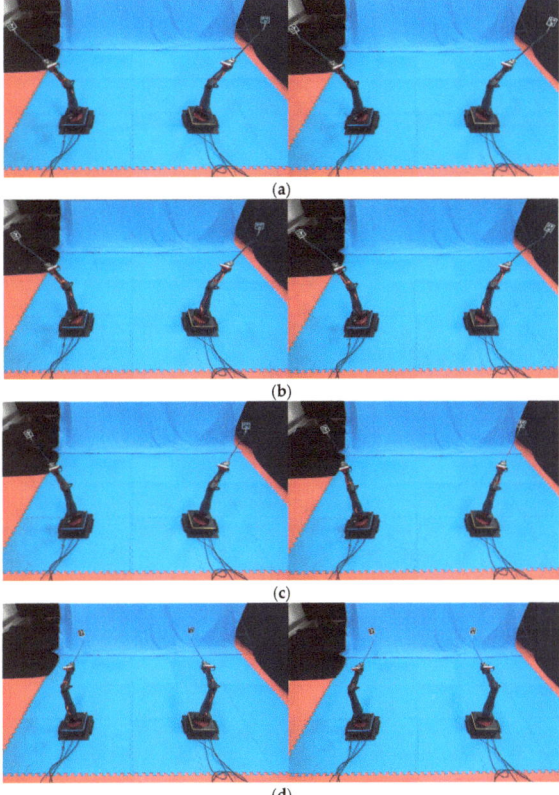

Figure 14. Dual-manipulator system under two control methods. (**a**) T = 0 s, (**b**) T = 1 s, (**c**) T = 5 s, (**d**) T = 50 s.

Figures 15 and 16 display the responses of the joint angles of the manipulators under the control laws in Equations (19) and (28), respectively. It implied that all the joint angles of the manipulators converge to the desired positions at about 50 s. It is clear that both controllers can effectively drive multiple manipulators to achieve the consensus.

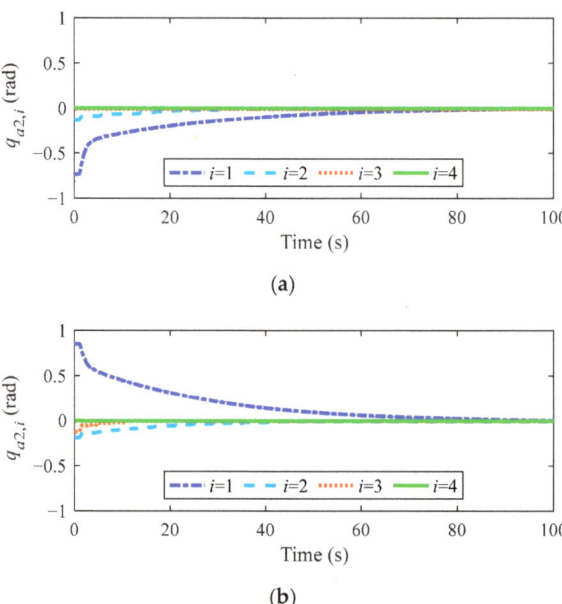

Figure 15. The experiment results under PBC. (**a**) Joint angles of manipulator 1. (**b**) Joint angles of manipulator 2.

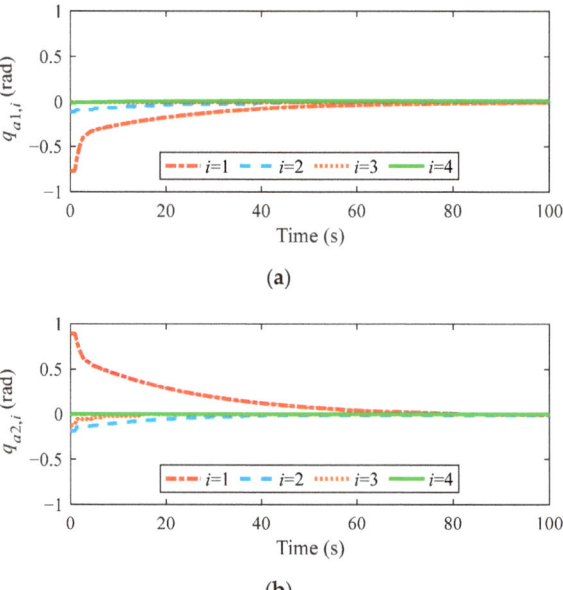

Figure 16. The experiment results under PBC with the feedback of VDFEB. (**a**) Joint angles of manipulator 1. (**b**) Joint angles of manipulator 2.

In order to emphasize the effectiveness of the VDFEB feedback-based vibration suppression in the proposed controller, the comparison of the deflection at the end of the flexible beam under the control laws in Equations (19) and (28) is conducted and Figure 17 shows the comparison results. It is clear that the control method based on the feedback of the VDFEB has an apparent effect on suppressing the vibration of the flexible beam. It can be seen intuitively that under the control law in Equation (28), the vibration amplitude of the flexible beam is smaller than that under the control law in Equation (19). At the same time, the amplitude decays faster.

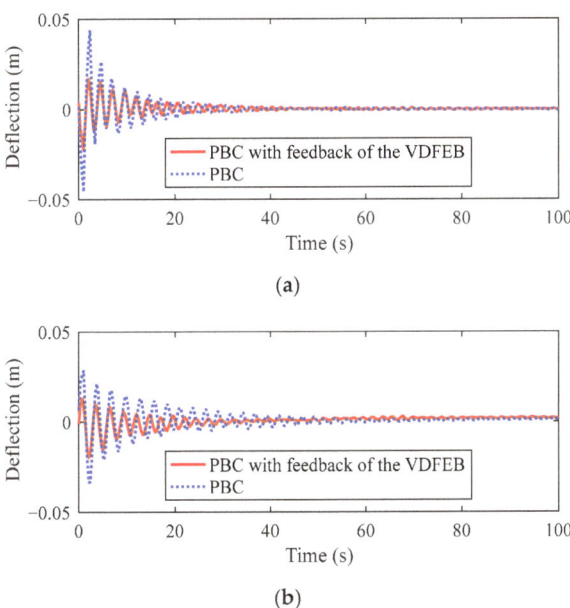

Figure 17. The deflections of the flexible beams. (**a**) Deflection of beam 1. (**b**) Deflection of beam 2.

6. Conclusions

This study presents a distributed PBC without or with the feedback of the VDFEB to achieve the leader–follower consensus of multiple manipulators holding flexible beams while simultaneously suppressing the vibration of the flexible beams. Lagrange's equations of the second kind and the assumed mode method are employed to describe the dynamics behaviors of the rigid–flexible coupling system. With the aid of the auxiliary Euler–Lagrange systems, the leader–follower distributed system can achieve consensus under undirected communication graphs. The feedback of the VDFEB is introduced into the PBC to complete the tracking tasks and the vibration suppression. Simulation results demonstrate that the multiple underactuated EL systems under the designed controller can achieve consensus more efficiently with a better performance of the vibration suppression in comparison with that under PBC. The experimental results show that both control methods can complete the consistency tracking task, and the distributed PBC with the feedback of the VDFEB has an obvious vibration suppression effect.

Author Contributions: Conceptualization, Y.C., Z.W. and T.C.; methodology, Y.C.; software, Y.C.; validation, Y.C., T.C. and E.Z.; formal analysis, Y.C. and T.C.; investigation, Y.C.; resources, T.C.; data curation, Y.C.; writing—original draft preparation, Y.C.; writing—review and editing, T.C., M.X., S.N. and Z.W.; visualization, Y.C.; supervision, T.C.; project administration, T.C.; funding acquisition, T.C. and Z.W. All authors have read and agreed to the published version of the manuscript.

Funding: This work was supported in part by the National Natural Science Foundation of China under Grants 12494562, 12472015 and 12232011, and in part by the Fundamental Research Funds for the Central Universities, No. NS2024004, and in part by the National Key Laboratory of Space Intelligent Control under Grant HTKJ2024KL502032.

Data Availability Statement: The datasets generated during and analyzed during the current study are available from the corresponding author upon reasonable request.

Conflicts of Interest: The authors declare no conflicts of interest.

References

1. Cao, Y.C.; Yu, W.W.; Ren, W.; Chen, G.R. An overview of recent progress in the study of distributed multi-agent coordination. *IEEE Trans. Ind. Inform.* **2012**, *9*, 427–438. [CrossRef]
2. Chen, T.; Shan, J.J.; Wen, H.; Xu, S.D. Review of attitude consensus of multiple spacecraft. *Astrodynamics* **2022**, *6*, 329–356. [CrossRef]
3. Liu, Y.; Jia, Y.M. Adaptive consensus control for multiple Euler-Lagrange systems with external disturbance. *Int. J. Control Autom. Syst.* **2017**, *15*, 205–211. [CrossRef]
4. Chen, T.; Shan, J.J.; Wen, H. *Distributed Attitude Consensus of Multiple Flexible Spacecraft*; Springer: Singapore, 2023; pp. 4–60.
5. Zhang, Y.H.; Jiang, Y.L.; Zhang, W.L.; Ai, X.L. Distributed coordinated tracking control for multi-manipulator systems under intermittent communications. *Nonlinear Dyn.* **2022**, *107*, 3573–3591. [CrossRef]
6. Meng, X.Z.; Mei, J.; Miao, Z.B.; Wu, A.G.; Ma, G.F. Fully distributed consensus of multiple Euler-Lagrange systems under switching directed graphs using only position measurements. *IEEE Trans. Autom. Control* **2023**, *69*, 1781–1788. [CrossRef]
7. Nuño, E. Consensus of Euler-Lagrange systems using only position measurements. *IEEE Trans. Control Netw. Syst.* **2016**, *5*, 489–498. [CrossRef]
8. Wang, Y.Q.; Yu, C.B. Distributed attitude and translation consensus for networked rigid bodies based on unit dual quaternion. *Int. J. Robust Nonlinear Control* **2017**, *27*, 3971–3989. [CrossRef]
9. Liu, Y.; Min, H.B.; Wang, S.C.; Ma, L.; Liu, Z.G. Consensus for multiple heterogeneous Euler–Lagrange systems with time-delay and jointly connected topologies. *J. Frankl. Inst.* **2014**, *351*, 3351–3363. [CrossRef]
10. Aldana, C.I.; Tabarez, L.; Nuño, E.; Romero, J.G. Leader-follower and leaderless pose consensus of robot networks with variable time-delays and without velocity measurements. *Int. J. Control* **2023**, *96*, 2885–2897. [CrossRef]
11. Ren, W. Distributed leaderless consensus algorithms for networked Euler–Lagrange systems. *Int. J. Control* **2009**, *82*, 2137–2149. [CrossRef]
12. Wang, S.M.; Zhang, H.W.; Baldi, S.; Zhong, R.X. Leaderless consensus of heterogeneous multiple Euler-Lagrange systems with unknown disturbance. *IEEE Trans. Autom. Control* **2022**, *68*, 2399–2406. [CrossRef]
13. Song, Y.F.; Li, Z.J.; Li, B.; Wen, G.X. Optimized leader-follower consensus control using combination of reinforcement learning and sliding mode mechanism for multiple robot manipulator system. *Int. J. Robust Nonlinear Control* **2024**, *34*, 5212–5228. [CrossRef]
14. Cong, Y.Z.; Du, H.B.; Li, X.L.; Jin, X.Z. Distributed bounded finite-time cooperative control algorithm for multiple nonlinear manipulators. *Int. J. Robust Nonlinear Control* **2024**, *34*, 8127–8143. [CrossRef]
15. Ghommam, J.; Bouterra, Y.; Mnif, F.; Poisson, G. Distributed backstepping control for synchronization of networked class of underactuated systems: A passivity approach. In Proceedings of the Mediterranean Conference on Control & Automation, Corfu, Greece, 20–23 June 2011; pp. 7–12.
16. Listmann, K.D.; Woolsey, C.A.; Adamy, J. Passivity-based coordination of multi-agent systems: A backstepping approach. In Proceedings of the European Control Conference, Budapest, Hungary, 23–26 August 2009; pp. 2450–2455.
17. Spong, M.W. Partial feedback linearization of underactuated mechanical systems. In Proceedings of the International Conference on Intelligent Robots and Systems, Munich, Germany, 12–16 September 1994; pp. 314–321.
18. Yan, Z.; Lai, X.Z.; Meng, Q.X.; Wu, M.; She, J.H.; Iwasaki, M. Modeling, analysis, and adaptive neural modified-backstepping control of an uncertain horizontal pendubot with double flexible joints. *Control Eng. Pract.* **2023**, *139*, 105647. [CrossRef]
19. Du, H.B.; Li, S.H. Attitude synchronization control for a group of flexible spacecraft. *Automatica* **2014**, *50*, 646–651. [CrossRef]
20. Cheikh, I.; Faqihi, H.; Benbrahim, M.; Kabbaj, M.N. Robust sliding mode based on partial feedback linearization Control for underactuated robot manipulator. In Proceedings of the International Conference on Digital Technologies and Applications, Ningbo, China, 30–31 May 2024; pp. 487–497.
21. Matouš, J.; Paliotta, C.; Pettersen, K.Y.; Varagnolo, D. The hand position concept for control of underactuated underwater vehicles. *IEEE Trans. Control Syst. Technol.* **2024**, *32*, 2223–2239. [CrossRef]
22. Olfati-Saber, R. Cascade normal forms for underactuated mechanical systems. In Proceedings of the IEEE Conference on Decision and Control, Sydney, Australia, 12–15 December 2000; pp. 2162–2167.

23. Olfati-Saber, R. Normal forms for underactuated mechanical systems with symmetry. *IEEE Trans. Autom. Control* **2002**, *47*, 305–308. [CrossRef]
24. Ortega, R.; Loria, A.; Nicklasson, P.J.; Sira-Ramirez, H. Euler-Lagrange Systems, Springer: London, UK, 1998; pp. 15–37.
25. Nuño, E.; Ortega, R. Achieving consensus of Euler–Lagrange agents with interconnecting delays and without velocity measurements via passivity-based control. *IEEE Trans. Control Syst. Technol.* **2017**, *26*, 222–232. [CrossRef]
26. Hatanaka, T.; Chopra, N.; Fujita, M.; Spong, M.W. *Passivity-Based Control and Estimation in Networked Robotics*; Springer: Cham, Switzerland, 2015; pp. 31–45.
27. Chen, T.; Shan, J.J.; Wen, H. Distributed passivity-based control for multiple flexible spacecraft with attitude-only measurements. *Aerosp. Sci. Technol.* **2019**, *94*, 105408. [CrossRef]
28. Mansour, T.; Konno, A.; Uchiyama, M. Modified PID control of a single-link flexible robot. *Adv. Robot.* **2008**, *22*, 433–449. [CrossRef]
29. Wang, E.M.; Wu, S.N.; Liu, Y.F.; Wu, Z.G.; Liu, X.D. Distributed vibration control of a large solar power satellite. *Astrodynamics* **2019**, *3*, 189–203. [CrossRef]
30. Li, D.; Zhong, L.; Zhu, W.; Xu, Z.; Tang, Q.; Zhan, W. A survey of space robotic technologies for on-orbit assembly. *Space Sci. Technol.* **2022**, *2022*, 1–13. [CrossRef]
31. Wu, Z.; Chen, Y.; Xu, W. A light space manipulator with high load-to-weight ratio: System development and compliance control. *Space Sci. Technol.* **2021**, *2021*, 1–12. [CrossRef]
32. Tayebi, J.; Chen, T.; Wang, H. Dynamics and control of flexible satellite using reaction sphere actuators. *Space Sci. Technol.* **2023**, *3*, 0077. [CrossRef]
33. Du, H.B.; Chen, M.Z.; Wen, G.H. Leader–following attitude consensus for spacecraft formation with rigid and flexible spacecraft. *J. Guid. Control Dyn.* **2016**, *39*, 944–951. [CrossRef]
34. Hu, J.; Hong, Y. Leader-following coordination of multi-agent systems with coupling time delays. *Phys. A Stat. Mech. Its Appl.* **2007**, *374*, 853–863. [CrossRef]
35. Zhu, Q. *Nonlinear Systems*; MDPI-Multidisciplinary Digital Publishing Institute: Beijing, China, 2023; pp. 19–47.
36. Ortega, R.; Van Der Schaft, A.; Maschke, B.; Escobar, G. Interconnection and damping assignment passivity-based control of port-controlled Hamiltonian systems. *Automatica* **2002**, *38*, 585–596. [CrossRef]
37. Ortega, R.; Garcia-Canseco, E. Interconnection and damping assignment passivity-based control: A survey. *Eur. J. Control* **2004**, *10*, 432–450. [CrossRef]
38. Xie, M.; Chen, T.; Ni, S.; Feng, C. Flexible payload transportation using cooperative space manipulators with statics compensation. *ISA Trans.* **2025**, in press. [CrossRef]
39. Lyu, B.; Liu, C.; Yue, X. Integrated predictor-observer feedback control for vibration mitigation of large-scale spacecraft with unbounded input time delay. *IEEE Trans. Aerosp. Electron. Syst.* **2025**, in press. [CrossRef]

Disclaimer/Publisher's Note: The statements, opinions and data contained in all publications are solely those of the individual author(s) and contributor(s) and not of MDPI and/or the editor(s). MDPI and/or the editor(s) disclaim responsibility for any injury to people or property resulting from any ideas, methods, instructions or products referred to in the content.

Article

Test Mass Capture Control for Drag-Free Satellite Based on State-Dependent Riccati Equation Method

Yingjie Chen, Yankai Wang, Ti Chen *, Zhengtao Wei and Javad Tayebi

State Key Laboratory of Mechanics and Control for Aerospace Structures, Nanjing University of Aeronautics and Astronautics, No. 29 Yudao Street, Nanjing 210016, China; yingjiechen@nuaa.edu.cn (Y.C.); wangyankai@nuaa.edu.cn (Y.W.); weizhengtao@nuaa.edu.cn (Z.W.); j.tayebi@nuaa.edu.cn (J.T.)
* Correspondence: chenti@nuaa.edu.cn

Abstract: The drag-free satellite plays an important role in the space-based gravitational wave observatory. The capture control of test mass after release is a crucial technology that can affect the success of the mission. The test mass must be released to the center of the electrostatic suspension cage accurately. This paper presents a nonlinear dynamic model of drag-free satellites in Lagrange formalism. A capture control scheme for test mass release phase is proposed based on the state-dependent Riccati equation (SDRE) strategy. To deal with the actuator saturation problem, a nonlinear saturation model is introduced to the dynamics of satellite, while the SDRE strategy is applied to the non-affine system. The effectiveness of the proposed methodology is verified by the numerical simulation for the drag-free satellite.

Keywords: drag-free satellite; test mass capture control; SDRE; non-affine system; actuator saturation

1. Introduction

The general theory of relativity is one of the greatest discoveries of the 20th century in physics. As direct proof of the general theory of relativity, gravitational waves (GWs) have aroused great desire for exploration. In February 2016, human beings directly detected GWs for the first time [1]. Limited by the length of the interference arm, the ground-based GW observatory can only observe GW signals with frequencies above 1 Hz. However, low-frequency GW signals (1 mHz up to 0.1 Hz) contain richer astronomical information. The SGO (Space-based Gravitational wave Observatory) with a large enough interference arm is able to detect low-frequency GWs. Many mission concepts have been proposed for the SGO, including the LISA (Laser Interferometer Space Antenna) mission [2,3], Taiji plan [4,5], TianQin mission [6], BBO (Big Bang Observer) [7], DECIGO (DECi hertz Interferometer Gravitational wave Observatory) [8,9], and LAGRANGE (Laser Gravitational wave Antenna at Geo-lunar Lagrange) [10], etc.

The nature of the SGO is a space-based Michelson interferometer based on the formation of drag-free satellites. The SGO detects GWs by measuring the relative distance between two free falling test masses (TMs) in two drag-free satellites [11]. To withstand the high vibration forces during launch, the TM is fixed by the cage and vent mechanism (CVM) [12]. In the release phase, the TM is released by the grabbing positioning and release mechanism (GPRM) [13]. Nevertheless, the GPRM inevitably introduces some errors caused by the asymmetry and adhesion [14], which will lead to the TM in unfavorable initial conditions. To perform science mode, the TM must be precisely captured in the center of the electrostatic suspension cage. However, these critical initial conditions together with the saturation of the electrostatic suspension force raise great challenges to capture the TM after it is released [15]. Consequently, there are many studies that focus on the release mechanism. In Ref. [16], a mathematical model for the retraction dynamics of the release tip in the GPRM was proposed to predict the dynamic conditions of the TM injection. Benedetti et al. [17] investigated the dynamics of the TM in free-falling conditions typical

---

Citation: Chen, Y.; Wang, Y.; Chen, T.; Wei, Z.; Tayebi, J. Test Mass Capture Control for Drag-Free Satellite Based on State-Dependent Riccati Equation Method. *Actuators* **2024**, *13*, 434. https://doi.org/10.3390/act13110434

Academic Editor: Liang Sun

Received: 14 September 2024
Revised: 20 October 2024
Accepted: 24 October 2024
Published: 27 October 2024

Copyright: © 2024 by the authors. Licensee MDPI, Basel, Switzerland. This article is an open access article distributed under the terms and conditions of the Creative Commons Attribution (CC BY) license (https://creativecommons.org/licenses/by/4.0/).

of space applications and built an analytical model to predict the imparted momentum in the case of conservative interaction forces. To give an estimation of the TM release velocity in the scope of the LISA-Pathfinder space mission, Bortoluzzi et al. [18] developed a model based on laboratory measurements for both the release tip retraction and the adhesion phenomenon. Moreover, they not only made a significant contribution to the establishment of the analytical model for the release mechanism but also developed an experimental technique aimed at measuring the momentum transfer that occurs when two free-falling bodies interacting with surface forces are impulsively separated [19]. These studies mentioned above only focus on the release mechanism itself or the interaction between the release mechanism and the TM and do not involve the capture controller design. The release mechanism is just a subsystem of the whole drag-free satellite, and the drag-free satellite is a high-dimensional system with 20 DoFs (Degree of Freedom). To consider the Capture Control Problem (CCP) of the TM after release, the dynamics of the whole drag-free satellite must be built and analyzed.

However, it is challenging to deal with the control problem of such a high-dimensional dynamic system. There are relatively few papers related to the CCP of the TM after release. At present, the existing controllers designed for the CCP are almost based on either the Sliding Mode Control (SMC) method or optimization techniques. For example, Montemurro et al. [20] developed a capture controller based on SMC for the LISA pathfinder mission, which is a science and technology demonstrator designed to pave the way for the LISA mission. For the CCP for the LISA mission, Capicchiano [21] designed a second order controller named Super Twisting SMC to avoid the possible collisions between the walls of the electrostatic suspension cage and the TM. In order to further improve the robustness of SMC, Lian et al. [22] designed a Radial Basis Function (RBF) neural network to compensate for the disturbance and combined this RBF neural network with the SMC method to propose an adaptive SMC for the CCP of the TM. On the other hand, under the framework of optimal control, Vidano et al. [23] proposed a capture controller for the LISA mission based on a Model Predictive Control (MPC) strategy. Due to the tuning difficulty and higher computational load of the MPC algorithm caused by the high dimensionality of the drag-free satellite, it is very difficult, even impossible, to design an MPC controller for the whole plant. Hence, in Ref. [23], the cost function defined for the MPC optimization problem only includes 12 DoFs related to the TM in the drag-free satellite, and then the sub-optimal control inputs are provided by solving this optimization problem. The time-optimal control for capturing the TM has also been considered. For instance, Gioia [24] proposed a time-optimal control law to reduce the TM capturing time. Furthermore, Lin et al. [25] not only proposed a minimum-time capture control method for the TM release phase of drag-free satellites but also gave an analytical solution for this optimal control problem based on Pontryagin's minimum principle and the simplified form of the high-dimensional nonlinear relative dynamics, which is the well-known Clohessy–Wiltshire equation [26,27]. These studies notes that, because of the complexity of high-dimensional systems or the limitations of the efficiency of optimization algorithms, these capture controllers must lose the accuracy of describing the nonlinearity of the system and adopt some simplified models. However, the whole drag-free satellite is a highly coupled nonlinear system, while the capture control requires high accuracy. Under such accuracy requirements, it is not appropriate to ignore some high-order nonlinear terms.

The other challenge in the CCP is the requirement of high control accuracy under the constraints of actuator saturation. Only little efforts have been made to address this issue. For example, in the Ref. [23], Vidano et al. considered the actuator saturation within the MPC framework, and the actuator saturation was treated as the constraints for the optimization problem. However, this method will cause the control inputs to jitter between saturation values.

The SDRE method is a systematic and effective way to design nonlinear feedback control for nonlinear systems and allows for the nonlinearities in the system states [28]. Hence, the SDRE strategy has become very popular within the control community over the last decade. Notably, to the best knowledge of the authors, the state-dependent Riccati equation (SDRE) method is never utilized for the CCP. To facilitate the design of an effective controller for drag-free satellites, this paper presents a 20 DoF dynamics model of drag-free satellites with the existence of external disturbances. Then, an SDRE controller considering the actuator saturation is proposed for the release phase of the drag-free satellite. Furthermore, when the actuator saturation is considered the constraints of the optimal problem, there will be a jitter phenomenon in the control input. In order to eliminate this jitter, a nonlinear saturation function is directly introduced to the dynamics of the drag-free satellite to model the actuator saturation. The main contributions of this study are as follows:

1. Compared with Refs. [20–25], the capture controller is designed based on the whole 20 DoF nonlinear dynamics model of the drag-free satellite using the SDRE strategy. And the feasible linear-like structure is proposed for fulfilling the SDRE controller design.
2. In contrast to the actuator saturation considered in Ref. [23], a nonlinear saturation model is introduced. Then, the actuator saturation is considered the non-affine term to the dynamics of drag-free satellites. The SDRE strategy is utilized to deal with this non-affine system.

The rest of this paper is organized as follows: In Section 2, the SDRE method is reviewed, and the dynamics model of drag-free satellites is presented. In Section 3, the control objective is presented, and the SDRE controller for the TM release phase is proposed. Section 4 provides the numerical simulation results to demonstrate the effectiveness of the proposed controller. The conclusions are summarized and analyzed in Section 5.

2. Problem Formulation

In this section, the mathematic notations used in this paper are introduced. Then, the traditional SDRE regulator is presented to show the SDRE integral servomechanism. The Euler angles are selected as the representation of attitude because of their advantages in small rotation cases compared to other method [29,30]. Finally, the dynamics model of the drag-free satellite is presented.

2.1. Mathematical Notations

2.1.1. Vectors and Coordinates

r denotes a three-dimensional vector. The Cartesian coordinate system is introduced by an orthonormal basis denoted as $\underline{e} = \begin{bmatrix} e_1 & e_2 & e_3 \end{bmatrix}^T$. Then, the coordinates of r with respect to $\underline{e}$ are given by $\underline{r}^{(e)} = \begin{bmatrix} r_1 & r_2 & r_3 \end{bmatrix}^T \in \mathbb{R}^3$, i.e., $r = \underline{r}^{(e)T}\underline{e}$. For a vector $a = \left(\underline{a}^{(e)}\right)^T \underline{e}$, the skew-symmetric matrix is defined as the following:

$$\underline{a}^{(e)\times} = \begin{bmatrix} 0 & -a_3 & a_2 \\ a_3 & 0 & -a_1 \\ -a_2 & a_1 & 0 \end{bmatrix} \quad (1)$$

where the coordinate $\underline{a}^{(e)}$ is denoted as $\underline{a}^{(e)} = \begin{bmatrix} a_1 & a_2 & a_3 \end{bmatrix}^T$.

2.1.2. Rotation Matrix

Assume that there are two different orthonormal bases $\underline{e}_1$ and $\underline{e}_2$ in $\mathbb{R}^3$, during the process of rotating from $\underline{e}_1$ to $\underline{e}_2$. The rotation matrix is defined as follows:

$$R_1^2 = \underline{e}_2 \cdot \underline{e}_1^T \quad (2)$$

Then, the different coordinates of the same vector r have the following relationship:

$$\underline{r}^{(e_2)} = R_1^2 \underline{r}^{(e_1)} \quad (3)$$

where $\underline{r}^{(e_1)}$ represents the coordinates of r with respect to $\underline{e}_1$, and $\underline{r}^{(e_2)}$ represents the coordinates of r with respect to $\underline{e}_2$.

And the time derivative of the rotation matrix is given by the following formula:

$$\dot{R}_1^2 = -\underline{\omega}_{21}^{(e_2)\times} R_1^2 \tag{4}$$

where $\underline{\omega}_{21}^{(e_2)}$ represents the coordinates of the angular velocity of $\underline{e}_2$ with respect to $\underline{e}_1$ in the basis $\underline{e}_1$.

2.1.3. Matrix Norms

For any matrix $A = [a_{ij}] \in \mathbb{R}^{n \times m}$, the ∞-norm defined on $\mathbb{R}^{n \times m}$ is given by $\|A\|_\infty = \max\limits_{1 \leq i \leq n} \sum\limits_{j=1}^{m} |a_{ij}|$. As a matrix norm, the following property holds for any matrices $A, B \in \mathbb{R}^{n \times m}$:

$$\|A + B\|_\infty \leq \|A\|_\infty + \|B\|_\infty \tag{5}$$

Significantly, the ∞-norm defined on $\mathbb{R}^{n \times m}$, $\mathbb{R}^{m \times q}$, and $\mathbb{R}^{n \times q}$ is mutually consistent, i.e., the following inequality holds for any $A \in \mathbb{R}^{n \times m}$ and $C \in \mathbb{R}^{m \times q}$:

$$\|AC\|_\infty \leq \|A\|_\infty \|C\|_\infty \tag{6}$$

2.2. SDRE Regulator

In this section, the traditional SDRE regulator for the optimal control problem of nonlinear system is reviewed [28].

Consider a nonlinear system, which is full-state observable, autonomous, and affine in the input, represented in the following form:

$$\dot{x}(t) = f(x) + B(x)u(t), \quad x(t_0) = x_0 \tag{7}$$

where $x \in \mathbb{R}^n$ is the state vector, $u \in \mathbb{R}^m$ represents the input vector, and the time variable $t \in [t_0, +\infty)$ with t_0 as the initial time. Note that $f : \mathbb{R}^n \to \mathbb{R}^n$ and $B : \mathbb{R}^n \to \mathbb{R}^{n \times m}$.

The infinite-time horizon optimal control problem (ITHOCP) is to seek the optimal input u^* such that the performance criterion

$$J = \frac{1}{2} \int_{t_0}^{\infty} \left[x^T(t) Q x(t) + u^T(t) R u(t) \right] dt, \tag{8}$$

is minimized, where $Q \geq 0$ and $R > 0$ denote the positive semi-definite matrix and positive definite matrix, respectively.

The suboptimal solution to this ITHOCP based on the SDRE method can be obtained through the following steps [28]:

1. Use direct factorization to bring a nonlinear system into the following linear-like structure that contains SDC (State-Dependent Coefficient) matrices:

$$\dot{x} = A(x)x + B(x)u \tag{9}$$

where the SDC matrix function $A(x) : \mathbb{R}^n \to \mathbb{R}^{n \times n}$, and $A(x)x$ is the SDC form of the nonlinear function $f(x)$ in Formula (4).

2. Solve the state-dependent Riccati equation

$$A^T(x)P(x) + P(x)A(x) - P(x)B(x)R^{-1}B^T(x)P(x) + Q = 0 \tag{10}$$

to obtain the positive semi-definite solution $P(x) \geq 0$.

3. Use $P(x)$ to obtain the nonlinear suboptimal feedback control law in the following equation.

$$u = -R^{-1}B^{\mathrm{T}}(x)P(x)x \tag{11}$$

2.3. SDRE Integral Servomechanism

To perform trajectory tracking control without steady-state error, the SDRE controller in the last subsection can be implemented as an integral servomechanism [31]. The process of tracking control is shown as follows:

1. The state vector x is decomposed as the following:

$$x = \begin{bmatrix} x_R^{\mathrm{T}} & x_N^{\mathrm{T}} \end{bmatrix}^{\mathrm{T}} \tag{12}$$

where x_R is the vector component of x to track a desired trajectory r_c, and the x_N represents the vector containing the remaining part of x.

2. The following new state vector is augmented with x_I:

$$\tilde{x}^{\mathrm{T}} = \begin{bmatrix} x_R^{\mathrm{T}} & x_N^{\mathrm{T}} & x_I^{\mathrm{T}} \end{bmatrix} \tag{13}$$

where x_I is the integral state of x_R, i.e., $x_I = \int_0^t x_R dt$.

3. The augmented system equation is given by the following:

$$\dot{\tilde{x}} = \tilde{A}(\tilde{x})\tilde{x} + \tilde{B}(\tilde{x})u \tag{14}$$

where

$$\tilde{A}(\tilde{x}) = \begin{bmatrix} A & 0 \\ I : 0 & 0 \end{bmatrix}, \tilde{B}(\tilde{x}) = \begin{bmatrix} B \\ 0 \end{bmatrix} \tag{15}$$

4. The positive semi-definite solution $\tilde{P}(\tilde{x}) \geq 0$ is obtained by solving the following state-dependent Riccati equation:

$$\tilde{A}^{\mathrm{T}}(\tilde{x})\tilde{P}(\tilde{x}) + \tilde{P}(\tilde{x})\tilde{A}(\tilde{x}) - \tilde{P}(\tilde{x})\tilde{B}(\tilde{x})\tilde{R}^{-1}\tilde{B}^{\mathrm{T}}(\tilde{x})\tilde{P}(\tilde{x}) + \tilde{Q} = 0 \tag{16}$$

5. The SDRE integral servo controller is given by:

$$u = -\tilde{R}^{-1}\tilde{B}^{\mathrm{T}}(\tilde{x})\tilde{P}(\tilde{x})\hat{x} \tag{17}$$

where

$$\hat{x} = \begin{bmatrix} x_R - r_c \\ x_N \\ x_I - \int_0^t r_c dt \end{bmatrix} \tag{18}$$

To arrive at the conclusion that the SDRE has a solution, the pointwise detectability condition must be satisfied. This step is accomplished by penalizing the integral states with the corresponding nonzero diagonal elements of $\tilde{Q}$ [32].

2.4. Dynamic Equation of Drag-Free Satellites

The drag-free satellite for the SGO mission mainly consists of five rigid bodies, i.e., Spacecraft (SC), Optical Assembly-1 (OA1), Optical Assembly-2 (OA2), TM1, and TM2, as shown in Figure 1. The OA1 and OA2 can rotate around the pivot axis that is mounted on the SC. Each TM is suspended inside the corresponding Optical Assembly (OA). The whole dynamics model of the drag-free satellite established in this paper is built from two systems. One system is called the operation platform (OP), which consists of the SC and two OAs. The other system is called the TM system, which consists of two TMs.

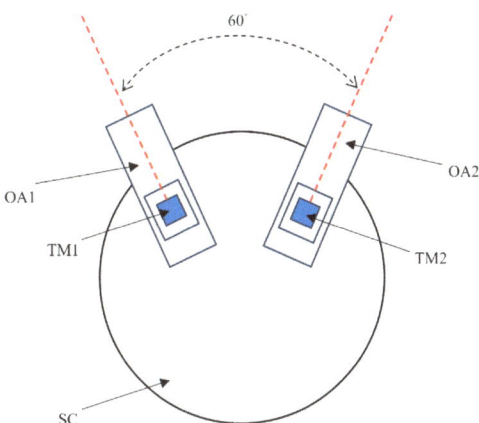

Figure 1. Drag-free satellite for SGO.

Essentially, the OP is a system of multiple rigid bodies. B_1, B_2, and B_3 represent the SC, OA1, and OA2, respectively. To describe the motion of B_i, an inertial reference frame (IRF) $\{O_I X_I Y_I Z_I\}$ and the body reference frame (BRF-i) $\{o_i x_i y_i z_i\}$ are introduced, as shown in Figure 2. Without a loss of generality, assume that BRF-i is aligned along the principal axes of inertia of the corresponding B_i.

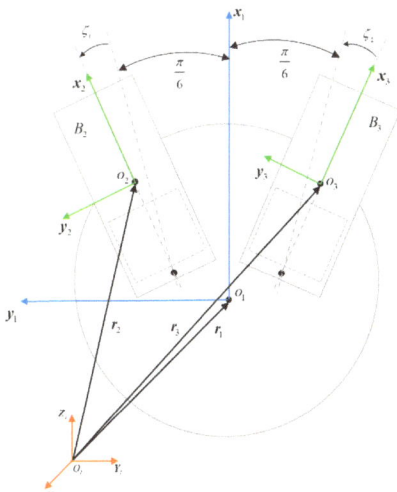

Figure 2. Reference system of OP.

The total kinetic energy of the OP system can be expressed as follows:

$$T = \sum_{i=1}^{3} \left[\frac{1}{2} m_i \dot{r}_i^{(I)T} \dot{r}_i^{(I)} + \frac{1}{2} \omega_i^T J_i \omega_i \right] \quad (19)$$

where the superscript "T" is the transpose operator, the mass of B_i is denoted as m_i, $r_i^{(I)} \in \mathbb{R}^3$ are the coordinates of the position of B_i in the IRF, and $\omega_i \in \mathbb{R}^3$ represent the coordinates of the angular velocity of B_i in BRF-i. Based on the assumption mentioned above, the inertia matrix of B_i is given by a diagonal matrix $J_i = \text{diag}(J_i^1, J_i^2, J_i^3)$.

By applying Lagrange's equation of the second kind, the dynamic equation of the OP is given in a matrix form as

$$M_{OP}(q_{OP})\ddot{q}_{OP} + C_{OP}(q_{OP}, \dot{q}_{OP})\dot{q}_{OP} + K_{OP}(q_{OP})q_{OP} = Q^a \quad (20)$$

where $q_{OP} = \begin{bmatrix} r_1^{(I)T} & \theta_{1I}^T & \zeta_1 & \zeta_2 \end{bmatrix}^T \in \mathbb{R}^8$ is the generalized coordinate vector, $M_{OP} \in \mathbb{R}^{8\times 8}$ represents the generalized inertia matrix, $C_{OP} \in \mathbb{R}^{8\times 8}$ is the Coriolis matrix, and $K_{OP} \in \mathbb{R}^{8\times 8}$ represents the generalized stiffness matrix. $r_1^{(I)}$ and θ_{1I} are the position vector and the Euler 123 angle of the SC with respect to the IRF, respectively, ζ_1 is the rotation angle of the OA1 with respect to the SC, and ζ_2 is the rotation angle of the OA2 with respect to the SC. Q^a is the generalized force vector.

As shown in Figure 3, two OA reference frames (ORF-j) $\{o_{OAj}x_{OAj}y_{OAj}z_{OAj}\}$ and two TM body reference frames (MRF-j) $\{o_{TMj}x_{TMj}y_{TMj}z_{TMj}\}$ are built for deriving the dynamic equations of TMs. Assume that MRF-i is aligned along the principal axes of inertia of the corresponding TM.

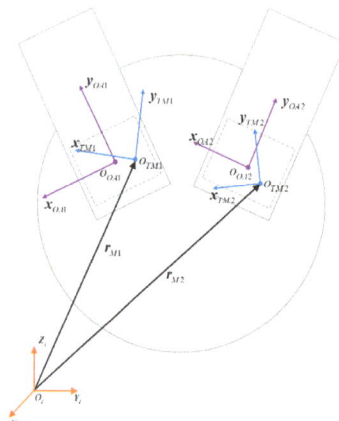

Figure 3. Reference system of TM.

The dynamic equation of j-th TM ($j = 1, 2$) can be obtained using the Newton–Euler equation [33] as follows:

$$\ddot{r}_{Mj}^{(I)} = -\mu \frac{r_{Mj}^{(I)}}{\left|r_{Mj}^{(I)}\right|^3} + \frac{1}{m_{Mj}} R_S^I R_{Oj}^S \left(F_{Ej}^{(Oj)} + F_{stj}^{(Oj)}\right) \quad (21)$$

$$J_{Mj}^{(Mj)} \dot{\omega}_{Mj}^{(Mj)} + \omega_{Mj}^{(Mj)\times} J_{Mj}^{(Mj)} \omega_{Mj}^{(Mj)} = R_{Oj}^{Mj} M_{Ej}^{(Oj)} + R_{Oj}^{Mj} M_{stj}^{(Oj)} \quad (22)$$

where $r_{Mj}^{(I)}$ represents the position coordinates of the mass center of the j-th TM in the IRF, μ is the standard gravitational parameter of the Sun, m_{Mj} represents the mass of the j-th TM, R_S^I is the rotation matrix from BRF-1 to the IRF, R_{Oj}^S represents the rotation matrix from ORF-j to BRF-1, $F_{Ej}^{(Oj)}$ represents the coordinates of the j-th electrostatic suspension force in ORF-j, and $F_{stj}^{(Oj)}$ represents the coordinates of the coupling force in ORF-j. $J_{Mj}^{(Mj)}$ is the inertia matrix of the j-th TM in MRF-j, $\omega_{Mj}^{(Mj)}$ represents the coordinates of the inertial angular velocity of the j-th TM in MRF-j, R_{Oj}^{Mj} is the rotation matrix from ORF-j to MRF-j, $M_{Ej}^{(Oj)}$ represents the coordinates of the electrostatic suspension torque in ORF-j, and $M_{stj}^{(Oj)}$ represents the coordinates of the coupling torque in ORF-j.

The dynamics model of two TMs can be rewritten in a matrix form as

$$M_{TM}(q_{TM})\ddot{q}_{TM} + C_{TM}(q_{TM}, \dot{q}_{TM})\dot{q}_{TM} + K_{TM}(q_{TM})q_{TM} = Q_{TM} \tag{23}$$

where $q_{TM} = \begin{bmatrix} r_{M1}^{(I)T} & \theta_{M1}^{T} & r_{M2}^{(I)T} & \theta_{M2}^{T} \end{bmatrix}^T \in \mathbb{R}^{12}$ is the generalized coordinate vector, $M_{TM} \in \mathbb{R}^{12 \times 12}$ represents the generalized inertia matrix, $C_{TM} \in \mathbb{R}^{12 \times 12}$ is the Coriolis matrix, $K_{TM} \in \mathbb{R}^{12 \times 12}$ denotes the generalized stiffness matrix, and $Q_{TM} \in \mathbb{R}^{12}$ is the generalized force.

The dynamic equation of the whole drag-free satellite reads

$$M(q)\ddot{q} + C(q)\dot{q} + K(q)q = Q \tag{24}$$

where $q = \begin{bmatrix} q_{OP}^T & q_{TM}^T \end{bmatrix}^T \in \mathbb{R}^{20}$ represents the generalized coordinates of whole drag-free satellite, $M(q) = \begin{bmatrix} M_{OP} & 0 \\ 0 & M_{TM} \end{bmatrix} \in \mathbb{R}^{20 \times 20}$ denotes the generalized inertia matrix, $C(q) = \begin{bmatrix} C_{OP} & 0 \\ 0 & C_{TM} \end{bmatrix} \in \mathbb{R}^{20 \times 20}$ is the Coriolis matrix of this dynamic system, $K(q) = \begin{bmatrix} K_{OP} & 0 \\ 0 & K_{TM} \end{bmatrix} \in \mathbb{R}^{20 \times 20}$ is the SDC matrix derived from the gravity, and the generalized force is denoted as $Q = \begin{bmatrix} (Q^a)^T & Q_{TM}^T \end{bmatrix}^T \in \mathbb{R}^{20}$.

3. Control Design

In this section, the control objective of the TM release phase is presented first. Then, to address the saturation of the electrostatic actuators and the micro propulsion system, an actuator saturation model is introduced. Finally, the capture controller based on the SDRE method is proposed.

3.1. Control Objective

Due to the error introduced by the GPRM in the release phase, the TM will deviate from the center of the electrostatic suspension cage. To perform science mode, the TM must be in the center and aligned with the ORF under required precision. During the capture process, the motion of the TM should be within the range of the electrostatic suspension cage [23]. Meanwhile, affected by the solar radiation pressure (SRP) and the reaction of the electrostatic actuators, the position and attitude of the SC and the rotation angles of OA will drift away. Hence, the SC should be controlled to keep the original position and attitude, and the rotation of OA should be stabilized in the nominal position. Furthermore, the actuator saturation of the electrostatic suspension is also considered as

$$-\bar{u}_F \leq F_{Ej}^{(Oj)} \leq \bar{u}_F, -\bar{u}_T \leq M_{Ej}^{(Oj)} \leq \bar{u}_T, \; j = 1, 2 \tag{25}$$

where $\bar{u}_F$ and $\bar{u}_T$ are the saturation of electrostatic force and torque, respectively.

3.2. Controller Design

In order to implement the capture control using the SDRE method, the dynamic model of the TM should be modified to the state-space representation. The state vector is defined as $x_{TM} = \begin{bmatrix} q_{TM}^T & \dot{q}_{TM}^T \end{bmatrix}^T \in \mathbb{R}^{24}$, and the input vector is given by $u_{TM} = \begin{bmatrix} (F_{E1}^{(O1)})^T & (F_{E2}^{(O2)})^T & (M_{E1}^{(O1)})^T & (M_{E2}^{(O2)})^T \end{bmatrix}^T \in \mathbb{R}^{12}$, where $F_{E1}^{(O1)}$ and $F_{E2}^{(O2)}$ are the electrostatic forces generated by the electrostatic actuators, and the $M_{E1}^{(O1)}$ and $M_{E2}^{(O2)}$ are the electrostatic torque generated by the electrostatic actuators. Then, the dynamics model of two TMs can be converted to the state-space representation as follows:

$$\dot{x}_{TM} = A_{TM}(x_{TM}, x_{OP})x_{TM} + B_{TM}(x_{TM}, x_{SP})u_{TM} \tag{26}$$

where $A_{TM}(x_{TM}, x_{OP}) = \begin{bmatrix} 0_{12} & I_{12} \\ -M_{TM}^{-1}K_{TM} & -M_{TM}^{-1}C_{TM} \end{bmatrix} \in \mathbb{R}^{24 \times 24}$ represents the system matrix, $B_{TM}(x_{TM}, x_{OP}) = \begin{bmatrix} 0_{12} & \left(M_{TM}^{-1}H_{TM}\right)^T \end{bmatrix}^T \in \mathbb{R}^{24 \times 12}$ is the input matrix, the I_{12} represents the 12-dimensional identity matrix, and $0_{12} \in \mathbb{R}^{12 \times 12}$ denotes the zero matrix. The matrix $H_{TM} \in \mathbb{R}^{12 \times 12}$ describes the relationship between the generalized force $\underline{Q}_{TM}$ and the control commands u_{TM}, i.e., $\underline{Q}_{TM} = H_{TM}(x_{TM}, x_{OP})u_{TM}$.

The commonly used saturation model is a piecewise function as shown in Refs. [34,35]. However, this function is not smooth mathematically. Hence, the model of actuator saturation in this paper is considered by using the hyperbolic tangent function [36], denoted as $\tanh(\cdot)$, i.e., $u = \bar{u}\tanh(v)$, where u is the input of the system subject to saturation type nonlinearity, $\bar{u}$ is the saturation value of u, and v is the virtual control input to be designed. In this way, the actuator output is written in the following matrix form:

$$u_{TM} = U_{TM}T(v_{TM}) \tag{27}$$

where $U_{TM} = \text{diag}(\bar{u}_{TM1}, \bar{u}_{TM2}, \cdots, \bar{u}_{TM12})$ is the saturation matrix with its k-th components, $\bar{u}_{TMk}$ is the saturation value of the k-th component of u_{TM}, $v_{TM} \in \mathbb{R}^{12}$ represents the virtual control input to be designed, and the saturation function is defined as $T(v_{TM}) : [v_{TM1}, v_{TM2}, \cdots, v_{TM12}]^T \mapsto [\tanh(v_{TM1}), \tanh(v_{TM2}), \cdots, \tanh(v_{TM12})]^T$.

By substituting Equation (27) into Equation (26), one can obtain the following system:

$$\dot{x}_{TM} = A_{TM}x_{TM} + B_{TM}U_{TM}T(v_{TM}) \tag{28}$$

Essentially, this system in the above equation is a non-affine system. The key step of the SDRE strategy for a non-affine system is the SDC parameterization of the non-affine term [37]. To implement the SDRE controller design, this system should be parameterized into an SDC form. Since $\frac{\tanh(v)}{v} \to 1$ as $v \to 0$, then the SDC form of Equation (28) can be obtained

$$\dot{x}_{TM} = A_{TM}x_{TM} + B_{TM}U_{TM}G(v_{TM})v_{TM} \tag{29}$$

where

$$G(v_{TM}) = \begin{cases} \text{diag}\left(\frac{\tanh(v_{TM1})}{v_{TM1}}, \frac{\tanh(v_{TM2})}{v_{TM2}}, \cdots, \frac{\tanh(v_{TM12})}{v_{TM12}}\right), & v_{TM} \neq 0 \\ I_{12}, & v_{TM} = 0 \end{cases} \tag{30}$$

In order to apply the SDRE integral servomechanism to system (29), the state vector x_{TM} is decomposed as $\bar{x}_{TM} = [x_{TMR}^T \ x_{TMN}^T]^T$, where $x_{TMR} = q_{TM}$ is desired to track a reference command r_{TM}, and x_{TMN} is the vector consisting of the remaining elements. Then, the augmented state can be written as

$$\tilde{x}_{TM} = [x_{TMR}^T \ x_{TMN}^T \ x_{TMI}^T]^T \tag{31}$$

where the state x_{TMI} is the integral state of x_{TMR}, i.e., $x_{TMI} = \int_0^t x_{TMR} dt$. The augmented system is given by

$$\dot{\tilde{x}}_{TM} = \tilde{A}_{TM}(\tilde{x}_{TM})\tilde{x}_{TM} + \tilde{B}_{TM}(\tilde{x}_{TM}, v_{TM})v_{TM} \tag{32}$$

where

$$\tilde{A}_{TM} = \begin{bmatrix} A_{TM} & 0_{24 \times 12} \\ I_{12} \vdots 0_{12 \times 12} & 0_{12 \times 12} \end{bmatrix} \in \mathbb{R}^{36 \times 36} \tag{33}$$

$$\widetilde{B}_{\text{TM}} = \begin{bmatrix} B_{\text{TM}} U_{\text{TM}} G(v_{\text{TM}}) \\ 0_{12 \times 12} \end{bmatrix} \in \mathbb{R}^{36 \times 12} \tag{34}$$

The SDRE integral servomechanism controller is obtained by

$$v_{\text{TM}} = -\widetilde{R}_{\text{TM}}^{-1} \widetilde{B}_{\text{TM}}(\widetilde{x}_{\text{TM}}, v_{\text{TM}}) P(\widetilde{x}_{\text{TM}}, v_{\text{TM}}) \begin{bmatrix} x_{\text{TMR}} - r_{\text{TM}} \\ x_{\text{TMN}} \\ x_{\text{TMI}} - \int_0^t r_{\text{TM}} dt \end{bmatrix} \tag{35}$$

where $P(\widetilde{x}_{\text{TM}}, v_{\text{TM}})$ is the positive semi-definite solution of the following state-dependent Riccati equation:

$$P(\widetilde{x}_{\text{TM}}, v_{\text{TM}}) \widetilde{A}_{\text{TM}}(\widetilde{x}_{\text{TM}}) + \widetilde{A}_{\text{TM}}^{\text{T}}(\widetilde{x}_{\text{TM}}) P(\widetilde{x}_{\text{TM}}, v_{\text{TM}}) + \widetilde{Q}_{\text{TM}} - P(\widetilde{x}_{\text{TM}}, v_{\text{TM}}) \widetilde{B}_{\text{TM}}(\widetilde{x}_{\text{TM}}, v_{\text{TM}}) \widetilde{R}_{\text{TM}}^{-1} \widetilde{B}_{\text{TM}}^{\text{T}}(\widetilde{x}_{\text{TM}}, v_{\text{TM}}) P(\widetilde{x}_{\text{TM}}, v_{\text{TM}}) = 0 \tag{36}$$

As for designing the controller of the OP system with an SDRE strategy, the state vector x_{OP} is defined as $x_{\text{OP}} = \begin{bmatrix} \underline{q}_{\text{OP}}^{\text{T}} & \underline{\dot{q}}_{\text{OP}}^{\text{T}} \end{bmatrix}^{\text{T}} \in \mathbb{R}^{16}$, and the input vector is given by $u_{\text{OP}} = \begin{bmatrix} \left(\underline{F}_S^{(S)}\right)^{\text{T}} & \left(\underline{M}_S^{(S)}\right)^{\text{T}} & M_{\text{OA1}} & M_{\text{OA2}} \end{bmatrix}^{\text{T}} \in \mathbb{R}^8$, where $\underline{F}_S^{(S)}$ represents the coordinates of the micro propulsion force in SRF, $\underline{M}_S^{(S)}$ represents the coordinates of the micro propulsion torque in SRF, and $M_{\text{OA}j}$ is the command torque of j-th OA. Then, the state-space representation of the OP is written in the following form:

$$\dot{x}_{\text{OP}} = A_{\text{OP}}(x_{\text{OP}}) x_{\text{OP}} + B_{\text{OP}}(x_{\text{OP}}) \underline{Q}^a \tag{37}$$

where $A_{\text{OP}}(x_{\text{OP}}) = \begin{bmatrix} 0 & I \\ -M_{\text{OP}}^{-1} K_{\text{OP}} & -M_{\text{OP}}^{-1} C_{\text{OP}} \end{bmatrix} \in \mathbb{R}^{16 \times 16}$ represents the system matrix, $B_{\text{OP}}(x_{\text{OP}}) = \begin{bmatrix} 0 \\ M_{\text{OP}}^{-1} \end{bmatrix} \in \mathbb{R}^{16 \times 8}$ is the input matrix, and $\underline{Q}^a$ denotes the generalized forces.

Note that $\underline{Q}^a = H_{\text{OP}}(x_{\text{OP}}) u_{\text{TM}} + D_{\text{OP}}(x_{\text{OP}}) u_{\text{TM}}$, where $H_{\text{OP}}(x_{\text{OP}}) \in \mathbb{R}^{8 \times 8}$ and $D_{\text{OP}}(x_{\text{OP}}) \in \mathbb{R}^{8 \times 12}$ are the SDC matrix, depending on the state of OP. It should be noted that u_{TM} will disturb the movement of the OP system. Fortunately, u_{TM} can be regarded as the matched disturbance to system (37) because the matrix H_{OP} is invertible. In this way, system (37) can be written as

$$\dot{x}_{\text{OP}} = A_{\text{OP}} x_{\text{OP}} + B_{\text{OP}} H_{\text{OP}} u'_{\text{OP}} \tag{38}$$

where $u'_{\text{OP}} = u_{\text{OP}} + H_{\text{OP}}^{-1} D_{\text{OP}} u_{\text{TM}}$ is considered as a new control input vector to system (38). The specific expression of $H_{\text{OP}}(x_{\text{OP}})$ and $D_{\text{OP}}(x_{\text{OP}})$ will be presented in the Appendix A.

The actuator saturation of the OP system is also considered by introducing the saturation model $u_{\text{OP}} = U_{\text{OP}} T(v_{\text{OP}})$, where v_{OP} is the virtual control input to be designed, $U_{\text{OP}} = \text{diag}(\overline{u}_{\text{OP1}}, \overline{u}_{\text{OP2}}, \cdots, \overline{u}_{\text{OP8}})$ is the saturation matrix, and the k-th components $\overline{u}_{\text{OP}k}$ represent the saturation value of k-th components of u_{OP}. Note that, because of the saturation of u_{OP} and u_{TM}, the virtual saturation of u'_{OP} is introduced as

$$u'_{\text{OP}} = U'_{\text{OP}} T(v'_{\text{OP}}) \tag{39}$$

where v'_{OP} is the virtual control input to be designed for system (38), the virtual saturation matrix is denoted as $U'_{\text{OP}} = \text{diag}(\overline{u}'_{\text{OP1}}, \overline{u}'_{\text{OP2}}, \cdots, \overline{u}'_{\text{OP8}})$, and the k-th component $\overline{u}'_{\text{OP}k}$ are the virtual saturation value of k-th component of u'_{OP}. Therefore, the following relationship holds by considering the actual saturation of u_{OP} and u_{TM} together with the virtual saturation of u'_{OP}

$$U'_{\text{OP}} T(v'_{\text{OP}}) = U_{\text{OP}} T(v_{\text{OP}}) + H_{\text{OP}}^{-1} D_{\text{OP}} U_{\text{TM}} T(v_{\text{TM}}) \tag{40}$$

In fact, the control law $u_{TM} = U_{TM}T(v_{TM})$ is already given by Equation (35); hence, once the $u'_{OP} = U'_{OP}T(v'_{OP})$ is determined, the virtual control input v_{OP} can be obtained by

$$v_{OP} = T^{-1}\left(U_{OP}^{-1}U'_{OP}T(v'_{OP}) - U_{OP}^{-1}H_{OP}^{-1}D_{OP}U_{TM}T(v_{TM})\right) \quad (41)$$

where $T^{-1}(\cdot): D \to \mathbb{R}^8$, $D = \{y \in \mathbb{R}^8 \mid \|y\|_\infty < 1\} \subset \mathbb{R}^8$, specifically, for any $y \in D$, $T^{-1}(y): [y_1 \; y_2 \; \cdots \; y_8] \mapsto [\text{artanh}(y_1) \; \text{artanh}(y_2) \; \cdots \; \text{artanh}(y_8)]$, and the artanh$(\cdot)$ represents the inverse hyperbolic tangent function. Hence, the term on the right side of (40) within the bracket of T^{-1} must satisfy the following condition to make v_{OP} to be solved:

$$\left(U_{OP}^{-1}U'_{OP}T(v'_{OP}) - U_{OP}^{-1}H_{OP}^{-1}D_{OP}U_{TM}T(v_{TM})\right) \in D \quad (42)$$

It should be noted that the virtual control input v'_{OP} would be determined by the controller designed in the last part of this subsection. Therefore, the virtual saturation matrix U'_{OP} is the only parameter to be designed to make (42) hold. From a physical perspective, the virtual saturation matrix U'_{OP} should be bounded, because of the saturation of the actual control input. The sufficient boundary is established in the following theorem:

Theorem 1. *if the inequality* $\|U'_{OP}\|_\infty < \frac{1}{\|U_{OP}^{-1}\|_\infty} - 10\|U_{TM}\|_\infty$ *is satisfied, then (42) holds.*

Proof of Theorem 1. The specific nonzero element of the matrix $H_{OP}^{-1}D_{OP} \in \mathbb{R}^{8\times 12}$ is denoted as a_{ij}, where the a_{ij} is either a constant less than 1 or a linear combination of $\sin \zeta_1$, $\sin \zeta_2$, $\cos \zeta_1$, and $\cos \zeta_2$, and the coefficients of every linear combination are all less than 1. Hence, it can be conservatively concluded that $\left\|H_{OP}^{-1}D_{OP}\right\|_\infty < 10$ by counting the maximum number of times $\sin \zeta_1$, $\sin \zeta_2$, $\cos \zeta_1$, $\cos \zeta_2$ and the constant appear in a row. Then, the following inequality holds:

$$\frac{1}{\left\|U_{OP}^{-1}\right\|_\infty} - 10\|U_{TM}\|_\infty \leq \frac{1}{\left\|U_{OP}^{-1}\right\|_\infty} - \left\|H_{OP}^{-1}D_{OP}\right\|_\infty \|U_{TM}\|_\infty \quad (43)$$

With this inequality (43), together with the inequality condition mentioned in Theorem 1, the following relationship can be obtained: $\|U'_{OP}\|_\infty < \frac{1}{\|U_{OP}^{-1}\|_\infty} - \left\|H_{OP}^{-1}D_{OP}\right\|_\infty \|U_{TM}\|_\infty$. Obviously, it is equivalent to $\|U'_{OP}\|_\infty + \left\|H_{OP}^{-1}D_{OP}\right\|_\infty \|U_{TM}\|_\infty < \frac{1}{\|U_{OP}^{-1}\|_\infty}$. According to the property of the ∞-norm shown in (6), $\left\|H_{OP}^{-1}D_{OP}U_{TM}\right\|_\infty \leq \left\|H_{OP}^{-1}D_{OP}\right\|_\infty \|U_{TM}\|_\infty$. Then, the inequality $\|U'_{OP}\|_\infty + \left\|H_{OP}^{-1}D_{OP}U_{TM}\right\|_\infty < \frac{1}{\|U_{OP}^{-1}\|_\infty}$ is obtained, and the following inequality holds:

$$\left\|U_{OP}^{-1}\right\|_\infty \left(\|U'_{OP}\|_\infty + \left\|H_{OP}^{-1}D_{OP}U_{TM}\right\|_\infty\right) < 1 \quad (44)$$

Finally, using the inequality (44) and the properties shown in the inequality (5, 6), the norm $\left\|U_{OP}^{-1}U'_{OP}T(v'_{OP}) - U_{OP}^{-1}H_{OP}^{-1}D_{OP}U_{TM}T(v_{TM})\right\|_\infty$ is bounded as shown in the following procedures:

$$\begin{aligned}
&\left\|U_{OP}^{-1}U'_{OP}T(v'_{OP}) - U_{OP}^{-1}H_{OP}^{-1}D_{OP}U_{TM}T(v_{TM})\right\|_\infty \\
&\leq \left\|U_{OP}^{-1}U'_{OP}T(v'_{OP})\right\|_\infty + \left\|U_{OP}^{-1}H_{OP}^{-1}D_{OP}U_{TM}T(v_{TM})\right\|_\infty \\
&\leq \left\|U_{OP}^{-1}\right\|_\infty \|U'_{OP}\|_\infty \|T(v'_{OP})\|_\infty + \left\|U_{OP}^{-1}\right\|_\infty \left\|H_{OP}^{-1}D_{OP}U_{TM}\right\|_\infty \|T(v_{TM})\|_\infty \\
&\leq \left\|U_{OP}^{-1}\right\|_\infty \|U'_{OP}\|_\infty + \left\|U_{OP}^{-1}\right\|_\infty \left\|H_{OP}^{-1}D_{OP}U_{TM}\right\|_\infty \\
&= \left\|U_{OP}^{-1}\right\|_\infty \left(\|U'_{OP}\|_\infty + \left\|H_{OP}^{-1}D_{OP}U_{TM}\right\|_\infty\right) < 1
\end{aligned} \quad (45)$$

Hence, $\left(U_{\text{OP}}^{-1}U'_{\text{OP}}T(v'_{\text{OP}}) - U_{\text{OP}}^{-1}H_{\text{OP}}^{-1}D_{\text{OP}}U_{\text{TM}}T(v_{\text{TM}})\right) \in D$, i.e., condition (42) holds. □

The virtual saturation matrix U'_{OP} can be selected based on Theorem 1. Now, the following statement is going to design the virtual control input v'_{OP}.

By substituting Equation (39) into Equation (38) and considering the SDC form of $u'_{\text{OP}} = U'_{\text{OP}}T(v'_{\text{OP}})$, i.e., $u'_{\text{OP}} = U'_{\text{OP}}G(v'_{\text{OP}})v'_{\text{OP}}$, then the state-space representation of the OP system that considers the actuator saturation is obtained in the following SDC form:

$$\dot{x}_{\text{OP}} = A_{\text{OP}}x_{\text{OP}} + B_{\text{OP}}H_{\text{OP}}U'_{\text{OP}}G(v'_{\text{OP}})v'_{\text{OP}} \tag{46}$$

where

$$G(v'_{\text{OP}}) = \begin{cases} \text{diag}\left(\frac{\tanh(v'_{\text{OP1}})}{v'_{\text{OP1}}}, \frac{\tanh(v'_{\text{OP2}})}{v'_{\text{OP2}}}, \ldots, \frac{\tanh(v'_{\text{OP8}})}{v'_{\text{OP8}}}\right), & v'_{\text{OP}} \neq 0 \\ I_8, & v'_{\text{OP}} = 0 \end{cases} \tag{47}$$

In order to control the OP system using an SDRE integral servomechanism, the state vector x_{OP} is decomposed as $\bar{x}_{\text{OP}} = \begin{bmatrix} x_{\text{OPR}}^{\text{T}} & x_{\text{OPN}}^{\text{T}} \end{bmatrix}^{\text{T}} = E_{\text{OP}}x_{\text{OP}}$, where the state $x_{\text{OPR}} = \begin{bmatrix} r_1^{(I)\text{T}} & \underline{\theta}_{1I}^{\text{T}} & \dot{r}_1^{(I)\text{T}} & \dot{\underline{\theta}}_{1I}^{\text{T}} \end{bmatrix}^{\text{T}} \in \mathbb{R}^{12}$ represents the state expected to track a designed trajectory r_{OP}, and x_{OPN} denotes the state vector composed of the remaining elements. $E_{\text{OP}} \in \mathbb{R}^{16 \times 16}$ is the elementary row transformation matrix from x_{OP} to $\bar{x}_{\text{OP}}$. Then, $\bar{x}_{\text{OP}}$ can be augmented to $\tilde{x}_{\text{OP}} = \begin{bmatrix} \bar{x}_{\text{OP}}^{\text{T}} & x_{\text{OPI}}^{\text{T}} \end{bmatrix}^{\text{T}} \in \mathbb{R}^{28}$, where $x_{\text{OPI}} \in \mathbb{R}^{12}$ is the integral state of x_{OPR}, i.e., $x_{\text{OPI}} = \int_0^t x_{\text{OPR}}dt$. Finally, the augmented system can be obtained as follows:

$$\dot{\tilde{x}}_{\text{OP}} = \tilde{A}_{\text{OP}}(\tilde{x}_{\text{OP}})\tilde{x}_{\text{OP}} + \tilde{B}_{\text{OP}}(\tilde{x}_{\text{OP}}, v'_{\text{OP}})v'_{\text{OP}} \tag{48}$$

where

$$\tilde{A}_{\text{OP}} = \begin{bmatrix} E_{\text{OP}}A_{\text{OP}}E_{\text{OP}}^{-1} & 0_{16 \times 12} \\ I_{12} \vdots 0_{12 \times 4} & 0_{12 \times 12} \end{bmatrix} \in \mathbb{R}^{28 \times 28} \tag{49}$$

$$\tilde{B}_{\text{OP}} = \begin{bmatrix} E_{\text{OP}}B_{\text{OP}}H_{\text{OP}}U'_{\text{OP}}G(v'_{\text{OP}}) \\ 0_{12 \times 8} \end{bmatrix} \in \mathbb{R}^{28 \times 8} \tag{50}$$

The SDRE control law is in the form of

$$v'_{\text{OP}} = -\tilde{R}_{\text{OP}}^{-1}\tilde{B}_{\text{OP}}^{\text{T}}(\tilde{x}_{\text{OP}}, v'_{\text{OP}})P(\tilde{x}_{\text{OP}}, v'_{\text{OP}})\begin{bmatrix} x_{\text{OPR}} - r_{\text{OP}} \\ x_{\text{OPN}} \\ x_{\text{OPI}} - \int_0^t r_{\text{OP}}dt \end{bmatrix} \tag{51}$$

where $P(\tilde{x}_{\text{OP}}, v'_{\text{OP}})$ is the positive semi-definite solution of the following state-dependent Riccati equation:

$$P(\tilde{x}_{\text{OP}}, v'_{\text{OP}})\tilde{A}_{\text{OP}}(\tilde{x}_{\text{OP}}) + \tilde{A}_{\text{OP}}^{\text{T}}(\tilde{x}_{\text{OP}})P(\tilde{x}_{\text{OP}}, v'_{\text{OP}}) + \tilde{Q}_{\text{OP}} - \\ P(\tilde{x}_{\text{OP}}, v'_{\text{OP}})\tilde{B}_{\text{OP}}(\tilde{x}_{\text{OP}}, v'_{\text{OP}})\tilde{R}_{\text{OP}}^{-1}\tilde{B}_{\text{OP}}^{\text{T}}(\tilde{x}_{\text{OP}}, v'_{\text{OP}})P(\tilde{x}_{\text{OP}}, v'_{\text{OP}}) = 0 \tag{52}$$

4. Simulation Results

In this section, the effectiveness of the designed controller is verified in a simulation environment including the following setup. The time step of the numerical solution process is set to 0.1 s. At each step, the SDRE becomes an algebraic Riccati equation. An iterative procedure based on Newton's method is used to solve such equations.

- Solar radiation pressure: the solar pressure acts on the spacecraft surface and pushes the spacecraft out of its nominal orbit, and it is the main disturbance source that can be described as a mathematical function [38].

- TM stiffness: the presence of this stiffness is determined by the interactions of the spacecraft's gravitational and electromagnetic fields with test mass. According to Ref. [39], this stiffness can be modeled by a linear combination of the position and attitude of the test mass relative to the cage frame.
- Electrostatic actuator saturation: the TM is controlled by electrostatic actuators, which can provide control forces and torques with saturations. According to Ref. [23], the saturation value of the electrostatic actuator is given by $\bar{u}_F = \begin{bmatrix} 998 & 1056 & 595 \end{bmatrix}^T \times 10^{-9}$ N, $\bar{u}_T = \begin{bmatrix} 11 & 16 & 9 \end{bmatrix}^T \times 10^{-9}$ Nm.
- The critical initial conditions of TMs: The TM is released by the GPRM. During this time interval, the test mass will move away from the cage center. According to Ref. [40], the initial conditions of the TM are selected as those shown in Table 1.

Table 1. The initial state of TM.

Initial State	Value
$r_{MO}^{(O)}$	$\begin{bmatrix} 50 & 50 & 50 \end{bmatrix}^T$ μm
θ_{MO}	$\begin{bmatrix} 3 & 3 & 3 \end{bmatrix}^T$ mrad
$\dot{r}_{MO}^{(O)}$	$\begin{bmatrix} 10 & 10 & 10 \end{bmatrix}^T$ μm/s
$\dot{\theta}_{MO}$	$\begin{bmatrix} 100 & 100 & 100 \end{bmatrix}^T$ μrad/s

where $r_{MO}^{(O)}$ is the initial relative position of the TM with respect to the center of the local cage frame, and θ_{MO} represents the initial Euler 123 angles of the TM with respect to the local cage frame, and $\dot{r}_{MO}^{(O)}$ and $\dot{\theta}_{MO}$ are the time derivatives of $r_{MO}^{(O)}$ and θ_{MO} at the initial time, respectively.

The parameters of the drag-free satellite shown in the Table 2 are selected as those in Ref. [41].

Table 2. Parameters of drag-free spacecraft.

Parameters	Value
SC Mass	1500 kg
SC Inertia	diag (800, 800, 1000) kgm^2
OA Mass	71 kg
OA Inertia	diag (6, 6, 17) kgm^2
OA Stiffness	90,000 Nm/rad
OA Damping	80 Nms/rad
OA Mounting Position	0.36 m
TM Mass	1.96 kg
TM Inertia	diag (0.6912, 0.6912, 0.6912) × 10^{-3} kgm^2

Once the TM is released, the control system is activated. The drag-free satellite is simulated by the nonlinear dynamic model (22) with a simulation horizon of 1000 s. The results in Figures 4 and 5 verify the validity of the SDRE controller for capturing the TM after it is released. The relative position and attitude of the TM with respect to the local cage frame can be controlled effectively. In Figures 6 and 7, the steady state of the TM is presented. The results further validate the control accuracy of the proposed controller. As shown in Figures 8 and 9, the control inputs are strictly within the range of saturation values because of the property of the selected saturation function. The performance is summarized in Table 3. The max settling time is about 400 s. The maximum overshoot in the relative position of the TM with respect to the cage center is 1.967 mm, and the maximum overshoot of the attitude response of the TM is 2.111×10^{-5} rad. Both of these two values are within the range of the sensor [23]. The most important index to evaluate the performance of release control is the steady state error. According to Ref. [23], the steady state error of the TM position should be within $\pm 2 \times 10^{-6}$ m, and the steady state error of the TM attitude should be within $\pm 5 \times 10^{-6}$ rad. As shown in Table 3, the max steady state error of the TM position is 7.280×10^{-8} m, and the max steady state error of the TM attitude is 2.735×10^{-10} rad. In Table 4, the maximum value of control input is presented, and it is shown that the saturation constraints are satisfied.

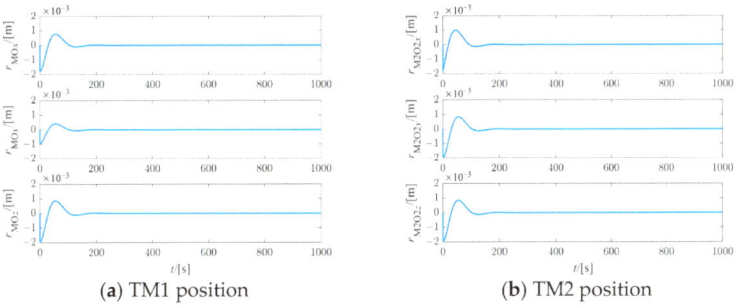

(a) TM1 position   (b) TM2 position

Figure 4. TM position varying with time during release control phase.

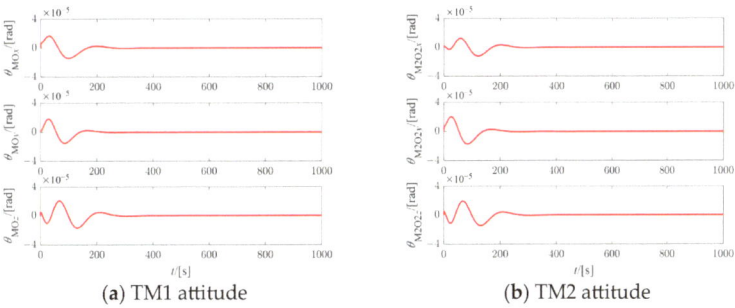

(a) TM1 attitude   (b) TM2 attitude

Figure 5. TM attitude varying with time during release control phase.

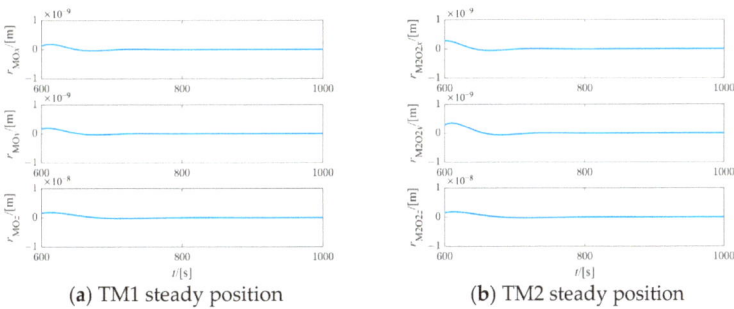

(a) TM1 steady position   (b) TM2 steady position

Figure 6. Steady position of TMs varying with time during release control phase.

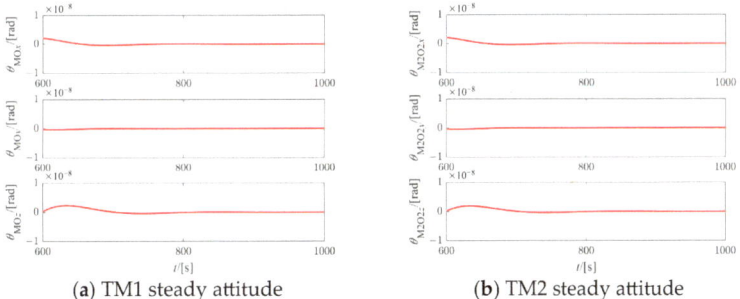

(a) TM1 steady attitude   (b) TM2 steady attitude

Figure 7. Steady attitude of TMs varying with time during release control phase.

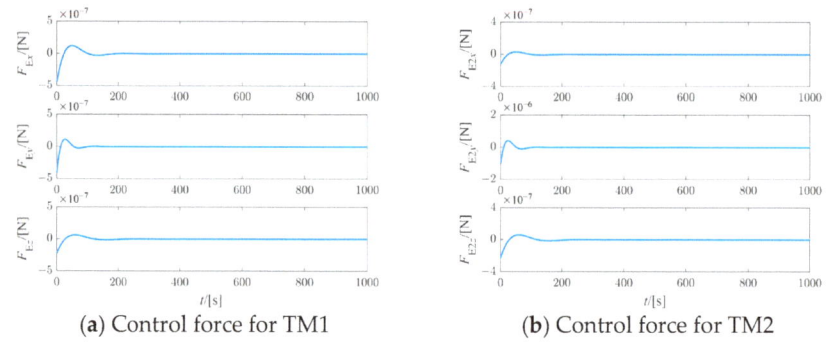

(a) Control force for TM1   (b) Control force for TM2

Figure 8. Control force for TMs varying with time during release control phase.

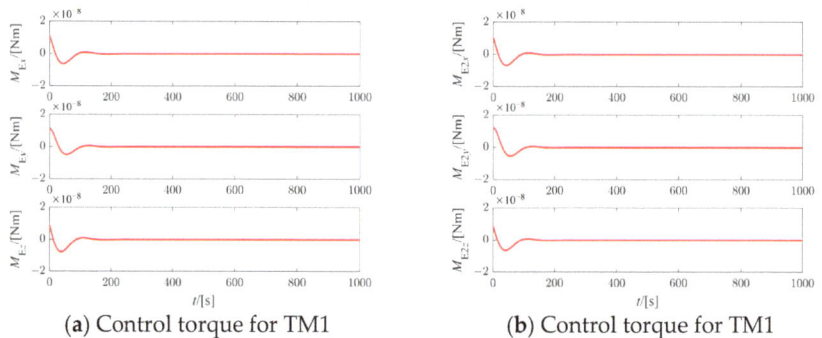

(a) Control torque for TM1   (b) Control torque for TM1

Figure 9. Control torque for TMs varying with time during release control phase.

Table 3. Performance of TM release control.

Output	Max Settling Time	Max Overshooting	Steady State Error
r_{MOx}	400 s	1.789 mm	$\pm 7.280 \times 10^{-8}$ m
r_{MOy}	400 s	0.946 mm	$\pm 4.491 \times 10^{-8}$ m
r_{MOz}	400 s	1.967 mm	$\pm 3.009 \times 10^{-8}$ m
θ_{MOx}	200 s	1.671×10^{-5} rad	$\pm 2.085 \times 10^{-9}$ rad
θ_{MOy}	200 s	1.824×10^{-5} rad	$\pm 2.696 \times 10^{-10}$ rad
θ_{MOz}	200 s	2.111×10^{-5} rad	$\pm 2.735 \times 10^{-10}$ rad

Table 4. The peak value of control input in TM release phase.

Control Input	Peak Value	Control Input	Peak Value
F_{Ex}	444 nN	F_{E2x}	118 nN
F_{Ey}	403 nN	F_{E2y}	1022 nN
F_{Ez}	226 nN	F_{E2z}	226 nN
M_{Ex}	10.889 nNm	M_{E2x}	10.351 nNm
M_{Ey}	11.167 nNm	M_{E2y}	12.250 nNm
M_{Ez}	8.826 nNm	M_{E2z}	8.649 nNm

Figures 10 and 11 show the states of spacecraft and the relative rotation angle of OA with respect to time. The solar radiation pressure is compensated to control the position and attitude of the spacecraft accurately, and, also, the rotation angle of OA is regulated to the equilibrium value.

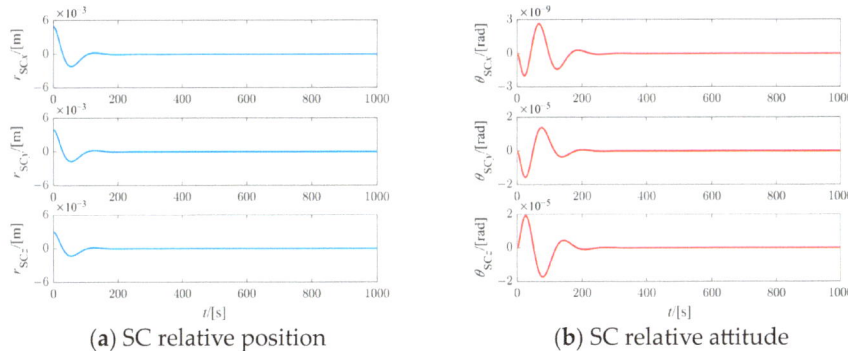

(a) SC relative position

(b) SC relative attitude

Figure 10. The relative position and attitude of SC varying with time during release control phase.

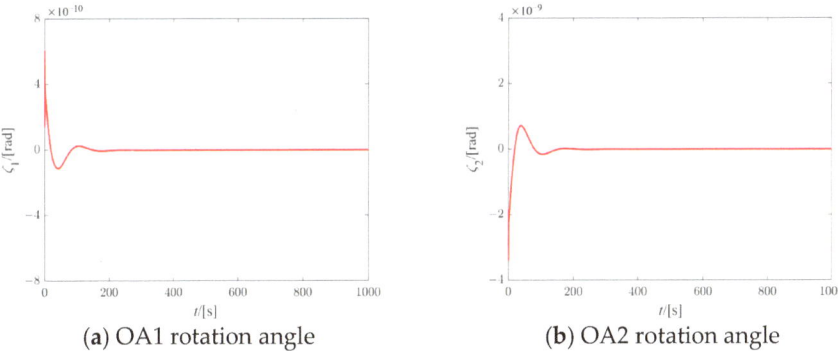

(a) OA1 rotation angle

(b) OA2 rotation angle

Figure 11. The rotation angles of OA varying with time during release control phase.

5. Conclusions

This paper presents a novel capture controller for the TM release phase based on the SDRE scheme. The whole control system consists of two SDRE controllers. One is for regulating the OP and compensating for the solar radiation pressure. The other one is designed for capturing the TM. This control strategy can eliminate the control inputs coupling between the OP loop and TM loop. Meanwhile, the actuator saturation of the electrostatic suspension cage is considered by the hyperbolic tangent function. This saturation function will lead to a non-affine system, and the SDRE controller for non-affine systems is implemented to solve saturation constraints. The simulation results demonstrate the validity of the proposed capture controller, and the strategy to solve the problem of the actuator saturation is verified to be effective. In this study, only the solar radiation pressure suffered by drag-free satellites were considered as the main perturbation because its amplitude is larger than other perturbations. However, the various perturbations on drag-free and attitude control systems, such as the actuator faults and the noise generated by the sensors and the actuators, should be considered in our future research.

Author Contributions: Conceptualization, Y.C. and T.C.; methodology, Y.C.; software, Y.C.; validation, Y.C., Y.W. and Z.W.; formal analysis, Y.C. and T.C.; investigation, Y.C.; resources, T.C.; data curation, T.C.; writing—original draft preparation, Y.C.; writing—review and editing, T.C., Y.W., Z.W. and J.T.; visualization, Y.C.; supervision, T.C.; project administration, T.C.; funding acquisition, T.C., Z.W. and J.T. All authors have read and agreed to the published version of the manuscript.

Funding: This work was supported by the National Key Research and Development Program of China (Grant No. 2022YFC2204800) and Jiangsu Funding Program for Excellent Postdoctoral Talent.

Data Availability Statement: The datasets generated during and analyzed during the current study are available from the corresponding author upon reasonable request.

Conflicts of Interest: The authors declare no conflicts of interest.

Appendix A

The specific expression of the matrix H_{OP} is as follows:

$$H_{OP}(x_{OP}) = \begin{bmatrix} R_S^I & 0_3 & 0_{3\times 1} & 0_{3\times 1} \\ 0_3 & R_S^I & 0_{3\times 1} & 0_{3\times 1} \\ 0_{1\times 3} & 0_{1\times 3} & 1 & 0 \\ 0_{1\times 3} & 0_{1\times 3} & 0 & 1 \end{bmatrix},$$

and the expression of the matrix D_{OP} is as follows:

$$D_{OP}(x_{OP}) = \begin{bmatrix} -R_S^I R_O^S & -R_S^I R_{O2}^S & 0_{3\times 3} & 0_{3\times 3} \\ -R_S^I \left(\underline{d}_{12}^{(S)} + \underline{d}_{22}^{1(S)}\right)^\times R_O^S & -R_S^I \left(\underline{d}_{13}^{(S)} + \underline{d}_{33}^{1(S)}\right)^\times R_{O2}^S & -R_S^I R_O^S & -R_S^I R_{O2}^S \\ \begin{bmatrix} 0 & -l_{22}^1 & 0 \end{bmatrix} & 0_{1\times 3} & 0_{1\times 3} & 0_{1\times 3} \\ 0_{1\times 3} & \begin{bmatrix} 0 & -l_{33}^1 & 0 \end{bmatrix} & 0_{1\times 3} & 0_{1\times 3} \end{bmatrix},$$

where $\underline{d}_{12}^{(S)}$ represents the coordinates of the vector from O_1 to the mounting point of the OA1 in BRF-1, $\underline{d}_{22}^{1(S)}$ represents the coordinates of the vector from the mounting point of the OA1 to the center of the corresponding electrostatic suspension cage in BRF-1, and l_{22}^1 is the modulus of $\underline{d}_{22}^{1(S)}$. Similarly, $\underline{d}_{13}^{(S)}$ represents the coordinates of the vector from O_1 to the mounting point of the OA2 in BRF-1, $\underline{d}_{33}^{1(S)}$ represents the coordinates of the vector from the mounting point of the OA2 to the center of the corresponding electrostatic suspension cage in BRF-1, and l_{33}^1 is the modulus of $\underline{d}_{33}^{1(S)}$.

References

1. Aasi, J.; Abbott, B.P.; Abbott, R.; Abbott, T.; Abernathy, M.R.; Ackley, K.; Adams, C.; Adams, T.; Addesso, P.; Adhikari, R.X.; et al. Advanced ligo. *Class. Quant. Grav.* **2015**, *32*, 074001.
2. Contaldi, C.R.; Pieroni, M.; Renzini, A.I.; Cusin, G.; Karnesis, N.; Peloso, M.; Ricciardone, A.; Tasinato, G. LISA Cosmology Working Group Maximum likelihood map making with the Laser Interferometer Space Antenna. *Phys. Rev. D* **2020**, *102*, 043502. [CrossRef]
3. Barausse, E.; Berti, E.; Hertog, T.; Hughes, S.A.; Jetzer, P.; Pani, P.; Sotiriou, T.P.; Tamanini, N.; Witek, H.; Yagi, K.; et al. Prospects for fundamental physics with LISA. *Gen. Relativ. Gravit.* **2020**, *52*, 1–33. [CrossRef]
4. Liu, H.; Luo, Z.; Jin, G. The Development of phasemeter for taiji space gravitational wave detection. *Microgav. Sci. Technol.* **2018**, *30*, 775–781. [CrossRef]
5. Ruan, W.-H.; Guo, Z.-K.; Cai, R.-G.; Zhang, Y.-Z. Taiji program: Gravitational-wave sources. *Int. J. Mod. Phys. A* **2020**, *35*, 2050075. [CrossRef]
6. Luo, J.; Chen, L.-S.; Duan, H.-Z.; Gong, Y.-G.; Hu, S.; Ji, J.; Liu, Q.; Mei, J.; Milyukov, V.; Sazhin, M.; et al. TianQin: A space-borne gravitational wave detector. *Class. Quant. Grav.* **2016**, *33*, 035010. [CrossRef]
7. Harry, G.M.; Fritschel, P.; A Shaddock, D.; Folkner, W.; Phinney, E.S. Laser interferometry for the Big Bang Observer. *Class. Quant. Grav.* **2006**, *23*, 4887–4894. [CrossRef]
8. Sato, S.; Kawamura, S.; Ando, M.; Nakamura, T.; Tsubono, K.; Araya, A.; Funaki, I.; Ioka, K.; Kanda, N.; Moriwaki, S.; et al. The Status of DECIGO. *J. Phys. Conf. Ser.* **2017**, *840*, 012010.
9. Geng, S.; Cao, S.; Liu, T.; Biesiada, M.; Qi, J.; Liu, Y.; Zhu, Z.-H. Gravitational-wave constraints on the cosmic opacity at $z \sim 5$: Forecast from space gravitational-wave antenna DECIGO. *Astrophys. J.* **2020**, *905*, 54. [CrossRef]
10. McKenzie, K.; Spero, R.; Klipstein, W.; de Vine, G.; Ware, B.; Vallisneri, M.; Cutler, C.; Ziemer, J.; Shaddock, D.; Skoug, R.; et al. LAGRANGE: A space-based gravitational-wave detector with geometric suppression of spacecraft noise. Presented at the Workshop on Gravitational Wave Mission Concepts, Linthicum, MD, USA, 20 December 2011.
11. Caprini, C.; Chala, M.; Dorsch, G.C.; Hindmarsh, M.; Huber, S.J.; Konstandin, T.; Kozaczuk, J.; Nardini, G.; No, J.M.; Rummukainen, K.; et al. Detecting gravitational waves from cosmological phase transitions with LISA: An update. *J. Cosmol. Astropart. Phys.* **2020**, *2020*, 024. [CrossRef]

12. Zanoni, C.; Bortoluzzi, D.; Conklin, J.W.; Köker, I.; Seutchat, B.; Vitale, S. Summary of the results of the LISA-Pathfinder Test Mass release. *J. Phys. Conf. Ser.* **2015**, *610*, 012022. [CrossRef]
13. Koker, I.; Rozemeijer, H.; Stary, F.; Reichenberger, K. Alignment and testing of the GPRM as part of the LTP caging mechanism. In Proceedings of the 15th European Space Mechanisms and Tribology Symposium, Noordwijk, The Netherlands, 25–27 September 2013.
14. Zanoni, C.; Bortoluzzi, D.; Conklin, J.W.; Köker, I.; Marirrodriga, C.G.; Nellen, P.M.; Vitale, S. Testing the injection of the LISA-pathfinder test mass into geodesic conditions. In Proceedings of the 15th European Space Mechanism and Tribology Symposium (ESMATS), Noordwijk, The Netherlands, 25–27 September 2013.
15. Schleicher, A.; Ziegler, T.; Schubert, R.; Brandt, N.; Bergner, P.; Johann, U.; Fichter, W.; Grzymisch, J. In-orbit performance of the LISA Pathfinder drag-free and attitude control system. *CEAS Space J.* **2018**, *10*, 471–485. [CrossRef]
16. Bortoluzzi, D.; Mäusli, P.A.; Antonello, R.; Nellen, P.M. Modeling and identification of an electro-mechanical system: The LISA grabbing positioning and release mechanism case. *Adv. Space Res.* **2011**, *47*, 453–465. [CrossRef]
17. Benedetti, M.; Bortoluzzi, D.; Vitale, S. A Momentum transfer measurement technique between contacting free-falling bodies in the presence of adhesion. *J. Appl. Mech.* **2008**, *75*, 011016. [CrossRef]
18. Bortoluzzi, D.; Conklin, J.W.; Zanoni, C. Prediction of the LISA-Pathfinder release mechanism in-flight performance. *Adv. Space Res.* **2013**, *51*, 1145–1156. [CrossRef]
19. Bortoluzzi, D.; De Cecco, M.; Vitale, S.; Benedetti, M. Dynamic measurements of impulses generated by the separation of adhered bodies under near-zero gravity conditions. *Exp. Mech.* **2008**, *48*, 777–787. [CrossRef]
20. Montemurro, F.; Fichter, W.; Schlotterer, M.; Vitale, S. Control Design of the test mass release mode for the lisa pathfinder mission. *AIP Conf. Proc.* **2006**, *873*, 583–587.
21. Capicchiano, L. Test Mass Release for LISA ESA Mission–Control Design and MonteCarlo Analysis. Doctoral Dissertation, Politecnico di Torino, Turin, Italy, 2020.
22. Lian, X.; Zhang, J.; Chang, L.; Song, J.; Sun, J. Test mass capture for drag-free satellite based on RBF neural network adaptive sliding mode control. *Adv. Space Res.* **2021**, *69*, 1205–1219. [CrossRef]
23. Vidano, S.; Novara, C.; Grzymisch, J.; Pagone, M. The LISA DFACS: Preliminary model predictive control for the test mass release phase. In Proceedings of the 71st International Astronautical Congress, Online, 12–14 October 2020.
24. Gioia, A. Time-Optimal Electrostatic Control and Capture of a Free-Falling Test Mass. Doctoral Dissertation, Politecnico di Milano, Milan, Italy, 2020.
25. Lin, M.; Zhang, J.; He, Y. Minimum-time control for the test mass release phase of drag-free spacecraft. *Space Sci. Technol.* **2024**, *4*, 0151.
26. Bai, S.; Wang, Y.; Liu, H.; Sun, X. Spacecraft fast fly-around formations design using the parallelogram configuration. In *Nonlinear Dynamics*; Springer Nature: Berlin/Heidelberg, Germany, 2024; pp. 1–22.
27. Zhou, H.; Jiao, B.; Dang, Z.; Yuan, J. Parametric formation control of multiple nanosatellites for cooperative observation of China Space Station. *Astrodynamics* **2024**, *8*, 77–95. [CrossRef]
28. Çimen, T. State-Dependent Riccati Equation (SDRE) Control: A Survey. *IFAC Proc. Vol.* **2008**, *41*, 3761–3775. [CrossRef]
29. Chen, T.; Shan, J.; Wen, H.; Xu, S. Review of attitude consensus of multiple spacecraft. *Astrodynamics* **2022**, *6*, 329–356. [CrossRef]
30. Sun, G.; Zhou, M.; Jiang, X. Non-cooperative spacecraft proximity control considering target behavior uncertainty. *Astrodynamics* **2022**, *6*, 399–411. [CrossRef]
31. Cloutier, J.; Stansbery, D. Control of a Continuously Stirred Tank Reactor Using an Asymmetric Solution of the State-Dependent Riccati Equation. In Proceedings of the Conference on Control Applications, Kohala Coast, HI, USA, 22–27 August 1999; pp. 893–898.
32. Stansbery, D.T.; Cloutier, J.R. Position and attitude control of a spacecraft using the state-dependent Riccati equation technique. In Proceedings of the 2000 American Control Conference. ACC (IEEE Cat. No.00CH36334), Chicago, IL, USA, 28–30 June 2000; Volume 3, pp. 1867–1871.
33. Vidano, S.; Novara, C.; Colangelo, L.; Grzymisch, J. The LISA DFACS: A nonlinear model for the spacecraft dynamics. *Aerosp. Sci. Technol.* **2020**, *107*, 106313. [CrossRef]
34. An, J.H.; Kim, H.S. Interval Type-2 Fuzzy-Model-Based Sampled-Data Control of an AUV Depth System with Input Saturation. *Actuators* **2024**, *13*, 71. [CrossRef]
35. Wu, J.; Li, B.; Li, J.; Li, M.; Yang, B. Global Stabilization of Control Systems with Input Saturation and Multiple Input Delays. *Actuators* **2024**, *13*, 306. [CrossRef]
36. Zhou, Y.; Liu, H.; Guo, H. L1 Adaptive Fault-Tolerant Control for Nonlinear Systems Subject to Input Constraint and Multiple Faults. *Actuators* **2024**, *13*, 258. [CrossRef]
37. Rafee Nekoo, S.; Geranmehr, B. Nonlinear observer based optimal control via state-dependent Riccati equation for a class of non-affine in control systems. *J. Control Eng. Appl. Inf.* **2014**, *16*, 5–13.
38. Georgevic, R.M. *Mathematical Model of the Solar Radiation Force and Torques Acting on the Components of a Spacecraft, Technical Report for NASA JPL*; Jet Propulsion Lab.: Pasadena, CA, USA, October 1971.
39. Merkowitz, S.M.; Haile, W.B.; Conkey, S.; Kelly, W.; Peabody, H. Self-gravity modelling for LISA. *Class. Quant. Grav.* **2005**, *22*, S395–S402. [CrossRef]

40. Bortoluzzi, D.; Armano, M.; Audley, H.; Auger, G.; Baird, J.; Binetruy, P.; Born, M.; Bortoluzzi, D.; Brandt, N.; Bursi, A.; et al. Injection of a Body into a Geodesic: Lessons Learnt from the LISA Pathfinder Case. In Proceedings of the 43rd Aerospace Mechanisms Symposium, Santa Clara, CA, USA, 4–6 June 2016; p. 20160008114.
41. Virdis, M. A Meteoroid Impact Recovery Control System for the LISA Gravitational Wave Observatory. Doctoral Dissertation, Politecnico di Torino, Turin, Italy, 2021.

Disclaimer/Publisher's Note: The statements, opinions and data contained in all publications are solely those of the individual author(s) and contributor(s) and not of MDPI and/or the editor(s). MDPI and/or the editor(s) disclaim responsibility for any injury to people or property resulting from any ideas, methods, instructions or products referred to in the content.

Article

Thrust Model and Trajectory Design of an Interplanetary CubeSat with a Hybrid Propulsion System

Alessandro A. Quarta

Department of Civil and Industrial Engineering, University of Pisa, I-56122 Pisa, Italy; alessandro.antonio.quarta@unipi.it

Abstract: This paper analyzes the performance of an interplanetary CubeSat equipped with a hybrid propulsion system (HPS), which combines two different types of thrusters in the same deep space vehicle, in a heliocentric transfer between two assigned (Keplerian) orbits. More precisely, the propulsion system of the CubeSat considered in this work consists of a combination of a (low-performance) photonic solar sail and a more conventional solar electric thruster. In particular, the characteristics of the solar electric thruster are modeled using a recent mathematical approach that describes the performance of the miniaturized engine that will be installed on board the proposed ESA's M-ARGO CubeSat. The latter will hopefully be the first interplanetary CubeSat to complete a heliocentric transfer towards a near-Earth asteroid using its own propulsion system. In order to simplify the design of the CubeSat attitude control subsystem, we assume that the orientation of the photonic solar sail is kept Sun-facing, i.e., the sail reference plane is perpendicular to the Sun-CubeSat line. That specific condition can be obtained, passively, by using an appropriate design of the shape of the sail reflective surface. The performance of an HPS-based CubeSat is analyzed by optimizing the transfer trajectory in a three-dimensional heliocentric transfer between two closed orbits of given characteristics. In particular, the CubeSat transfer towards the near-Earth asteroid 99942 Apophis is studied in detail.

Keywords: hybrid propulsion system; photonic solar sail; solar electric thruster; interplanetary CubeSat; M-ARGO CubeSat; preliminary trajectory design; asteroid 99942 Apophis

Citation: Quarta, A.A. Thrust Model and Trajectory Design of an Interplanetary CubeSat with a Hybrid Propulsion System. *Actuators* **2024**, *13*, 384. https://doi.org/10.3390/act13100384

Academic Editors: Ti Chen, Junjie Kang, Shidong Xu, Shuo Zhang and Dongdong Li

Received: 29 August 2024
Revised: 24 September 2024
Accepted: 25 September 2024
Published: 1 October 2024

Copyright: © 2024 by the authors. Licensee MDPI, Basel, Switzerland. This article is an open access article distributed under the terms and conditions of the Creative Commons Attribution (CC BY) license (https://creativecommons.org/licenses/by/4.0/).

1. Introduction

The trajectory analysis and the transfer performance of an interplanetary spacecraft are closely related to the specific characteristics of the propulsion system installed on board [1–3]. This crucial aspect of spacecraft design is particularly important when considering a small spacecraft, such as a CubeSat, whose recent use in interplanetary applications is revealing its high potential [4–7] thanks to successful missions such as the pioneering NASA's Mars Cube One (MarCO) [8,9] and the more recent Light Italian CubeSat for Imaging of Asteroids (LICIACube) of the Italian Space Agency (ASI) [10]. In the context of the design of a small-spacecraft-based interplanetary mission [11–13], the choice of the type of propulsion system is usually a compromise between conflicting requirements [14,15] such as, for example, the weight limitation (or the volume reserved to the thruster subsystem), and the possibility of having a continuous, steerable thrust vector of sufficiently high magnitude.

From the viewpoint of the trajectory design, in fact, a continuous propulsive acceleration vector can be used for a long period of time in order to cover complex interplanetary orbits, or to complete deep space missions that require a high (or very-high) velocity change [16] without an excessive propellant expenditure when the propellant mass flow rate is low enough. A steerable thrust vector with a magnitude that can be varied within a prescribed (and sufficiently wide) range can be achieved by the use of a typical electric thruster, the use of which in small spacecraft has recently become a viable option [14,15,17] due to technological advances in the miniaturization of space vehicle components and subsystems, as detailed in Refs. [18–21]. However, the use of an electric thruster within a trajectory design process introduces the usual constraint due to the finite amount of propellant mass that can be stored on board of the interplanetary (small) spacecraft.

This last constraint can be overcome by considering a more advanced (and in some ways more exotic) propulsion system, such as the Electric Solar Wind Sail (E-sail) proposed by Dr. Pekka Janhunen [22,23], which deflects the charged particles of the solar wind using an artificial electric field generated by a series of long conducting tethers [24,25], or a more well-known photonic solar sail, which is a propellantless thruster that converts solar radiation pressure into thrust using a large (usually highly reflective) membrane with a metalized film coating [26,27]. Currently, starting from the successful Japanese mission Interplanetary Kite-craft Accelerated by Radiation Of the Sun (IKAROS) in 2010 [28,29], which first tested the concept of a solar sail-based propulsion system in interplanetary space, among the different propellantless thrusters proposed in the literature only the solar sail concept seems to have the technological maturity to be effectively employed in interplanetary robotic missions whose launch can be planned in the near future. Very hopefully, the upcoming flight tests of a scaled-down version of the E-sail in a Moon-centered high-elliptic orbit [30–32], will bring this fascinating propulsion system into the list of those actually usable in scientific missions to interplanetary space.

The use of a solar sail as the primary propulsion system, however, poses some additional constraints that must be carefully considered during the spacecraft trajectory design. In fact, the propulsive capabilities of a photonic solar sail are limited by the intrinsic impossibility of pointing the thrust vector towards the Sun and by the potential difficulty of continuously varying the attitude of a large space structure in order to follow a given guidance law [33]. Furthermore, taking into account the current technological level in solar sail design [34], the maximum magnitude of the propulsive acceleration vector given by that propellantless propulsion system is typically small when compared to the (local) Sun's gravitational acceleration.

In this context, a possible solution is to employ a hybrid propulsion system (HPS) that combines, in the same spacecraft, two different types of space thrusters using different propellants or, more generally, different "thrust sources" in the case where propellantless propulsion systems are considered. The HPS concept is different from the more recent multimode propulsion system [35], which uses two different types of spacecraft engines that, however, share the same propellant type. In this case, the interesting review by Rovey et al. [36] represents an excellent starting point to recover the bibliography and technical information on this useful concept of (multimode) propulsion system. Assuming an HPS-propelled small spacecraft, this paper analyzes the transfer performance of an interplanetary CubeSat equipped with a combination of a (low-performance) photonic solar sail and a miniaturized electric thruster. The combined use of a solar sail and an electric thruster is not a new idea in the context of the heliocentric trajectory design, since such an interesting proposal (more precisely, the use of a solar sail and a nuclear electric propulsion system) was advanced by Dr. Giovanni Vulpetti over fifty years ago in one of his many pioneering articles on the use of photonic solar sail [37]. More precisely, Ref. [37] discussed the potential of a deep space probe equipped with an HPS to reach (and explore) the outer regions of the Solar System. The same idea of using a solar sail and an electric thruster was then taken up several times in the scientific literature [38,39], also by the author from the trajectory optimization point of view [40–42]. In this context, the interested reader can appreciate the elegant approach proposed by Ceriotti and McInnes [43] for the use of an HPS-equipped spacecraft for the generation of non-Keplerian orbits that can be used to observe the medium-high latitude areas of the Earth's surface. In particular, the latter mission application is a sort of evolution of the concept of the "pole sitter" [44,45] proposed by Dr. Gregory Matloff [46] about twenty years ago.

In this paper, the propulsive characteristics of the HPS are described with the sole purpose of obtaining a simple mathematical model that can be used to simulate (and optimize by defining a suitable performance index) the CubeSat transfer trajectory in a typical heliocentric mission scenario. In other words, the design of the spacecraft from the point of view of its subsystems and the integration of the HPS within the space vehicle structure are not part of the scope of this work and, for this reason, will not be covered in this paper. Consequently, the presence of an HPS installed on board the interplanetary CubeSat will be schematized by describing the part of the total propulsive acceleration due to the miniaturized solar electric thruster and the part related to the presence of the photonic solar sail. Note that the presence of a complex system such as the HPS would require, in reality, the careful choice of the spacecraft attitude variation law during interplanetary flight, in order to avoid the propellant expelled by the solar electric thruster hitting the solar sail, thus degrading [47,48] the reflective characteristics of the film that covers the sail membrane. In order to simplify the design of the small spacecraft attitude control subsystem, we assume that the orientation of the (low-performance) photonic solar sail is kept Sun-facing, i.e., the sail nominal plane is perpendicular to the Sun-CubeSat line. That specific condition is obtained, passively, by using an appropriate design of the shape of the reflective surface. Indeed, as McInnes [49] pointed out, a Sun-facing attitude is obtained passively using a slightly conical (or, in general, an axially symmetric [50]) sail shape with the apex pointing towards the Sun.

The propulsive characteristics of the HPS considered in this work are obtained by analyzing the literature data regarding two interplanetary CubeSats equipped with a solar electric thruster or a photonic solar sail [51]. In particular, the characteristics of the miniaturized solar electric thruster are modeled using a recent mathematical description proposed by the author [52], which schematize the performance of the space engine that will be installed on board the proposed ESA's CubeSat M-ARGO (which is the acronym for The Miniaturised Asteroid Remote Geophysical Observer); see the artistic concept in Figure A1. In fact, M-ARGO will hopefully be the first interplanetary CubeSat to complete a heliocentric orbit transfer using the solar electric thruster (in this specific case, a single radiofrequency gridded ion thruster) installed on board [53,54]. In particular, the M-ARGO's planned heliocentric orbital transfer will start from the second collinear Lagrangian point of the Sun-Earth system, where the (piggyback) CubeSat will be parked by the launch system and will allow the small spacecraft to complete a rendezvous with a near-Earth asteroid. The asteroid to be reached has yet to be selected from a set of possible targets [55].

More precisely, the mathematical model of the solar electric propulsion system used in this paper coincides with the one described in Ref. [52], which is in turn a simplified version of the surrogate, and elegant, model proposed by Topputo et al. [54] a few years ago for the preliminary trajectory design of M-ARGO. This simplified thrust model is summarized in Appendix A, while the interested reader can find all the useful details of the mathematical model and some interesting mission applications in the two related references [52,54]. On the other hand, the propulsive characteristics of the solar sail were chosen using data available in the literature regarding NASA's mission Near-Earth Asteroid Scout (NEA Scout) [56–58], an artistic representation of which is shown in Figure A3. In fact, NEA Scout was one of the ten CubeSat deployed during the maiden flight of the Space Launch System in November 2022, and that interplanetary mission was designed to approach asteroid 2020 GE and to take some pictures of this very small celestial body which approaches the orbit of Earth. Unfortunately, the ground station was unable to establish contact with the solar sail-based CubeSat, so the NEA Scout was considered lost in December 2022. The thrust model of the photonic solar sail that constitutes the second part of the HPS is briefly described in Appendix B, where the thermo-optical characteristics of the reflective film have been obtained by the interesting work of Heaton et al. [59]. The latter reference has been also used by Pezent et al. [60] to obtain high-fidelity trajectories of NEA Scout CubeSat.

In addition to the two appendices that briefly describe, as already mentioned, the thrust models of the miniaturized electric thruster and the photonic solar sail, and to Section 4 that contains, as usual, the conclusions of the work, this paper has two other sections.

In particular, the next section describes the mission scenario and schematizes the (total) propulsive acceleration vector by using the simple mathematical models summarized in Appendixes A and B. Furthermore, the next section briefly describes the approach used to study the optimal transfer trajectory of the HPS-propelled spacecraft in a selected heliocentric mission scenario. In this context, a detailed description of the optimization model from the mathematical point of view has been avoided, since the general form of the latter has been repeatedly employed and described in detail in several (and also recent) works of the author [61]. In this regard, the bibliographical references to which the interested reader can refer are indicated in the References section. Finally, Section 3 contains the results of the numerical simulations and some related comments about the performance of the HPS in a specific heliocentric mission scenario. In particular, an orbit-to-orbit three-dimensional transfer from the Earth to asteroid 99942 Apophis is simulated, and the numerical results are studied as a function of the design characteristics of the HPS-propelled CubeSat.

2. Problem Description and Mathematical Model

The main purpose of this section is to present the mathematical model of the HPS thrust vector. This model allows us to describe, through a compact analytical relationship, the total thrust vector T and the time variation in the mass dm/dt of an interplanetary CubeSat equipped with an HPS. The thrust model uses the equations presented in Appendixes A and B to describe the performance of the miniaturized electric thruster and the photonic solar sail in a Sun-facing configuration, respectively.

The HPS thrust model is employed to analyze the CubeSat transfer performance in a typical interplanetary mission scenario. More precisely, this work considers a three-dimensional orbit-to-orbit heliocentric transfer, that is, a transfer between two closed Keplerian orbits of assigned characteristics. In particular, the orbital elements of the CubeSat heliocentric parking orbit (or target orbit) coincide with those of the Keplerian orbit of the Earth (or the asteroid 99942 Apophis) around the Sun. Consequently, the HPS thrust model is employed to analyze a sort of ephemeris-free transfer to this interesting near-Earth asteroid that will closely approach our planet on 13 April 2029. That transfer will be studied by minimizing the flight time, without considering any potential planetary gravity assist maneuver [62], as briefly described in the last part of this section.

2.1. HPS Analytical Thrust Model

The thrust vector T given by the HPS can be considered as the (vector) sum of the part T_e due to the miniaturized electric thruster, which is analyzed in detail in Appendix A, and the part T_s due to the photonic solar sail in a Sun-facing configuration, which is discussed in Appendix B. Therefore, using Equations (A1) and (A9) to express the two vectors T_e and T_s, respectively, one obtains the (simplified) analytical form of the total thrust vector given by the HPS

$$T = T_e + T_s \equiv \tau \, T_{e_{\max_\oplus}} F \, \hat{t}_e + \eta \, P_\oplus \, A \left(\frac{r_\oplus}{r}\right)^2 \hat{r} \qquad (1)$$

where $F = F(r_\oplus/r)$ is a dimensionless auxiliary function defined as

$$F \triangleq \frac{d_{N_1}(r/r_\oplus)^4 + d_{N_2}(r/r_\oplus)^3 + d_{N_3}(r/r_\oplus)^2 + d_{N_4}(r/r_\oplus) + d_{N_5}}{(r/r_\oplus)^3 + d_{D_1}(r/r_\oplus)^2 + d_{D_2}(r/r_\oplus) + d_{D_3}} \qquad (2)$$

which has been introduced here to obtain a more compact expression of T. The meaning and values of the terms appearing on the right-hand side of Equation (1) are indicated and discussed in the two appendices, while the variation in the auxiliary function F with r is schematized in Figure 1. Note that $F = 1$ when the solar distance is equal to 1 astronomical unit, that is, when $r = r_\oplus$. Recall that $\hat{t}_e$ is the unit vector that defines the direction of the thrust given by the miniaturized electric thruster, while $\hat{r}$ is the Sun-CubeSat unit vector which coincides with the direction of the thrust given by the Sun-facing (photonic) solar sail. Therefore, the control terms in Equation (1) are the unit vector $\hat{t}_e$ and the throttle function $\tau \in [0, 1]$, so that there are three scalar control variables (recall that $\|\hat{t}_e\| = 1$).

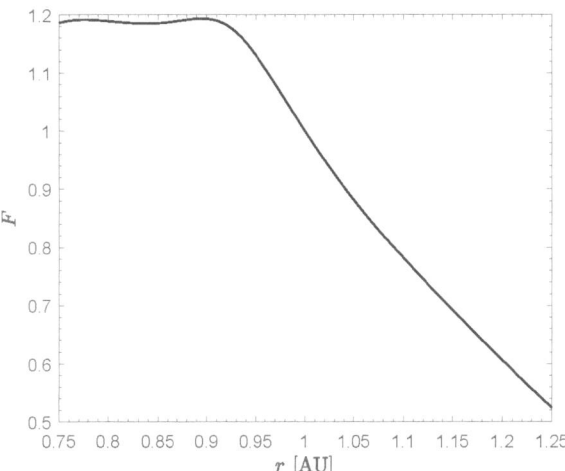

Figure 1. Variation with $r \in [0.75, 1.25]$ AU of the auxiliary function F defined in Equation (2).

According to Equation (1), the maximum value of the HPS-induced thrust magnitude is obtained when $\tau = 1$ and $\hat{t}_e = \hat{r}$, that is, when the solar electric propulsion system gives an outward radial thrust with a full throttle level. In this case, one has

$$\max(\|T\|) = \tau\, T_{e\max_\oplus} F + \eta\, P_\oplus A \left(\frac{r_\oplus}{r}\right)^2 \tag{3}$$

which is shown in Figure 2 as a function of the solar distance $r \in [0.75, 1.25]$ AU. The minimum value of the thrust vector magnitude, instead, is $\min(\|T\|) = 0$, because at a given solar distance r, the value of $\|T_s\|$ is always less than the maximum value of $\|T_e\|$; compare Figure A2 with Figure A4. However, the condition $\min(\|T\|) = 0$ is obtained by balancing the solar sail-induced outward thrust with an inward radial thrust (i.e., a case in which $\hat{t}_e = -\hat{r}$) given by the electric propulsion system. This condition requires a propellant expenditure, so the presence of a coasting arc in the HPS-based CubeSat requires a total mass variation when the photonic solar sail is constrained in a Sun-facing configuration. The components of the thrust vector T are conveniently calculated in a typical Radial–Transverse–Normal (RTN) orbital reference frame, in which $\hat{i}_R \equiv \hat{r}$ is the radial unit vector, $\hat{i}_N = (\hat{r} \times v)/\|\hat{r} \times v\|$ is the normal unit vector (where v is the CubeSat inertial velocity vector), and $\hat{i}_T = \hat{i}_N \times \hat{i}_R$ is the transverse unit vector; see Figure 3 of Ref. [61]. Note that the plane $(\hat{i}_R, \hat{i}_T)$ coincides with the orbital plane of the CubeSat osculating orbit, while the direction of $\hat{i}_N$ coincides with the direction of the specific angular momentum vector. Bearing in mind the expression of the total thrust vector given by Equation (1), the three components $\{T_R, T_T, T_N\}$ of T in the RTN reference frame are written as

$$T_R = T \cdot \hat{i}_R = \tau\, T_{e\max_\oplus} F \cos\alpha + \eta\, P_\oplus A \left(\frac{r_\oplus}{r}\right)^2 \tag{4}$$

$$T_T = T \cdot \hat{i}_T = \tau\, T_{e\max_\oplus} F \sin\alpha \cos\delta \tag{5}$$

$$T_N = T \cdot \hat{i}_N = \tau\, T_{e\max_\oplus} F \sin\alpha \sin\delta \tag{6}$$

where $\alpha \in [0, 180]$ deg is the angle between the Sun-CubeSat line and the direction of the unit vector $\hat{t}_e$, while $\delta \in [0, 360]$ deg is the angle between the direction of $\hat{i}_T$ and the projection of $\hat{t}_e$ into the plane $(\hat{i}_T, \hat{i}_N)$. Note that $\{\alpha, \delta\}$ are the two angles that define the direction of $\hat{t}_e$, i.e., the direction of the electric thruster-induced thrust vector T_e in the RTN reference frame; see also Figure 6 of Ref. [61]. Indeed, the unit vector $\hat{t}_e$ can be written as a function of angles $\{\alpha, \delta\}$ as

$$\hat{t}_e = \cos\alpha\, \hat{i}_R + \sin\alpha\, \cos\delta\, \hat{i}_T + \sin\alpha\, \sin\delta\, \hat{i}_N \tag{7}$$

so that, alternatively, one can consider $\{\tau, \alpha, \delta\}$ as the three dimensionless control terms. In this context, Equation (7) can be used to obtain the HPS thrust bubble, that is, the surface plot that shows the variation in the radial component T_R of T as a function of T_T and T_N when $\tau = 1$, at an assigned solar distance. The thrust bubble is shown in Figure 3 for three values of the Sun-CubeSat distance, namely $r \in \{0.75, 1, 1.25\}$ AU. In particular, the cases of $r \in \{0.75, 1.25\}$ AU indicate the two scenarios in which the solar distance reaches the boundaries of the range interval in which the mathematical model of the electric thruster is valid; see the discussion in Appendix A. Note the evident size reduction in the thrust bubble when the distance of the CubeSat from the Sun increases.

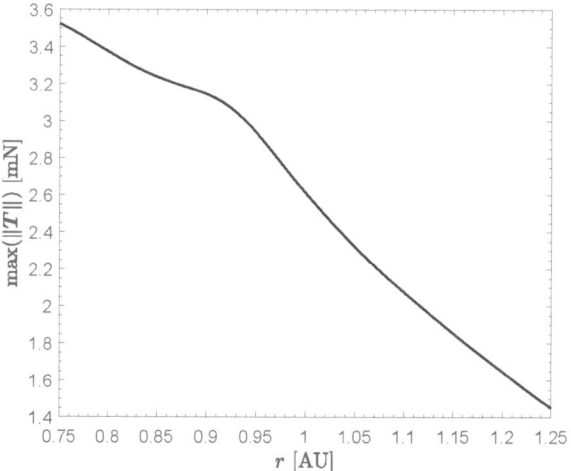

Figure 2. Variation with $r \in [0.75, 1.25]$ AU of the maximum magnitude of the HPS-induced thrust vector T.

The CubeSat propulsive acceleration vector a_p is easily obtained from Equation (1) by introducing the spacecraft mass m, whose initial (assigned) value is $m(t_0) = m_0$, where $t_0 \triangleq 0$ is the initial time instant. Using the expression of the total thrust given by Equation (1), one has

$$a_p = \frac{T}{m} = \tau \frac{T_{e_{\max_\oplus}} F}{m} \hat{t}_e + \frac{\eta P_\oplus A}{m}\left(\frac{r_\oplus}{r}\right)^2 \hat{r} \tag{8}$$

The variation in the CubeSat mass during the flight is due exclusively to the propellant expelled by the miniaturized electric thruster whose specific impulse I_{sp} is a function of the solar distance r; see the lower part of Figure A2. Bearing in mind the mathematical model described in Appendix A, the time variation in the mass can be written as

$$\frac{dm}{dt} = -\frac{\|T_e\|}{g_0\, I_{sp}} \equiv -\frac{\tau\, T_{e_{\max_\oplus}} F}{g_0\, I_{sp}} \tag{9}$$

where $g_0 = 9.80665 \, \text{m/s}^2$ is the standard gravity, F is the auxiliary dimensionless function defined in Equation (2), while I_{sp} is the (electric thruster) specific impulse given by Equation (A3) as a function of $r \in [0.75, 1.25]$ AU. Note that the absolute value of the term on the right-hand side of Equation (9) coincides with the propellant mass flow rate.

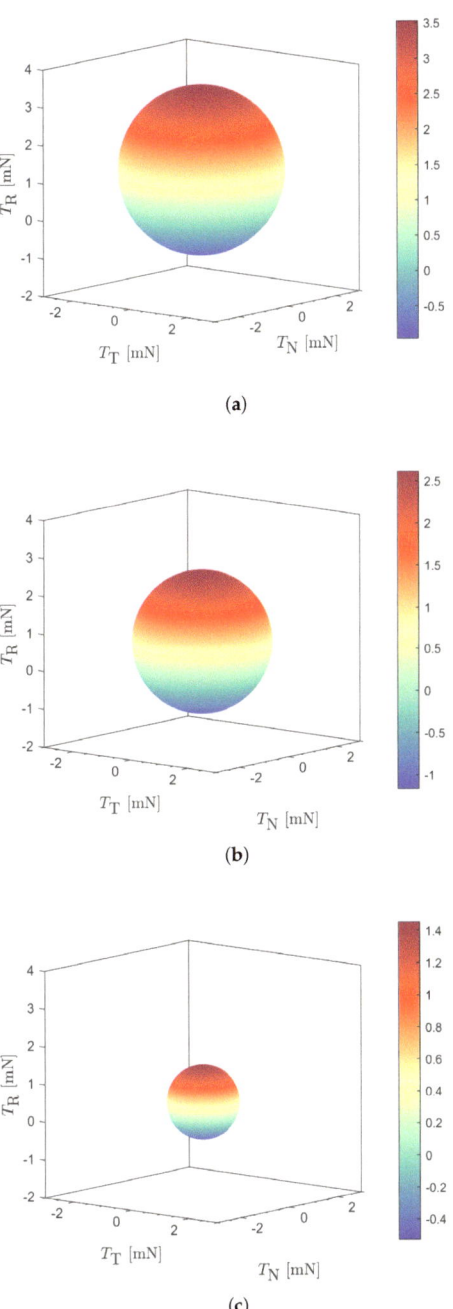

Figure 3. Thrust bubble (when $\tau = 1$) as a function of the Sun-CubeSat distance r. The ticks in the color bar are in millinewtons. (**a**) Case of $r = 0.75$ AU; (**b**) Case of $r = 1$ AU; (**c**) Case of $r = 1.25$ AU.

Accordingly, the variation in the propellant mass flow rate with the solar distance is shown in Figure 4 when the throttle function is $\tau = 1$.

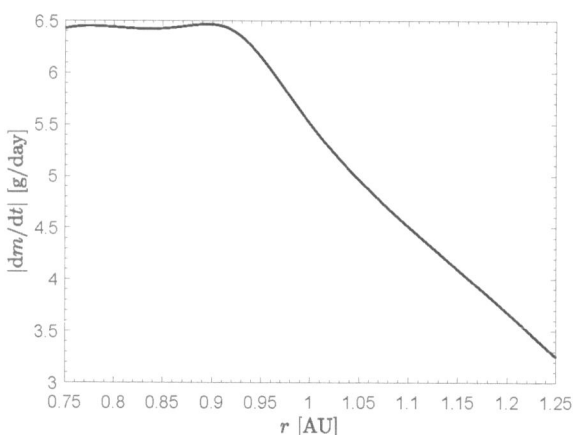

Figure 4. Variation with $r \in [0.75, 1.25]$ AU of propellant mass flow rate when the throttle function is $\tau = 1$; see the right side of Equation (9).

To summarize, the simplified thrust model of the HPS, in which the photonic solar sail has a Sun-facing attitude, is given by Equations (8) and (9). In this model, the (dimensionless) control terms are $\{\tau, \hat{t}_e\}$ or $\{\tau, \alpha, \delta\}$ if an RTN reference frame is used to express the components of the thrust vector; see Equation (7).

2.2. Description of the Trajectory Optimization Process

The HPS analytical thrust model given by Equations (8) and (9) has been implemented in an optimization routine to simulate the rapid transfer trajectory of the CubeSat in a typical orbit-to-orbit mission scenario, which approximates the transfer towards the near-Earth asteroid 99942 Apophis. To this end, an in-house routine based on the classical calculus of variation [63,64] and Pontryagin's maximum principle [65–67] has been adapted to handle the case of a CubeSat equipped with an HPS. The details of the mathematical approach are illustrated in Ref. [61]. In particular, the heliocentric dynamics of the CubeSat are described by using the modified equinoctial orbital elements [68], while the values of the dimensionless throttle function τ and the two thrust angles $\{\alpha, \delta\}$ are obtained, as a function of the costates, by maximizing at any time instant the (scalar) Hamiltonian function. More precisely, the expression of the Hamiltonian function is consistent with Equation (21) of Ref. [61]. The differential equations of the optimization process have been integrated by using a PECE solver based on the Adams–Bashforth method, while the boundary value problem has been solved by adapting a recent procedure proposed by the author [69]. The numerical results of the trajectory optimization process are illustrated in the next section as a function of the value of the initial mass of the CubeSat.

3. Results of Numerical Simulations and Parametric Study

The thrust model of the HPS-propelled CubeSat has been used to simulate a three-dimensional heliocentric transfer from the orbit of the Earth to that of asteroid 99942 Apophis, without considering perturbations and ephemeris constraints. In this scenario, the orbital data of the parking (i.e., the Earth) and the target (i.e., the asteroid) orbits are retrieved from the well-known JPL Horizon system. The numerical simulations have been performed by considering a reference value of the initial CubeSat mass of $m_0 = 30$ kg. Note that the launch mass of the proposed ESA's M-ARGO CubeSat is roughly 22.6 kg, while the initial mass of NASA's NEA Scout was 14 kg.

In this case, the CubeSat heliocentric (rapid) transfer trajectory is shown in Figure 5, while Figure 6 summarizes the time variation in the two thrust angles α and δ.

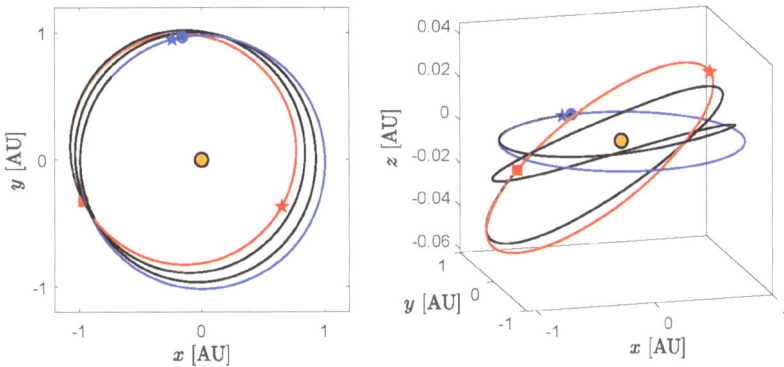

Figure 5. Ecliptic projection and isometric view of the rapid transfer trajectory towards asteroid 99942 Apophis when the initial CubeSat mass is $m_0 = 30$ kg. The z-axis of the isometric view is exaggerated to highlight the three-dimensionality of the trajectory. Black line → CubeSat transfer trajectory; blue line → Earth's orbit; red line → asteroid's orbit; filled star → perihelion; blue dot → starting point; red square → arrival point; orange dot → the Sun.

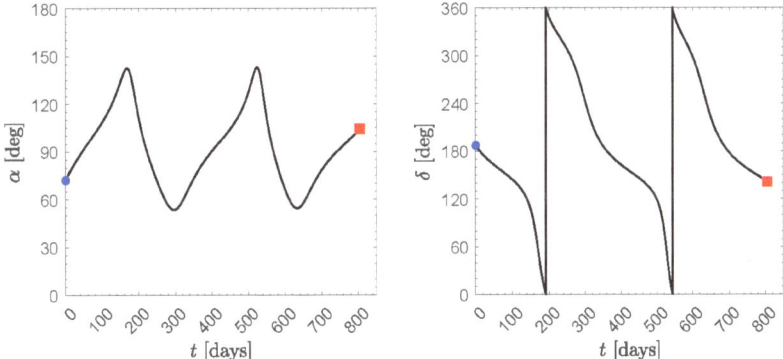

Figure 6. Time variation in the thrust angles α and δ along the rapid transfer trajectory towards asteroid 99942 Apophis when the initial CubeSat mass is $m_0 = 30$ kg. Blue dot → starting point; red square → arrival point.

In this reference case, the miniaturized electric thruster gives always the maximum thrust magnitude (i.e., $\tau = 1$ during all the flight), the rapid transfer does not contain any coasting arc and requires a flight time of about 803 days, while the required propellant mass is slightly less of 4.6 kg. Finally, Figure 7 shows the time variation in the CubeSat mass and the distance from the Sun during the transfer. Note that the solar distance value remains within the allowed range $r \in [0.75, 1.25]$ AU throughout the interplanetary flight. The numerical procedure has been repeated by varying the initial CubeSat mass in the range $m_0 \in [26, 35]$ kg, with a step of 1 kg, to perform a sort of parametric study of the transfer performance as a function of m_0. The results are shown in Figure 8, which clearly indicates that the transfer performances, in terms of both the flight time and the required propellant mass, depend on the value of the initial CubeSat mass. In this respect, a reduction in the value of m_0 of about 2 kg (with respect to the reference value of 30 kg) allows the flight time to be reduced of roughly 50 days.

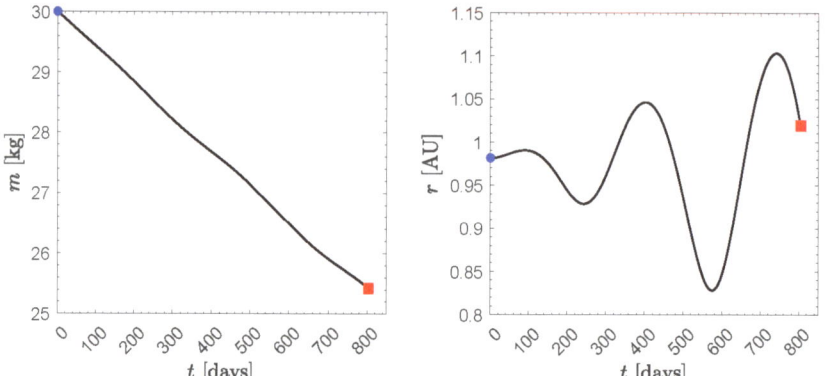

Figure 7. Time variation in the mass m and solar distance r along the rapid transfer trajectory towards asteroid 99942 Apophis when the initial CubeSat mass is $m_0 = 30$ kg. Blue dot → starting point; red square → arrival point.

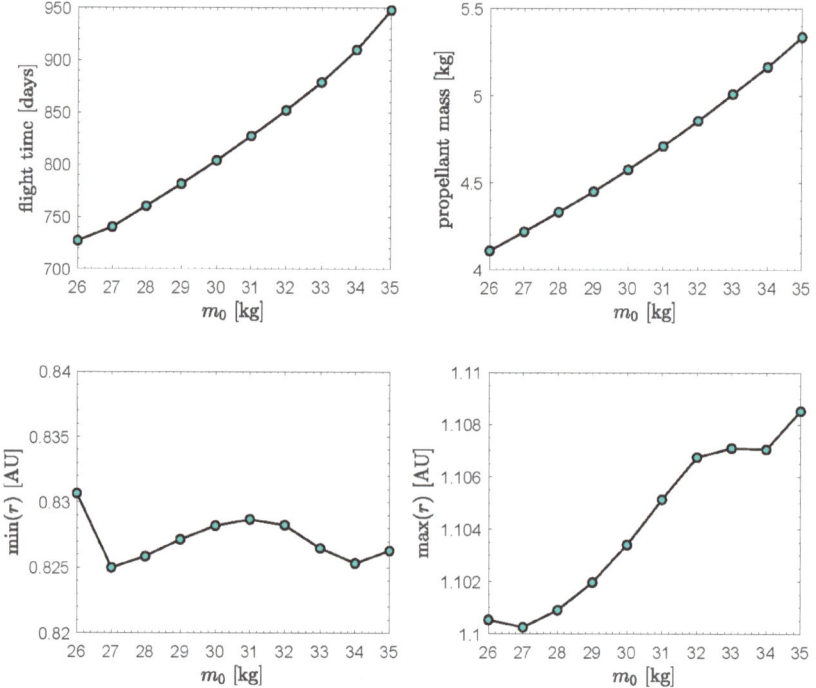

Figure 8. Results of the parametric study of the Earth–Apophis, orbit-to-orbit, rapid transfer as a function of the value of the initial CubeSat mass m_0 (step of 1 kg, green dots).

4. Conclusions

In this paper, the thrust model of a CubeSat equipped with an HPS consisting of a miniaturized electric thruster and a photonic solar sail in a Sun-facing configuration has been analyzed. The proposed mathematical model considers the performance of the miniaturized electric thruster that will be installed on the first interplanetary CubeSat capable of a heliocentric transfer using its own propulsion system, while the performance of the photonic solar sail has been schematized by using the classical optical force model and the design characteristics of the sail installed on NASA's NEA scout. The HPS model gives an analytical expression of both the thrust vector and the propellant mass flow rate, which can be used to study the CubeSat trajectory in a typical heliocentric mission scenario. In this context, the paper discussed the performance in an orbit-to-orbit transfer from the Earth to asteroid 99942 Apophis. In particular, the optimal performance is calculated in terms of minimum transfer time. However, the proposed thrust model can be used to simulate CubeSat transfers in other interesting mission scenarios where, for example, the performance index to be optimized is a suitable combination of flight time and required propellant mass.

The potential extension of this work can be achieved in two subsequent steps. First, a mathematical model of the CubeSat subsystems mass can be used to determine the actual impact of an HPS on the initial mass of the spacecraft. This will allow us to correctly compare the performance of a conventional spacecraft with a single propulsion system with that of a vehicle equipped with an HPS. Second, the constraint on the Sun-facing condition can be relaxed in order to evaluate the effect of the photonic solar sail attitude variation on the overall interplanetary transfer performance.

Funding: This research received no external funding.

Institutional Review Board Statement: Not applicable.

Informed Consent Statement: Not applicable.

Data Availability Statement: The original contributions presented in the study are included in the article; further inquiries can be directed to the corresponding author.

Conflicts of Interest: The author declares no conflict of interest.

Appendix A. Mathematical Model of the Electric Thruster-Induced Propulsive Acceleration

This appendix reports the thrust model of the (miniaturized) electronic thruster that constitutes a part of the HPS considered in this work. More precisely, this thrust model is derived from the recent work of the author [52], which describes a simplified mathematical approach for the description of the propulsive characteristics of the electric thruster installed on board the proposed ESA's M-ARGO CubeSat, whose artistic representation is shown in Figure A1. In turn, the mathematical model described in Ref. [52] is an extension of the surrogate engine model recently proposed by Topputo et al. [54] for the preliminary design of the M-ARGO interplanetary trajectory.

In particular, the simplified thrust model described in Ref. [52] gives the expression of the thrust vector T_e and the specific impulse I_{sp} of the M-ARGO's miniaturized electric thruster as a function of the Sun-CubeSat distance $r \in [0.75, 1.25]$ AU. The constraints on the minimum (i.e., 0.75 AU) and maximum (i.e., 1.25 AU) solar distance are related to the original surrogate model of Topputo et al. [54] (which, in turn, is linked to the admissible value of the thruster input power), and will also be considered in this work during the selection of the potential mission scenario.

According to Equations (12) and (16) of Ref. [52], and bearing in mind that the electric thruster-induced thrust vector is assumed to be freely steerable during the flight, one obtains the compact expressions of T_e

$$T_e = \tau \, T_{e_{max_\oplus}} \frac{d_{N_1}(r/r_\oplus)^4 + d_{N_2}(r/r_\oplus)^3 + d_{N_3}(r/r_\oplus)^2 + d_{N_4}(r/r_\oplus) + d_{N_5}}{(r/r_\oplus)^3 + d_{D_1}(r/r_\oplus)^2 + d_{D_2}(r/r_\oplus) + d_{D_3}} \hat{t}_e \quad \text{(A1)}$$

where $r_\oplus \triangleq 1$ AU is a reference distance which is used to normalize the results of the best-fit procedure described in Ref. [52], $T_{e_{max_\oplus}} \triangleq 1.8897$ mN is the maximum value of the magnitude of T_e when the solar distance is 1 astronomical unit, $\hat{t}_e$ is the electric thruster-induced thrust unit vector, and $\tau \in [0, 1]$ is the dimensionless throttle parameter which models the local thrust magnitude variation between the minimum (when $\tau = 0$) and the maximum (when $\tau = 1$) value. In essence, the dimensionless parameter τ models the actual behaviour of a typical electric thruster in which the admissible thrust magnitude levels are given by the classical throttle Table [61]. In Equation (A1), the coefficients appearing in the fraction on the right-hand side are obtained through a best-fit procedure. Their values are given by Equation (13) of Ref. [52], viz.

Figure A1. Artistic representation of the ESA's M-ARGO CubeSat approach to a potential near-Earth asteroid. Image: © ESA.

$$d_{N_1} = -1.6239, \ d_{N_2} = 6.6115, \ d_{N_3} = -9.7377, \ d_{N_4} = 6.1927, \ d_{N_5} = -1.4378,$$
$$d_{D_1} = -2.4888, \ d_{D_2} = 2.0463, \ d_{D_3} = -0.5527 \quad (A2)$$

The approximate expression of the specific impulse is given by Equation (14) of Ref. [52] as a function of the solar distance r. The function $I_{sp} = I_{sp}(r)$ is written in a compact form as

$$I_{sp} = I_{sp_\oplus} \frac{e_{N_1}(r/r_\oplus)^4 + e_{N_2}(r/r_\oplus)^3 + e_{N_3}(r/r_\oplus)^2 + e_{N_4}(r/r_\oplus) + e_{N_5}}{(r/r_\oplus)^3 + e_{D_1}(r/r_\oplus)^2 + e_{D_2}(r/r_\oplus) + e_{D_3}} \quad (A3)$$

where $I_{sp_\oplus} \triangleq 3022.6\,\text{s}$ is a reference value of the specific impulse (which coincides with the value of I_{sp} at a solar distance of 1 astronomical unit), while the best-fit coefficients on the fraction at the right-hand side of the equation are

$$e_{N_1} = -0.3556, \ e_{N_2} = 2.2133, \ e_{N_3} = -4.0643, \ e_{N_4} = 2.9771, \ e_{N_5} = -0.7599,$$
$$e_{D_1} = -2.5148, \ e_{D_2} = 2.0994, \ e_{D_3} = -0.5740 \quad (A4)$$

The r-variation of the maximum thrust magnitude $\max(\|T_e\|)$ and the specific impulse I_{sp} are shown in Figure A2, which is consistent with the graph reported in Figure 4 of Ref. [52]. Equations (A1) and (A3) are used in Section 2 to obtain a simplified version of the HPS thrust vector model. In this regard, the next appendix describes the mathematical approach used to evaluate the contribution of the solar sail to the total propulsive acceleration of the CubeSat.

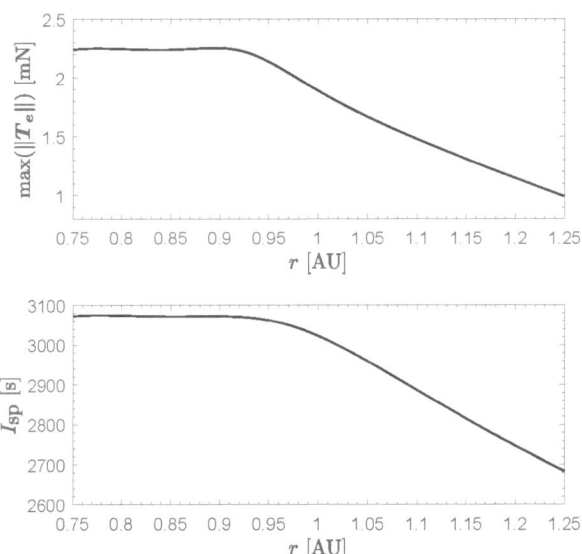

Figure A2. Miniaturized electric thruster: variation of max $(\|T_e\|)$ and I_{sp} with $r \in [0.75, 1.25]$ AU.

Appendix B. Mathematical Model of the Solar Sail-Induced Propulsive Acceleration

This appendix describes the mathematical model used to schematize the thrust vector T_s due to the presence of the photonic solar sail. In this context, the solar sail is considered as a flat surface, that is, the billowing of the reflective membrane [70] is neglected in evaluating the expression of T_s. The latter is a fairly common assumption in the preliminary design of a solar sail-based trajectory, while recent literature [71] discusses an approach based on the use of finite element analysis to evaluate the impact of sail billowing during the interplanetary flight.

From the point of view of the effect of the thermo-optical characteristics of the reflective membrane, the classical optical force model illustrated by McInnes [72] has been employed. In this respect, using the formulation proposed in Ref. [73] the compact expression of the solar sail-induced thrust vector is

$$T_s = P_\oplus A \left(\frac{r_\oplus}{r}\right)^2 \left\{ b_1 \left(\hat{n} \cdot \hat{r}\right) \hat{r} + \left[b_2 \left(\hat{n} \cdot \hat{r}\right)^2 + b_3 \left(\hat{n} \cdot \hat{r}\right)\right] \hat{n} \right\} \tag{A5}$$

where $P_\oplus \simeq 4.56 \times 10^{-6}\,\text{N/m}^2$ is the solar radiation pressure exerted on a perfectly absorbing surface at 1 astronomical unit of distance from the Sun, A is the sail reflective surface, $r_\oplus = 1$ AU is a reference distance, $\hat{r}$ is the Sun-spacecraft radial unit vector, and $\hat{n}$ is the unit vector normal to the sail reflective membrane surface in direction opposite to the Sun. In Equation (A5), the dimensionless terms $\{b_1, b_2, b_3\}$ are the so called "sail force coefficients", whose values depend on the thermo-optical characteristics of the film which coves the sail membrane. The expressions of the sail force coefficients are [73].

$$b_1 \triangleq 1 - \rho s \quad ; \quad b_2 \triangleq 2\rho s \quad ; \quad b_3 \triangleq B_{fr}\rho(1-s) + (1-\rho)\frac{\epsilon_{fr} B_{fr} - \epsilon_b B_b}{\epsilon_{fr} + \epsilon_b} \tag{A6}$$

where s is the fraction of photons that are specularly reflected, B_{fr} (or B_b) is the non-Lambertian coefficient of the front (or back) sail membrane, ϵ_{fr} (or ϵ_b) is the membrane front (or back) emissivity, and $\rho < 1$ the reflection coefficient.

In this work, the photonic solar sail performances are modeled by using the literature data referring to the NASA's NEA Scout CubeSat; see the artistic image in Figure A3.

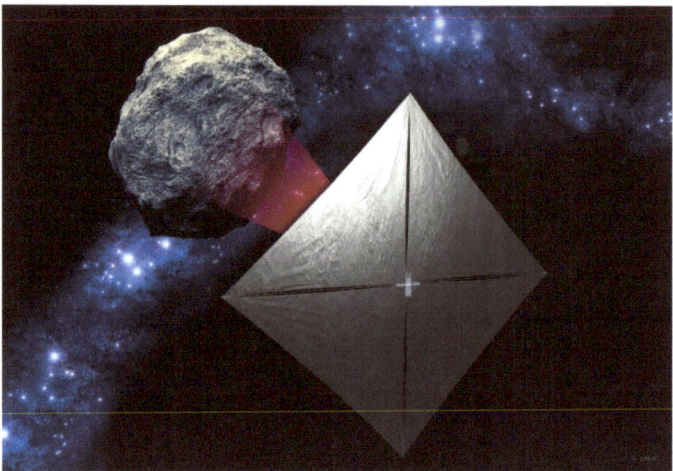

Figure A3. Artistic concept of the NASA's Near-Earth Asteroid Scout (NEA Scout) approaching the target asteroid. The solar sail-based CubeSat failed to make contact with ground station after launch, and the mission NEA Scout was considered lost in December 2022. Image credit: NASA.

In this case, the thermo optical coefficients of that (ill-fated) solar sail-based CubeSat are indicated in Refs. [59,60], viz.

$$\rho = 0.91 \quad , \quad s = 0.89 \quad , \quad B_{fr} = 0.79 \quad , \quad B_b = 0.67 \quad , \quad \epsilon_{fr} = 0.025 \quad , \quad \epsilon_b = 0.27 \tag{A7}$$

so that, according to Equation (A6), the sail force coefficients are

$$b_1 = 0.1901 \quad , \quad b_2 \simeq 1.6198 \quad , \quad b_3 = 0.0299 \tag{A8}$$

In the special case in which the sail attitude is Sun-facing, that is, the direction of $\hat{n}$ coincides with that of $\hat{r}$, one has $\hat{n} \cdot \hat{r} = 1$ and Equation (A5) simplifies as

$$T_s = P_\oplus A \left(\frac{r_\oplus}{r}\right)^2 (b_1 + b_2 + b_3)\,\hat{r} \equiv \eta\, P_\oplus A \left(\frac{r_\oplus}{r}\right)^2 \hat{r} \tag{A9}$$

where $\eta < 2$ is a dimensionless coefficient that quantifies the effect of the actual thermo-optical characteristics of the sail film with respect to the ideal case of a complete (and specular) reflection in which $\eta = 2$. In this case, the value of η is

$$\eta \triangleq b_1 + b_2 + b_3 = 1.8398 \tag{A10}$$

Equation (A9) is used to describe the (Sun-facing) photonic solar sail-induced thrust vector. In this respect, a value of the sail area consistent with the NEA Scout case has been assumed, that is, $A = 86\,\mathrm{m}^2$. Accordingly, the variation with r of the magnitude of the solar sail thrust vector T_s is drawn in Figure A4. Note that at $r = r_\oplus$ one has $\|T_s\| \simeq 0.7215\,\mathrm{mN}$. In fact, the NEA Scout propulsive acceleration magnitude at 1 astronomical unit of distance from the Sun was estimated in about $0.05\,\mathrm{mm/s}^2$ and, bearing in mind that the CubeSat mass was 14 kg, one obtains $0.05 \times 14 \simeq 0.7\,\mathrm{mN}$, as expected.

Finally, note that $\|T_s\|$ is of the same order of magnitude as the maximum propulsive thrust generated by the miniaturized electric thruster; compare the upper part of Figure A2 with the graph in Figure A4. This is an important aspect of the analysis, because it indicates that the contribution of the photon solar sail in obtaining the total thrust vector is not negligible.

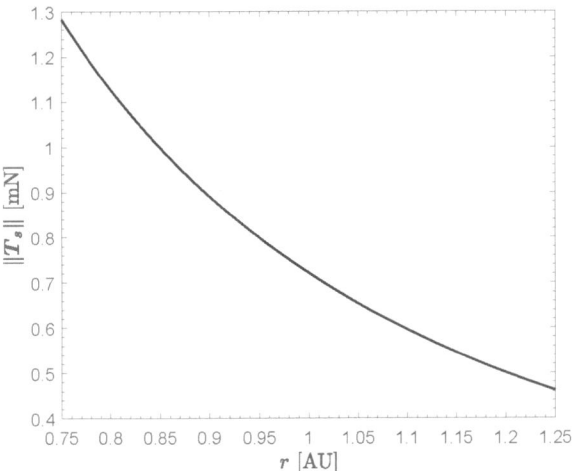

Figure A4. Photonic solar sail in a Sun-facing configuration: variation of the thrust magnitude $\|T_s\|$ with the solar distance $r \in [0.75, 1.25]$ AU, according to Equation (A9). The solar sail characteristics (in terms of sail area and sail force coefficients) are consistent with the system installed onboard of the NASA's NEA Scout CubeSat.

References

1. Brophy, J.R.; Noca, M. Electric Propulsion for Solar System Exploration. *J. Propuls. Power* **1998**, *14*, 700–707. [CrossRef]
2. Burke, L.M.; Falck, R.D.; McGuire, M.L. *Interplanetary Mission Design Handbook: Earth-to-Mars Mission Opportunities 2026 to 2045*; techreport NASA/TM—2010-216764; National Aeronautics and Space Administration: Cleveland, OH, USA, 2010.
3. Urrios, J.; Pacheco-Ramos, G.; Vazquez, R. Optimal planning and guidance for Solar System exploration using Electric Solar Wind Sails. *Acta Astronaut.* **2024**, *217*, 116–129. [CrossRef]
4. Staehle, R.; Blaney, D.; Hemmati, H.; Jones, D.; Klesh, A.; Liewer, P.; Lazio, J.; Wen-Yu Lo, M.; Mouroulis, P.; Murphy, N.; et al. Interplanetary CubeSat Architecture and Missions, In Proceedings of the AIAA SPACE 2012 Conference & Exposition, Pasadena, CA, USA, 11–13 September 2012. [CrossRef]
5. Schoolcraft, J.; Klesh, A.; Werne, T., MarCO: Interplanetary Mission Development on a CubeSat Scale. In *Space Operations: Contributions from the Global Community*; Springer International Publishing: Berlin, Germany, 2017; Chapter 10, pp. 221–231. [CrossRef]
6. Malphrus, B.K.; Freeman, A.; Staehle, R.; Klesh, A.T.; Walker, R. Interplanetary CubeSat missions. In *CubeSat Handbook: From Mission Design to Operations*; Cappelletti, C., Battistini, S., Malphrus, B.K., Eds.; Academic Press: London, UK, 2021; Chapter 4, pp. 85–121. [CrossRef]
7. Freeman, A.; Malphrus, B.K.; Staehle, R. CubeSat science instruments. In *CubeSat Handbook: From Mission Design to Operations*; Cappelletti, C., Battistini, S., Malphrus, B.K., Eds.; Academic Press: London, UK, 2021; Chapter 3, pp. 67–83. [CrossRef]
8. Schoolcraft, J.; Klesh, A.T.; Werne, T. MarCO: Interplanetary mission development on a cubesat scale. In Proceedings of the SpaceOps 2016 Conference, Daejeon, Republic of Korea, 16–20 May 2016. [CrossRef]
9. Klesh, A.T. MARCO: Flight results from the first interplanetary CubeSat mission. In Proceedings of the 70th Annual International Astronautical Congress, Washington, DC, USA, 21–25 October 2019.
10. Gutierrez, G.; Riccobono, D.; Bruno, E.; Bonariol, T.; Vigna, L.; Reverberi, G.; Fazzoletto, E.; Cotugno, B.; Vitiello, A.; Saita, G.; et al. LICIACube: Mission Outcomes of Historic Asteroid Fly-By Performed by a CubeSat. In Proceedings of the 2024 IEEE Aerospace Conference, Big Sky, MT, USA, 2–9 March 2024. [CrossRef]
11. Franzese, V.; Topputo, F. Celestial Bodies Far-Range Detection with Deep-Space CubeSats. *Sensors* **2023**, *23*, 4544. [CrossRef]
12. Cervone, A.; Topputo, F.; Franzese, V.; Pérez-Silva, A.R.; Leon, B.B.; Garcia, B.D.; Minacapilli, P.; Rosa, P.; Bay, G.; Radu, S. The path towards increasing RAMS for novel complex missions based on CubeSat technology. *CEAS Space J.* **2023**, *16*, 203–224. [CrossRef]
13. Mitchell, A.M.; Panicucci, P.; Franzese, V.; Topputo, F.; Linares, R. Improved detection of a Near-Earth Asteroid from an interplanetary CubeSat mission. *Acta Astronaut.* **2024**, *223*, 685–692. [CrossRef]
14. Lemmer, K. Propulsion for CubeSats. *Acta Astronaut.* **2017**, *134*, 231–243. [CrossRef]
15. Alnaqbi, S.; Darfilal, D.; Swei, S.S.M. Propulsion Technologies for CubeSats: Review. *Aerospace* **2024**, *11*, 502. [CrossRef]
16. O'Reilly, D.; Herdrich, G.; Kavanagh, D.F. Electric Propulsion Methods for Small Satellites: A Review. *Aerospace* **2021**, *8*, 22. [CrossRef]

17. Yeo, S.H.; Ogawa, H.; Kahnfeld, D.; Schneider, R. Miniaturization perspectives of electrostatic propulsion for small spacecraft platforms. *Prog. Aerosp. Sci.* **2021**, *126*, 100742. [CrossRef]
18. Grönland, T.A.; Rangsten, P.; Nese, M.; Lang, M. Miniaturization of components and systems for space using MEMS-technology. *Acta Astronaut.* **2007**, *61*, 228–233. [CrossRef]
19. Alvara, A.N.; Lee, L.; Sin, E.; Lambert, N.; Westphal, A.J.; Pister, K.S. BLISS: Interplanetary exploration with swarms of low-cost spacecraft. *Acta Astronaut.* **2024**, *215*, 348–361. [CrossRef]
20. Shukla, H.; Velidi, G. Energetic material characterization and ignition study of MEMS based micro-thruster for multi spacecrafts missions. *FirePhysChem* **2024**, *4*, 122–130. [CrossRef]
21. Kabirov, V.; Semenov, V.; Torgaeva, D.; Otto, A. Miniaturization of spacecraft electrical power systems with solar-hydrogen power supply system. *Int. J. Hydrogen Energy* **2023**, *48*, 9057–9070. [CrossRef]
22. Janhunen, P. Electric sail for spacecraft propulsion. *J. Propuls. Power* **2004**, *20*, 763–764. [CrossRef]
23. Janhunen, P.; Sandroos, A. Simulation study of solar wind push on a charged wire: Basis of solar wind electric sail propulsion. *Ann. Geophys.* **2007**, *25*, 755–767. [CrossRef]
24. Janhunen, P.; Toivanen, P.K.; Polkko, J.; Merikallio, S.; Salminen, P.; Haeggström, E.; Seppänen, H.; Kurppa, R.; Ukkonen, J.; Kiprich, S.; et al. Electric solar wind sail: Toward test missions. *Rev. Sci. Instrum.* **2010**, *81*, 111301. [CrossRef]
25. Bassetto, M.; Niccolai, L.; Quarta, A.A.; Mengali, G. A comprehensive review of Electric Solar Wind Sail concept and its applications. *Prog. Aerosp. Sci.* **2022**, *128*, 1–27. [CrossRef]
26. Gong, S.; Macdonald, M. Review on solar sail technology. *Astrodynamics* **2019**, *3*, 93–125. [CrossRef]
27. Fu, B.; Sperber, E.; Eke, F. Solar sail technology—A state of the art review. *Prog. Aerosp. Sci.* **2016**, *86*, 1–19. [CrossRef]
28. Tsuda, Y.; Mori, O.; Funase, R.; Sawada, H.; Yamamoto, T.; Saiki, T.; Endo, T.; Kawaguchi, J. Flight status of IKAROS deep space solar sail demonstrator. *Acta Astronaut.* **2011**, *69*, 833–840. [CrossRef]
29. Tsuda, Y.; Mori, O.; Funase, R.; Sawada, H.; Yamamoto, T.; Saiki, T.; Endo, T.; Yonekura, K.; Hoshino, H.; Kawaguchi, J. Achievement of IKAROS—Japanese deep space solar sail demonstration mission. *Acta Astronaut.* **2013**, *82*, 183–188. [CrossRef]
30. Mughal, M.R.; Praks, J.; Vainio, R.; Janhunen, P.; Envall, J.; Näsilä, A.; Oleynik, P.; Niemelä, P.; Nyman, S.; Slavinskis, A.; et al. Aalto-1, multi-payload CubeSat: In-orbit results and lessons learned. *Acta Astronaut.* **2021**, *187*, 557–568. [CrossRef]
31. Dalbins, J.; Allaje, K.; Iakubivskyi, I.; Kivastik, J.; Komarovskis, R.O.; Plans, M.; Sünter, I.; Teras, H.; Ehrpais, H.; Ilbis, E.; et al. ESTCube-2: The Experience of Developing a Highly Integrated CubeSat Platform. In Proceedings of the 2022 IEEE Aerospace Conference (AERO), Big Sky, MT, USA, 5–12 March 2022; pp. 1–16. [CrossRef]
32. Palos, M.F.; Janhunen, P.; Toivanen, P.; Tajmar, M.; Iakubivskyi, I.; Micciani, A.; Orsini, N.; Kütt, J.; Rohtsalu, A.; Dalbins, J.; et al. Electric Sail Mission Expeditor, ESME: Software Architecture and Initial ESTCube Lunar Cubesat E-sail Experiment Design. *Aerospace* **2023**, *10*, 694. [CrossRef]
33. Boni, L.; Bassetto, M.; Niccolai, L.; Mengali, G.; Quarta, A.A.; Circi, C.; Pellegrini, R.C.; Cavallini, E. Structural response of Helianthus solar sail during attitude maneuvers. *Aerosp. Sci. Technol.* **2023**, *133*, 1–9. [CrossRef]
34. Zhao, P.; Wu, C.; Li, Y. Design and application of solar sailing: A review on key technologies. *Chin. J. Aeronaut.* **2023**, *36*, 125–144. [CrossRef]
35. Berg, S.P.; Rovey, J.L. Assessment of Multimode Spacecraft Micropropulsion Systems. *J. Spacecr. Rocket.* **2017**, *54*, 592–601. [CrossRef]
36. Rovey, J.L.; Lyne, C.T.; Mundahl, A.J.; Rasmont, N.; Glascock, M.S.; Wainwright, M.J.; Berg, S.P. Review of multimode space propulsion. *Prog. Aerosp. Sci.* **2020**, *118*, 100627. [CrossRef]
37. Vulpetti, G. Missions to the heliopause and beyond by staged propulsion spacecrafts. In Proceedings of the 43rd International Astronautical Congress, Washington, DC, USA, 28 August–5 September 1992.
38. Baig, S.; McInnes, C.R. Artificial Three-Body Equilibria for Hybrid Low-Thrust Propulsion. *J. Guid. Control. Dyn.* **2008**, *31*, 1644–1655. [CrossRef]
39. Li, T.; Wang, Z.; Zhang, Y. Multi-objective trajectory optimization for a hybrid propulsion system. *Adv. Space Res.* **2018**, *62*, 1102–1113. [CrossRef]
40. Mengali, G.; Quarta, A.A. Tradeoff performance of hybrid low-thrust propulsion system. *J. Spacecr. Rocket.* **2007**, *44*, 1263–1270. [CrossRef]
41. Mengali, G.; Quarta, A.A. Trajectory design with hybrid low-thrust propulsion system. *J. Guid. Control. Dyn.* **2007**, *30*, 419–426. [CrossRef]
42. Mengali, G.; Quarta, A.A. Trajectory analysis and optimization of Hesperides mission. *Universe* **2022**, *8*, 364. [CrossRef]
43. Ceriotti, M.; McInnes, C.R. Hybrid solar sail and solar electric propulsion for novel Earth observation missions. *Acta Astronaut.* **2011**, *69*, 809–821. [CrossRef]
44. Ceriotti, M.; McInnes, C.R. An Earth pole-sitter using hybrid propulsion. In Proceedings of the AIAA/AAS Astrodynamics Specialist Conference, Toronto, Canada, 2–5 August 2010. [CrossRef]
45. Ceriotti, M.; McInnes, C.R. Generation of optimal trajectories for Earth hybrid pole sitters. *J. Guid. Control. Dyn.* **2011**, *34*, 847–859. [CrossRef]
46. Matloff, G.L. The solar photon thruster as a terrestrial pole sitter. *Ann. N. Y. Acad. Sci.* **2004**, *1017*, 468–474. [CrossRef]

47. Dachwald, B.; Seboldt, W.; Macdonald, M.; Mengali, G.; Quarta, A.; McInnes, C.; Rios-Reyes, L.; Scheeres, D.; Wie, B.; Gorlich, M.; et al. Potential Solar Sail Degradation Effects on Trajectory and Attitude Control. In Proceedings of the AIAA Guidance, Navigation, and Control Conference and Exhibit, San Francisco, CA, USA, 15–18 August, 2005. [CrossRef]
48. Dachwald, B.; Macdonald, M.; McInnes, C.R.; Mengali, G.; Quarta, A.A. Impact of optical degradation on solar sail mission performance. *J. Spacecr. Rocket.* **2007**, *44*, 740–749. [CrossRef]
49. McInnes, C.R. Orbits in a Generalized Two-Body Problem. *J. Guid. Control. Dyn.* **2003**, *26*, 743–749. [CrossRef]
50. Mengali, G.; Quarta, A.A. Optimal control laws for axially symmetric solar sails. *J. Spacecr. Rocket.* **2005**, *42*, 1130–1133. [CrossRef]
51. Li, Y.; Cheah, K.H., Solar sail as propellant-less micropropulsion. In *Space Micropropulsion for Nanosatellites*; Cheah, K.H., Ed.; Elsevier: Amsterdam, The Netherlands, 2022; Chapter 10, pp. 273–284. [CrossRef]
52. Quarta, A.A. Continuous-Thrust Circular Orbit Phasing Optimization of Deep Space CubeSats. *Appl. Sci.* **2024**, *14*, 7059. [CrossRef]
53. Franzese, V.; Topputo, F.; Ankersen, F.; Walker, R. Deep-Space Optical Navigation for M-ARGO Mission. *J. Astronaut. Sci.* **2021**, *68*, 1034–1055. [CrossRef]
54. Topputo, F.; Wang, Y.; Giordano, C.; Franzese, V.; Goldberg, H.; Perez-Lissi, F.; Walker, R. Envelop of reachable asteroids by M-ARGO CubeSat. *Adv. Space Res.* **2021**, *67*, 4193–4221. [CrossRef]
55. Franzese, V.; Giordano, C.; Wang, Y.; Topputo, F.; Goldberg, H.; Gonzalez, A.; Walker, R. Target selection for M-ARGO interplanetary cubesat. In Proceedings of the 71st International Astronautical Congress, Virtual Event, 12–14 October 2020.
56. McNutt, L.; Johnson, L.; Clardy, D.; Castillo-Rogez, J.; Frick, A.; Jones, L. Near-earth asteroid scout. In Proceedings of the AIAA SPACE 2014 Conference and Exposition, San Diego, CA, USA, 4–7 August 2014.
57. Johnson, L.; Castillo-Rogez, J.; Lockett, T. Near Earth asteroid Scout: Exploring asteroid 1991VG using a Smallsat. In Proceedings of the 70th International Astronautical Congress, Washington, DC, USA, 21–25 October 2019.
58. Lockett, T.R.; Castillo-Rogez, J.; Johnson, L.; Matus, J.; Lightholder, J.; Marinan, A.; Few, A. Near-Earth Asteroid Scout Flight Mission. *IEEE Aerosp. Electron. Syst. Mag.* **2020**, *35*, 20–29. [CrossRef]
59. Heaton, A.; Miller, K.; Ahmad, N. Near earth asteroid Scout solar sail thrust and torque model. In Proceedings of the 4th International Symposium on Solar Sailing (ISSS 2017), Kyoyo, Japan, 17–20 January 2017.
60. Pezent, J.; Sood, R.; Heaton, A. High-fidelity contingency trajectory design and analysis for NASA's near-earth asteroid (NEA) Scout solar sail Mission. *Acta Astronaut.* **2019**, *159*, 385–396. [CrossRef]
61. Quarta, A.A.; Mengali, G.; Bassetto, M. Rapid orbit-to-orbit transfer to asteroid 4660 Nereus using Solar Electric Propulsion. *Universe* **2023**, *9*, 459. [CrossRef]
62. Fan, Z.; Huo, M.; Quarta, A.A.; Mengali, G.; Qi, N. Improved Monte Carlo Tree Search-based Approach to Low-thrust Multiple Gravity assist Trajectory Design. *Aerosp. Sci. Technol.* **2022**, *130*, 1–8. [CrossRef]
63. Bryson, A.E.; Ho, Y.C. *Applied Optimal Control*; Hemisphere Publishing Corporation: New York, NY, USA, 1975; Chapter 2, pp. 71–89. ISBN: 0-891-16228-3.
64. Stengel, R.F. *Optimal Control and Estimation*; Dover Books on Mathematics, Dover Publications, Inc.: New York, NY, USA, 1994; pp. 222–254.
65. Ross, I.M. *A Primer on Pontryagin's Principle in Optimal Control*; Collegiate Publishers: San Francisco, CA, USA, 2015; Chapter 2, pp. 127–129.
66. Prussing, J.E. *Optimal Spacecraft Trajectories*; Oxford University press: Oxford, UK, 2018; Chapter 4, pp. 32–40.
67. Prussing, J.E. *Spacecraft Trajectory Optimization*; Cambridge University Press: Cambridge, UK, 2010; Chapter 2, pp. 16–36. [CrossRef]
68. Walker, M.J.H.; Ireland, B.; Owens, J. A set of modified equinoctial orbit elements. *Celest. Mech.* **1985**, *36*, 409–419. Erratum in *Celest. Mech.* **1986**, *38*, 391–392. [CrossRef]
69. Quarta, A.A. Initial costate approximation for rapid orbit raising with very low propulsive acceleration. *Appl. Sci.* **2024**, *14*, 1124. [CrossRef]
70. Boni, L.; Bassetto, M.; Quarta, A.A. Characterization of a solar sail membrane for Abaqus-Based Simulations. *Aerospace* **2024**, *11*, 151. [CrossRef]
71. Boni, L.; Mengali, G.; Quarta, A.A. Finite Element Analysis of Solar Sail Force Model with Mission Application. *Proc. Inst. Mech. Eng. Part G J. Aerosp. Eng.* **2019**, *233*, 1838–1846. [CrossRef]
72. McInnes, C.R. *Solar Sailing: Technology, Dynamics and Mission Applications*; Springer-Praxis Series in Space Science and Technology; Springer-Verlag: Berlin, Germany, 1999; pp. 46–54. 119–120. [CrossRef]
73. Mengali, G.; Quarta, A.A. Optimal three-dimensional interplanetary rendezvous using nonideal solar sail. *J. Guid. Control. Dyn.* **2005**, *28*, 173–177. [CrossRef]

Disclaimer/Publisher's Note: The statements, opinions and data contained in all publications are solely those of the individual author(s) and contributor(s) and not of MDPI and/or the editor(s). MDPI and/or the editor(s) disclaim responsibility for any injury to people or property resulting from any ideas, methods, instructions or products referred to in the content.

Article

Cooperative Integrated Guidance and Control for Active Target Protection in Three-Player Conflict

Xiaopeng Gong, Wanchun Chen and Zhongyuan Chen *

School of Astronautics, Beihang University, Beijing 100191, China; gxpjzbg@163.com (X.G.); wanchun_chen@buaa.edu.cn (W.C.)
* Correspondence: zhongyuan_buaa@163.com

Abstract: This paper addresses the active target protection problem in a three-player (Target–Attacker–Defender, TAD) conflict by proposing a cooperative integrated guidance and control (IGC) strategy. Unlike previous studies that have designed guidance and control loops separately, this work establishes an IGC model by linearizing both the translational motion and the rotational motion of the vehicles, thereby generating actuator commands directly. This model integrates the kinematics and short-period dynamics, providing a more comprehensive and accurate representation of the vehicles' characteristics. Based on the linearization and order reduction, differential game theory and the sweep method are employed to derive and analytically solve the Riccati differential equation, yielding an optimal control strategy with an explicit expression. The theoretical rigor of the proposed approach is ensured through a proof of optimality sufficiency. Furthermore, factors influencing the computational accuracy of the Riccati equation solution, including the singular values of the control matrix and condition numbers of the solution matrix, are analyzed. Taking into account the dynamic response and limitations of the actuators, numerical simulations demonstrate the effectiveness and superiority of the proposed IGC strategy in intercepting the attacker and protecting the target compared to traditional separated guidance and control designs.

Keywords: target–attacker–defender (TAD) game; integrated guidance and control (IGC); differential game; Riccati differential equation; active protection; three-player conflict

Citation: Gong, X.; Chen, W.; Chen, Z. Cooperative Integrated Guidance and Control for Active Target Protection in Three-Player Conflict. *Actuators* **2024**, *13*, 245. https://doi.org/10.3390/act13070245

Academic Editors: Ti Chen, Junjie Kang, Shidong Xu, Shuo Zhang and Dongdong Li

Received: 28 May 2024
Revised: 25 June 2024
Accepted: 26 June 2024
Published: 28 June 2024

Copyright: © 2024 by the authors. Licensee MDPI, Basel, Switzerland. This article is an open access article distributed under the terms and conditions of the Creative Commons Attribution (CC BY) license (https:// creativecommons.org/licenses/by/ 4.0/).

1. Introduction

Differential game theory, a branch of mathematical control theory, explores the strategic interactions in dynamic settings where players' decisions influence the outcome over time [1]. One of the pivotal applications of differential games is in pursuit–evasion scenarios, which involve multiple agents with conflicting objectives. These scenarios are not only academically intriguing but have substantial practical implications in areas such as military tactics, security systems, and autonomous vehicle navigation [2–4].

Among the various configurations of pursuit–evasion games, three-player conflicts involving a Target, Attacker, and Defender (TAD) present a particularly complex and rich problem space [5–7]. In the classic two-player pursuit–evasion game, the scenario typically involves an attacker pursuing a target whose primary strategy is passive evasion. Instead, in TAD configurations the target benefits from the active protection provided by the defender, whose role is to intercept or obstruct the attacker's efforts.

As a multi-agent system, with the inherent dynamics and interdependencies, the TAD conflict has garnered increasing attention due to its relevance to real-world scenarios in which valuable assets are actively defended [8]. Garcia et al. analyzed the game's dynamics, focusing on the regions where the attacker prevails and the defender's optimal interception strategies [9]. Their subsequent work in [10] developed optimal strategies for the attacker while ensuring that solutions were aligned with the Hamilton–Jacobi–Isaacs equation. A barrier analysis was introduced in [11] to delineate winning regions for both the attacker and the target–defender team, facilitating strategic decision-making through explicit policy and geometric analysis. Subsequent research explored stage-based strategies for a fast nonsuicidal attacker while adjusting tactics based on capture radii [12]. In [13], the authors presented a geometric interception strategy using the Apollonius circle, offering a practical approach for a slower defender to intercept a faster attacker.

The insights gained from TAD scenarios significantly enhance active protection systems for aerial vehicles, improving their effectiveness and survivability in hostile environments. The work in [14] optimized control inputs for the target and the defender, allowing effective interception even with slower defenders. The guidance law for defense missiles of nonmaneuverable aircraft was developed in [15], demonstrating superiority in non-coplanar engagements through high-fidelity simulations. Similar work was presented in [16], focusing on an air combat scenarios in which a defending missile aims to maximize the separation between a target aircraft and an attacking missile at the point of interception.

However, the aforementioned modeling approach in aerial pursuit–evasion games often simplifies the dynamics by directly using heading angles as control inputs, overlooking critical factors such as turning radii and the dynamic response of the system to commands, such as oscillations or delays. Additionally, these models tend to ignore the physical limitations of the maneuverability and actuators, which play a crucial role in the realistic implementation of control strategies. The oversimplified approach does not adequately leverage the general principles and knowledge of dynamics, and neglects the characteristics of aerial vehicles. As a result, the obtained strategies usually lack realistic feasibility. In fact, although there are nonlinear factors such as actuator saturation, aerial vehicles typically still exhibit pronounced linear characteristics, which form the classical foundation for the design of guidance and control systems [17,18]. It is essential for the design of guidance and control systems to include considerations of these linear characteristics along with the vehicle's dynamic response processes.

This need for realistic modeling and design has led to the development of research that focuses on model linearization and order reduction. The prevailing research trend gravitates towards one-sided optimal control or differential games. The distinction lies in that optimal control theory-based guidance laws necessitate prior knowledge of the adversary's control tactics. Although the reliance on prior information for one-sided optimization can be reduced through information sharing between the target and the defender [19], there are still challenges, such as the impracticality of deploying numerical optimization algorithms onboard [20,21]. Conversely, the differential game has attracted broader interest for its independence from presumptions regarding the opponent's strategy [22]. The differential game can obtain the game strategy of the two opponents by finding the saddle point solution; under the condition of accurate modeling, this can guarantee the optimality of the strategy against the opponent's arbitrary maneuver [23,24]. Considering the drawback that the control of linear quadratic differential game guidance law may exceed operational bounds, the bounded differential game was proposed and verified in a two-dimensional plane and three-dimensional space [25,26]. In addition, in order to be closer to the real combat scenarios, recent studies have accounted for constraints such as limited operational capabilities [27], state estimation under imperfect information through Kalman filtering [28], relative intercept angle limitations on attack protocols [24,29,30], cooperative multi-vehicle strategy against an active defense target [24,31], and weapon–target allocation strategies [32].

However, traditional active protection research has focused on designing guidance laws that generate acceleration commands without considering how these commands are effectively tracked by the autopilot. Such approaches overlook the control loop required to generate actuator commands. With the introduction of advanced guidance laws and increased confrontation, oversimplification can lead to significant distortion [33]. Moreover, in the endgame phase of engagements, where the relative motion between vehicles changes rapidly, the assumption of spectral separation between the guidance and control loops becomes untenable [34,35]. Therefore, integrated guidance and control (IGC) was proposed to address these issues by merging the guidance and control loops, thereby mitigating the negative impacts of their coupling [36,37]. IGC ensures that the commands sent to actuators are both feasible and effective within the physical constraints of the system. In the view of engineering designers, IGC is expected to significantly improve missile performance, resulting in lower weight and enhanced lethality [18]. In recent years, the development of IGC has progressed rapidly, with research increasingly focusing on a variety of advanced methods, including sliding mode control [38,39], back-stepping [40], robust control [41], and small-gain theorem [42], among others.

Building on the previously discussed challenges and advancements in three-player conflicts and IGC, this article aims to delve deeper into an IGC design specifically tailored for aerial vehicle active protection scenarios. We present an IGC model for active protection distinguished by the linearization of both the center of mass motion and rotation around the center of mass. Our approach leverages differential game theory and the sweep method to derive and solve the Riccati differential equation analytically. The main contributions of this paper are summarized as follows:

1. **Advanced IGC Modeling for Active Protection**: An advanced modeling approach for active protection is introduced that uniquely combines linearization of both translational and rotational dynamics around the center of mass. Although the linearization technique is a well-established practice in aerial vehicle design, our contribution lies in applying this approach to the IGC design in a three-player conflict scenario. This modeling framework allows for a more comprehensive and accurate representation of the characteristics of aerial vehicles compared to previous studies.
2. **Analytical Derivation and Solution of Riccati Equation**: Through the application of differential game theory and the sweep method, this study derives and solves the Riccati differential equation, providing an analytical expression for the optimal control strategy and in turn presenting possibilities for real-time onboard calculation.
3. **Theoretical Rigor in Solution Analysis**: The theoretical rigor of the proposed approach is provided through a proof of optimality sufficiency. Additionally, we examine factors that influence the computational accuracy of the Riccati equation solution using singular values of the control matrix and condition numbers of the solution matrix.

The rest of the paper is organized as follows. In Section 2, we describe the three-player conflict in the IGC framework, focusing on model linearization and order reduction. Section 3 is dedicated to the derivation of the IGC active protection strategy. In Section 4, we conduct a theoretical analysis of the IGC strategy. Following this, Section 5 presents simulations to demonstrate the practical applicability and effectiveness of the strategy. Finally, Section 6 concludes the paper by summarizing the findings.

2. Problem Formulation

2.1. Nonlinear Engagement Model

In the TAD three-player conflict, the adversarial attacker aims at the target, which is protected by a defender aiming to intercept the incoming threat mid-course. In general, the target is weak in maneuvering and has difficulty avoiding being hit by the attacker through simple maneuvering strategies; thus, the target adopts an active defense strategy in order to survive.

As shown in Figure 1, the engagement is refined into a geometric representation within an inertial coordinate system X_IOY_I. In the aerial conflict, each entity's flight state is characterized by its flight path angle (γ), flight velocity (V), and actual acceleration (a), denoted by the respective subscripts A, T, and D for attacker, target, and defender. The engagement is divided into two distinct but interrelated processes: the attacker's pursuit of the target (AT) and the defender's interception of the attacker to protect the target (AD). In these engagements, r stands for the distance between the respective vehicles, LOS denotes the line-of-sight, and λ represents the line-of-sight angle. Specifically, $\dot{\gamma}$, $\dot{\lambda}$, and $\dot{r}$ denote the respective derivatives of γ, λ, and r. Initial conditions at the start of the engagement are indicated by the subscript 0.

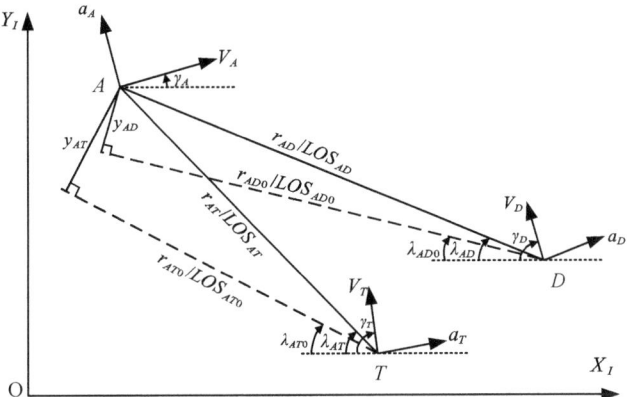

Figure 1. Three-body confrontation engagement geometry.

As shown in Figure 1, the nonlinear engagement model of attacker–target and attacker–defender can be represented as

$$
\begin{aligned}
\dot{r}_{Ai} &= -V_i \cos(\gamma_i - \lambda_{Ai}) - V_A \cos(\gamma_A + \lambda_{Ai}) \\
\dot{\lambda}_{Ai} &= \frac{-V_i \sin(\gamma_i - \lambda_{Ai}) + V_A \sin(\gamma_A + \lambda_{Ai})}{r_{Ai}},
\end{aligned}
\quad (1)
$$

where A stands for the attacker and i represents the target or defender, i.e., $i \in \{T, D\}$.

The rate of change of the flight path angle for each vehicle can be expressed as

$$
\dot{\gamma}_j = \frac{a_j}{V_j} = \frac{F_j(\alpha_j, \delta_j)}{m_j V_j}, \quad j \in \{A, T, D\}, \quad (2)
$$

where m_j represents the mass of the vehicle j and F_j is the normal aerodynamic force acting on the airframe, which is a function of the angle of attack α_j and the fin deflection δ_j from the actuator.

Unlike previous studies focusing solely on translational motion in active protection, the IGC design incorporates rotational kinematics and dynamics around the center of mass involving the pitch angle θ and pitch rate q. The angular acceleration $\dot{q}_j$ is generated by the torque applied to the airframe divided by the moment of inertia I_j. Thus, it is possible to obtain

$$
\begin{aligned}
\dot{\alpha} &= q - \frac{F_j(\alpha_j, \delta_j)}{m_j V_j} \\
\ddot{\theta} &= \dot{q} = \frac{M_j(\alpha_j, \delta_j)}{I_j}.
\end{aligned}
\quad (3)
$$

Although the nonlinear differential Equations (1)–(3) can be solved numerically, adopting an analytical approach provides clearer insights and understanding of the engagement dynamics. To facilitate this, we linearize the equations around an operating condition, which allows for analysis and design based on linear systems theory.

2.2. Linearization and Order Reduction

To facilitate analysis of the engagement dynamics in the terminal phase, the model is simplified based on the following set of assumptions:

- Constant Velocity: The velocities of the entities are assumed to be constant.

- Co-planar Movement: The three entities are considered to move within the same plane; the defender is launched by the target, meaning that LOS_{MT0} and LOS_{MD0} coincide.
- Linear Plant Models: The dynamics of each entity are approximated as linear systems.
- Linearization of Collision Triangles: It is assumed that the two collision triangles can be linearized along their respective initial lines of sight (LOS).

The dynamics model of each vehicle can be represented by a linear equation of arbitrary order:

$$\begin{aligned} \dot{x}_j &= A_j x_j + B_j u_j, \quad j \in \{A, T, D\} \\ a_j &= C_j x_j + d_j u_j, \quad j \in \{A, T, D\} \end{aligned} \tag{4}$$

where x_j represents the internal state variables of each vehicle, a_j is the acceleration perpendicular to the initial line-of-sight direction, and u_j is the corresponding control input.

In previous studies, vehicles described by Equation (4) have generally been simplified to ideal dynamics or first-order delay dynamics. Ideal dynamics implies that $A_j = B_j = C_j = 0$ and $d_j = 1$. First-order delay dynamics implies that $A_j = 1/\tau_j$, $B_j = 1/\tau_j$, $C_j = 1/\tau_j$, and $d_j = 0$ with the time constant of τ_j. In both cases, u_j represents the command acceleration. We further extend the scope of Equation (4) to cover the short-period dynamics described by Equation (3) for the target and the defender [43], which means that

$$A_i = \begin{bmatrix} Z_\alpha^{(i)} & 1 \\ M_\alpha^{(i)} & 0 \end{bmatrix} \quad B_i = \begin{bmatrix} Z_\delta^{(i)} \\ M_\delta^{(i)} \end{bmatrix}$$

$$C_i = \begin{bmatrix} -Z_\alpha^{(i)} V^{(i)} & 0 \end{bmatrix} \quad d_i = -Z_\delta^{(i)} V^{(i)} \quad i \in \{T, D\}, \tag{5}$$

where the numerical coefficients are defined by

$$\begin{aligned} Z_\alpha^{(i)} &= -\frac{1}{m_i V_i} \frac{\partial F_i(\alpha_i, \delta_i)}{\partial \alpha_i} & Z_\delta^{(i)} &= -\frac{1}{m_i V_i} \frac{\partial F_i(\alpha_i, \delta_i)}{\partial \delta_i}, \\ M_\alpha^{(i)} &= \frac{1}{I_i} \frac{\partial M_i(\alpha_i, \delta_i)}{\partial \alpha_i} & M_\delta^{(i)} &= \frac{1}{I_i} \frac{\partial M_i(\alpha_i, \delta_i)}{\partial \delta_i}. \end{aligned} \tag{6}$$

The target and defender are modeled as IGC configurations and the attacker is modeled with first-order lag dynamics. Choosing the state variables as $x_{Ai} = [y_{Ai}, \dot{y}_{Ai}, \alpha_i, q_i, a_A]^\top$, the equations of motion for the attacker to engage the target or defender can be expressed as

$$\dot{x}_{Ai} = A_{Ai} x + \tilde{B}_i u_i + \tilde{B}_A u_A, i \in \{T, D\} \tag{7}$$

where

$$A_{Ai} = \begin{bmatrix} 0 & 1 & 0 & 0 & 0 \\ 0 & 0 & \mp Z_\alpha^{(i)} V_i & 0 & \mp 1 \\ 0 & 0 & Z_\alpha^{(i)} & 1 & 0 \\ 0 & 0 & M_\alpha^{(i)} & 0 & 0 \\ 0 & 0 & 0 & 0 & -1/\tau_A \end{bmatrix} \tag{8}$$

$$\tilde{B}_T = \begin{bmatrix} 0 & -Z_\delta^{(T)} V_T & Z_\delta^{(T)} & M_\delta^{(T)} & 0 \end{bmatrix}^\top \tag{9}$$

$$\tilde{B}_D = \begin{bmatrix} 0 & Z_\delta^{(D)} V_D & Z_\delta^{(D)} & M_\delta^{(D)} & 0 \end{bmatrix}^\top \tag{10}$$

$$\tilde{B}_A = \begin{bmatrix} 0 & 0 & 0 & 0 & 1/\tau_A \end{bmatrix}^\top. \tag{11}$$

The symbols $\mp$ in Equation (8) correspond to $i = T$ and $i = D$, respectively. For the target and the defender, the control input u_i is defined as the fin deflection of the actuator δ_i. For the attacker, the control input u_A is the commanded acceleration a_A^c.

The time-to-go between the attacker–target pair and the defender–attacker pair is denoted by t_{go1} and t_{go2}, respectively. The time-to-go can be calculated by $t_{go1} = t_{f1} - t$ and $t_{go2} = t_{f2} - t$, where the interception time is defined as

$$t_{f1} = \frac{r_{AT0}}{[V_A \cos(\gamma_{A0} + \lambda_{AT0}) + V_T \cos(\gamma_{T0} - \lambda_{AT0})]},$$
$$t_{f2} = \frac{r_{AD0}}{[V_A \cos(\gamma_{A0} + \lambda_{AD0}) + V_D \cos(\gamma_{D0} - \lambda_{AD0})]}. \quad (12)$$

We assume that the engagement of the attacker with the defender precedes the engagement of the attacker with the target, i.e., t_{f1} and t_{f2} satisfy $t_{f1} - t_{f2} > 0$ in the timeline. This is because once the attacker hits or misses the target, it means that the game is over and the defender is no longer needs to continue the engagement.

Therefore, unlike previous studies that only considered guidance loops, the linearization encompasses two aspects: linearization of the airframe and linearization of engagement geometry.

1. Linearization of the airframe: Linearization of the airframe dynamics is typically performed around a trim condition, which represents a steady-state flight condition; in our case, this corresponds to flight at a constant velocity and a small angle of attack. This approach is widely accepted in aerospace control systems design, especially for short-period dynamics, as deviations from this trim condition during maneuvers are usually small enough to maintain the validity of the linear model.
2. Linearization of engagement geometry: For the engagement geometry, we employ linearization around a nominal collision course. This approach is commonly used and justified in terminal guidance problems, particularly in the endgame phase, where the relative geometry changes are relatively small and can be approximated linearly.

This situation can be formulated as a three-player differential game between two opposing sides with different objectives. The cost function that we utilize incorporates both miss distance and energy consumption through a linear–quadratic (LQ) form, striking a balance between the two factors. This also means that it is a soft constraint, ensuring that the control effort does not exceed the maneuverability. The specific formulation of the cost function is as follows:

$$\min_{[u_T, u_D]} \max_{u_A} J = \frac{1}{2}\alpha_{AT} y_{AT}^2\left(t_{f1}\right) + \frac{1}{2}\alpha_{AD} y_{AD}^2\left(t_{f2}\right) + \frac{1}{2}\int_0^{t_{f1}} \beta_T u_T^2 dt + \frac{1}{2}\int_0^{t_{f2}} \beta_D u_D^2 dt - \frac{1}{2}\int_0^{t_{f1}} u_A^2 dt \quad (13)$$

where α_{AT} and α_{AD} are non-negative weight coefficients related to miss distance and the coefficients β_U and β_D reflect the maneuverability of the target and defender relative to the attacker. When the target has weaker maneuverability compared to the attacker, β_T takes a larger value; $\beta_T \to \infty$ indicates that the target has no maneuverability. When the defender has stronger maneuverability compared to the attacker, β_D takes a smaller value, with $\beta_D \to 0$ signifying strong maneuverability of the defender.

It is still difficult to obtain the analytical solution of the above linear optimal control problem, so reducing the model's order is essential. In the context of guidance analysis, the zero-effort miss (ZEM) is a commonly used concept, which physically represents the miss distance of the interceptor if no maneuver is performed, i.e., the control command is zero from the current moment. From the perspective of signal processing, this is equivalent to the zero-input response of the linear system. Using the terminal projection transformation of the linear system, the ZEM z_1 and z_2 can be expressed as

$$z_1(t) = L\Phi_{AT}(t_{f1}, t)x_{AT},$$
$$z_2(t) = L\Phi_{AD}(t_{f2}, t)x_{AD}, \quad (14)$$

where $L = \begin{bmatrix} 1 & 0 & 0 & 0 & 0 \end{bmatrix}$ is a coefficient matrix and $\Phi_{Ai}(t_f, t)$ is the state transition matrix.

The derivatives of z_1 and z_2 with respect to time in compact matrix form are

$$\dot{z} = \begin{bmatrix} \dot{z}_1(t) \\ \dot{z}_2(t) \end{bmatrix} = G_1 \begin{bmatrix} u_T \\ u_D \end{bmatrix} + G_2 u_A \quad (15)$$
$$= G_1 u_{T/D} + G_2 u_A,$$

where

$$G_1 = \begin{bmatrix} \Phi_{AT,y}(t_{go1}) \\ \Phi_{AD,y}(t_{go2}) \end{bmatrix} \cdot [\tilde{B}_T \quad \tilde{B}_D]$$
$$G_2 = \begin{bmatrix} \Phi_{AT,y}(t_{go1}) \\ \Phi_{AD,y}(t_{go2}) \end{bmatrix} \cdot \tilde{B}_A \quad (16)$$
$$\Phi_{AT,y}(t_{go1}) = L\Phi_{AT}(t_{f1}, t)$$
$$\Phi_{AD,y}(t_{go2}) = L\Phi_{AD}(t_{f2}, t).$$

Correspondingly, the cost function of the reduced-order differential game problem is provided by

$$\min_{u_{T/D}} \max_{u_A} J_z = \frac{1}{2} z^\top(t_f) Q z(t_f) + \frac{1}{2} \int_{t_0}^{t_f} \left(u_{T/D}^\top R_1 u_{T/D} + u_A^\top R_2 u_A \right) dt, \quad (17)$$

where

$$Q = \begin{bmatrix} -\alpha_{AT} & 0 \\ 0 & \alpha_{AD} \end{bmatrix} \quad R_1 = \begin{bmatrix} \beta_T & 0 \\ 0 & \beta_D \end{bmatrix} \quad R_2 = -1$$

3. Derivation of IGC Active Protection Strategy

The differential game approach is applied to solve for the optimal control of each vehicle. The Hamiltonian of the linearized system is

$$H = \frac{1}{2} \left(u_{T/D}^\top R_1 u_{T/D} + u_A^\top R_2 u_A \right) + \lambda^\top (G_1 u_{T/D} + G_2 u_A), \quad (18)$$

where λ stands for the introduced costates. According to the calculus of variations, by equating the derivatives of the Hamiltonian with respect to the players' controllers to zero we obtain

$$\frac{\partial H}{\partial u_{T/D}} = R_1 u_{T/D} + G_1^\top \lambda = 0 \quad \Rightarrow u_{T/D} = -R_1^{-1} G_1^\top \lambda,$$
$$\frac{\partial H}{\partial u_A} = R_2 u_A + G_2^\top \lambda = 0 \quad \Rightarrow u_A = -R_2^{-1} G_2^\top \lambda. \quad (19)$$

It follows that in order to solve for the expression of the optimal control we need to obtain an expression for the costate λ. From the adjoint equation and the transversality condition, we have

$$\dot{\lambda} = -\frac{\partial H}{\partial z} = 0 \quad \lambda(t_f) = \begin{bmatrix} -\alpha_{AT} z_1(t_{f1}) \\ \alpha_{AD} z_2(t_{f2}) \end{bmatrix} = \begin{bmatrix} \lambda_1(t) \\ \lambda_2(t) \end{bmatrix}. \quad (20)$$

As shown in Equation (20), the expression for λ includes $z_1(t_{f1})$ and $z_2(t_{f2})$. However, unlike previous studies such as [22] which tackled more idealized dynamic models, the complexity inherent to the IGC problem precludes a straightforward analytical solution. Therefore, to address these challenges we adopt the sweep method to solve for the costates. We assume that $\lambda(t)$ can be written in the form of linear feedback of reduced-order states as

$$\lambda(t) = P(t)z(t), \quad (21)$$

i.e.,

$$\begin{bmatrix} \lambda_1(t) \\ \lambda_2(t) \end{bmatrix} = \begin{bmatrix} P_{11}(t) & P_{12}(t) \\ P_{21}(t) & P_{22}(t) \end{bmatrix} \begin{bmatrix} z_1(t) \\ z_2(t) \end{bmatrix}, \quad (22)$$

with the transversality condition

$$\lambda(t_f) = Qz(t_f). \tag{23}$$

Taking the derivative of λ, as shown in Equation (21), yields

$$\dot{\lambda} = \dot{P}z + P\dot{z} = 0. \tag{24}$$

Next, incorporating the reduced-order dynamics of the system, represented by Equation (15), results in

$$\dot{P}z + P\dot{z} = \dot{P}z + P\left[-G_1(t)R_1^{-1}G_1^\top(t)P(t) - G_2(t)R_2^{-1}G_2^\top(t)P(t)\right]z(t) = 0. \tag{25}$$

Upon rearranging and simplifying the terms, we obtain the Riccati differential equation

$$\dot{P}(t) = P(t)\left[G_1(t)R_1^{-1}G_1^\top(t)P(t) + G_2(t)R_2^{-1}G_2^\top(t)P(t)\right]. \tag{26}$$

Equation (26) characterizes the evolution of the matrix $P(t)$ over time. The terms within the brackets represent the combined influence of the system dynamics and the control effort costs on the evolution of $P(t)$. However, the $P(t)$ matrix in Equation (26) has four elements and the variables are coupled, which makes it impossible to solve it directly. In order to find the analytical expression for $P(t)$, the matrix $K(t)$ is introduced to satisfy

$$K(t) = P^{-1}(t) \quad K(t)P(t) = I, \tag{27}$$

where I is the identity matrix. Therefore, the derivative of $K(t)$ can be expressed in terms of $P(t)$:

$$\dot{K}(t)P(t) + K(t)\dot{P}(t) = 0. \tag{28}$$

This leads to the fully decoupled matrix differential equation for $K(t)$:

$$\begin{aligned}
\dot{K}(t) &= -K(t)\dot{P}(t)P^{-1}(t) \\
&= -\underbrace{K(t)P(t)}_{I} G_1(t)R_1^{-1}G_1^\top(t) - \underbrace{K(t)P(t)}_{I} G_2(t)R_2^{-1}G_2^\top(t) \\
&= -G_1(t)R_1^{-1}G_1^\top(t) - G_2(t)R_2^{-1}G_2^\top(t).
\end{aligned} \tag{29}$$

This decoupled differential equation for $K(t)$ is more tractable and its terms purely depend on the matrices $G_1(t)$, $G_2(t)$, R_1, and R_2, which are typically known from the system dynamics and cost function. The terminal condition for the matrix $P(t)$ can be determined from the costate's terminal condition stated in (23):

$$P(t_f) = Q = \begin{bmatrix} -\alpha_{AT} & 0 \\ 0 & \alpha_{AD} \end{bmatrix}. \tag{30}$$

Correspondingly, the terminal condition for $K(t)$ is provided by

$$K(t_f) = P^{-1}(t_f) = \begin{bmatrix} -1/\alpha_{AT} & 0 \\ 0 & 1/\alpha_{AD} \end{bmatrix}. \tag{31}$$

This setup allows for backward integration from t_f to the initial time while using $K(t_f)$ to calculate $K(t)$, providing a method for determining $P(t)$ across the engagement timeline.

In light of the computational challenges posed by backward integration and the specific representation of the state transition matrix in terms of time-to-go, it is useful to reverse the time in the differential equation for the matrix $K(t)$. Moreover, reversing the time allows for dynamic adjustment of control inputs based on changes in the engagement through state feedback, rather than relying on a precomputed offline $P(t)$. This adjustment leads to

$$\frac{dK}{dt_{go}} = -\frac{dK}{dt} = G_1(t)R_1^{-1}G_1^\top(t) + G_2(t)R_2^{-1}G_2^\top(t). \tag{32}$$

Simultaneously, the terminal conditions in (31) are converted into initial conditions for the integration process in terms of t_{go}. Thus, the problem is transformed into an initial value problem for system ordinary differential equations in a decoupled form. The decoupling facilitates solving for $K(t)$ directly through indefinite integration, which in turn allows for determining the solution $P(t)$ of the Riccati differential Equation (26).

Furthermore, in order to obtain the analytical expression of the optimal control it is necessary to solve for the relevant elements in Equation (16). Because A_{Ai} in (7) is time-invariant, the state transition matrix can be solved by $\Phi_{Ai} = \mathcal{L}^{-1}\left[(sI - A_{Ai})^{-1}\right]$. To find the inverse of $(sI - A_{Ai})$, we can augment the given matrix with the identity matrix and perform Gaussian elimination. By performing elementary row and column operations on the augmented matrix, we obtain

$$L \cdot (sI - A_{Ai})^{-1} = \begin{bmatrix} \dfrac{1}{s} & \dfrac{1}{s^2} & \pm \dfrac{V_i Z_\alpha^{(i)}}{s(M_\alpha^{(i)} + s(-s + Z_\alpha^{(i)}))} & \pm \dfrac{V_i Z_\alpha^{(i)}}{s^2(M_\alpha^{(i)} + s(-s + Z_\alpha^{(i)}))} & \mp \dfrac{\tau_A}{s^2 + s^3 \tau_A} \end{bmatrix}. \tag{33}$$

Thus, $\Phi_{AT,y}(t_{go1})$ and $\Phi_{AD,y}(t_{go2})$ in G_1 and G_2 can be given by

$$\begin{aligned}\Phi_{AT,y}(t_{go1}) &= \mathcal{L}^{-1}\left[L \cdot (sI - A_{AT})^{-1}\right], \\ \Phi_{AD,y}(t_{go2}) &= \mathcal{L}^{-1}\left[L \cdot (sI - A_{AD})^{-1}\right]. \end{aligned} \tag{34}$$

In order to solve the inverse Laplace transform of the above equation, the following transformation is provided:

$$\frac{Z_\alpha}{M_\alpha + Z_\alpha s - s^2} = -\frac{Z_\alpha}{\kappa_1} \cdot \frac{\kappa_1}{(s + \kappa_2)^2 + \kappa_1^2} \tag{35}$$

$$\frac{Z_\alpha}{s(M_\alpha + Z_\alpha s - s^2)} = \frac{Z_\alpha}{M_\alpha}\left[\frac{1}{s} - \frac{s + \kappa_2}{(s + \kappa_2)^2 + \kappa_1^2} + \frac{Z_\alpha}{2a} \cdot \frac{\kappa_1}{(s + \kappa_2)^2 + \kappa_1^2}\right] \tag{36}$$

$$\frac{Z_\alpha}{s^2(M_\alpha + Z_\alpha s - s^2)} = \frac{Z_\alpha}{M_\alpha}\left[\frac{1}{s^2} - \frac{Z_\alpha}{M_\alpha s} + \frac{Z_\alpha}{M_\alpha}\frac{s + \kappa_2}{(s + \kappa_2)^2 + \kappa_1^2} - \frac{1 + \frac{Z_\alpha^2}{2M_\alpha}}{\kappa_1} \cdot \frac{\kappa_1}{(s + \kappa_2)^2 + \kappa_1^2}\right] \tag{37}$$

where $\kappa_1 = \sqrt{-M_\alpha - \frac{Z_\alpha^2}{4}}$, $\kappa_2 = -\frac{Z_\alpha}{2}$. It should be noted that Equations (35)–(37) are not a partial fraction decomposition of the original fraction, as we have not found the roots of the denominator. The rational fractions in Equations (35)–(37) allow us to utilize the following inverse Laplace transform:

$$\mathcal{L}^{-1}\left[\frac{a}{(s+b)^2 + a^2}\right] = e^{-bt}\sin at \quad \mathcal{L}^{-1}\left[\frac{s+b}{(s+b)^2 + a^2}\right] = e^{-bt}\cos at. \tag{38}$$

Let $\kappa_1^{(T)} = \sqrt{-M_\alpha^{(T)} - \frac{(Z_\alpha^{(T)})^2}{4}}$, $\kappa_2^{(T)} = -\frac{Z_\alpha^{(T)}}{2}$ and $\kappa_1^{(D)} = \sqrt{-M_\alpha^{(D)} - \frac{(Z_\alpha^{(D)})^2}{4}}$, $\kappa_2^{(D)} = -\frac{Z_\alpha^{(D)}}{2}$; then, the first row of the state transition matrix used to compute the ZEM $\Phi_{AT,y}(t, t_{f1})$ and $\Phi_{AD,y}(t, t_{f2})$ can be derived from the inverse Laplace transform lookup table, which can be represented as

$$\Phi_{AT,y}\left(t, t_{f1}\right) = \Phi_{AT,y}\left(t_{go1}\right)$$

$$= \begin{bmatrix} 1 \\ t_{go1} \\ \frac{Z_a^{(T)} V_T}{M_a^{(T)}}\left(1 - e^{-\kappa_2^{(T)} t_{go1}} \cos\left(\kappa_1^{(T)} t_{go1}\right) + \frac{e^{-\kappa_2^{(T)} t_{go1}} Z_a^{(T)}}{2\kappa_1^{(T)}} \sin\left(\kappa_1^{(T)} t_{go1}\right)\right) \\ \frac{Z_a^{(T)} V_T}{M_a^{(T)}}\left(t_{go1} - \frac{Z_a^{(T)}}{M_a^{(T)}} + \frac{Z_a^{(T)}}{M_a^{(T)}} e^{-\kappa_2^{(T)} t_{go1}} \cos\left(\kappa_1^{(T)} t_{go1}\right) - \frac{1 + \frac{(Z_a^{(T)})^2}{2 M_a^{(T)}}}{\kappa_1^{(T)}} e^{-\kappa_2^{(T)} t_{go1}} \sin\left(\kappa_1^{(T)} t_{go1}\right)\right) \\ -\tau_A\left(t_{go1} - \tau_A + e^{-\frac{t_{go1}}{\tau_A}} \tau_A\right) \end{bmatrix}^T \quad (39)$$

and

$$\Phi_{AD,y}\left(t, t_{f2}\right) = \Phi_{AD,y}\left(t_{go2}\right)$$

$$= \begin{bmatrix} 1 \\ t_{go2} \\ -\frac{Z_a^{(D)} V_D}{M_a^{(D)}}\left(1 - e^{-\kappa_2^{(D)} t_{go2}} \cos\left(\kappa_1^{(D)} t_{go2}\right) + \frac{e^{-\kappa_2^{(D)} t_{go2}} Z_a^{(D)}}{2\kappa_1^{(D)}} \sin\left(\kappa_1^{(D)} t_{go2}\right)\right) \\ -\frac{Z_a^{(D)} V_D}{M_a^{(D)}}\left(t_{go2} - \frac{Z_a^{(D)}}{M_a^{(D)}} + \frac{Z_a^{(D)}}{M_a^{(D)}} e^{-\kappa_2^{(D)} t_{go2}} \cos\left(\kappa_1^{(D)} t_{go2}\right) - \frac{1 + \frac{(Z_a^{(D)})^2}{2 M_a^{(D)}}}{\kappa_1^{(D)}} e^{-\kappa_2^{(D)} t_{go2}} \sin\left(\kappa_1^{(D)} t_{go2}\right)\right) \\ \tau_A\left(t_{go2} - \tau_A + e^{-\frac{t_{go2}}{\tau_A}} \tau_A\right) \end{bmatrix}^T \quad (40)$$

Substituting $\Phi_{AT,y}(t_{go1})$ and $\Phi_{AD,y}(t_{go2})$ into G_1 and G_2 in (32), it can be found that the elements of $\frac{dK}{dt_{go}}$ share the following form and that the indefinite integral can be expressed in terms of elementary functions:

$$\int \left(n + tp - re^{tq}\cos(mt) + ve^{tq}\sin(mt)\right)^2 dt$$

$$= \frac{1}{12}\begin{pmatrix} 4t(3n^2 + 3npt + p^2 t^2) + \frac{3e^{2qt}(r^2+v^2)}{q} \\ + \frac{3e^{2qt}(2mrv + q(r^2 - v^2))\cos(2mt)}{m^2 + q^2} - \frac{3e^{2qt}(2qrv + m(-r^2 + v^2))\sin(2mt)}{m^2 + q^2} \\ - \frac{24e^{qt}(m^2 r(p + nq + pqt) + q^2 r(nq + p(-1 + qt)) + m^3(n + pt)v + mq(nq + p(-2 + qt))v)\cos(mt)}{(m^2 + q^2)^2} \\ + \frac{24e^{qt}(-m^3 r(n + pt) - mqr(nq + p(-2 + qt)) + m^2(p + nq + pqt)v + q^2(nq + p(-1 + qt))v)\sin(mt)}{(m^2 + q^2)^2} \end{pmatrix} \quad (41)$$

This indefinite integral can be obtained by expanding the square and employing general integration techniques.

At this point, we can obtain the expression of $K(t)$ directly through an indefinite integral, thereby obtaining the solution of the Riccati equation.

4. Theoretical Analysis of IGC Active Protection Strategy

4.1. Saddle Point Sufficiency Proof

In the development of IGC strategies, establishing the optimality of a control solution is crucial. This not only ensures the effectiveness of the control in achieving its goals, thereby minimizing miss distance and control effort, it provides theoretical validation that the proposed solution indeed performs better than any other strategy under the given model assumptions. To this end, proving that a solution to the Riccati equation serves as a sufficient condition for the existence of an optimal saddle point solution is essential. Such a proof confirms that the derived control law is not just a feasible solution but the best possible strategy within the framework of the IGC problem [44].

Theorem 1. *The existence of a solution to the Riccati equation (Equation (26)) is a sufficient condition for the existence of a saddle point solution, and the control derived from the Riccati equation is optimal.*

Proof. Consider the reduced-order active protection problem defined by

$$\min_{u_{T/D}} \max_{u_A} J_z = \frac{1}{2} z^\top(t_f) Q z(t_f) + \frac{1}{2} \int_{t_0}^{t_f} \left(u_{T/D}^\top R_1 u_{T/D} + u_M^\top R_2 u_M \right) dt \qquad (42)$$

subject to the dynamics

$$\dot{z}(t) = G_1 u_{T/D} + G_2 u_M. \qquad (43)$$

To analyze this, we construct the identity

$$S = z_0^\top P(t_0) z_0 - z^\top(t_f) Q_f z(t_f) + \int_0^{t_f} \frac{d}{dt}(z^\top P z) \, dt \equiv 0. \qquad (44)$$

Using the chain rule for differentiation, we obtain

$$\frac{d}{dt}(z^\top P z) = (\dot{z}^\top P z) + (z^\top \dot{P} z) + (z^\top P \dot{z}). \qquad (45)$$

Substituting and expanding within S, we obtain

$$S = z_0^\top P(0) z_0 - z^\top(t_f) Q_f z(t_f) - \int_0^{t_f} \left(z^\top Q z + u_{T/D}^\top R_1 u_{T/D} + u_M^\top R_2 u_M \right) dt$$
$$+ \int_0^{t_f} \left(u_{T/D} + R_1^{-1} G_1^\top P z \right)^\top R_1 \left(u_{T/D} + R_1^{-1} G_1^\top P z \right) dt \qquad (46)$$
$$- \int_0^{t_f} \left(u_M + R_2^{-1} G_2^\top P z \right)^\top R_2 \left(u_M + R_2^{-1} G_2^\top P z \right) dt.$$

From this, we can deduce that the performance index J_z satisfies

$$2 J_Z = z_0^\top P(0) z_0 + \int_0^{t_f} \left(u_{T/D} + R_1^{-1} G_1^\top P z \right)^\top R_1 \left(u_{T/D} + R_1^{-1} G_1^\top P z \right) dt$$
$$- \int_0^{t_f} \left(u_M + R_2^{-1} G_2^\top P z \right)^\top R_2 \left(u_M + R_2^{-1} G_2^\top P z \right) dt \qquad (47)$$
$$= z_0^\top P(0) z_0 + \int_0^{t_f} (u_{T/D} - u_{T/D}^*)^\top R_1 (u_{T/D} - u_{T/D}^*) dt - \int_0^{t_f} (u_M - u_M^*)^\top R_2 (u_M - u_M^*) dt.$$

The expression clearly shows that $J_Z(u_{T/D}^*, u_M) \leq J_Z(u_{T/D}^*, u_M^*) \leq J_Z(u_{T/D}, u_M^*)$, indicating that the performance index is minimized for the optimal controls $u_{T/D}^*$ and u_M^*, confirming the saddle point nature of the solution. This completes the proof. □

4.2. Analysis of the Invertibility of Matrix $K(t)$

The invertibility of the matrix $K(t)$ is critical in solving the Riccati equation and ensuring the computation accuracy of the proposed strategy. The ill-conditioning of $K(t)$ significantly impacts the numerical stability, especially during matrix inversion operations. Ill-conditioning refers to scenarios where small variations in inputs cause large changes in outputs, leading to significant numerical errors in calculations such as matrix inversion. An ill-conditioned $K(t)$ can undermine the reliability of computing its inverse.

It can be seen through Equation (32) that $K(t)$ are influenced by the matrices G_1 and G_2, which incorporate control inputs from $\tilde{B}_T$, $\tilde{B}_D$, and $\tilde{B}_M$. Significant differences in the control matrices can be amplified through quadratic operations in Equation (32), leading to wide disparities in the magnitudes of the elements in $K(t)$ and exacerbating its ill-conditioning.

In practical terms, control inputs such as fin deflection $u_{T/D}$ and commanded accelerations u_A differ fundamentally in scale and system impact. For example, a single-degree change in fin deflection might equate to a 20 m/s^2 change in acceleration. The difference in control efficiency affects how sensitive the system state is to increments in $u_{T/D}$ and u_M.

In multiple-input–multiple-output (MIMO) systems, the disparity in control effectiveness across different actuators can be quantitatively expressed through the singular values of their control matrices. The largest singular value of control matrices offers a measure of the "control efficiency", reflecting the sensitivity of the system's state to changes in respective control inputs. This is because the largest singular value in a control matrix provides critical insight into the maximum stretch factor that the matrix imparts to any vector it multiplies.

To illustrate the relationship between matrix scaling and the largest singular values, we consider a scenario where the short-period dynamics parameters for the target and defender are taken from [45] and [17], respectively. We define the scaled control matrices as follows:

$$\tilde{B}_T^* = B^* \cdot \tilde{B}_T$$
$$\tilde{B}_D^* = B^* \cdot \tilde{B}_D \quad (48)$$
$$\tilde{B}_A^* = B^* \cdot \tilde{B}_A$$

where B^* is a scaling factor. The sensitivity of the system's state to different control inputs is demonstrated by comparing the largest singular values of the control matrices under various scaling factors.

Figure 2 presents the variation of the largest singular value $\bar{\sigma}$ for each control matrix under different scaling factors. Given that the defender's control input typically exhibits the highest control efficiency, we use the largest singular value of $\tilde{B}_D$ as the benchmark. The graph shows that the largest singular values approximate a linear relationship on a logarithmic scale. For the flight characteristics considered here, when the scaling factor for $\tilde{B}_T$ is about 10 and for $\tilde{B}_A$ about 250 the relationships satisfy

$$\bar{\sigma}\left(\tilde{B}_T^*\right) \approx \bar{\sigma}\left(\tilde{B}_D\right) \approx \bar{\sigma}\left(\tilde{B}_A^*\right). \quad (49)$$

This finding suggests that the magnitude differences in control matrices must be considered when applying the IGC active protection strategy, as they directly affect the ill-conditioning of $K(t)$ and potentially the robustness of the control solution.

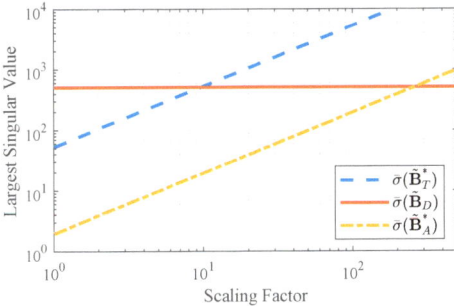

Figure 2. Comparison of largest singular values of different control matrices.

Furthermore, in order to check the accuracy of the calculation for $K^{-1}(t)$ during the simulation, we introduce the condition number as a real-time monitor. The condition number of a matrix K is defined as

$$\text{cond}(K) = \|K\| \|K^{-1}\|, \quad (50)$$

where $\|K\|$ is a norm of K. The condition number quantifies the sensitivity of the solution of a system of linear equations to changes in the input or errors in the data. A low condition number indicates that the matrix is well-conditioned, meaning that it is stable and reliable for numerical calculations. Conversely, a high condition number suggests that the matrix is ill-conditioned, which can lead to significant numerical errors in computations, especially inverses. Matrices with very high condition numbers may result in large errors in computed solutions, indicating that even small errors in data can lead to unreliable or incorrect results.

In Section 5, we demonstrate through specific examples how the methodologies proposed in this paper effectively reduce the condition number of matrix $K(t)$. By improving the conditioning of $K(t)$, we enhance the reliability and accuracy of the matrix inversion operations critical to the IGC active protection strategies.

4.3. Simplified IGC Active Protection Strategy

In the development of the IGC active protection strategy, while modeling both the target and the defender within the IGC framework is feasible, it tends to increase computational complexity and numerical instability. A more streamlined approach can be adopted by focusing the IGC model primarily on the defender.

In general, the defender exhibits more significant agility and dynamic response than the target in active protection scenarios. By concentrating the IGC model on the defender, we can effectively capture and manage these dynamics while reducing the model's overall complexity. This simplified strategy simplifies calculations while addressing the more pronounced coupling effects between guidance and control loops in high-speed and highly maneuverable defenders.

The state vector for this simplified model is chosen as

$$x = [y_{AT}, \dot{y}_{AT}, a_T, y_{AD}, \dot{y}_{AD}, \alpha_D, q_D, a_A]^\top. \tag{51}$$

This state vector includes the relative positions, velocities, and accelerations of the attacker and target as well as the angle of attack and pitch rate relevant to the defender.

The linearized dynamics model representing the three-player conflict can be expressed as

$$\dot{x} = \bar{A}x + \bar{B}_T u_T + \bar{B}_D u_D + \bar{B}_A u_A, \tag{52}$$

where

$$\bar{A} = \begin{bmatrix} 0 & 1 & 0 & & & & & 0 \\ 0 & 0 & 1 & & [0]_{3\times 4} & & & -1 \\ 0 & 0 & -1/\tau_T & & & & & 0 \\ \hline & & & 0 & 1 & [0]_{1\times 2} & & 0 \\ & [0]_{4\times 3} & & 0 & 0 & C_D & & 1 \\ & & & [0]_{2\times 1} & [0]_{2\times 1} & A_D & & 0 \\ \hline & [0]_{1\times 3} & & & [0]_{1\times 4} & & & -1/\tau_M \end{bmatrix} \tag{53}$$

$$\bar{B}_T = \begin{bmatrix} 0 & 0 & 1/\tau_T & [0]_{1\times 4} & 0 \end{bmatrix}^\top \tag{54}$$

$$\bar{B}_D = \begin{bmatrix} [0]_{1\times 3} & 0 & -d_D & B_D^\top & 0 \end{bmatrix}^\top \tag{55}$$

$$\bar{B}_A = \begin{bmatrix} [0]_{1\times 3} & [0]_{1\times 4} & 1/\tau_M \end{bmatrix}^\top. \tag{56}$$

This model simplifies the integration of guidance and control for the defender by isolating and directly addressing the most dynamically significant elements of the system. The derivation follows procedures similar to those previously discussed; for the sake of brevity, it is not repeated here.

5. Simulation Results and Analysis

This section describes the simulation setup and presents the results of two specific simulation examples to validate the proposed IGC approach within the context of aerial vehicle active protection scenarios. The results are compared to traditional methods where guidance and control loops are designed separately.

5.1. Simulation Setup

The simulation is designed to evaluate the effectiveness of the IGC strategy proposed in this paper. The defender model is based on the one described in [17]. The fin deflection commands generated by the autopilot (AP) or IGC are transmitted to the actuator. In aerial vehicle systems, the actuator is a critical component that converts the autopilot's commands into physical movements, such as the deflection of tail fins, to achieve the desired motion. Considering the need for rapid response, actuators in aerial vehicles are typically high-bandwidth devices which exceed the bandwidth of the flight control loop itself. Although most modern actuators are electromechanical, hydraulic options are used in specific applications.

For the initial design and analysis phases, it is common to model the actuator dynamics using a simplified second-order transfer function. While this model does not capture the full complexity of the actuator hardware, it provides a reasonable approximation for understanding how the actuator's response characteristics can influence the performance of the guidance and control loops. The second-order modeling is expressed by the transfer function

$$\frac{\delta(s)}{\delta_c(s)} = \frac{\omega_a^2}{s^2 + 2\zeta_a\omega_a s + \omega_a^2}, \tag{57}$$

where $\delta(s)$ is the actuator output, $\delta_c(s)$ is the command from the autopilot, ω_a is the natural frequency, and ζ_a is the damping ratio of the actuator. Reasonably designing the parameters of actuators is also crucial for the stability of a control system. A series of curves comparing different settings of the second-order response and the characteristics of these responses are shown in Figure 3.

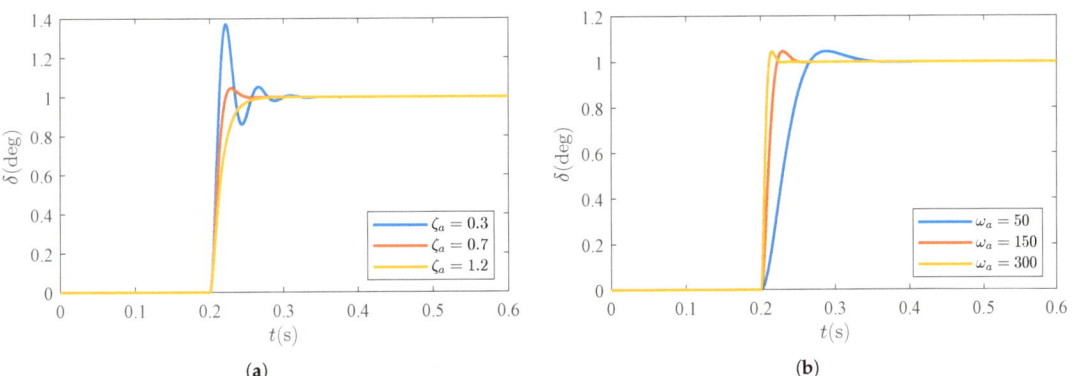

Figure 3. Step response of a second-order actuator model: (a) setting $\omega_a = 150$ and (b) setting $\zeta_a = 0.7$.

For subsequent simulations, the parameters chosen for the actuator model are $\omega_a = 150$ rad/s and $\zeta_a = 0.7$. These parameters are selected to balance responsiveness and stability, ensuring that the actuator can adequately follow the control commands without inducing undue oscillations or delays in the vehicle's response.

5.2. Simulation Scenario I

In this simulation example, the attacker directly employs the Proportional Navigation (PN) guidance law to engage the target. The defender's performance is evaluated under two different configurations: an IGC design, and a traditional approach in which guidance and control loops are designed separately, i.e., PN+AP. In the traditional setup, the defender uses a PN guidance law coupled with a classic three-loop autopilot to track acceleration commands. The initial conditions for this simulation are detailed in Table 1:

Table 1. Initial parameters for simulation scenario I.

Parameter	Value
(x_{T0}, y_{T0})	(0, 1000) m
(x_{D0}, y_{D0})	(0, 1000) m
(x_{A0}, y_{A0})	(2000, 1200) m
τ_A	0.5 s
u_A^{max}	40 g

The engagement trajectories of the vehicles are depicted in Figure 4. Under the IGC design, the defender successfully intercepts the attacker at 2.55 s with a miss distance of only 0.16 m. This high-precision interception effectively neutralizes the threat, preventing any damage to the target. In stark contrast, the defender utilizing the traditional AP+PN design results in a significantly larger miss distance of 28.67 m. This considerable deviation results in the attacker successfully hitting the target, demonstrating the limitations of the traditional approach in high-stakes scenarios requiring precise interception.

In addition, the defender employing the IGC design follows a more direct and efficient flight path. This path not only demonstrates the system's enhanced responsiveness to dynamic threats, it indicates a lower demand for the defender's maneuvering capabilities; conversely, the trajectory of the defender using the traditional PN+AP design exhibits more curvature and requires greater maneuvering effort, which can lead to higher fuel consumption and greater stress on the defender.

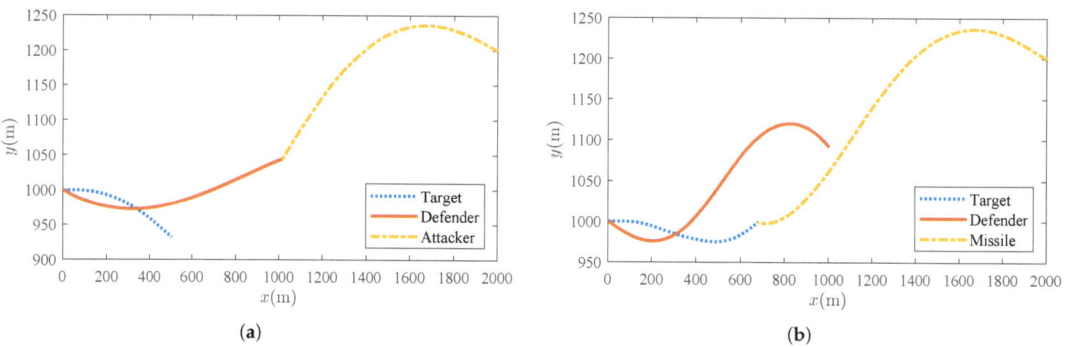

Figure 4. Trajectory when the attacker adopts the PN guidance law: (**a**) trajectory when the defender employs IGC and (**b**) trajectory when the defender employs PN+AP.

The ZEM is a critical metric in three-player aerial engagements, providing insight into the effectiveness of the active protection strategy by measuring the predicted miss distance (assuming that no further maneuvers are performed). An analysis of the ZEM for both the IGC strategy and the PN+AP reveals significant differences in performance and stability.

For the defender utilizing the IGC design, the ZEM z_2 exhibits a stable and smooth convergence towards zero, as shown in Figure 5. This stable convergence indicates that the defender effectively locks the attacker within the interception triangle, maintaining a consistent trajectory that ensures successful interception. Conversely, the ZEM values for the defender employing the traditional PN+AP display considerable fluctuations, highlighting instability and a lack of control precision. The erratic nature of the ZEM curve when using PN+AP suggests that the defender frequently adjusts their path, reacting to the attacker's maneuvers rather than anticipating, which leads to a failure to intercept.

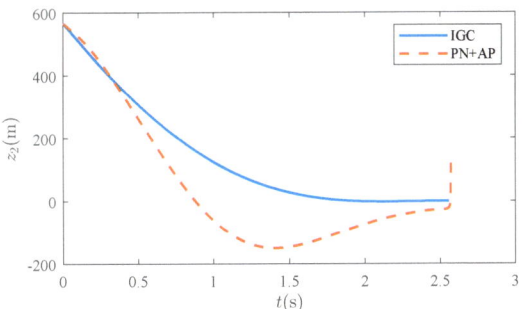

Figure 5. ZEM z_2.

In the proposed IGC design, computing the inverse of matrix $K(t)$ is crucial for solving the Riccati differential equation in order to derive the solution matrix $P(t)$. The condition number of matrix $K(t)$ serves as a critical indicator of its invertibility and the potential for numerical instability. A low condition number suggests that the matrix is not sensitive to small perturbations or numerical errors, ensuring the reliability of the matrix inversion operation.

Throughout the engagement process, the condition number of matrix $K(t)$ maintained values typically below 10, as depicted in Figure 6. This indicates that matrix $K(t)$ remained non-singular and well-conditioned, ensuring the accuracy and reliability of the inversion computations required for solving $P(t)$. This stability in the condition number affirms the method's robustness against potential numerical issues, particularly when real-time onboard calculations are required for dynamic aerial engagements.

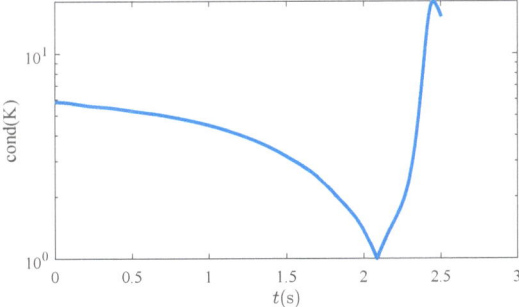

Figure 6. Condition number of $K(t)$.

Our approach based on the differential game framework does not necessitate prior knowledge of the adversary's exact strategies or their functional form; however, it does demand specific battlefield information, such as the attacker's time constant τ_A and acceleration a_A. Although τ_A can be inferred from open-source information or engineering experience, and a_A can be estimated in real-time using filtering algorithms, these methods introduce inevitable errors. In order to assess potential disadvantages, we conducted Monte Carlo simulations considering deviation in estimates of τ_A and a_A. In addition, we introduced varying levels of noise interference in the measurement of the Line-of-Sight (LOS) rotation rate. The defender's kill radius was set to 5 m, meaning that an interception is successful if the miss distance is less than 5 m. The attacker's initial position was varied between 1500–3000 m laterally and 200–1800 m longitudinally.

We compiled the defender's interception success rates under different noise levels, as shown in Table 2. The results indicate that when the measurement noise and model deviation are below 25% the proposed IGC strategy's performance remains largely unaffected, demonstrating the method's robustness to noise interference.

Table 2. Success rate under noise and deviation.

Noise and Deviation	0	10%	15%	25%
Success Rate	100%	100%	100%	100%

These findings address concerns about sensor noise while providing a more comprehensive evaluation of our IGC strategy's performance under realistic conditions. In addition, they demonstrate the practical applicability of our approach in scenarios where sensor imperfections and countermeasures are present.

5.3. Simulation Scenario II

The vehicles' simulation parameters are detailed in Table 3. Faced with the threat of interception, the attacker employs weave maneuvers to evade the defender. If the attacker succeeds in evading the defender, it then switches to the proportional navigation (PN) guidance law to strike the target.

Table 3. Initial parameters of the vehicles.

Initial Parameters	Value
(x_{T0}, y_{T0})	(0, 1000) m
(x_{D0}, y_{D0})	(0, 1000) m
(x_{A0}, y_{A0})	(3500, 800) m
τ_A	0.5 s
u_A^{max}	20 g

With the implementation of the IGC design, the trajectories of the three vehicles are illustrated in Figure 7a. The defender successfully intercepts the attacker with a miss distance of 0.33 m, demonstrating the effectiveness of the IGC approach in protecting the target.

To highlight the superiority of the proposed IGC approach, we compare it against a traditional scheme where the guidance and control loops are designed separately. In this traditional configuration, the defender employs a PN guidance law coupled with a three-loop autopilot to follow the acceleration commands. The trajectory for this separation design scheme is depicted in Figure 7b. In contrast to the IGC approach, the defender using the PN+AP scheme fails to intercept the attacker, resulting in a significant miss distance of 49.64 m. Consequently, this failure allows the attacker to successfully hit the target.

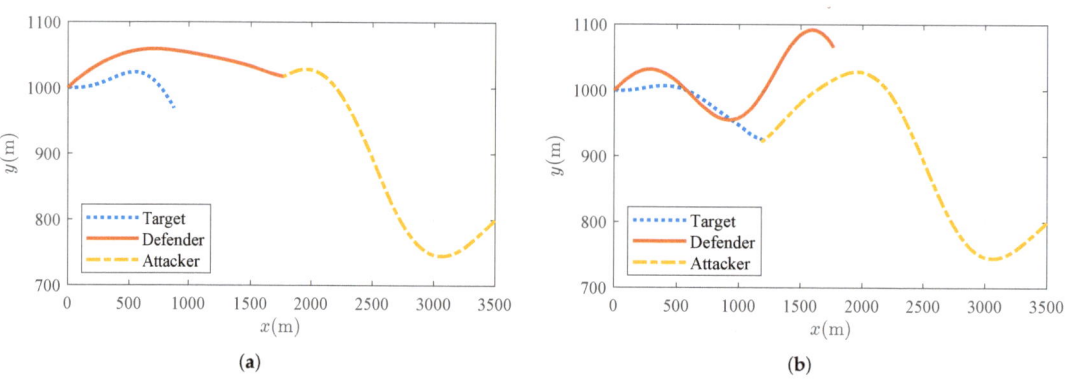

Figure 7. Trajectory when attacker performs weave maneuver: (**a**) trajectory when the defender employs IGC and (**b**) trajectory when the defender employs PN+AP.

The ZEM z_2 comparison is illustrated in Figure 8. The defender utilizing IGC shows markedly less influence from the attacker's evasive maneuvers, successfully locking the attacker within the interception triangle. In contrast, the defender using the separate PN+AP approach experiences significant fluctuations in z_2, ultimately failing to intercept the attacker.

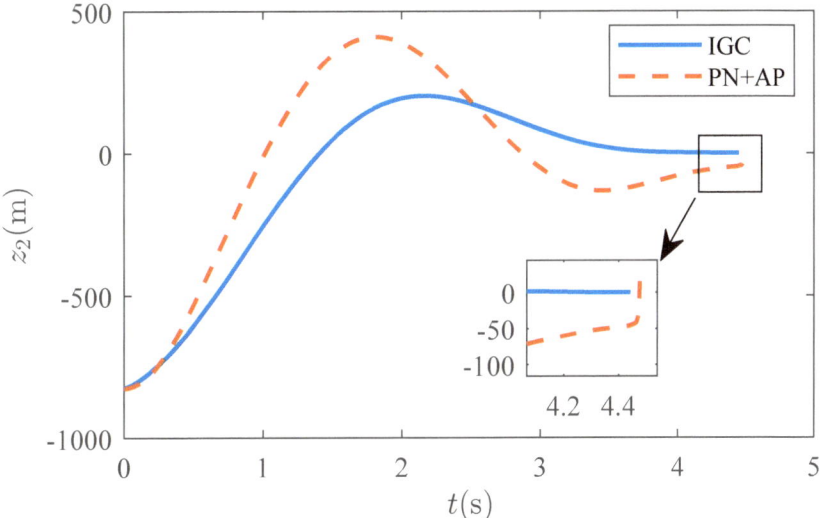

Figure 8. ZEM z_2.

The acceleration curves of the flight vehicles under both the integrated and separated guidance and control schemes are presented in Figure 9a,b. The simulation with the IGC design ends at 4.44 s, whereas the simulation with the PN+AP approach extends to 5.94 s. Notably, the attacker's acceleration profile before evading the defender remains consistent in both figures until it switches to a PN guidance law heading to the target.

The acceleration of the defender under these two scenarios is particularly noteworthy. With the IGC approach, the defender experiences brief initial oscillations before quickly stabilizing. Throughout the engagement, the defender maintains relatively low overload levels generally not exceeding 5 g, with a maximum peak below 10 g. In contrast, under the traditional separated design the defender's demand for maneuverability significantly increases, with peak overloads reaching up to 30 g.

The comparison in acceleration clearly illustrates the advantages of integrating guidance and control loops in the IGC design. By considering the dynamic characteristics of the vehicle comprehensively, the IGC approach effectively reduces the demand for maneuverability in high-dynamic environments, leading to more stable and efficient flight performance. Such an integrated design enhances interception efficacy while minimizing the stress on the vehicle's structural and operational limits.

The fin deflection of the defender is depicted in Figure 10, showing a smooth variation that accounts for actuator response delays. This controlled adjustment ensures compatibility with realistic hardware capabilities, demonstrating the practical applicability of the IGC approach.

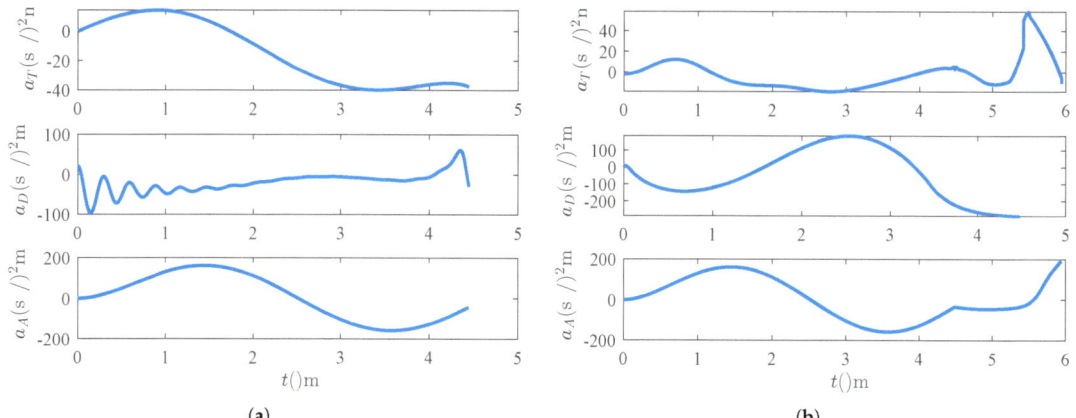

Figure 9. Acceleration when attacker performs weave maneuver: (**a**) acceleration when the defender employs IGC and (**b**) trajectory when the defender employs PN+AP.

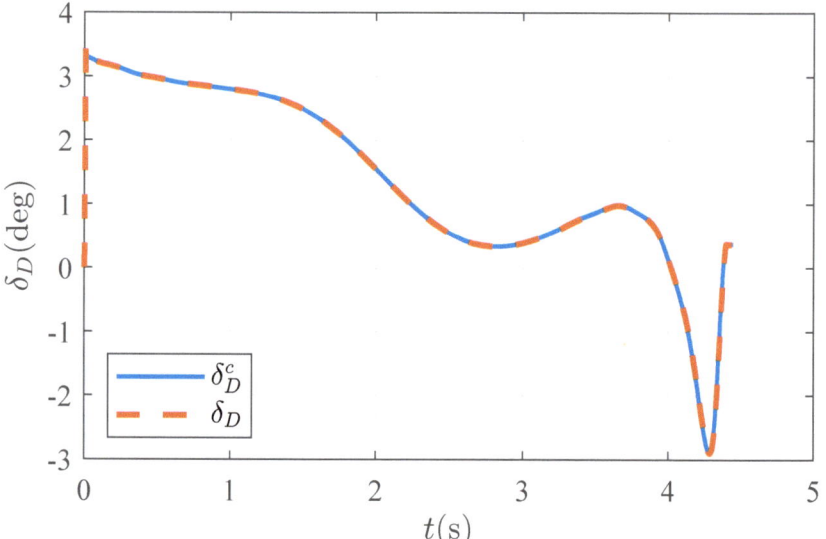

Figure 10. Fin deflection.

To ensure the numerical accuracy during the inversion operation of matrix $K(t)$, we analyzed the changes in the elements of matrix $K(t)$ and its condition number. As displayed in Figure 11, the variation curves of each element confirm that the matrix does not exhibit any singularity, validating the feasibility of obtaining matrix $P(t)$ through inversion. In addition, we compared the theoretical solution obtained from the linearized model with the actual values observed in the nonlinear model simulations. The close agreement between the two sets of results shown in Figure 11 serves as empirical evidence supporting the validity and feasibility of the linearization method employed in this work.

Further, as shown in Figure 12, the condition number of matrix $K(t)$ generally remains below 10 and does not exceed 100. This indicates that matrix $K(t)$ is not ill-conditioned, ensuring that the precision of the inversion operation is safeguarded.

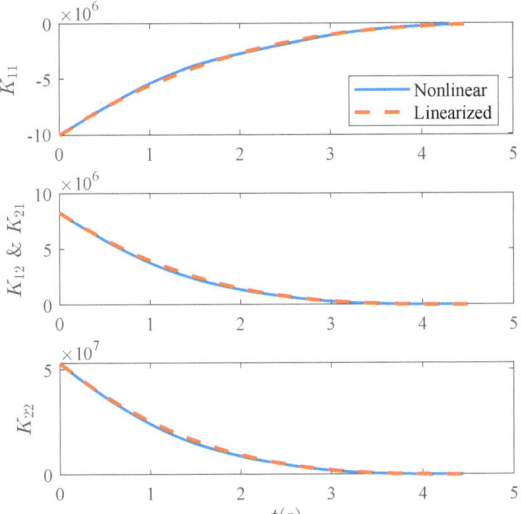

Figure 11. Matrix $K(t)$.

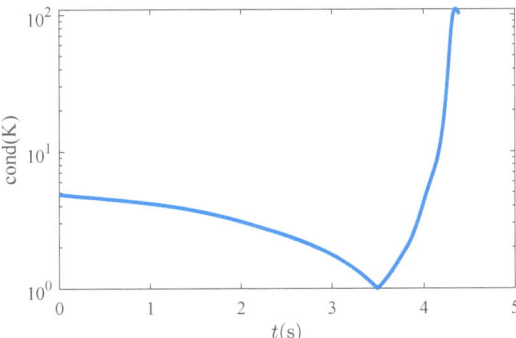

Figure 12. Condition number of $K(t)$.

6. Conclusions

This paper has proposed a cooperative IGC strategy for active target protection in a three-player conflict scenario. The IGC model was established by linearizing both the translational and rotational dynamics of the target and defender vehicles. Differential game theory and the sweep method were applied to derive and solve the Riccati differential equation analytically, providing an optimal control solution. The theoretical optimality of the derived strategy was proven and the computational accuracy of the solution was analyzed in terms of the control matrix singular values and solution matrix condition numbers. Compared to traditional separated guidance and control designs, the IGC strategy demonstrates superior performance, smoother control trajectories, and reduced maneuverability demands on the defender, leading to enhanced interception efficacy and minimized stress on the vehicle's structural and operational limits. The proposed IGC strategy provides a promising framework for active target protection in aerial vehicle engagements. Future work could explore extending the model to incorporate more complex vehicle dynamics, multi-vehicle cooperation, and robustness against uncertainties in system parameters.

Author Contributions: Conceptualization, X.G. and W.C.; methodology, X.G.; software, Z.C.; validation, X.G.; formal analysis, X.G.; investigation, X.G.; resources, Z.C.; data curation, W.C.; writing—original draft preparation, X.G.; writing—review and editing, Z.C.; visualization, X.G.; supervision, Z.C.; project administration, Z.C. and W.C.; funding acquisition, Z.C. All authors have read and agreed to the published version of the manuscript.

Funding: This work was supported by the China Postdoctoral Science Foundation (Grant No. 2021M700321) and the Fundamental Research Funds for the Central Universities (Grant No. YWF-23-Q-1041).

Data Availability Statement: All data used during the study appear in the submitted article.

Acknowledgments: The study described in this paper was supported by the China Postdoctoral Science Foundation (Grant No. 2021M700321) and the Fundamental Research Funds for the Central Universities (Grant No. YWF-23-Q-1041). The authors fully appreciate the financial support.

Conflicts of Interest: The authors declare no conflicts of interest.

References

1. Isaacs, R. *Differential Games: A Mathematical Theory with Applications to Warfare and Pursuit, Control and Optimization*; Dover Publications: Mineola, NY, USA, 1999.
2. Wei, X.; Yang, J. Optimal Strategies for Multiple Unmanned Aerial Vehicles in a Pursuit/Evasion Differential Game. *J. Guid. Control Dyn.* **2018**, *41*, 1799–1806. [CrossRef]
3. Hayoun, S.Y.; Shima, T. On guaranteeing point capture in linear n -on-1 endgame interception engagements with bounded controls. *Automatica* **2017**, *85*, 122–128. [CrossRef]
4. Buzikov, M.; Galyaev, A. The Game of Two Identical Cars: An Analytical Description of the Barrier. *J. Optim. Theory Appl.* **2023**, *198*, 988–1018. [CrossRef]
5. Singh, S.K.; Reddy, P.V. Dynamic Network Analysis of a Target Defense Differential Game with Limited Observations. *IEEE Trans. Control Netw. Syst.* **2023**, *10*, 308–320. [CrossRef]
6. Wang, K.; Zhou, S.; Yao, Y.; Sun, Q.; Wang, Y. A target defence–intrusion game with considering the obstructive effect of target. *IET Control Theory Appl.* **2024**. [CrossRef]
7. Liang, L.; Deng, F.; Wang, J.; Lu, M.; Chen, J. A Reconnaissance Penetration Game With Territorial-Constrained Defender. *IEEE Trans. Autom. Control* **2022**, *67*, 6295–6302. [CrossRef]
8. Weintraub, I.E.; Pachter, M.; Garcia, E. An Introduction to Pursuit-evasion Differential Games. In Proceedings of the 2020 American Control Conference (ACC), Denver, CO, USA, 1–3 July 2020; pp. 1049–1066. [CrossRef]
9. Garcia, E.; Casbeer, D.W.; Pachter, M. Pursuit in the Presence of a Defender. *Dyn. Games Appl.* **2019**, *9*, 652–670. [CrossRef]
10. Garcia, E.; Casbeer, D.W.; Pachter, M. The Complete Differential Game of Active Target Defense. *J. Optim. Theory Appl.* **2021**, *191*, 675–699. [CrossRef]
11. Liang, L.; Deng, F.; Peng, Z.; Li, X.; Zha, W. A differential game for cooperative target defense. *Automatica* **2019**, *102*, 58–71. [CrossRef]
12. Liang, L.; Deng, F.; Lu, M.; Chen, J. Analysis of Role Switch for Cooperative Target Defense Differential Game. *IEEE Trans. Autom. Control* **2021**, *66*, 902–909. [CrossRef]
13. Nayak, S.P.; Rajawat, A.P.; Kothari, M. Inverse Geometric Guidance Strategy for a Three-Body Differential Game. In Proceedings of the AIAA Scitech 2021 Forum, Reston, VA, USA, 11–15 & 19–21 January 2021; AIAA SciTech Forum, pp. 1–17. [CrossRef]
14. Garcia, E.; Casbeer, D.W.; Pachter, M. Cooperative Strategies for Optimal Aircraft Defense from an Attacking Missile. *J. Guid. Control Dyn.* **2015**, *38*, 1510–1520. [CrossRef]
15. Harini Venkatesan, R.; Sinha, N.K. A New Guidance Law for the Defense Missile of Nonmaneuverable Aircraft. *IEEE Trans. Control Syst. Technol.* **2015**, *23*, 2424–2431. [CrossRef]
16. Garcia, E.; Casbeer, D.W.; Fuchs, Z.E.; Pachter, M. Cooperative Missile Guidance for Active Defense of Air Vehicles. *IEEE Trans. Aerosp. Electron. Syst.* **2018**, *54*, 706–721. [CrossRef]
17. Zarchan, P. *Tactical and Strategic Missile Guidance*, 6th ed.; Progress in Astronautics and Aeronautics; American Institute of Aeronautics and Astronautics: Reston, VA, USA, 2012; Volume 239. [CrossRef]
18. Yanushevsky, R. *Modern Missile Guidance*, 2nd ed.; Taylor et Francis/CRC Press: Boca Raton, FL, USA; London, UK; New York, NY, USA, 2019.
19. Shaferman, V.; Oshman, Y. Stochastic Cooperative Interception Using Information Sharing Based on Engagement Staggering. *J. Guid. Control Dyn.* **2016**, *39*, 2127–2141. [CrossRef]
20. Prokopov, O.; Shima, T. Linear Quadratic Optimal Cooperative Strategies for Active Aircraft Protection. *J. Guid. Control Dyn.* **2013**, *36*, 753–764. [CrossRef]
21. Shima, T. Optimal Cooperative Pursuit and Evasion Strategies Against a Homing Missile. *J. Guid. Control Dyn.* **2011**, *34*, 414–425. [CrossRef]

22. Perelman, A.; Shima, T.; Rusnak, I. Cooperative Differential Games Strategies for Active Aircraft Protection from a Homing Missile. *J. Guid. Control Dyn.* **2011**, *34*, 761–773. [CrossRef]
23. Alkaher, D.; Moshaiov, A. Game-Based Safe Aircraft Navigation in the Presence of Energy-Bleeding Coasting Missile. *J. Guid. Control Dyn.* **2016**, *39*, 1539–1550. [CrossRef]
24. Liu, F.; Dong, X.; Li, Q.; Ren, Z. Cooperative differential games guidance laws for multiple attackers against an active defense target. *Chin. J. Aeronaut.* **2022**, *35*, 374–389. [CrossRef]
25. Rubinsky, S.; Gutman, S. Three-Player Pursuit and Evasion Conflict. *J. Guid. Control Dyn.* **2014**, *37*, 98–110. [CrossRef]
26. Rubinsky, S.; Gutman, S. Vector Guidance Approach to Three-Player Conflict in Exoatmospheric Interception. *J. Guid. Control Dyn.* **2015**, *38*, 2270–2286. [CrossRef]
27. Qi, N.; Sun, Q.; Zhao, J. Evasion and pursuit guidance law against defended target. *Chin. J. Aeronaut.* **2017**, *30*, 1958–1973. [CrossRef]
28. Shaferman, V.; Shima, T. Cooperative Multiple-Model Adaptive Guidance for an Aircraft Defending Missile. *J. Guid. Control Dyn.* **2010**, *33*, 1801–1813. [CrossRef]
29. Shaferman, V.; Shima, T. Cooperative Differential Games Guidance Laws for Imposing a Relative Intercept Angle. *J. Guid. Control Dyn.* **2017**, *40*, 2465–2480. [CrossRef]
30. Saurav, A.; Kumar, S.R.; Maity, A. Cooperative Guidance Strategies for Aircraft Defense with Impact Angle Constraints. In Proceedings of the AIAA Scitech 2019 Forum, Reston, VA, USA, 7–11 January 2019. [CrossRef]
31. Liang, H.; Wang, J.; Liu, J.; Liu, P. Guidance strategies for interceptor against active defense spacecraft in two-on-two engagement. *Aerosp. Sci. Technol.* **2020**, *96*, 105529. [CrossRef]
32. Shalumov, V.; Shima, T. Weapon–Target-Allocation Strategies in Multiagent Target–Missile–Defender Engagement. *J. Guid. Control Dyn.* **2017**, *40*, 2452–2464. [CrossRef]
33. Chen, Z.; Chen, W.; Liu, X.; Cheng, J. Three-dimensional fixed-time robust cooperative guidance law for simultaneous attack with impact angle constraint. *Aerosp. Sci. Technol.* **2021**, *110*, 106523. [CrossRef]
34. Li, G.; Liu, L.; Liu, J.; Wu, Y.; Zhao, J. Three-dimensional low-order fixed-time integrated guidance and control for STT missile with strap-down seeker. *J. Frankl. Inst.* **2023**, *360*, 9788–9811. [CrossRef]
35. Li, Z.; Zhang, X.; Zhang, H.; Zhang, F. Three-dimensional approximate cooperative integrated guidance and control with fixed-impact time and azimuth constraints. *Aerosp. Sci. Technol.* **2023**, *142*, 108617. [CrossRef]
36. Williams, D.; Richman, J.; Friedland, B. Design of an integrated strapdown guidance and control system for a tactical missile. In Proceedings of the Guidance and Control Conference, Reston, VA, USA, 15–August 1983. [CrossRef]
37. Santoso, F.; Garratt, M.A.; Anavatti, S.G. State-of-the-Art Integrated Guidance and Control Systems in Unmanned Vehicles: A Review. *IEEE Syst. J.* **2021**, *15*, 3312–3323. [CrossRef]
38. Yao, C.; Liu, Z.; Zhou, H.; Gao, C.; Li, J.; Zhang, Z. Integrated guidance and control for underactuated fixed-trim moving mass flight vehicles. *Aerosp. Sci. Technol.* **2023**, *142*, 108680. [CrossRef]
39. Shima, T.; Idan, M.; Golan, O.M. Sliding-Mode Control for Integrated Missile Autopilot Guidance. *J. Guid. Control Dyn.* **2006**, *29*, 250–260. [CrossRef]
40. Xingling, S.; Honglun, W. Back-stepping active disturbance rejection control design for integrated missile guidance and control system via reduced-order ESO. *ISA Trans.* **2015**, *57*, 10–22. [CrossRef] [PubMed]
41. Liu, C.; Jiang, B.; Zhang, K. Adaptive Fault-Tolerant H-Infinity Output Feedback Control for Lead-Wing Close Formation Flight. *IEEE Trans. Syst. Man, Cybern. Syst.* **2019**, *50*, 2804–2814. [CrossRef]
42. Yan, H.; Hou, M. A Small-Gain Approach for Three-Dimensional Integrated Guidance and Control in Pursuit-Evasion Games. In Proceedings of the 2020 Chinese Control And Decision Conference (CCDC), Hefei, China, 22–24 August 2020; pp. 4069–4076. [CrossRef]
43. Stevens, B.L.; Lewis, F.L.; Johnson, E.N. *Aircraft Control and Simulation: Dynamics, Controls Design, and Autonomous Systems*, 3rd ed.; John Wiley & Sons: Hoboken, NJ, USA, 2016.
44. Green, M.; Limebeer, D.J.N. *Linear Robust Control*, dover ed.; Dover Publications Inc.: Mineola, NY, USA, 2012.
45. Yechout, T.R.; Morris, S.L.; Bossert, D.E.; Hallgren, W.F.; Hall, J.K. *Introduction to Aircraft Flight Mechanics: Performance, Static Stability, Dynamic Stability, Classical Feedback Control, and State-Space Foundations*, 2nd ed.; AIAA Education Series; American Institute of Aeronautics and Astronautics Inc.: Reston, VA, USA, 2014.

Disclaimer/Publisher's Note: The statements, opinions and data contained in all publications are solely those of the individual author(s) and contributor(s) and not of MDPI and/or the editor(s). MDPI and/or the editor(s) disclaim responsibility for any injury to people or property resulting from any ideas, methods, instructions or products referred to in the content.

Article

Intelligent Reduced-Dimensional Scheme of Model Predictive Control for Aero-Engines

Zhen Jiang [1], Xi Wang [1], Jiashuai Liu [1,*], Nannan Gu [2] and Wei Liu [3]

1. School of Energy and Power Engineer, Beihang University, Beijing 102206, China; jiangzhen@buaa.edu.cn (Z.J.); xwang@buaa.edu.cn (X.W.)
2. School of Electronics and Information Engineering, Ningbo University of Technology, Ningbo 315211, China; gunannan@nbut.edu.cn
3. AECC Sichuan Gas Turbine Research Establishment, Chengdu 610500, China; gte@cgte.com.cn
* Correspondence: ljsbuaa@buaa.edu.cn

Abstract: Model Predictive Control (MPC) has many advantages in controlling an aero-engine, such as handling actuator constraints, but the computational burden greatly obstructs its application. The current multiplex MPC can reduce computational complexity, but it will significantly decrease the control performance. To guarantee real-time performance and good control performance simultaneously, an intelligent reduced-dimensional scheme of MPC is proposed. The scheme includes a control variable selection algorithm and a control sequence coordination strategy. A constrained optimization problem with low computational complexity is first constructed by using only one control variable to define a reduced-dimensional control sequence. Therein, the control variable selection algorithm provides an intelligent mode to determine the control variable that has the best control effect at the current sampling instant. Furthermore, a coordination strategy is adopted in the reduced-dimensional control sequence to consider the interaction of control variables at different predicting instants. Finally, an intelligent reduced-dimensional MPC controller is designed and implemented on an aero-engine. Simulation results demonstrate the effectiveness of the intelligent reduced-dimensional scheme. Compared with the multiplex MPC, the intelligent reduced-dimensional MPC controller enhances the control quality significantly by 34.06%; compared with the standard MPC, the average time consumption is decreased by 64.72%.

Keywords: model predictive control; computational complexity; reduced-dimensional; control variable selection; aero-engine

1. Introduction

As the power source of flight, the aero-engine plays an important role in the modern aircraft. During operation, the engine needs to not only meet the power requirement of the flight mission but also to ensure its own safety [1]. These all impose strict demands on the engine's control system. With the increasing demands, traditional control techniques cannot handle the tasks well [2,3]. Therefore, the engineering community of aero-engine control is urged to seek more advanced control techniques. Model Predictive Control (MPC), a model-based optimal control method, has attracted engineers' attention [4–6]. MPC emerged in the 1970s and then rapidly flourished in process control [7,8]. MPC has an online prediction model, which can estimate unmeasurable performance parameters. This is lacking in other techniques (e.g., robust control, adaptive control) [9,10]. Additionally, MPC solves a constrained optimization problem, which can achieve the control objective and constrains management simultaneously. The constraints consider the physical limitations of the actuators, whereas other techniques (e.g., anti-windup) require additional design to handle them [11]. Therefore, researchers try to apply MPC to aero-engines [12–15].

Although MPC can enhance control performance, it has real-time implementation issues. The main reason is that optimization in MPC leads to a huge computational burden [16]. The MPC controller mainly includes an optimization problem and a mathematical algorithm. At each sampling instant, a nonlinear optimization problem can be constructed according to the control objective and the constraint conditions. Then, the mathematical algorithm is employed to solve this optimization problem. Since there have been many well-developed mathematical algorithms to be chosen, this paper focuses on the optimization problem [17,18].

Among the existing methods, Bemporad aimed to perform the solution of the online optimization problem offline and proposed the explicit MPC [19–21]. Based on multiparameter programming theory, the explicit MPC controller is designed offline in each region of the state space. When implementing online, the corresponding control parameters are searched according to the current state. The method of offline calculation and online query greatly reduces the online computational complexity of the MPC controller and thus improves the real-time property. Gu improved the control system's real-time performance by applying the multiparameter quadratic programming explicit model predictive control on a turboshaft engine [22], while Feng designed the explicit MPC controller for a turbofan engine and demonstrated that it meets the requirement of real-time property by a hardware-in-the-loop test [23]. However, this explicit implementation must satisfy that the optimization problem can be simplified into a multiparameter programming. In addition, the control parameters of all regions should be stored in the explicit MPC controller, placing high demands on the memory of the control system. What is more, the explicit MPC controller that has been designed offline may fail when the engine deteriorates.

From handling the online optimization problem itself, there are two directions to reduce the computational complexity. The first is to decrease the order of the optimization problem. Normally, the 2-norm is used to construct a quadratic cost function as the performance index, resulting in the optimization problem being nonlinear [24]. Using $1/\infty$-norm instead of 2-norm can construct a linear cost function which reduces the optimization problem's order and can be solved by simpler linear programming. Genceli employed the 1-norm to form the cost function and thus solved the control law by online linear programming [25]. Kerrigan utilized $1/\infty$-norm in a robustly stable MPC problem and found that the ∞-norm has higher real-time performance, while the 1-norm has higher solution accuracy [26]. However, compared with the 2-norm, the $1/\infty$-norm results in obvious differences in the solution results, which hinders further development in this direction.

Another direction is to lower the scale of the optimization problem by reducing the dimension of the control sequence. In reality, the computational complexity of the optimization problem relates to the cube of the control sequence's length [27,28]. Different from standard MPC that updates all the control inputs at the same time, multiplexed MPC (mMPC), which is proposed by Ling, is employed to update only one control variable at a time and all the control variables sequentially and cyclically in the implementation process, resulting in the computation speed up [29,30]. Richter applied mMPC to a large commercial turbofan engine and demonstrated the computational savings of this method [31]. Then, Pang employed this method to control a gas turbine engine, and all the results showed that the time consumption can be greatly reduced in comparison with the standard MPC [32,33]. Although it has shown superiority in reducing computational complexity, mMPC finds a suboptimal solution to the original optimization problem actually, and the control variable's update mode is fixed and inflexible. Each control variable has the same possibility to be used to control the system despite their different regulating abilities. Therefore, the control quality actually witnesses a significant decrease. In the implementation in an aero-engine, the mMPC controller has a much bigger control error than the standard MPC ones [31,32].

Aiming at the defect of existing methods, a novel MPC intelligent reduced-dimensional scheme is proposed in this paper to realize great real-time performance and control performance simultaneously. The main contributions are as follows:

(1) Different from mMPC, a selection algorithm is designed in the scheme to determine the control variable with the best control effect at each sampling instant, which is an intelligent update mode and helps to enhance the control performance;

(2) To search for a better sub-optimal solution, a coordination strategy is developed in the scheme, which considers the interaction of the control variables at different predicting instants in the control sequence;
(3) By constructing an optimization problem with low computational complexity, the intelligent reduced-dimensional scheme guarantees the superiority in time consumption.

The remainder of this paper is organized as follows. Section 2 details the methodology of the proposed intelligent reduced-dimensional scheme. Section 3 shows the simulation results to demonstrate the effectiveness of the proposed method. Finally, Section 4 concludes this paper.

2. MPC Intelligent Reduced-Dimensional Scheme

2.1. MPC Optimization Problem with Low Computational Complexity

Consider a linear state-space model with forms of discrete-time and small deviation as the predictive model:

$$\begin{cases} x_{k+1} = Ax_k + Bu_k \\ y_k = Cx_k + Du_k \\ u_{k+1} = u_k + \Delta u_k \end{cases} \tag{1}$$

where $x \in \mathbb{R}^n$ is the state vector, $u \in \mathbb{R}^m$ is the input vector, the output vector $y \in \mathbb{R}^r$ contains the controlled parameters $y_{ctrl} \in \mathbb{R}^s$ and the constrained parameters $y_{con} \in \mathbb{R}^t$, and k is the sampling instant. The system matrices A, B, C and D have dimensions of n by n, n by m, r by n and r by m, respectively. Note that the symbol Δ, which represents a small deviation, is omitted for simplifying the expression.

In Equation (1), the third formula introduces integral action to the MPC, and Δu_k actually represents the increment of input vector with a small deviation form, i.e., $(\Delta)u_{k+1} = (\Delta)u_k + (\Delta)\Delta u_k$ [31].

After augmenting the state vector with the input vector, Equation (1) can be rewritten as

$$\begin{cases} x_{a,k+1} = A_a x_{a,k} + B_a \Delta u_k \\ y_{ctrl,k} = C_{a,ctrl} x_{a,k} \\ y_{con,k} = C_{a,con} x_{a,k} \end{cases} \tag{2}$$

where $x_{a,k}^T = \begin{bmatrix} x_k^T & u_k^T \end{bmatrix} \in \mathbb{R}^{1 \times (n+m)}$, $A_a = \begin{bmatrix} A & B \\ 0 & I \end{bmatrix} \in \mathbb{R}^{(n+m) \times (n+m)}$, $B_a = \begin{bmatrix} 0 \\ I \end{bmatrix} \in \mathbb{R}^{(n+m) \times m}$, $C_{a,ctrl} = \begin{bmatrix} C_{ctrl} & D_{ctrl} \end{bmatrix} \in \mathbb{R}^{s \times (n+m)}$, $C_{a,con} = \begin{bmatrix} C_{con} & D_{con} \end{bmatrix} \in \mathbb{R}^{t \times (n+m)}$.

For the MPC, the computation complexity of its optimization problem depends on the dimension of Δu_k in Equation (2) to a great extent. Therefore, the input vector is encouraged to lower the dimension from m to 1 to minimize the computation complexity.

Without loss of generality, consider the ith control variable $\Delta u_{i,k}$ as the available one and $\Delta u_{j,k} = 0, j \neq i$. $\Delta u_{i,k}$ can be obtained by removing all the zero elements from Δu_k. Then, the first formula in Equation (2) is revised as

$$\begin{aligned} \Delta u_{i,k} &= E_i \Delta u_k \\ x_{a,k+1} &= A_a x_{a,k} + (B_a E_i^T) \Delta u_{i,k} \end{aligned} \tag{3}$$

where $E_i \in \mathbb{R}^{1 \times m}$ is a transfer matrix, and its ith element is 1, while the others are zero.

It should be noted that the other control variables keep their previous values when selecting u_i at sampling instant k.

Then, a reduced-dimensional control sequence in the MPC can be defined as

$$\Delta \tilde{U} = [\Delta u_{i,k}^T, \Delta u_{i,k+1}^T, \ldots, \Delta u_{i,k+N_c-1}^T]^T \tag{4}$$

where N_c is the control horizon.

According to Equations (2)–(4), the predicted output vector over the prediction horizon can be denoted as

$$Y_{ctrl} = P_{ctrl}x_{a,k} + H_{ctrl}\Delta\widetilde{U}$$
$$Y_{con} = P_{con}x_{a,k} + H_{con}\Delta\widetilde{U} \quad (5)$$

where $Y_{ctrl} = [y_{ctrl,k+1}^T, y_{ctrl,k+2}^T, \ldots, y_{ctrl,k+Np}^T]^T$, $Y_{con} = [y_{con,k+1}^T, y_{con,k+2}^T, \ldots, y_{con,k+Np}^T]^T$, Np is the prediction horizon. The $P_{ctrl} \in \mathbb{R}^{sNp \times (n+m)}$, $P_{con} \in \mathbb{R}^{tNp \times (n+m)}$, $H_{ctrl} \in \mathbb{R}^{sNp \times Nc}$ and $H_{con} \in \mathbb{R}^{tNp \times Nc}$ are

$$P_{ctrl} = \begin{bmatrix} C_{a,ctrl}A_a \\ C_{a,ctrl}A_a^2 \\ \vdots \\ C_{a,ctrl}A_a^{Nc} \\ \vdots \\ C_{a,ctrl}A_a^{Np} \end{bmatrix}, H_{ctrl} = \begin{bmatrix} C_{a,ctrl}B_aE_i^T & 0 & 0 & \cdots & 0 \\ C_{a,ctrl}A_aB_aE_i^T & C_{a,ctrl}B_aE_i^T & 0 & \cdots & 0 \\ \vdots & \vdots & \vdots & \vdots & \vdots \\ C_{a,ctrl}A_a^{Nc-1}B_aE_i^T & C_{a,ctrl}A_a^{Nc-2}B_aE_i^T & C_{a,ctrl}A_a^{Nc-3}B_aE_i^T & \cdots & C_{a,ctrl}B_aE_i^T \\ \vdots & \vdots & \vdots & \vdots & \vdots \\ C_{a,ctrl}A_a^{Np-1}B_aE_i^T & C_{a,ctrl}A_a^{Np-2}B_aE_i^T & C_{a,ctrl}A_a^{Np-3}B_aE_i^T & \cdots & C_{a,ctrl}A_a^{Np-Nc}B_aE_i^T \end{bmatrix}$$

$$P_{con} = \begin{bmatrix} C_{a,con}A_a \\ C_{a,con}A_a^2 \\ \vdots \\ C_{a,con}A_a^{Nc} \\ \vdots \\ C_{a,con}A_a^{Np} \end{bmatrix}, H_{con} = \begin{bmatrix} C_{a,con}B_aE_i^T & 0 & 0 & \cdots & 0 \\ C_{a,con}A_aB_aE_i^T & C_{a,con}B_aE_i^T & 0 & \cdots & 0 \\ \vdots & \vdots & \vdots & \vdots & \vdots \\ C_{a,con}A_a^{Nc-1}B_aE_i^T & C_{a,con}A_a^{Nc-2}B_aE_i^T & C_{a,con}A_a^{Nc-3}B_aE_i^T & \cdots & C_{a,con}B_aE_i^T \\ \vdots & \vdots & \vdots & \vdots & \vdots \\ C_{a,con}A_a^{Np-1}B_aE_i^T & C_{a,con}A_a^{Np-2}B_aE_i^T & C_{a,con}A_a^{Np-3}B_aE_i^T & \cdots & C_{a,con}A_a^{Np-Nc}B_aE_i^T \end{bmatrix} \quad (6)$$

Equation (5) is considered to have a high prediction accuracy, and with the help of it, a constrained optimization problem under the reduced-dimensional control sequence Equation (4) can be constructed as follows.

A common quadratic performance index that contains the control error and the control energy consumption is defined as the cost function of the optimization problem.

$$J = e^T Q e + \Delta\widetilde{U}^T R \Delta\widetilde{U} \quad (7)$$

where $e = r_{ctrl} - Y_{ctrl}$ is the error vector, and $r_{ctrl} = [r_{k+1}^T \; r_{k+2}^T \; \cdots \; r_{k+Np}^T]^T$ is the command vector. Q and R are two weight matrices to balance the control error and the control energy consumption. In other words, larger coefficients in Q denote smaller control errors of the controlled parameters, while larger coefficients in R indicate less energy consumption, which reduces the system's response speed when applying the ith control variable.

$$Q = \begin{bmatrix} \widetilde{Q} & & \\ & \ddots & \\ & & \widetilde{Q} \end{bmatrix} \in \mathbb{R}^{sNp \times sNp}, \widetilde{Q} = \begin{bmatrix} Q_1 & & \\ & \ddots & \\ & & Q_s \end{bmatrix} \in \mathbb{R}^{s \times s}, R = \begin{bmatrix} R_i & & \\ & \ddots & \\ & & R_i \end{bmatrix} \in \mathbb{R}^{Nc \times Nc} \quad (8)$$

where $Q_i, i = 1, 2, \ldots, s$ is the weight for each controlled parameter, and R_i is the weight for the ith control variable.

The constraint conditions of the output parameters and the control variable can be represented in the form of $\Delta\widetilde{U}$, as follows.

$$\begin{cases} H_{con}\Delta\widetilde{U} \leq Y_{ub} \\ -H_{con}\Delta\widetilde{U} \leq Y_{lb} \\ \Delta\widetilde{U} \leq dU_{ub} \\ -\Delta\widetilde{U} \leq dU_{lb} \\ L_{con}\Delta\widetilde{U} \leq U_{ub} \\ -L_{con}\Delta\widetilde{U} \leq U_{lb} \end{cases} \quad (9)$$

with

$$\begin{cases} Y_{ub} = [I_{t,1}, I_{t,2}, \ldots, I_{t,Np}]^T y_{con,ub} - P_{con}x_{a,k} \\ Y_{lb} = -[I_{t,1}, I_{t,2}, \ldots, I_{t,Np}]^T y_{con,lb} + P_{con}x_{a,k} \\ dU_{ub} = [du_{i,ub}, du_{i,ub}, \ldots, du_{i,ub}]^T \\ dU_{lb} = -[du_{i,lb}, du_{i,lb}, \ldots, du_{i,lb}]^T \\ U_{ub} = [u_{i,ub}, u_{i,ub}, \ldots, u_{i,ub}]^T \\ U_{lb} = -[u_{i,lb}, u_{i,lb}, \ldots, u_{i,lb}]^T \end{cases} \quad (10)$$

$$L_{con} = \begin{bmatrix} 1 & 0 & 0 & \cdots & 0 \\ 1 & 1 & 0 & \cdots & 0 \\ 1 & 1 & 1 & \cdots & 0 \\ \vdots & \vdots & \vdots & & \vdots \\ 1 & 1 & 1 & 1 & 1 \end{bmatrix} \in \mathbb{R}^{Nc \times Nc} \quad (11)$$

where the subscripts *ub* and *lb* denote the upper limit and the lower limit.

Since Equation (2) has a form of small deviation, $y_{con,ub} = [y_{con,1,ub}, y_{con,2,ub}, \ldots, y_{con,t,ub}]^T$ and $y_{con,lb} = [y_{con,1,lb}, y_{con,2,lb}, \ldots, y_{con,t,lb}]^T$ are the constrained parameters' increment upper and lower limit, respectively. $du_{i,ub}$ and $du_{i,lb}$ are the *i*th control variable increment rate's upper and lower limit, respectively. $u_{i,ub}$ and $u_{i,lb}$ are the *i*th control variable increment magnitude's upper and lower limit, respectively.

Finally, a constrained optimization problem with low computation complexity is summarized as

$$\min J = e^T Q e + \Delta \tilde{U}^T R \Delta \tilde{U}$$
$$s.t. \ M\Delta\tilde{U} \leq C \quad (12)$$

where $M = \begin{bmatrix} H_{con} \\ -H_{con} \\ I \\ -I \\ L_{con} \\ -L_{con} \end{bmatrix}, C = \begin{bmatrix} Y_{ub} \\ Y_{lb} \\ dU_{ub} \\ dU_{lb} \\ U_{ub} \\ U_{lb} \end{bmatrix}.$

Some well-developed methods can cope with the constructed optimization problem. Among them, the interior point method is accessible. For the solution $\Delta\tilde{U}$, the first element $\Delta u_{i,k}$ is utilized to update the *i*th control variable and control the aero-engine at sampling instant *k*.

2.2. Control Variable Selection Algorithm

In fact, the control effect on the engine during the regulating process is accumulated by the control effect at each sampling instant, and the control effect at each sampling instant depends on the regulating ability of the selected control variable. Therefore, it is extremely necessary to consider which control variable is to be chosen at each sampling instant for better control results. However, the existing method, such as mMPC, does not pay enough attention to this selection issue. It changes the control variable sequentially and cyclically as the sampling instant increases, as shown in Figure 1. Essentially, it is an inflexible and non-intelligent selection mode. Since the regulating ability of each control variable is evidently not equivalent, an intelligent mode is needed to choose the suitable control variable at each sampling instant.

To achieve the intelligent mode, a control variable selection algorithm is designed to select the control variable that has the best control effect at the current sampling instant. According to the system information of the current sampling instant and the control target, the control effect of each control variable can be predicted for selection. The control variable selection algorithm is conducted as follows.

Firstly, the value range of each control variable at the first predicting instant is calculated. For example, at sampling instant *k*, the *i*th control variable at predicting instant *k* (i.e.,

$\Delta u_{i,k}$) should be able to meet the requirements of the constraints, that is, the limits of the control variable increment rate, the limits of the control variable increment magnitude and the limits of the constrained parameters' increment.

$$du_{i,lb} \leq \Delta u_{i,k} \leq du_{i,ub}$$
$$u_{i,lb} \leq \Delta u_{i,k} \leq u_{i,ub} \quad (13)$$

$$y_{con,lb} \leq y_{con,k+1} = C_{a,con}A_a x_{a,k} + C_{a,con}B_a E_i^T \Delta u_{i,k} \leq y_{con,ub} \quad (14)$$

According to Equations (13) and (14), the value range of $\Delta u_{i,k}$ can be calculated as

$$[\Delta u_{i,k,\min} \quad \Delta u_{i,k,\max}] \quad (15)$$

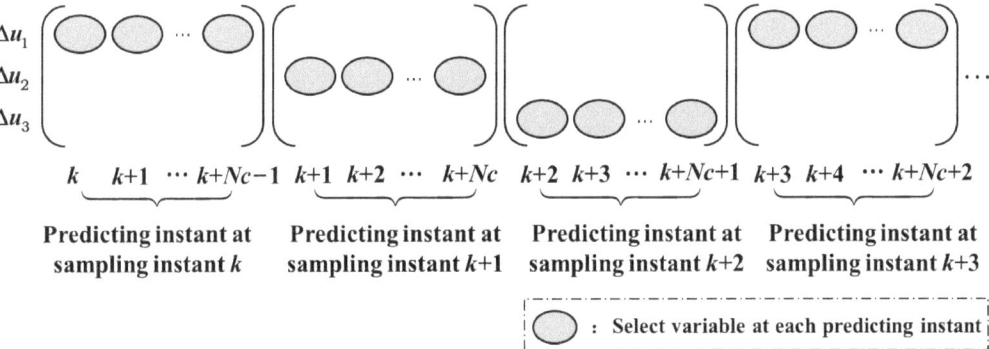

Figure 1. Schematic of the update mode of 3 control variables in mMPC.

Then, the value range of the controlled parameters at predicting instant $k + 1$ can be computed as

$$\begin{cases} y_{ctrl,k+1,\min}(\Delta u_{i,k}) = \min(C_{a,ctrl}A_a x_{a,k} + C_{a,ctrl}B_a E_i^T \Delta u_{i,k,\min}, C_{a,ctrl}A_a x_{a,k} + C_{a,ctrl}B_a E_i^T \Delta u_{i,k,\max}) \\ y_{ctrl,k+1,\max}(\Delta u_{i,k}) = \max(C_{a,ctrl}A_a x_{a,k} + C_{a,ctrl}B_a E_i^T \Delta u_{i,k,\min}, C_{a,ctrl}A_a x_{a,k} + C_{a,ctrl}B_a E_i^T \Delta u_{i,k,\max}) \end{cases} \quad (16)$$

By comparing r_{k+1} with $[y_{ctrl,k+1,\min}(\Delta u_{i,k}) \quad y_{ctrl,k+1,\max}(\Delta u_{i,k})]$, the possible value of $\Delta u_{i,k}$ can be obtained.

$$\Delta u_{i,k} = \begin{cases} \Delta u_{i,k,\min} \text{ or } \Delta u_{i,k,\max}, & if\ r_{k+1} < y_{ctrl,k+1,\min}(\Delta u_{i,k}) \\ \frac{r_{k+1} - C_{a,ctrl}A_a x_{a,k}}{C_{a,ctrl}B_a E_i^T}, & if\ y_{ctrl,k+1,\min}(\Delta u_{i,k}) < r_{k+1} < y_{ctrl,k+1,\max}(\Delta u_{i,k}) \\ \Delta u_{i,k,\max} \text{ or } \Delta u_{i,k,\min}, & if\ r_{k+1} > y_{ctrl,k+1,\max}(\Delta u_{i,k}) \end{cases} \quad (17)$$

In Equation (17), the first formula represents the value of $\Delta u_{i,k}$ that corresponds to $y_{ctrl,k+1,\min}(\Delta u_{i,k})$ and so does the third formula.

Subsequently, a merit function that considers the minimum control error and the possible control energy consumption is used to evaluate the control effect of $\Delta u_{i,k}$.

$$\widetilde{J}(\Delta u_{i,k}) = \begin{cases} \widetilde{Q}(r_{k+1} - y_{ctrl,k+1,\min})^2 + R_i \Delta u_{i,k}^2, & if\ r_{k+1} < y_{ctrl,k+1,\min}(\Delta u_{i,k}) \\ R_i \Delta u_{i,k}^2, & if\ y_{ctrl,k+1,\min}(\Delta u_{i,k}) < r_{k+1} < y_{ctrl,k+1,\max}(\Delta u_{i,k}) \\ \widetilde{Q}(r_{k+1} - y_{ctrl,k+1,\max})^2 + R_i \Delta u_{i,k}^2, & if\ r_{k+1} > y_{ctrl,k+1,\max}(\Delta u_{i,k}) \end{cases} \quad (18)$$

In Equation (18), $\Delta u_{i,k}$ takes the corresponding value in Equation (17). The second formula indicates the case in which the control objective is in the range of the control variable's regulating ability.

Finally, the merit functions of m control variables are calculated separately, and the control variable with the smallest merit function value is taken as the selected one.

The flowchart of the control variable selection algorithm at the current sampling instant can be illustrated as in Figure 2.

2.3. Control Sequence Coordination Strategy

Despite only the first element $\Delta u_{i,k}$ in the control sequence being fed to the engine, MPC actually predicts all the other elements, i.e., $\Delta u_{i,k+1}, \ldots, \Delta u_{i,k+Nc-1}$, by solving the constructed optimization problem. In the solution process, the result of each element will affect the others. For example, if the second element is another variable $\Delta u_{j,k+1}$, the solution result of $\Delta u_{i,k}$ changes accordingly. Among the elements, each one represents the predictive value of the control variable at the corresponding sampling instant. According to Section 2.2, each sampling instant can select the best variable to be the control action. Assume that the variable at sampling instant $k + 1$ is Δu_j; it is better to employ Δu_j as the second element in the control sequence of sampling instant k. This is true for the other sampling instants ($k + 2, \ldots, k + Nc - 1$). Under this arrangement, $\Delta u_{i,k}$ can obtain a better solution to enhance the control performance since each element is selected properly, whereas all the elements are the same in mMPC, which neglects the interaction among the predicting instants in the control sequence.

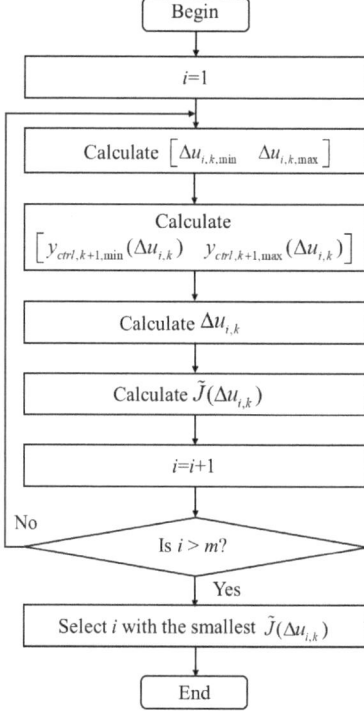

Figure 2. Flowchart of the control variable selection algorithm.

The arrangement is called a control sequence coordination strategy. The strategy is conducted as follows to optimally select the variables at subsequent predicting instants in the control sequence.

Based on Section 2.2, the first element and its potential value in the control sequence can be determined as $\Delta u_{i,k}$. Assuming that the second element is $\Delta u_{j,k+1}$, it should also satisfy the constraints of control variable and constrained parameters.

$$du_{j,lb} \leq \Delta u_{j,k+1} \leq du_{j,ub} \\ u_{j,lb} \leq \Delta u_{j,k+1} \leq u_{j,ub} \tag{19}$$

$$y_{con,lb} \leq y_{con,k+2} = C_{a,con} A_a x_{a,k+1} + C_{a,con} B_a E_j^T \Delta u_{j,k+1} \leq y_{con,ub} \tag{20}$$

In Equation (20), $x_{a,k+1}$ can be obtained with the help of Equations (2) and (3).

Similarly, the value range of $\Delta u_{j,k+1}$ and the value range of the controlled parameters at predicting instant $k+2$ can be derived as

$$[\Delta u_{j,k+1,\min} \quad \Delta u_{j,k+1,\max}] \tag{21}$$

$$\begin{cases} y_{ctrl,k+2,\min}(\Delta u_{j,k+1}) = \min(C_{a,ctrl} A_a x_{a,k+1} + C_{a,ctrl} B_a E_j^T \Delta u_{j,k+1,\min}, C_{a,ctrl} A_a x_{a,k+1} + C_{a,ctrl} B_a E_j^T \Delta u_{j,k+1,\max}) \\ y_{ctrl,k+2,\max}(\Delta u_{j,k+1}) = \max(C_{a,ctrl} A_a x_{a,k+1} + C_{a,ctrl} B_a E_j^T \Delta u_{j,k+1,\min}, C_{a,ctrl} A_a x_{a,k+1} + C_{a,ctrl} B_a E_j^T \Delta u_{j,k+1,\max}) \end{cases} \tag{22}$$

Then, the possible value of $\Delta u_{j,k+1}$ is

$$\Delta u_{j,k+1} = \begin{cases} \Delta u_{j,k+1,\min} \text{ or } \Delta u_{j,k+1,\max}, & \text{if } r_{k+2} < y_{ctrl,k+2,\min}(\Delta u_{j,k+1}) \\ \frac{r_{k+2} - C_{a,ctrl} A_a x_{a,k+1}}{C_{a,ctrl} B_a E_j^T}, & \text{if } y_{ctrl,k+2,\min}(\Delta u_{j,k+1}) < r_{k+2} < y_{ctrl,k+2,\max}(\Delta u_{j,k+1}) \\ \Delta u_{j,k+1,\max} \text{ or } \Delta u_{j,k+1,\min}, & \text{if } r_{k+2} > y_{ctrl,k+2,\max}(\Delta u_{j,k+1}) \end{cases} \tag{23}$$

In the same way, to evaluate the control effect of $\Delta u_{j,k+1}$, the merit function can be constructed as

$$\widetilde{J}(\Delta u_{j,k+1}) = \begin{cases} \widetilde{Q}(r_{k+2} - y_{ctrl,k+2,\min})^2 + R_j \Delta u_{j,k+1}^2, & \text{if } r_{k+2} < y_{ctrl,k+2,\min}(\Delta u_{j,k+1}) \\ R_j \Delta u_{j,k+1}^2, & \text{if } y_{ctrl,k+2,\min}(\Delta u_{j,k+1}) < r_{k+2} < y_{ctrl,k+2,\max}(\Delta u_{j,k+1}) \\ \widetilde{Q}(r_{k+2} - y_{ctrl,k+2,\max})^2 + R_j \Delta u_{j,k+1}^2, & \text{if } r_{k+2} > y_{ctrl,k+2,\max}(\Delta u_{j,k+1}) \end{cases} \tag{24}$$

By computing the m merit functions, the second element and its potential value in the control sequence can be determined. Then, accordingly, the variable that yields the optimal control effect at each predicting instant can be selected to form a coordinated control sequence, as illustrated in Figure 3.

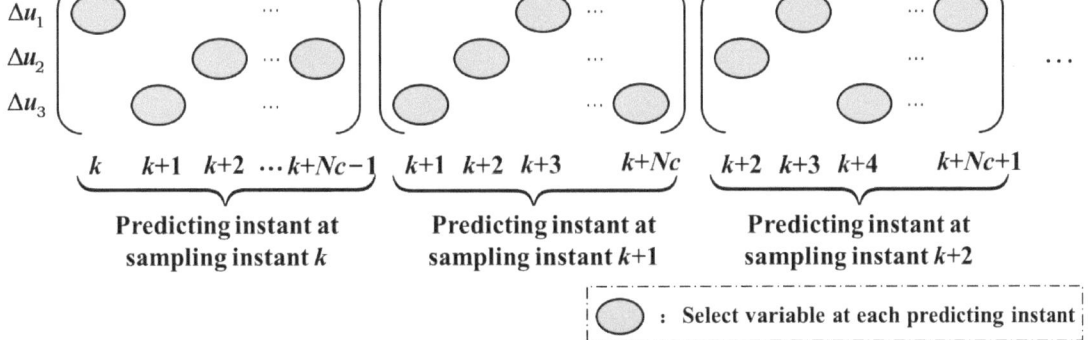

Figure 3. Schematic of the update mode of 3 control variables in the proposed scheme.

2.4. Intelligent Reduced-Dimensional MPC

Denote the coordinated control sequence as

$$\Delta \widehat{U} = [\Delta u_{i,k}, \Delta u_{j,k+1}, \Delta u_{l,k+2}, \ldots]^T \quad (25)$$

where i, j and l separately represent the selected control variable at their corresponding predicting instants.

Based on Equation (25), a new constrained optimization problem can be constructed as follows. Firstly, Equation (5) can be rewritten as

$$\begin{aligned} Y_{ctrl} &= P_{ctrl} x_{a,k} + \widehat{H}_{ctrl} \Delta \widehat{U} \\ Y_{con} &= P_{con} x_{a,k} + \widehat{H}_{con} \Delta \widehat{U} \end{aligned} \quad (26)$$

where $\widehat{H}_{ctrl}$ has a difference from H_{ctrl} in that the transfer matrix E_i^T in each column should be changed according to Equation (25). The same is true for the matrix $\widehat{H}_{con}$.

The cost function is also changed as

$$\widehat{J} = e^T Q e + \Delta \widehat{U}^T \widehat{R} \Delta \widehat{U} \quad (27)$$

where $\widehat{R} = \begin{bmatrix} R_i & & & \\ & R_j & & \\ & & R_l & \\ & & & \ddots \end{bmatrix} \in \mathbb{R}^{Nc \times Nc}$.

Then, for the constrained parameters and control variables, the corresponding equations about limitation are

$$\begin{cases} \widehat{H}_{con} \Delta \widehat{U} \leq Y_{ub} \\ -\widehat{H}_{con} \Delta \widehat{U} \leq Y_{lb} \end{cases} \\ \begin{cases} \Delta \widehat{U} \leq d\widehat{U}_{ub} \\ -\Delta \widehat{U} \leq d\widehat{U}_{lb} \end{cases} \\ \begin{cases} \widehat{L}_{con} \Delta \widehat{U} \leq \widehat{U}_{ub} \\ -\widehat{L}_{con} \Delta \widehat{U} \leq \widehat{U}_{lb} \end{cases} \quad (28)$$

where

$$\begin{cases} d\widehat{U}_{ub} = [du_{i,ub}, du_{j,ub}, du_{l,ub}, \ldots]^T \\ d\widehat{U}_{lb} = -[du_{i,lb}, du_{j,lb}, du_{l,lb}, \ldots]^T \\ \widehat{U}_{ub} = [u_{i,ub}, u_{j,ub}, u_{l,ub}, \ldots]^T \\ \widehat{U}_{lb} = -[u_{i,lb}, u_{j,lb}, u_{l,lb}, \ldots]^T \end{cases} \quad (29)$$

$\widehat{L}_{con} \in \mathbb{R}^{Nc \times Nc}$ is a lower triangular matrix with diagonal elements all equal to 1. Except for the diagonal, for any two variables in the control sequence $\{\Delta u_i(k_1) \; \Delta u_i(k_2) | k_1 < k_2, \; i = 1, 2, \cdots,$ $\widehat{L}_{con}(k_2, k_1) = 1$.

Finally, the constrained optimization problem in the intelligent reduced-dimensional MPC can be abstracted as

$$\begin{aligned} \min \widehat{J} &= e^T Q e + \Delta \widehat{U}^T \widehat{R} \Delta \widehat{U} \\ s.t. \; \widehat{M} \Delta \widehat{U} &\leq \widehat{C} \end{aligned} \quad (30)$$

where $\widehat{M} = \begin{bmatrix} \widehat{H}_{con} \\ -\widehat{H}_{con} \\ I \\ -I \\ \widehat{L}_{con} \\ -\widehat{L}_{con} \end{bmatrix}, \widehat{C} = \begin{bmatrix} Y_{ub} \\ Y_{lb} \\ \widehat{U}_b \\ \widehat{U}_a \\ \widehat{U}_{ub} \\ \widehat{U}_{lb} \end{bmatrix}.$

In summary, the proposed intelligent reduced-dimensional scheme has three key points. Firstly, reduce the dimension of control variables from m to 1 to lower the computational complexity of the optimization problem in MPC; secondly, determine the control variable with the best control effect at the current sampling instant by the control variable selection algorithm; finally, define a coordinated reduced-dimensional control sequence as Equation (25) with the help of the control variable coordination strategy and thus construct the constrained optimization problem Equation (30). By solving the optimization problem and applying the solution to the aero-engine, a receding horizon control process can operate. The flowchart of the proposed method is shown in Figure 4.

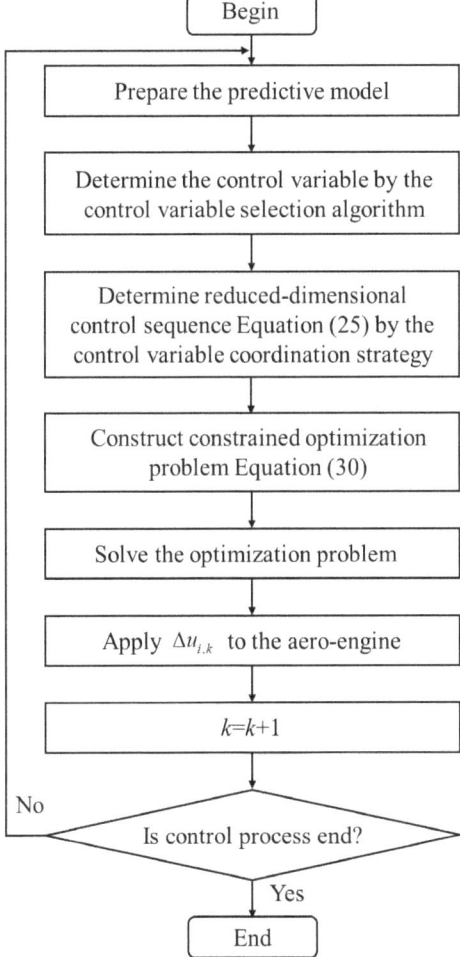

Figure 4. Flowchart of the intelligent reduced-dimensional MPC.

3. Simulation and Discussion

3.1. Simulation Cases

A twin-shaft turbofan engine with core driven fan stage (CDFS) is selected as the controlled object. The engine has components of intake, fan, CDFS, high-pressure compressor (HPC), combustor, high-pressure turbine (HPT), low-pressure turbine (LPT), duct and nozzle. By referring to the modeling method in [34,35] and using the Simulink Toolbox for the Modeling and Analysis of Thermodynamics Systems (T-MATS), the component-level model of this engine is constructed. In the model, all the co-operation equations are formulated based on aero-thermodynamics principles, flow continuity equations, pressure balance equations and rotor dynamics equations and are solved with the help of the Newton-Raphson method. All the components' maps and performance data are acquired from GasTurb [36].

The input of the component-level model includes the flight condition [H, Ma] and the following control variables: the fuel flow W_f, the nozzle area A_8 and the variable guide vane angle of LPT α_{LPT}, i.e., $\Delta u = [\Delta W_f \quad \Delta A_8 \quad \Delta \alpha_{LPT}]^T$. The output parameters are composed of the low-pressure rotor speed n_L, the high-pressure rotor speed n_H, the HPC exhaust pressure P_3 and the LPT exhaust temperature T_5. Among them, n_L is selected as the controlled parameter due to the strong correlation with thrust, while n_H, P_3 and T_5 are considered the constrained parameters for safety operation, i.e., $y_{ctrl} = n_L$ and $y_{con} = [n_H \quad P_3 \quad T_5]^T$.

To demonstrate the effectiveness of the proposed intelligent reduced-dimensional scheme, simulations are conducted at two flight conditions: [H, Ma] = [0, 0] and [H, Ma] = [11,000 m, 1.2]. In these conditions, the configuration of the engine is listed in Tables 1 and 2, wherein T_s = 20 ms is the sampling period.

Table 1. The configuration of the engine at [H, Ma] = [0, 0].

Parameters		Initial Steady-State Value	Magnitude Limit	Rate Limit	Unit
Output	n_L	98			%
	n_H	99.7	$n_H \leq 102$		%
	P_3	2618.4	$P_3 \geq 2500$		kPa
	T_5	1198.2	$T_5 \leq 1300$		K
Control variable	W_f	0.768	$0.1 \leq W_f \leq 1$	$0.03/T_s$	kg/s
	A_8	100	$90 \leq A_8 \leq 110$	$0.5/T_s$	%
	α_{LPT}	1	$0 \leq \alpha_{LPT} \leq 5$	$0.2/T_s$	°

Table 2. The configuration of the engine at [H, Ma] = [11,000 m, 1.2].

Parameters		Initial Steady-State Value	Magnitude Limit	Rate Limit	Unit
Output	n_L	98			%
	n_H	98.6	$n_H \leq 102$		%
	P_3	1425.4	$P_3 \geq 1300$		kPa
	T_5	1189	$T_5 \leq 1300$		K
Control variable	W_f	0.429	$0.1 \leq W_f \leq 1$	$0.03/T_s$	kg/s
	A_8	100	$90 \leq A_8 \leq 110$	$0.5/T_s$	%
	α_{LPT}	1	$0 \leq \alpha_{LPT} \leq 5$	$0.2/T_s$	°

A set value of Δn_L = 2% is given as the control objective of both cases. A method is adopted to determine the reference trajectory based on the set value and the current value of the controlled parameter [37], wherein α is a softening factor.

$$r_{k+j} = \alpha^j y_{ctrl,k} + (1 - \alpha^j)\Delta n_L, j = 1, 2, \ldots, N_p \qquad (31)$$

For each case, an intelligent reduced-dimensional linear MPC controller, denoted as "irdMPC" and shown in Figure 5, can be designed according to Section 2 and implemented to achieve the control targets. Concurrently, both the standard MPC controller and the multiplex MPC controller, denoted as "sMPC" and "mMPC", respectively, are designed for comparative analysis.

3.2. Control Effect Comparison

Firstly, a simulation is conducted at the flight condition [H, Ma] = [0, 0] to illustrate the effectiveness of the proposed method in control performance. During the simulation, the designed MPC controllers, which are "irdMPC", "sMPC" and "mMPC", adopt the same parameter settings, i.e., prediction horizon Np, control horizon Nc, weight matrices Q and R, in order to intuitively reflect the differences in control quality. By trial-and-error simulations, they can be well tuned. The controlled parameter's response under the three MPC controllers is depicted in Figure 6. The constrained parameters are depicted in Figure 7.

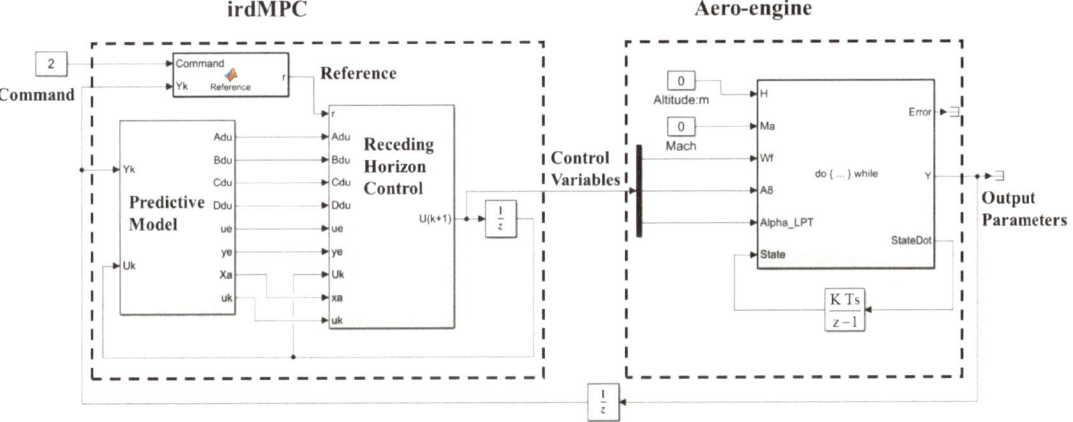

Figure 5. Control structure of "irdMPC" in Simulink environment.

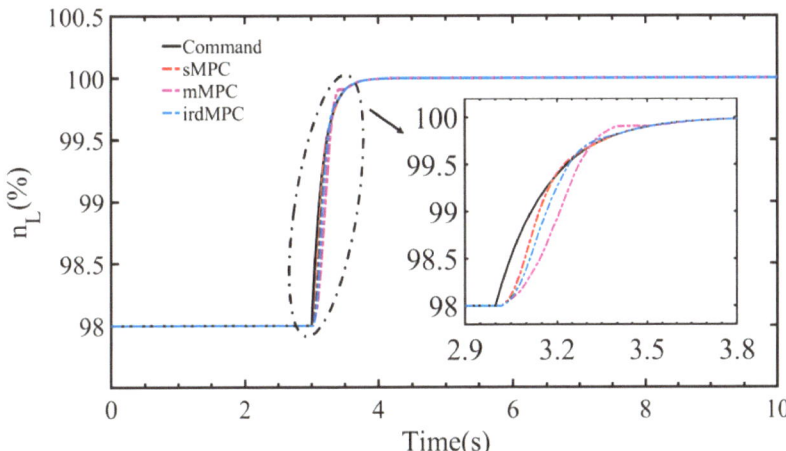

Figure 6. Controlled parameters' response under "sMPC", "mMPC" and "irdMPC".

Although all the MPC controllers can accomplish effective command tracking and constrains management, they actually have different performances. From Figure 6, it can be seen that "mMPC" witnesses a noticeable overshoot and long tracking response time. Thus, it has a poorer control outcome in command tracking than the others. "irdMPC" outperforms "mMPC", but it is still slightly inferior to "sMPC". The reason is that "irdMPC" and "mMPC" degrade their regulating ability by reducing the dimension of the control sequence. Nevertheless, "irdMPC" can track the reference trajectories well and has smaller tracking errors than "mMPC", demonstrating the effectiveness of the intelligent reduced-dimensional scheme in achieving good control quality. In terms of constraint management, the constrain parameters of the three MPC controllers reach different values after the transition process, as shown in Figure 7. Similarly, "irdMPC" is closer to "sMPC" than "mMPC", which represents that "irdMPC" derives a better operating state than "mMPC". Moreover, "mMPC" has a certain oscillation during transition in Figure 7b.

The parameters' response originates from the control variables' changes. During the process, the control variables' changes are shown in Figure 8. During transition, the selected control variable in "irdMPC" at each sampling instant is represented as "o" and is displayed in Figure 9.

In Figure 8, the control variables have various changes in the MPC controllers. Unlike "sMPC", only one control variable is optimized in "mMPC" and "irdMPC" at each sampling instant. Moreover, the control variable to be optimized among them is also different. Thus, the results of solving constraints' optimization problems, namely the changes of control variables, are different, resulting in their corresponding control outcomes. From Figure 9, it can be seen that the intelligent reduced-dimensional scheme selects W_f for adjustment in the initial stage of transition, while it selects the three variables separately in the middle stage and A_8 in the final stage. This intelligent update mode makes the variables of "irdMPC" nearer to those of "sMPC" and thus leads to a better parameter response than "mMPC". Although the α_{LPT} change is different from the others, it has only a small impact on the overall results since the regulating ability of α_{LPT} is the lowest among the variables.

For further comparison, the Root Mean Squared Error (RMSE) of the controlled parameter n_L is calculated by Equation (32) to quantify the control errors of the controllers. Table 3 shows the RMSE of the three MPC controllers.

$$\text{RMSE} = \sqrt{\frac{\sum_{i=1}^{N}(r_i - y_{ctrl,i})^2}{N}} \tag{32}$$

where N is the number of sample points.

Table 3. RMSE comparison of "sMPC", "mMPC" and "irdMPC".

	sMPC	mMPC	irdMPC
RMSE ($\times 10^{-2}$)	3.8146	8.046	5.3057

By comparing the controllers' RMSE, it can be seen that the multiplex MPC controller has much larger RMSE compared with the standard one. The RMSE of the "mMPC" is 2.11 times and differs by 4.2314×10^{-2} from that of "sMPC". While "irdMPC" greatly decreases the control errors, it has an RMSE of 5.3057×10^{-2} and a 34.06% improvement in control performance over "mMPC". Compared with "mMPC", "irdMPC" significantly reduces the RMSE deviation from 4.2314×10^{-2} to 1.4911×10^{-2}, which indicates the superiority of the proposed intelligent reduced-dimensional scheme to the existing method, i.e., the multiplex MPC, in control performance.

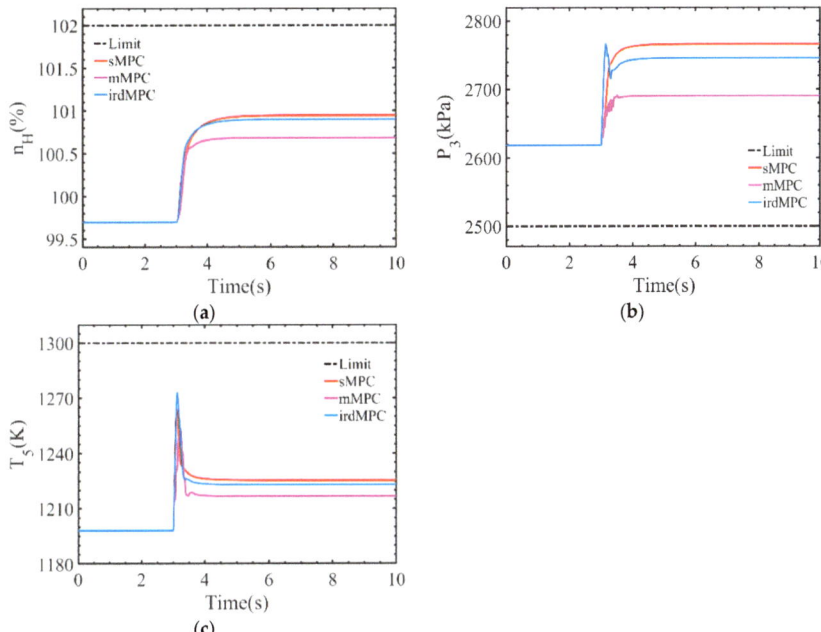

Figure 7. Constrained parameters' response under "sMPC", "mMPC" and "irdMPC". (**a**) n_H response curve. (**b**) P_3 response curve. (**c**) T_5 response curve.

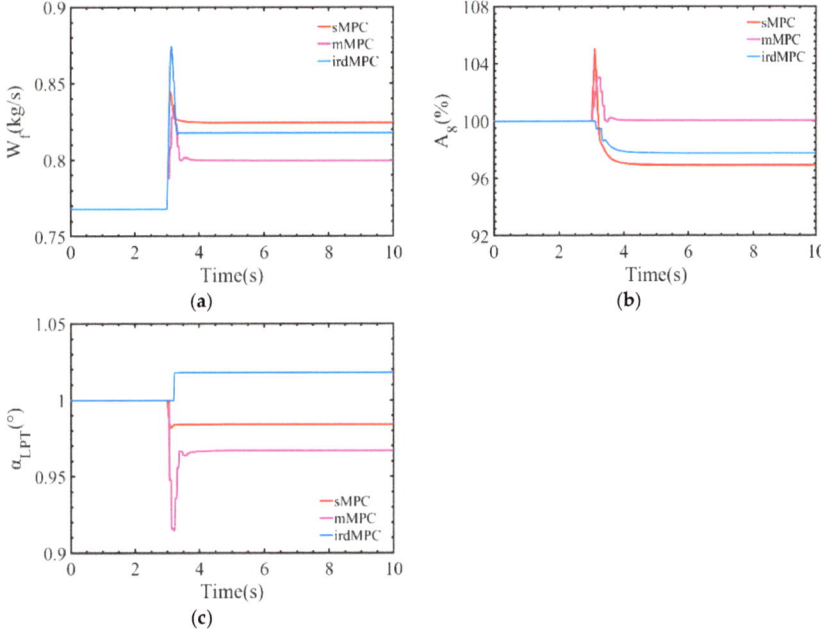

Figure 8. Control variables' changes under "sMPC", "mMPC" and "irdMPC". (**a**) W_f change curve. (**b**) A_8 change curve. (**c**) α_{LPT} change curve.

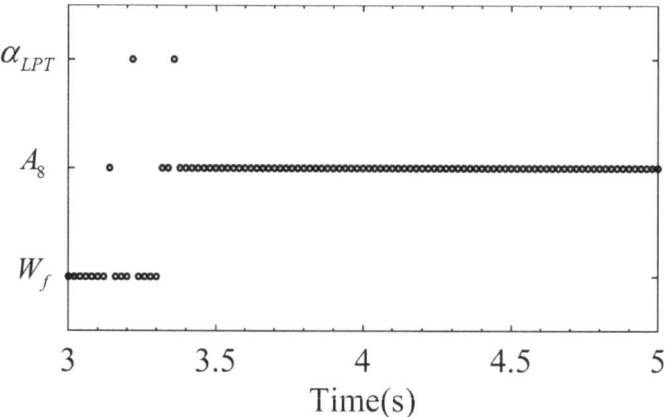

Figure 9. Selected control variable at each sampling instant during transition in "irdMPC".

3.3. Time Consumption Comparison

To illustrate the effectiveness of the proposed method in real-time performance, some simulations of different control horizons are conducted at the flight condition [H, Ma] = [0, 0]. During the simulation, the time required for the optimization solving processes in "irdMPC", "sMPC" and "mMPC" are recorded and compared. Since the MPC controllers adopt the same configuration and the simulations implement the same test environment, the results make sense. Table 4 lists the average time of the optimization problem solving, which is defined as Equation (33) and denoted as "t_{avg}".

$$t_{avg} = \frac{\sum_{i=1}^{N} t_i}{N} \qquad (33)$$

Table 4. Average time comparison of the "sMPC", "mMPC" and "irdMPC".

Nc	t_{avg} (ms)		
	sMPC	mMPC	irdMPC
3	6.3461	4.858	4.9351
4	7.4579	4.9323	5.1989
5	9.1945	5.0723	5.4687
6	11.111	5.2522	5.6339
7	13.2113	5.3438	5.8783
8	15.2596	5.5492	6.008
9	17.4286	5.6789	6.148

In Table 4, the prediction horizon Np is equal to 10. It can be seen that the average time consumption of these MPC controllers increases as the control horizon increases. The increase in the control horizon leads to an increase in the dimension of the control sequence. Therefore, the computational complexity increases, resulting in the time required for the optimization solutions increasing, while under the same control horizon, "irdMPC" has a lower dimension of the control sequence compared with "sMPC" and thus has lower computational complexity and a shorter average time consumption. The average time consumption of "irdMPC" is 4.9351 ms when the control horizon is 3, while that of the "sMPC" is 6.3461 ms. The t_{avg} of "sMPC" is 1.286 times that of "irdMPC". When the control horizon is 9, the t_{avg} of "irdMPC" is 6.148 ms, while that of "sMPC" is 17.4286 ms. The t_{avg} of "sMPC" becomes 2.835 times that of "irdMPC". As the dimension of the control sequence increases, "irdMPC" reduces the t_{avg} of "sMPC" at most by 64.72%.

Compared with "mMPC", "irdMPC" has a slightly bigger time consumption as a result of the control variable selection algorithm and control sequence coordinate strategy embedded in the scheme bringing extra computational tasks. For the t_{avg} of "irdMPC" and "mMPC", the ratio fluctuates between 1.016 and 1.1, which is small and acceptable considering a significant improvement in control performance. Furthermore, this ratio does not have a growing trend as the control horizon increases.

In the case of $Nc = 5$, some single-step time consumption results are listed in Table 5 to show details about the real-time performance comparison.

Table 5. Single-step time comparison of "sMPC", "mMPC" and "irdMPC".

Sampling Instant	t (ms)		
	sMPC	mMPC	irdMPC
$k+1$	9.1861	5.2361	4.8048
$k+2$	8.3069	5.3285	6.2879
$k+3$	9.2412	4.9081	6.1662
$k+4$	9.6713	4.9276	5.6472
$k+5$	13.01	5.7536	4.8898
$k+6$	10.3904	5.104	6.2612
$k+7$	7.2827	5.2765	5.421
$k+8$	7.7749	4.976	5.3001
$k+9$	9.7422	4.8451	5.6359

Table 5 records the time consumption of solving optimization problems at nine sampling instants. It can be seen that the results of these sampling instants are close to the values in Table 4. For each sampling instant, "sMPC" has larger solving time than "mMPC" and "irdMPC". Owing to extra computational tasks, the time consumption of "irdMPC" is slightly bigger than that of "mMPC" at most of the sampling instants. Nevertheless, their results have a similar level, and there exist two sampling instants at which "irdMPC" has smaller results.

The above results and analysis demonstrate the superiority of the intelligent reduced-dimensional MPC controller in time consumption over the standard MPC controller. In a fixed sampling period during the control process, the intelligent reduced-dimensional MPC controller has the potential to operate under a bigger control horizon.

3.4. Further Verification

Under another flight condition [H, Ma] = [11,000 m, 1.2], simulations are implemented to further illustrate the effectiveness of the proposed method. Similar results can be seen in Figure 10 and Table 6.

Table 6. Simulation results for real-time performance at [H, Ma] = [11,000 m, 1.2].

Nc	t_{avg} (ms)		
	sMPC	mMPC	irdMPC
3	6.3336	5.0491	5.1505
4	8.0574	5.0809	5.3599
5	9.8516	5.1691	5.603
6	11.8859	5.4101	5.7942
7	14.5266	5.4756	5.9671
8	16.6374	5.6881	6.1405
9	19.3425	5.8344	6.2934

Through the results under two flight conditions, the ability of the proposed intelligent reduced-dimensional scheme in achieving both good control performance and real-time performance is demonstrated, and the prospect of the designed intelligent reduced-dimensional MPC controller in regulating the aero-engine is validated.

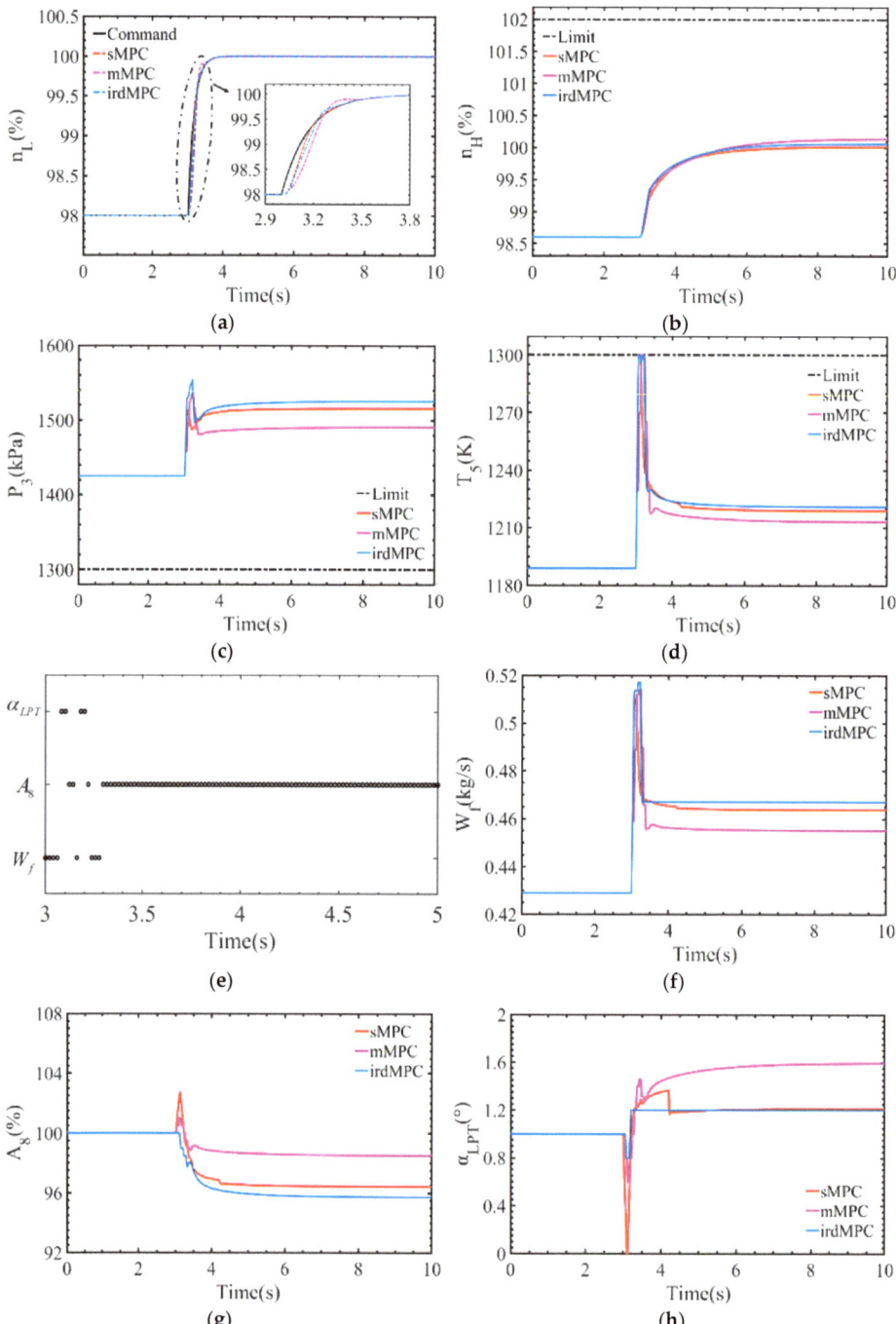

Figure 10. Simulation results for control performance at [H, Ma] = [11,000 m, 1.2]. (**a**) n_L response curve. (**b**) n_H response curve. (**c**) P_3 response curve. (**d**) T_5 response curve. (**e**) Selected control variable at each sampling instant during transition in "irdMPC". (**f**) W_f change curve. (**g**) A_8 change curve. (**h**) α_{LPT} change curve.

4. Conclusions

For good control performance and real-time performance simultaneously, an intelligent reduced-dimensional scheme of model predictive control is proposed in this paper. By reducing the dimension of the control vector in the control sequence to 1, a reduced-dimensional optimization problem with low computational complexity is constructed. In the proposed scheme, a novel selection algorithm is proposed to intelligently select the control variable with the best control effect as the available one in the optimization problem. In addition, the scheme embeds a coordination strategy to take into account the interaction among the control variables at different predicting instants in order to obtain a better optimization solution.

Simulation examples are implemented to illustrate the effectiveness of the proposed method. The results are as follows: (1) by applying the intelligent reduced-dimensional scheme, the control error of the multiplex MPC is greatly improved by 34.06%; (2) the intelligent reduced-dimensional scheme guarantees superiority in computational complexity. Compared to the standard MPC, it can save average time consumption by up to 64.72% when the control horizon increases to nine. All the results demonstrate that the proposed method has not only a better control outcome than the multiplex MPC but also less computation time than the standard MPC and thus helps the implementation of the MPC controller in the aero-engine.

The limitation of the proposed intelligent reduced-dimensional scheme is that this methodology is currently suitable for linear MPC, so the extension to nonlinear MPC needs further research. In addition, only numerical simulations are conducted in this paper to illustrate the effectiveness of the method. For further verification, the hardware-in-the-loop experiment will be considered in future research.

Author Contributions: Conceptualization, Z.J. and X.W.; methodology, Z.J. and J.L.; software, Z.J. and N.G.; validation, Z.J. and W.L; formal analysis, N.G.; investigation, W.L.; resources, X.W.; data curation, Z.J. and J.L.; writing—original draft preparation, Z.J.; writing—review and editing, X.W., N.G. and W.L.; visualization, Z.J.; supervision, X.W.; project administration, X.W. All authors have read and agreed to the published version of the manuscript.

Funding: This research received no external funding.

Data Availability Statement: The data that support the findings of this study are available from the corresponding author upon reasonable request.

Conflicts of Interest: Author Wei Liu was employed by the company China Gas Turbine Establishment. The remaining authors declare that the research was conducted in the absence of any commercial or financial relationships that could be construed as a potential conflict of interest.

References

1. Wang, X.; Yang, S.; Zhu, M.; Kong, X.X. *Aeroengine Control Principles*; Science Press: Beijing, China, 2021.
2. Garg, S. *Introduction to Advanced Engine Control Concepts*; NASA Glenn Research Center: Cleveland, OH, USA, 2007.
3. Zhu, Y.; Pan, M.; Zhou, W.; Huang, J. Intelligent direct thrust control for multivariable turbofan engine based on reinforcement and deep learning methods. *Aerosp. Sci. Technol.* **2022**, *131*, 107972. [CrossRef]
4. Akpan, V.A.; Hassapis, G.D. Nonlinear model identification and adaptive model predictive control using neural networks. *ISA Trans.* **2011**, *50*, 177–194. [CrossRef] [PubMed]
5. Bing, Y.U.; Zhouyang, L.I.; Hongwei, K.E.; Zhang, T. Wide-range model predictive control for aero-engine transient state. *Chin. J. Aeronaut.* **2022**, *35*, 246–260.
6. Chen, Q.; Sheng, H.; Zhang, T. A novel direct performance adaptive control of aero-engine using subspace-based improved model predictive control. *Aerosp. Sci. Technol.* **2022**, *128*, 107760. [CrossRef]
7. Richalet, J.; Rault, A.; Testud, J.L.; Papon, J. Model predictive heuristic control. *Automatica* **1978**, *14*, 413–428. [CrossRef]
8. Cutler, C.R.; Ramaker, B.L. Dynamic matrix control—A computer control algorithm. In Proceedings of the Joint Automatic Control Conference, San Francisco, CA, USA, 13–15 August 1980; p. 72.
9. Liu, J.; Wang, X.; Liu, X.; Zhu, M.; Pei, X.; Zhang, S.; Dan, Z. μ-Synthesis control with reference model for aeropropulsion system test facility under dynamic coupling and uncertainty. *Chin. J. Aeronaut.* **2023**, *36*, 246–261. [CrossRef]
10. Liu, J.; Wang, X.; Liu, X.; Zhu, M.; Pei, X.; Dan, Z.; Zhang, S. μ-Synthesis-based robust L1 adaptive control for aeropropulsion system test facility. *Aerosp. Sci. Technol.* **2023**, *140*, 108457. [CrossRef]

11. Liu, J.; Wang, X.; Liu, X.; Pei, X.; Dan, Z.; Zhang, S.; Yang, S.; Zhang, L. An anti-windup design with local sector and H2/H∞ optimization for flight environment simulation system. *Aerosp. Sci. Technol.* **2022**, *128*, 107787. [CrossRef]
12. Richter, H. *Advanced Control of Turbofan Engines*; Springer Science & Business Media: Berlin/Heidelberg, Germany, 2011.
13. Brunell, B.J.; Viassolo, D.E.; Prasanth, R. Model adaptation and nonlinear model predictive control of an aircraft engine. In Proceedings of the Turbo Expo: Power for Land, Sea, and Air, Vienna, Austria, 14–17 June 2004; pp. 673–682.
14. Wang, Y.; Huang, J.; Zhou, W.; Lu, F.; Hu, W. Neural network-based model predictive control with fuzzy-SQP optimization for direct thrust control of turbofan engine. *Chin. J. Aeronaut.* **2022**, *35*, 59–71. [CrossRef]
15. Wang, Y.; Zheng, Q.G.; Du, Z.Y.; Zhang, H. Research on nonlinear model predictive control for turboshaft engines based on double engines torques matching. *Chin. J. Aeronaut.* **2020**, *33*, 176–186. [CrossRef]
16. Nikolaidis, T.; Li, Z.; Jafari, S. Advanced constraints management strategy for real-time optimization of gas turbine engine transient performance. *Appl. Sci.* **2019**, *9*, 5333. [CrossRef]
17. Boyd, S.P.; Vandenberghe, L. *Convex Optimization*; Cambridge University Press: Cambridge, UK, 2004.
18. Kochenderfer, M.J.; Wheeler, T.A. *Algorithms for Optimization*; MIT Press: Cambridge, MA, USA, 2019.
19. Bemporad, A.; Morari, M.; Dua, V.; Pistikopoulos, E.N. The explicit solution of model predictive control via multiparametric quadratic programming. In Proceedings of the 2000 American Control Conference, ACC (IEEE Cat. No. 00CH36334), Chicago, IL, USA, 28–30 June 2000; IEEE: Piscataway, NJ, USA, 2000; Volume 2, pp. 872–876.
20. Tøndel, P.; Johansen, T.A.; Bemporad, A. An algorithm for multi-parametric quadratic programming and explicit MPC solutions. *Automatica* **2003**, *39*, 489–497. [CrossRef]
21. Alessio, A.; Bemporad, A. A survey on explicit model predictive control. In *Nonlinear Model Predictive Control: Towards New Challenging Applications*; Springer: Berlin/Heidelberg, Germany, 2009; pp. 345–369.
22. Gu, N.; Wang, X.; Zhu, M. Multi-Parameter Quadratic Programming Explicit Model Predictive Based Real Time Turboshaft Engine Control. *Energies* **2021**, *14*, 5399. [CrossRef]
23. Feng, C.; Du, X.; Yang, B. Design of Turbofan Engine Controller Based on Explicit Predictive Control. *J. Propuls. Technol.* **2022**, *6*, 043.
24. Lewis, F.L.; Vrabie, D.; Syrmos, V.L. *Optimal Control*; John Wiley & Sons: Hoboken, NJ, USA, 2012.
25. Genceli, H.; Nikolaou, M. Robust stability analysis of constrained l1-norm model predictive control. *AIChE J.* **1993**, *39*, 1954–1965. [CrossRef]
26. Kerrigan, E.C.; Maciejowski, J.M. Robustly stable feedback min-max model predictive control. In Proceedings of the 2003 American Control Conference, Denver, CO, USA, 4–6 June 2003; IEEE: Piscataway, NJ, USA, 2003; Volume 4, pp. 3490–3495.
27. Ling, K.V.; Maciejowski, J.; Richards, A.; Wu, B.F. Multiplexed model predictive control. *Automatica* **2012**, *48*, 396–401. [CrossRef]
28. Wang, X.Q.; Ho, W.K.; Ling, K.V. Computational load comparison of multiplexed and standard model predictive control. In Proceedings of the 2018 4th International Conference on Mechatronics and Robotics Engineering, Valenciennes, France, 7–11 February 2018; pp. 17–22.
29. Ling, K.V.; Maciejowski, J.; Wu, B.F. Multiplexed model predictive control. *IFAC Proc. Vol.* **2005**, *38*, 574–579. [CrossRef]
30. Ling, K.V.; Ho, W.K.; Wu, B.F.; Lo, A.; Yan, H. Multiplexed MPC for multizone thermal processing in semiconductor manufacturing. *IEEE Trans. Control Syst. Technol.* **2009**, *18*, 1371–1380. [CrossRef]
31. Richter, H.; Singaraju, A.V.; Litt, J.S. Multiplexed predictive control of a large commercial turbofan engine. *J. Guid. Control Dyn.* **2008**, *31*, 273–281. [CrossRef]
32. Pang, S.; Jafari, S.; Nikolaidis, T.; Qiuhong, L.I. Reduced-dimensional MPC controller for direct thrust control. *Chin. J. Aeronaut.* **2022**, *35*, 66–81. [CrossRef]
33. Pang, S.; Li, Q.; Ni, B. Improved nonlinear MPC for aircraft gas turbine engine based on semi-alternative optimization strategy. *Aerosp. Sci. Technol.* **2021**, *118*, 106983. [CrossRef]
34. Liu, Z.W.; Wang, Z.X.; Huang, H.C.; Cai, J.H. Numerical simulation on performance of variable cycle engines. *J. Aero-Space Power* **2010**, *25*, 1310–1315.
35. Wang, Y.; Zhang, P.P.; Li, Q.H.; Huang, X.H. Research and validation of variable cycle engine modeling method. *J. Aerosp. Power* **2014**, *29*, 2643–2651.
36. Kurzke, J. *GasTurb 12: Design and Off-Design Performance of Gas Turbines*; Mtu: Munich, Germany, 2012.
37. Du, X. *Application of Sliding Mode Control and Model Predictive Control to Limit Management for Aero-Engines*; Northwestern Polytechnical University: Xi'an, China, 2016.

Disclaimer/Publisher's Note: The statements, opinions and data contained in all publications are solely those of the individual author(s) and contributor(s) and not of MDPI and/or the editor(s). MDPI and/or the editor(s) disclaim responsibility for any injury to people or property resulting from any ideas, methods, instructions or products referred to in the content.

Article

Vibration Suppression of a Flexible Beam Structure Coupled with Liquid Sloshing via ADP Control Based on FBG Strain Measurement

Chunyang Kong [1,2], Dangjun Zhao [1,2,*] and Buge Liang [1,2]

1. School of Automation, Central South University, Changsha 410083, China; kong_cy@csu.edu.cn (C.K.); liangbuge@csu.edu.cn (B.L.)
2. Hunan Provincial Key Laboratory of Optic-Electronic Intelligent Measurement and Control, Changsha 410083, China
* Correspondence: zhao_dj@csu.edu.cn

Abstract: In this study, an adaptive dynamic programming (ADP) control strategy based on the strain measurement of a fiber Bragg grating (FGB) sensor array is proposed for the vibration suppression of a complicated flexible-sloshing coupled system, which usually exists in aerospace engineering, such as launch vehicles with a large amount of liquid propellant as well as a flexible beam structure. To simplify the flexible-sloshing coupled dynamics model, the equivalent spring-mass-damper (SMD) model of liquid sloshing is employed, and a finite-element method (FEM) dynamic model for the beam structure coupled with the liquid sloshing is mathematically established. Then, a strain-based vibration dynamic model is derived by employing a transformation matrix based on the relationship between displacement and strain of the beam structure. To facilitate the design of a strain-based control, a tracking differentiator is designed to provide the strains' derivative signals as partial states' estimations. Feeding the system with the strain measurements and their derivatives' estimations, an ADP controller with an action-dependent heuristic dynamic programming structure is proposed to suppress the vibration of the flexible-sloshing coupled system, and the corresponding Lyapunov stability of the closed-loop system is theoretically guaranteed. Numerical results show the proposed method can effectively suppress coupled vibration depending on limited strain measurements irrespective of external disturbances.

Keywords: fiber Bragg grating (FBG) sensors; finite element method (FEM); Euler–Bernoulli beam; spring-mass-damper model; adaptive dynamic programming (ADP); tracking differentiator

1. Introduction

Flexible-sloshing coupling problems are crucial in many fields [1], such as aerospace, architecture, and ocean engineering. Particularly in aerospace fields [2], modern spacecraft typically carry a large amount of liquid fuel and are also equipped with large flexible structures such as solar panels, communication antennae, and space manipulators. During attitude and orbit motion, the spacecraft is easily disturbed by liquid sloshing and flexible appendage vibrations. For flexible-sloshing coupling systems, elastic vibration can trigger liquid sloshing, resulting in the production of sloshing forces. Additionally, the coupling of sloshing and elastic vibration exerts a profound influence on the performance of the systems [3]. Therefore, the coupling dynamic and control of elastic vibration and liquid sloshing have received significant attention. In response to this issue, researchers continue to study the influence of structural elasticity and liquid sloshing on the control system. Specifically, they have (1) established equivalent beam models for elastic vibration [4]; (2) established equivalent mechanical models for liquid sloshing, including the spring-mass-damper model [5,6], single pendulum model [7], etc.; (3) modified equivalent models through ground modal analysis [8] and experimentation [9].

For elastic vibration, specific devices such as displacement sensors and accelerators can be employed to obtain vibration motion parameters, while the accelerators and gyroscopes installed on structures primarily serve to provide motion parameters for control systems. In these systems, error signals resulting from vibration-induced deformations can influence the accuracy of control performance. In contrast, fiber Bragg grating (FBG) sensors exhibit advantages, including lighter weight, stronger resistance to electromagnetic interference, easier integration, and higher accuracy. Therefore, they can be more conveniently integrated with control systems [10–12]. In addition, the exploration of sloshing dynamics encompasses theoretical analysis, numerical simulation, experimental methods, and theoretical and numerical synthesis methods. In the 1960s, NASA provided an analytical solution for the sloshing dynamics of several fixed-shape tanks [13]. Meanwhile, the advancement of computational fluid dynamics (CFD) technology has enabled more precise modeling of sloshing dynamics through FEM, the boundary element method, and finite difference method. Nevertheless, due to the constraints of onboard computer speed, memory, and the intricate nature of sloshing dynamics, a simple and efficient method is still required to replace complex calculation tasks in practical applications. Therefore, equivalent mechanical models are commonly employed in engineering to depict liquid sloshing by capturing sloshing patterns inside the tank using the motion of a rigid body [13].

In engineering practice, some novel vibration control devices, such as the tuned mass damper (TMD) [14] and negative stiffness mechanisms [15] (KDamper [16,17]), have been widely used. In addition, in the field of control algorithms, the classical control theory has been applied with great maturity. However, the classical controller can no longer satisfy the demand for stability in flexible-sloshing coupling systems. In response to practical needs [18,19] and theoretical challenges [20], several control methods have received great attention, including positive position feedback (PPF) control [21,22], independent modal space control (IMSC) [23,24], sliding mode control [25], boundary control [26,27], adaptive control [28], and intelligent control [29–32]. Among them, the PPF method is used to select the appropriate zeros and poles of the second-order filter to ensure the stability margin of closed-loop systems. The PPF method was first proposed in [21], and a modified PPF vibration active control method based on an adaptive controller was proposed in [22]. Vibration active control based on PPF can be numerically simulated to prove the effectiveness of this algorithm. The IMSC method involves discretizing the control object into various modal features of different orders and controlling the discrete modes. Active vibration control methods based on IMSC, as proposed in [23,24], can effectively suppress vibration. In [25], a sliding mode controller that only uses boundary information is proposed for the stabilization problem of an Euler–Bernoulli beam system. Boundary control is recognized as a highly practical approach for vibration control of flexible structures. In [26], a novel barrier Lyapunov function was employed to design a vibration boundary controller for an Euler–Bernoulli beam with boundary output constraints. This approach successfully suppressed beam vibrations without violating the constraints. Additionally, for the problem of vibration attenuation of Euler–Bernoulli beam systems with imprecise system parameters, external disturbances, asymmetric input saturation, and output constraint, the boundary controllers in [27,28] were designed by constructing adaptive laws. Furthermore, intelligent control has received significant attention due to its robust self-learning and self-adaptive ability. Ref. [29] conducted a study on the delay feedback control of cantilever vibration based on a genetic algorithm. Ref. [30] proposed a decomposed parallel fuzzy control with the adaptive neuro-fuzzy concept. Additionally, [31,32] proposed active vibration control algorithms based on reinforcement learning (RL). In particular, the adaptive dynamic programming (ADP) method, which leverages a critic-action network structure based on RL, has gained widespread recognition [33]. This approach boasts two key advantages: (1) ADP operates as a data-driven learning control method, eliminating the need for mathematical models; (2) the parameters of ADP can be adaptively updated over time when the system is disturbed. Drawing from these advantages, the ADP method is considered in this paper for vibration and sloshing suppression control.

In summary, this paper explores the control of a beam attached to a tank. To construct the flexible-sloshing coupling dynamic model, the vibration of the beam is analyzed using an Euler–Bernoulli beam and FEM, while the liquid sloshing in the tank is equivalent to a spring-mass-damper model. Drawing on the benefits of FBG sensors, a vibration and sloshing suppression control algorithm based on FBG strain information is developed in this paper. A strain-based vibration dynamic model is established by employing a transformation matrix that relies on the relationship between displacement and strain. Additionally, a tracking differentiator is designed to provide real-time first-order derivatives of the strain. Finally, based on the strain-based vibration dynamic model, an ADP method based on FBG strain information is proposed to suppress elastic vibration and sloshing. The main innovations of this paper are as follows:

(1) Compared with complex dynamic models, the Euler–Bernoulli beam model and spring-mass-damper equivalent model provide a simpler and more convenient way to construct the flexible-sloshing coupling dynamic model. The development of a strain-based vibration dynamic model facilitates the full utilization of FBG sensors' information.

(2) Compared with control methods that require motion parameters, the control method proposed in this paper can effectively suppress elastic vibration and sloshing even when only partial strain information is applied. Furthermore, the utilization of FBG strain information allows for direct measurement, eliminating the need for estimation of vibration parameters. This controller's advantages in practicality make it highly suitable for engineering applications.

The rest of the paper is organized as follows. In Section 2, the flexible-sloshing coupling dynamic model and the strain-based vibration dynamic equation are described. Section 3 presents the ADP controller based on FBG strain information and corresponding theoretical analysis. Numerical simulations are provided in Section 4. Finally, Section 5 concludes this paper.

2. Preliminaries and System Descriptions

2.1. Flexible-Sloshing Coupling Dynamic Model

As illustrated in Figure 1, to investigate the flexible-sloshing coupling dynamic model, a combined structure consisting of a tank attached to a beam is considered. The vibration of the beam stimulates the liquid sloshing within the tank, which in turn generates a sloshing force that impacts the tank and alters the vibration of the beam.

Assumption 1. *The flexible-sloshing coupling dynamic model is applicable to the case of small amplitude vibration and sloshing. The beam is slender and uniform, i.e., the shear deformation is not considered* [34]. *In addition, the spring-mass-damper model must satisfy the conditions that the liquid in the tank is incompressible, non-viscous, and non-rotating* [35].

According to Figure 1, the base coordinate system $O^B X^B Y^B Z^B$ and the tank coordinate system $O^S X^S Y^S Z^S$ are established. The origins O^B and O^S are located at the end of the beam and the center of the tank's bottom, respectively. Both coordinate systems are right-hand systems. Note that the base coordinate system is a global coordinate system, and the tank coordinate system changes with the vibration of the beam. M_{SB} is the rotation transformation matrix from the tank coordinate system to the base coordinate system, which is defined as

$$M_{SB} = \begin{bmatrix} \cos\theta^B_{O^S} & \sin\theta^B_{O^S} & 0 \\ -\sin\theta^B_{O^S} & \cos\theta^B_{O^S} & 0 \\ 0 & 0 & 1 \end{bmatrix}, \tag{1}$$

where $\theta^B_{O^S}$ is the rotation angle of the beam's position relative to the origin of the tank coordinate system in the base coordinate system, which varies with time.

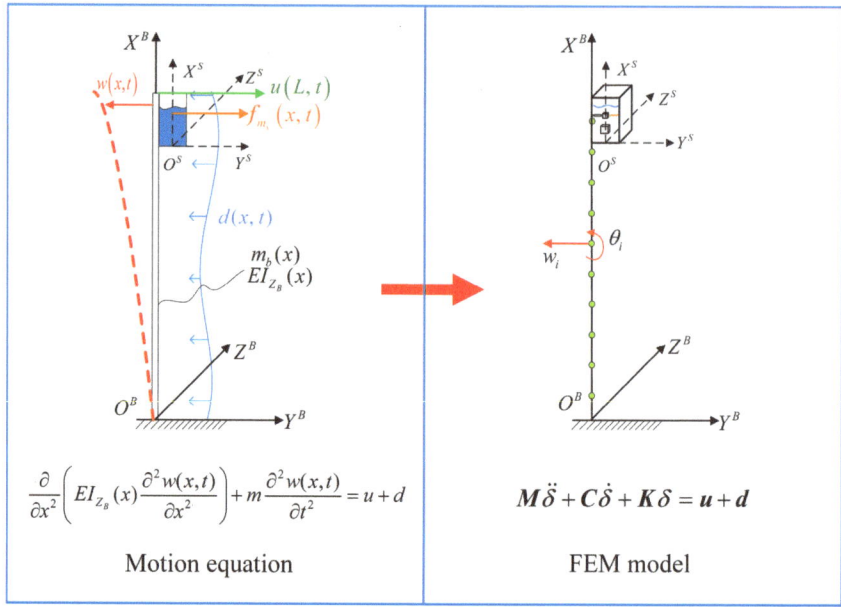

Figure 1. Flexible-sloshing coupling system including spring-mass-damper model and Euler–Bernoulli cantilever beam's FEM model.

Further, $m_b(x)$ is the mass of the beam; $EI_{Z_B}(x)$ is the bending stiffness with respect to $O_B Z_B$ axis; $w(x,t)$ is the transverse displacement of the beam related to time t and the position on the beam x; L is the length of the beam; u is control input acts on the end of the structure; $d(x,t)$ represents the unknown disturbance force applied to the beam; $f_{m_s}(x,t)$ denotes the sloshing force in the tank coordinate system. Note that d and f_{m_s} are distributed load and concentrated load, respectively.

Assumption 2. *For the unknown disturbance $d(t)$, we assume that there exists a constant $\bar{d}$, such that $|d(t)| < \bar{d}$, $\forall t \in [0, \infty)$. The time-varying $d(t)$ has finite energy and, thus, is bound, i.e., $d(t) \in L_\infty$ [26].*

Remark 1 [36]. *When the force f_d acting on the beam is distributed load, the unit force of f_d is provided as*

$$f_d^e = \frac{f_d}{12}\begin{bmatrix} 6l & l^2 & 6l & -l^2 \end{bmatrix},$$

and the unit force of concentrated load f_c is

$$f_c^e = f_c [N_1(x_0), N_2(x_0), N_3(x_0), N_4(x_0)]^T,$$

where $N_i(x)$ is the Hermite shape function, and x_0 is the coordinate of position where f_c acts on the beam. We define $\xi = x_0/l$, and $N_i(x_0)$ is provided as

$$N_1(x_0) = 1 - 3\xi^2 + 2\xi^3, \ N_2(x_0) = (\xi - 2\xi^2 + \xi^3)l$$
$$N_3(x_0) = 3\xi^2 - 2\xi^3, \ N_4(x_0) = (-\xi^2 + \xi^3)l$$

□

The analytical solution of the Euler–Bernoulli cantilever beam's motion equation is difficult to obtain in the case of non-uniform inertia and stiffness properties. So the FEM is often used to model the motion of the beam, and the second-order differential Equation [4] is obtained:

$$M\ddot{\delta} + C\dot{\delta} + K\delta = u + d, \qquad (2)$$

which includes n beam nodes and can be considered as a $2n$ degree-of-freedom system. Each node consists of two degrees of freedom, namely translation w and rotational θ, i.e.,

$$\{\delta_i\} = \begin{Bmatrix} w_i \\ \theta_i \end{Bmatrix} = \begin{Bmatrix} w_i \\ w_i' \end{Bmatrix}, i = 1, \ldots, n. \qquad (3)$$

M, C, and K are the mass, damping, and stiffness matrices, respectively. M and K are assembled by the following elemental matrices:

$$M^e = \frac{l}{420}\begin{bmatrix} 156 & 22l & 54 & -13l \\ 22 & 4l^2 & 13l & -3l^2 \\ -54 & 13l & 156 & -22l \\ -13l & -3l^2 & -22l & 4l^2 \end{bmatrix}, K^e = \frac{EI}{l^3}\begin{bmatrix} 12 & 6l & -12 & 6l \\ 6l & 4l^2 & -6l & 2l^2 \\ -12 & -6l & 12 & -6l \\ 6l & 2l^2 & -6l & 4l^2 \end{bmatrix}, \qquad (4)$$

where l is the length of the beam's element. Additionally, the corresponding damping matrix is $C = \alpha_1 M + \alpha_2 K$.

By introducing the influence of the sloshing force f_{m_s}, Equation (2) becomes

$$\overline{M}\ddot{\delta} + \overline{C}\dot{\delta} + \overline{K}\delta = \overline{F}, \qquad (5)$$

where $\overline{M}$, $\overline{C}$, and $\overline{K}$ are mass, damping, and stiffness matrices, and $\overline{F} = u + d + f_{m_s}$ represents the force driving the coupling system, including control force u, disturbance d, and sloshing force f_{m_s}.

As shown in Figure 2, the spring-mass-damper equivalent mechanical model is employed in this paper to estimate the liquid sloshing force. The liquid sloshing dynamic model in the tank can be regarded as a spring-mass-damper system [6], allowing for the calculation of the sloshing force f_{m_s}. The mass of the liquid in the tank is m_p, and the spring-mass-damper system divides the liquid into sloshing mass m_s and non-sloshing mass m_n. On the translation plane, along the two axes perpendicular to the central axis of the tank, the sloshing mass is connected to the tank through springs and dampers. Therefore, the sloshing mass is a secondary damping vibration perpendicular to the symmetry axis.

The mass of the liquid m_p is

$$m_p = \rho d^2 h, \qquad (6)$$

where ρ is the density of the liquid, d is the length of the bottom side of the tank, and h is the height of the liquid level.

According to Equation (6), h can be calculated as

$$h = \frac{m_p}{\rho d^2}. \qquad (7)$$

For the first-order sloshing mode of the rectangular tank, according to Section 5.2 of [37], the following equivalent parameters are established.

$$\begin{aligned} k &= \frac{8 m_p g \tanh^2(3\pi h/d)}{9h} \\ m_s &= m_p \frac{8}{\pi^3} \frac{\tanh(3\pi h/d)}{9h/d} \\ m_n &= m_p - m_s \\ l_s &= \frac{\tanh(3\pi h/2d)}{3\pi h/2d} \\ l_n &= -\frac{1}{m_n/m_p}\frac{m_s l_s}{m_p} \end{aligned} \qquad (8)$$

where l_s and l_n are the distances of sloshing and non-sloshing mass from the free liquid surface, respectively. k is elastic constant, and the relation between it and natural frequency ω_{ns} is $k = \omega_{ns}^2$.

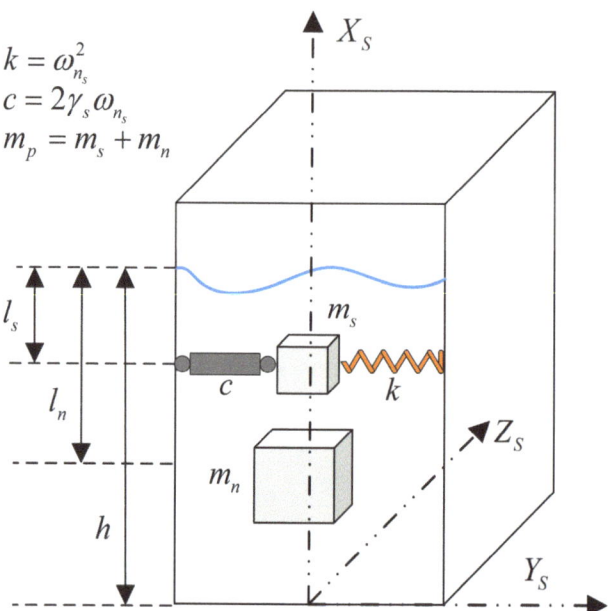

Figure 2. Spring-mass-damper model.

Note that the energy dissipation model in the oscillating system is approximately provided by the equivalent second-order damping ratio γ_s. Due to the lack of effective data for the tank design, according to [38], the equivalent second-order damping ratio of all sloshing modes is presumed to be $\gamma_s = 0.08$.

According to [35], the dynamic equation of liquid sloshing is as follows:

$$\begin{aligned} \dot{s}_{m_s}^S &= v_{m_s}^S \\ \dot{v}_{m_s}^B &= \frac{1}{m_s} A_s \begin{bmatrix} \Delta \dot{s}_{m_s}^S \\ v_{m_s}^S \end{bmatrix} + E_x \left[\dot{v}^B + \ddot{\varphi}_{m_s}^S \right] + E_{yz} g \\ \dot{v}_{m_s}^S &= \dot{v}_{m_s}^B - \dot{v}^B - \ddot{\varphi}_{m_s}^S \end{aligned} \quad (9)$$

where Δ represents the perturbation from the nominal reference. E_x only selects the axial acceleration component and E_{yz} only selects the normal component. The sloshing mass acceleration relative to the body in the axial direction is zero. A_s is

$$A_s = \begin{bmatrix} 0 & 0 & 0 & 0 & 0 & 0 \\ 0 & -\omega_{ns}^2 & 0 & 0 & -2\gamma_s \omega_{ns} & 0 \\ 0 & 0 & -\omega_{ns}^2 & 0 & 0 & -2\gamma_s \omega_{ns} \end{bmatrix}, \quad (10)$$

and the rotation term $\ddot{\varphi}_{m_s}^S$ is

$$\ddot{\varphi}_{m_s}^S = 2\omega^B \times v_{m_s}^S + \dot{\omega}^S \times s_{m_s}^S + \omega^S \times \omega^S \times s_{m_s}^S. \quad (11)$$

The force generated by the sloshing liquid is opposite to the force exerted on the liquid by the translation and rotation of the body, and the axial acceleration of the sloshing mass relative to the body must be considered. Assuming that most of the acceleration occurs along the x^B axis, this component acts on the force applied by the sloshing mass such that:

$$f_{m_s}^B = -m_s \dot{v}_{m_s}^B - m_s E_x \dot{v}^B. \tag{12}$$

Note that $f_{m_s}^B$ acts on the beam as a concentrated load.

2.2. Strain-Based Vibration Dynamic Model

In this paper, the strain information is used to design the controller for the system (5). To acquire the strain-based vibration dynamic model, the relationship [39] between the strain and the node displacement of the beam is established as

$$\begin{bmatrix} \varepsilon_1 \\ \varepsilon_2 \\ \vdots \\ \vdots \\ \varepsilon_{2n-1} \\ \varepsilon_{2n} \end{bmatrix} = \begin{bmatrix} T_1 \\ T_2 \\ & \ddots \\ & & \ddots \\ & & & T_{2n-1} \\ & & & T_{2n} \end{bmatrix} \begin{bmatrix} w_1 \\ \theta_1 \\ w_2 \\ \theta_2 \\ \vdots \\ w_{n-1} \\ \theta_{n-1} \\ w_n \\ \theta_n \end{bmatrix}. \tag{13}$$

The element transfer matrix [40] is

$$\begin{bmatrix} T_{2i-1} \\ T_{2i} \end{bmatrix} = \frac{1}{l^2} \begin{bmatrix} -6 & -4l & 6 & -2l \\ -6+12l & l(-4+6l) & 6-12l & l(-2+6l) \end{bmatrix} \times \frac{h_t}{2}, \ i=1,\ldots,n, \tag{14}$$

where h_t is the height of the beam's cross-section.

Thus, Equation (13) is written as

$$\varepsilon = T\delta. \tag{15}$$

According to Equation (15), Equation (5) is rewritten as

$$T^{-T}\overline{F} = T^{-T}\overline{M}T^{-1}\ddot{\varepsilon} + T^{-T}\overline{C}T^{-1}\dot{\varepsilon} + T^{-T}\overline{K}T^{-1}\varepsilon. \tag{16}$$

The dynamic model can be obtained as

$$M_\varepsilon \ddot{\varepsilon} + C_\varepsilon \dot{\varepsilon} + K_\varepsilon \varepsilon = F_\varepsilon, \tag{17}$$

where M_ε, C_ε, and K_ε are the mass, damper, and stiffness matrices, which satisfy $M_\varepsilon = T^{-T}\overline{M}T^{-1}$, $C_\varepsilon = T^{-T}\overline{C}T^{-1}$, and $K_\varepsilon = T^{-T}\overline{K}T^{-1}$. Additionally, force F_ε is provided as

$$F_\varepsilon = T^{-T}\overline{F} = T^{-T}\left(u + d + f_{m_s}\right).$$

By defining variables $X_\varepsilon = \begin{bmatrix} \varepsilon^T, & \dot{\varepsilon}^T \end{bmatrix}^T$, the state equation with strain and the first-order derivatives of the strain as state variables can be obtained as follows:

$$\begin{aligned} \dot{X}_\varepsilon &= \begin{bmatrix} 0 & I \\ -M_\varepsilon^{-1}K_\varepsilon & -M_\varepsilon^{-1}C_\varepsilon \end{bmatrix} X_\varepsilon + \begin{bmatrix} 0 \\ M_\varepsilon^{-1} \end{bmatrix} \left(u_\varepsilon + d_\varepsilon + f_{\varepsilon m_s}\right). \\ &= A_\varepsilon X_\varepsilon + B_\varepsilon F_\varepsilon \end{aligned} \tag{18}$$

In engineering applications, due to the installation of FBG sensors at regular intervals on the structure, only the strain values at some locations on the structure can be measured. Therefore, Equation (18) is rewritten as

$$\dot{X}_{\varepsilon r} = A_{\varepsilon r} X_{\varepsilon r} + B_{\varepsilon r} F_{\varepsilon r}, \tag{19}$$

where $X_{\varepsilon r}$ is the reduced state vector, and $A_{\varepsilon r}$ and $B_{\varepsilon r}$ are the system matrices.

The dimension of the dynamic model (18) is $2n$, and n represents the number of elements divided by FEM. To ensure accuracy, the value of n is relatively high, which will make real-time calculations on the onboard computer challenging. Therefore, the dimension of the system (19) is set as $2r$, and $r < n$ is the number of FBG sensors, such that the dimension of the system is reduced.

Remark 2. *The above method of associating the dimension of the system with the actual number of FBG sensors can better meet the requirements of engineering. This method has the same effect as the dynamic model reduction technology, which includes two main types: the physical reduce-order model and the agent model. The agent model is used to approximate the original complex structure through mathematical expressions with less computation, and the characteristics of the complex structure are obtained in the form of solving mathematical expressions, including the artificial neural network model. On the other hand, a neural network-based ADP control method is adopted in this paper. As a type of agent model, a neural network takes partial strain information as the input, which can better reflect the response characteristics of the whole structure. Moreover, the proposed ADP method in this paper is an optimal control method, which enables the elastic vibration and sloshing suppression control of the beam on the basis of only partial strain information through reasonable parameters design of the ADP method.*

On the other hand, FBG can only measure the strain value ε and cannot directly measure the first-order derivative of the strain $\dot{\varepsilon}$. Hence, it is necessary to introduce a tracking differentiator to estimate the state vector $X_{\varepsilon r}$. According to references [12,41], the tracking differentiator is designed as follows:

$$\begin{aligned} \dot{\hat{\varepsilon}}_0 &= \hat{\varepsilon}_1 + k_1(\varepsilon_m - \hat{\varepsilon}_0) \\ \dot{\hat{\varepsilon}}_1 &= \hat{\varepsilon}_2 + k_2(\varepsilon_m - \hat{\varepsilon}_0) \\ \dot{\hat{\varepsilon}}_2 &= k_3(\varepsilon_m - \hat{\varepsilon}_0) \end{aligned}, \tag{20}$$

where $k_i (i = 1, 2, 3)$ are the designed parameters; ε_m is the strain value measured by FBG; $\hat{\varepsilon}_i (i = 1, 2, 3)$ is the states of the differentiator. The real-time designed differentiator provides estimations of strain ε and its first-order derivative $\dot{\varepsilon}$ from the measurement ε_m.

Note that, according to [12], the estimation errors $e = \varepsilon_m - \hat{\varepsilon}$ of the tracking differentiator (20) exponentially converge into a region of hyperball when given an appropriate choice of $k_i (i = 1, 2, 3)$. This paper does not describe the proof process, and the detailed process is shown in [12]. In summary, based on the characteristics of the proposed differentiator, the states $\hat{\varepsilon}_0$ and $\hat{\varepsilon}_1$ are used as the estimations of the strain measurement ε_m as well as its first-order derivative $\dot{\varepsilon}$.

Substituting the estimations of tracking differentiator into Equation (19), we can obtain

$$\begin{aligned} \dot{\hat{X}}_{\varepsilon r} &= A_{\varepsilon r} \hat{X}_{\varepsilon r} + B_{\varepsilon r} \left(u_{\varepsilon r} + d_{\varepsilon r} + f_{\varepsilon r m_s} \right) \\ &= A_{\varepsilon r} \hat{X}_{\varepsilon r} + B_{\varepsilon r} F_{\varepsilon r} \end{aligned}, \tag{21}$$

where $\hat{X}_{\varepsilon r} = \left[\hat{\varepsilon}^T, \dot{\hat{\varepsilon}}^T \right]^T$. According to the strain-based vibration dynamic model (21), the intelligent controller is designed in the next section.

3. Control Strategy Design

3.1. Design of ADP Control Method Based on Strain Information of FBG

This paper proposed an ADP control method based on FBG strain information and neural networks. In contrast to previous work using motion parameters or modal parameters, this method can enable control performance using only strain information. The basic strategy of ADP [42] entails utilizing function approximators, such as linear function and neural network approximators, to construct critic networks to approximate the performance index function in the Hamilton–Jacobi–Bellman (HJB) function. For Equation (21), action-dependent heuristic dynamic programming (ADHDP) is adopted, and the system diagram is shown in Figure 3.

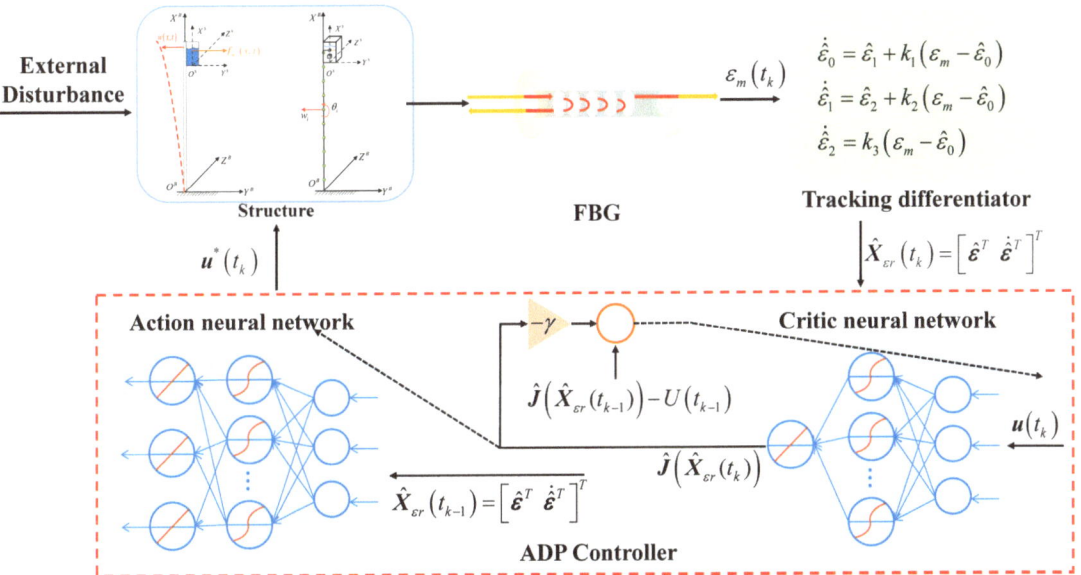

Figure 3. Vibration control for flexible-sloshing system based on FBG information and ADHDP.

First, the cost function with a discount factor γ is defined as follows:

$$J(t) = \int_{i=t}^{\infty} \gamma^{i-t} U(\hat{X}_{\varepsilon r}(t), u(t)), \tag{22}$$

where $0 < \gamma < 1$ is produced for the infinite horizon problem. U is the utility function, and the quadratic performance index [33] is the most commonly used; that is,

$$U(\hat{X}_{\varepsilon r}, u) = \hat{X}_{\varepsilon r}^T Q \hat{X}_{\varepsilon r} + u^T R u, \tag{23}$$

where Q and R are state and input utility weight matrices and positive-definite diagonal matrices.

The goal of the proposed method is to determine the appropriate control input $u(t)$ to minimize $J(t)$. We define $J^*(t)$ as the optimal cost function, which is shown as

$$J^*(t) = \min_{u(t)} \int_{i=t}^{\infty} \gamma^{i-t} U(\hat{X}_{\varepsilon r}(t), u(t)). \tag{24}$$

Based on the optimal control theory, $J^*(t)$ satisfies the following Bellman equation:

$$J^*(t) = \min_{u(t)} \{ U(\hat{X}_{\varepsilon r}(t), u(t)) + \gamma J^*(t+1) \}. \tag{25}$$

ADHDP adopts an action-critic network structure to obtain the approximate solution of the Bellman equation. The input of the critic network is the state and control input of the system, and the output $\hat{J}(t)$ is the estimation of $J^*(t)$. The action network's input and output are, respectively, the state of the system and the approximate optimate control input $u^*(t)$. In addition, both the action and critic networks adopt the three-layer neural network, which contains the input, hidden, and output layers. A hyperbolic tangent transfer function [33] is used as the activation function Φ. For any variable z, the hyperbolic tangent transfer function is defined as follows:

$$f_h(z) = \frac{1 - e^{-z}}{1 + e^{-z}}$$

Remark 3 [43]. *The approximation error of the neural network can be arbitrarily small as long as there are enough hidden layer neurons when the weights of the input layer–hidden layer are randomly initialized and kept constant. Consequently, in the learning process of the critic and action neural network, this paper only updates the weights of the hidden output layer. The update rules adopt the gradient descent method, which will be described later.*

The design of the critic neural networks is

$$\begin{aligned} \hat{J}(t) &= \omega_c^2 \Phi(h_c(t)) \\ h_c(t) &= \omega_c^1 [\hat{X}_{er}(t), u(\hat{X}_{er})] \end{aligned} \quad , \tag{26}$$

where ω_c^1 and ω_c^2 are the weight matrices of the hidden input and hidden output layers.

To train the critic network through the backpropagation method, the prediction error $e_c(t)$ is defined as follows:

$$e_c(t) = \gamma \hat{J}(\hat{X}_{er}(t)) - [\hat{J}(\hat{X}_{er}(t - \Delta t)) - U(\hat{X}_{er}(t), u(t))]. \tag{27}$$

Then, the error function to be minimized by the critic network is

$$E_c(t) = \frac{1}{2} e_c^2(t). \tag{28}$$

The output of the action network, namely the optimal control strategy, is

$$\begin{aligned} u^* &= \omega_a^2 \Phi(h_a(\hat{X}_{er})) \\ h_a(\hat{X}_{er}) &= \omega_a^1 \hat{X}_{er}(t) \end{aligned} \quad , \tag{29}$$

where ω_a^1 and ω_a^2 are the weight matrices of the hidden input and hidden output layers.

The prediction error $e_a(t)$ is defined as the difference between the estimated cost function $\hat{J}(t)$ and the desired ultimate object function U_c, which is backpropagated to the network to train the action network. The expression of $e_a(t)$ is

$$e_a(t) = \hat{J}(t) - U_c. \tag{30}$$

The error function to be minimized by the action network is

$$E_a(t) = \frac{1}{2} e_a^2(t). \tag{31}$$

The purpose of the controller proposed in this paper is to make the whole structure stable; that is, the FBG strain value is zero as a result. Hence, we can set the desired ultimate object function to $U_c = 0$.

Note that, according to Remark 3, ω_c^1 and ω_a^1 are randomly initialized and kept constant, and ω_c^2 and ω_a^2 are updated based on the gradient descent method. Therefore, the expression of weight updating policy by the chain rule is

$$\omega_c^2(t+\Delta t) = \omega_c^2(t) - \beta_c \left(\frac{\partial E_c(t)}{\partial \omega_c^2(t)} \right)$$

$$\frac{\partial E_c(t)}{\partial \omega_c^2(t)} = \frac{\partial E_c(t)}{\partial \hat{J}(t)} \frac{\partial \hat{J}(t)}{\partial \omega_c^2(t)} \tag{32}$$

$$\omega_a^2(t+\Delta t) = \omega_a^2(t) - \beta_a \left(\frac{\partial E_a(t)}{\partial \omega_a^2(t)} \right)$$

$$\frac{\partial E_a(\hat{X}_{er}(t))}{\partial \omega_a^2(t)} = \frac{\partial E_a(t)}{\partial \hat{J}(\hat{X}_{er}, u^*)} \frac{\partial \hat{J}(\hat{X}_{er}, u^*)}{\partial u^*} \frac{\partial u^*}{\partial \omega_a^2(t)} \tag{33}$$

where $\beta_c > 0$ and $\beta_a > 0$ denote the learning rates of the critic and action networks, respectively.

3.2. Stability Analysis

This section will analyze the Lyapunov stability of the above control system. Firstly, the following assumptions and lemmas are held.

Assumption 3. *Let ω_c^{2*} and ω_a^{2*} are the optimal weights of the hidden output layer in the critic and action neural network. Both of them are bound, i.e., $\|\omega_c^{2*}\| \leq \omega_{cm}$, $\|\omega_a^{2*}\| \leq \omega_{am}$, where ω_{cm} and ω_{am} are positive, satisfying the following equations:*

$$\omega_c^{2*} = \underset{\omega_c^2}{\arg\min} \|\gamma \hat{J}(t) - [\hat{J}(t-\Delta t) - U(t)]\|,$$

$$\omega_a^{2*} = \underset{\omega_a^2}{\arg\min} \|\hat{J}(t)\|.$$

Lemma 1 [44]. *Assumption 3 is held. The outputs of the critic and action networks are (26) and (29). The weights ω_c^1 and ω_a^1 of the critic and action networks are initialized and remain unchanged after initialization. The weights ω_c^2 and ω_a^2 are updated based on (32) and (33). Then, the errors between the optimal weights, ω_c^{2*} and ω_a^{2*}, and the weights ω_c^2 and ω_a^2, obtained based on the above update rules, are uniformly ultimately bound.*

The proof process of Lemma 1 can be seen in [44] and is not repeated in this paper. According to Lemma 1, when the neural network settings and weight updating rules are used in this paper, the estimation errors of the critic and action networks are uniformly ultimately bound [44]; that is, the method proposed in this paper can obtain the optimal control law.

Theorem 1. *For the dynamic model (21), the control law (29) is designed. If Assumption 3 and Lemma 1 are satisfied, the flexible-sloshing coupling control system is uniformly ultimately bounded.*

Proof of Theorem 1. Consider the Lyapunov function candidate defined as follows:

$$V = \frac{1}{2} \hat{X}_{er}^T Q \hat{X}_{er}. \tag{34}$$

The time derivative of (34) is

$$\dot{V} = \hat{X}_{er}^T Q \dot{\hat{X}}_{er}. \tag{35}$$

Substituting (21) into (35) yields

$$\begin{aligned}\dot{V} &= \hat{X}_{\varepsilon r}^T Q(A_{\varepsilon r}\hat{X}_{\varepsilon r} + B_{\varepsilon r}F_{\varepsilon r}) \\ &= \hat{X}_{\varepsilon r}^T Q\Big(-\hat{X}_{\varepsilon r} + (A_{\varepsilon r} + I)\hat{X}_{\varepsilon r} + B_{\varepsilon r}T^{-T}\Big(u + d + f_{m_s}\Big)\Big)\end{aligned} \quad (36)$$

According to the control law (29), (36) is rewritten as

$$\begin{aligned}\dot{V} \leq\ &-\hat{X}_{\varepsilon r}^T Q \hat{X}_{\varepsilon r} \\ &+\|\hat{X}_{\varepsilon r}\|\|Q\|\Big(\|(A_{\varepsilon r}+I)\|\|\hat{X}_{\varepsilon r}\| + \|B_{\varepsilon r}\|\|T^{-T}\|\Big(\|\omega_a^2\|\|\Phi(h_a(\hat{X}_{\varepsilon r}))\| + \|f_{m_s}\| + \|d\|\Big)\Big) \\ =\ &-2V + \|Q\|\|(A_{\varepsilon r}+I)\|\|\hat{X}_{\varepsilon r}\|^2 + \|\hat{X}_{\varepsilon r}\|\|Q\|\|B_{\varepsilon r}\|\|T^{-T}\|\Big(\|f_{m_s}\| + \|d\|\Big) \\ &+\|\hat{X}_{\varepsilon r}\|\|Q\|\|B_{\varepsilon r}\|\|T^{-T}\|\|\omega_a^2\|\|\Phi(h_a(\hat{X}_{\varepsilon r}))\| \\ =\ &-2V + \Theta\end{aligned} \quad (37)$$

where Θ satisfies

$$\begin{aligned}\Theta =\ &\|Q\|\|(A_{\varepsilon r}+I)\|\|\hat{X}_{\varepsilon r}\|^2 + \|\hat{X}_{\varepsilon r}\|\|Q\|\|B_{\varepsilon r}\|\|T^{-T}\|\Big(\|f_{m_s}\| + \|d\|\Big) \\ &+\|\hat{X}_{\varepsilon r}\|\|Q\|\|B_{\varepsilon r}\|\|T^{-T}\|\|\omega_a^2\|\|\Phi(h_a(\hat{X}_{\varepsilon r}))\|\end{aligned}.$$

In this paper, the flexible-sloshing coupling system is simplified according to the actual physical system, and the parameters in the actual physical system are bound, so the parameters of the proposed systems are bound; that is, $\|A_{\varepsilon r}\|$, $\|B_{\varepsilon r}\|$, $\|T^{-T}\|$, $\|d\|$, and $\|\hat{X}_{\varepsilon r}\|$ and are bound. The weight is bound according to Assumption 3 and Lemma 1. Based on the definition of the hyperbolic tangent transfer function, $\Phi(h_a(\hat{X}_{\varepsilon r})) \in [-1, 1]$, then $\|\Phi(h_a(\hat{X}_{\varepsilon r}))\|$ is bound; Q is a positive-definite diagonal matrix and $\|Q\|$ is bound. In summary, we can see that $\Theta > 0$ is bound. Therefore, the system (21) based on the proposed control law (29) is uniformly ultimately bound [45]. □

4. Numerical Simulations

The following numerical simulation demonstrates the effectiveness of the proposed ADP control strategy with strain information compared with the PD controller. An aluminum alloy cantilever beam and a cuboid tank are considered in this simulation. This tank is lightweight and thin-walled, so its weight can be ignored. The liquid in the tank is water. The parameters of the combined structure are shown in Table 1.

Table 1. Parameters of the structure.

	Parameters	Value
Aluminum alloy beam	Density (kg/m^2)	2690
	Elastic modulus (Pa)	6.98e10
	Length (m)	0.972
	Width (m)	0.02
	Height (m)	0.003
Cuboid tank	Height of tank(m)	0.1
	Length of bottom (m)	0.04
	Height of liquid level (m)	0.04
Water	Density (kg/m^2)	1000

The initial conditions in this simulation are provided as $w(0) = 0$, $s_{m_s}^S(0) = 0$, and $X_{\varepsilon r}(0) = [0, 0]^T$. The parameters of the tracking differentiator are set as $k_1 = 1/\delta^2$, $k_2 = 2/\delta^2$, and $k_3 = 3/\delta^2$, where δ satisfies $\delta = 0.015$. The damping matrix is set as $C = 0.001M + 0.001K$. The simulation time is 0.001s, and the total simulation time is 15s. A sinusoidal disturbance $d(t) = 0.1(\sin(5\pi t) + \cos(10\pi t))$ is applied. The dynamic responses of the structure are examined in the following cases.

Free Case: There is no control input in this simulation, i.e., $u(t) = 0$, and the spatial time representation is shown in Figure 4.

ADP Case 1: The ADP controller applies 10 FBG sensors' information. The number of hidden layer nodes is 30, the learning rates are $l_c = l_a = 10$, and the state and input utility weight matrices are $Q = diag\{ones(20,1)\}$ and $R = [1]$. The spatial time representation is shown in Figure 5.

ADP Case 2: The ADP controller applies 20 sensors' information. The number of hidden layer nodes and learning rates are set the same as ADP Case 1. The state and input weight matrices are $Q = diag\{ones(40,1)\}$ and $R = [1]$. The spatial time representation is shown in Figure 6.

PD Case: For comparison, the spatial time representation of the displacement with the PD controller is shown in Figure 7. The PD controller is designed as $u(t) = -k_p w(L,t) - k_d \dot{w}(L,t)$, and the parameters are set as $k_p = 20$, $k_d = 0.1$.

As shown in Figure 4a, it can be observed that there are significant vibrations along the structure subjected to the sinusoidal disturbance $d(t)$. In Figure 4b–d, both the ADP and PD controllers can suppress the vibration of the coupling systems when the system is subjected to external disturbance $d(t)$. Furthermore, it is evident that the displacement under ADP control is smaller than that under PD control.

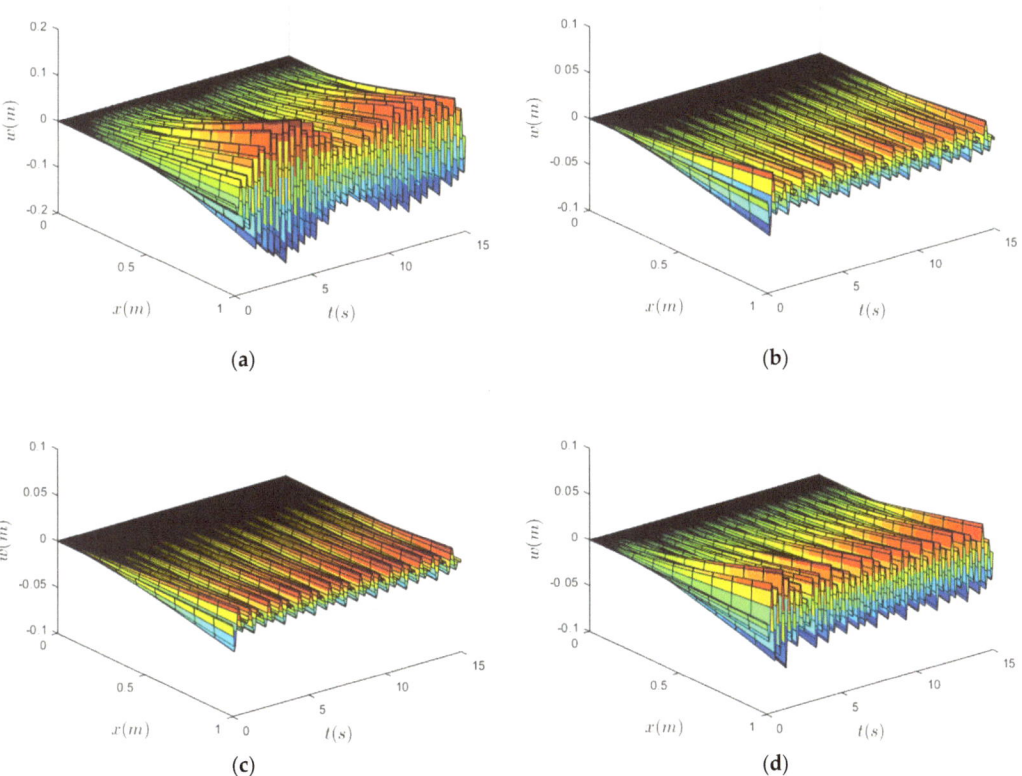

Figure 4. Displacement of the structure in (**a**) free case; (**b**) ADP case 1 with 10 sensors; (**c**) ADP case 2 with 20 sensors; (**d**) PD case.

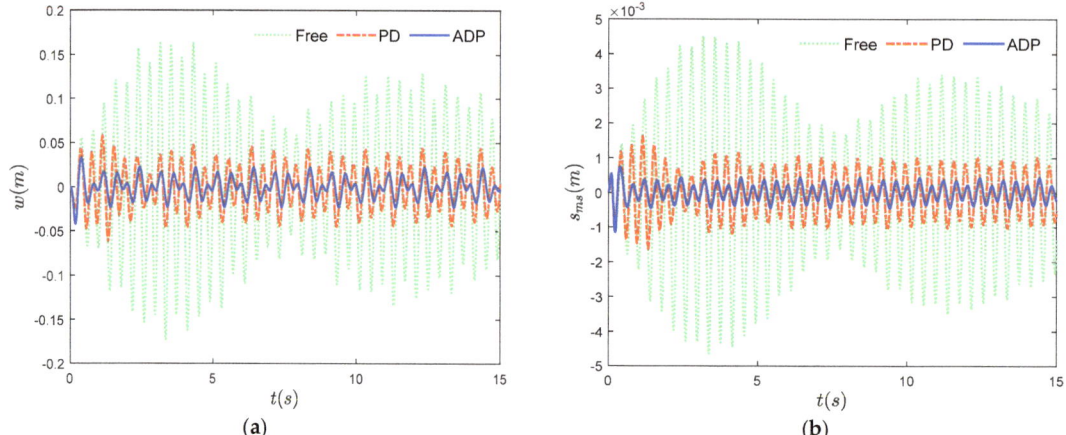

Figure 5. Results of displacement. (**a**) Displacement of the structure's free end in free case, PD case, and ADP case 1; (**b**) sloshing displacement in free case, PD case, and ADP case 1.

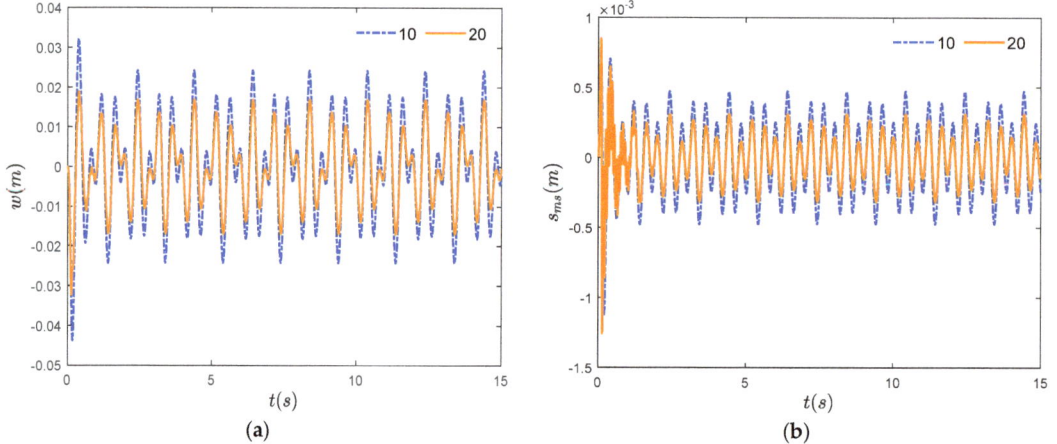

Figure 6. Results of displacement. (**a**) Displacement of the structure's free end for ADP control with 10 and 20 sensors; (**b**) sloshing displacement for ADP control with 10 and 20 sensors.

For further analysis, the displacements of the structure's free end and sloshing in the free case, PD case, and ADP case 1 are shown in Figure 5. The displacement errors (between PD case and free case, between ADP case 1 and free case) of the structure's free end and sloshing are shown in Figure 7a,b. As illustrated in Figures 5a and 7a, the ADP controller with 10 sensors and a PD controller can suppress the vibration at the small neighborhood of its equilibrium position. Moreover, the output position in ADP case 1 is smaller than that in the PD case. Similarly, Figures 5b and 7b demonstrate that both the ADP controller with 10 sensors and the PD controller can suppress the sloshing in the tank subject to disturbance, with the ADP case 1 producing smaller sloshing displacement than the PD case. The simulation results in Figures 5 and 7 indicate that the ADP method has better performance of elastic vibration and sloshing suppression in comparison to the PD controller.

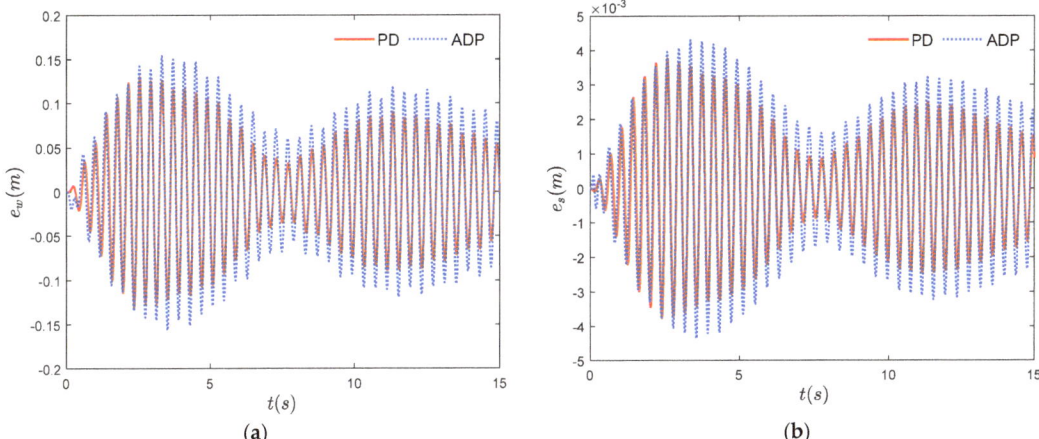

Figure 7. Error results of displacement relative to free case. (**a**) Displacement of the structure's free end in PD case and ADP case 1; (**b**) sloshing displacement in PD case and ADP case 1.

To analyze the performance of the ADP controller using varying numbers of sensors, the displacement of the structure's free end and sloshing are shown in Figure 6a,b, respectively. The displacement errors of the structure's free end and sloshing between ADP case 1 and ADP case 2 are shown in Figure 8a,b. We can observe that the displacement of the structure's free end and sloshing in ADP case 2 is smaller than that in ADP case 1. As the amount of sensor information increases, the performance of the ADP controller becomes more effective. However, the displacement of sloshing and control input within 2 s changes drastically under ADP control using 20 sensors' information. Finally, the ADP and PD control inputs are shown in Figure 9.

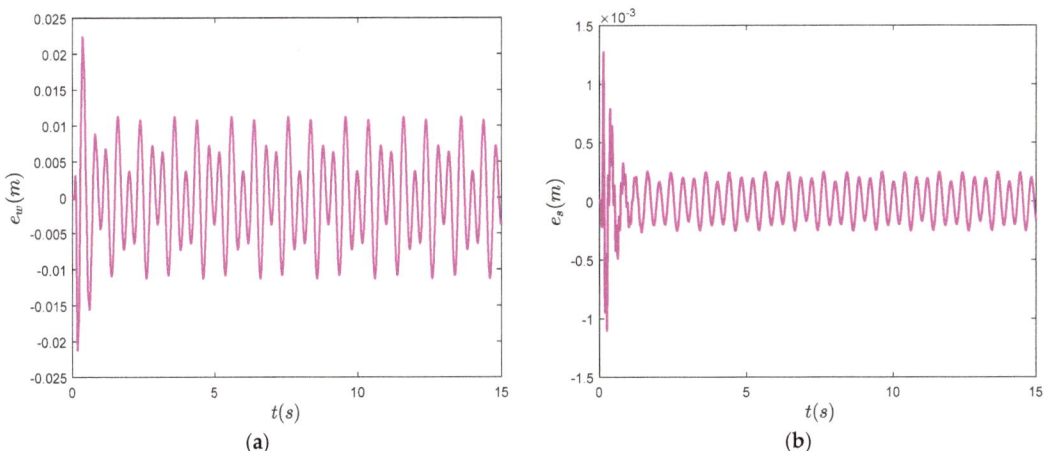

Figure 8. Error results of displacement between ADP case 1 and ADP case 2. (**a**) Displacement of the structure's free end; (**b**) sloshing displacement.

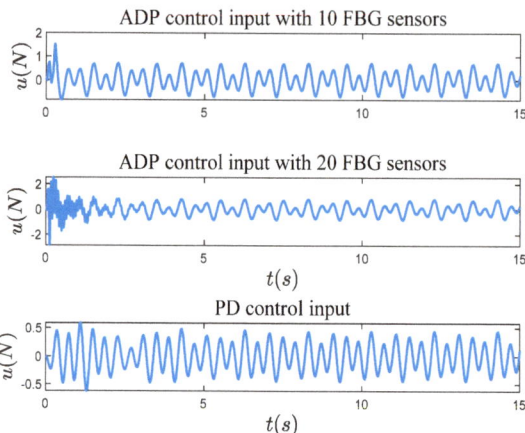

Figure 9. ADP control inputs with 10 and 20 sensors and PD control input.

In conclusion, the simulation results have demonstrated the effectiveness of the ADP method in suppressing the elastic vibration and sloshing of the coupling system when disturbance is exerted. Compared with other methods, the designed ADP controller only requires partial strain information and does not need the motion parameters of the coupling system. Additionally, as the amount of sensor information applied increases, the performance of the ADP controller becomes more effective. However, the displacement of sloshing undergoes significant changes, which means that the selection of the appropriate number of sensors depends on the compromise between the final performance of the ADP control and the initial violent sloshing during control.

5. Conclusions

In this paper, the elastic vibration and sloshing suppression control problem is investigated via the ADP algorithm based on strain information measured from FBG. The controller is designed using the strain information measured by FBG, with its first-order derivative estimated by a tracking differentiator. The Euler–Bernoulli theory and spring-mass-damper equivalent mechanical model are employed to establish the flexible-sloshing coupling dynamic model. Additionally, the strain-based vibration dynamic model is derived through the relationship between the strain and the node displacement based on the FEM. Furthermore, the controller is designed based on ADHDP structure with neural networks. The main advantages of the proposed method can be concluded as follows:

(1) The usages of the Euler–Bernoulli beam model and the spring-mass-damper equivalent model provide a simpler and more convenient way for constructing the flexible-sloshing coupling dynamic model.
(2) The development of a strain-based vibration dynamic model facilitates the full utilization of FBG sensors' strain information.
(3) This controller can effectively suppress elastic vibration and sloshing with only partial strain information, eliminating the process of estimating the vibration motion parameters. The control strategy indicates important implications in engineering applications.

In the future, we will conduct further research on three-dimensional vibration suppression control and the control performance of different numbers of sensors.

Author Contributions: Conceptualization, D.Z.; methodology, C.K.; software, C.K.; validation, C.K. and D.Z.; formal analysis, C.K.; investigation, C.K.; resources, D.Z.; data curation, C.K.; writing—original draft preparation, C.K.; writing—review and editing, D.Z. and B.L.; visualization, C.K.; supervision, D.Z. and B.L.; project administration, D.Z. and B.L.; funding acquisition, D.Z. and B.L. All authors have read and agreed to the published version of the manuscript.

Funding: This research received no external funding.

Data Availability Statement: No new data were created or analyzed in this study. Data sharing is not applicable to this article.

Conflicts of Interest: The authors declare no conflict of interest.

References

1. Guo, S.-S.; Kim, J. Some Recent Developments in the Vibration Control and Structure Health Monitoring. *Actuators* **2023**, *12*, 11. [CrossRef]
2. He, G.; Cao, D. Dynamic Modeling and Attitude-Vibration Cooperative Control for a Large-Scale Flexible Spacecraft. *Actuators* **2023**, *12*, 167. [CrossRef]
3. Dai, J.; Qin, Y.; Wang, C.; Zhu, J.; Zhu, J. Research on Stability Control Technology of Hazardous Chemical Tank Vehicles Based on Electromagnetic Semi-Active Suspension. *Actuators* **2023**, *12*, 333. [CrossRef]
4. Su, W.; King, C.K.; Clark, S.R.; Griffin, E.D.; Suhey, J.D.; Wolf, M.G. Dynamic Beam Solutions for Real-Time Simulation and Control Development of Flexible Rockets. *J. Spacecr. Rocket.* **2017**, *54*, 403–416. [CrossRef]
5. Peterson, L.D.; Crawley, E.F.; Hansman, R.J. Nonlinear fluid slosh coupled to the dynamics of a spacecraft. *AIAA J.* **1989**, *27*, 1230–1240. [CrossRef]
6. Frosch, J.A.; Vallely, D.P. Saturn AS-501/S-IC flight control system design. *J. Spacecr. Rocket.* **1967**, *4*, 1003–1009. [CrossRef]
7. Unruh, J.F.; Kana, D.D.; Dodge, F.T.; Fey, T.A. Digital data analysis techniques for extraction of slosh model parameters. *J. Spacecr. Rocket.* **1986**, *23*, 171–177. [CrossRef]
8. Guo, N.; Yang, Z.; Wang, L.; Ouyang, Y.; Zhang, X. Dynamic model updating based on strain mode shape and natural frequency using hybrid pattern search technique. *J. Sound Vib.* **2018**, *422*, 112–130. [CrossRef]
9. Mooij, E.; Gransden, D.I. The Effect of Sloshing on the Controllability of a Conventional Aeroelastic Launch Vehicle. In *AIAA Scitech 2019 Forum*; AIAA SciTech Forum; American Institute of Aeronautics and Astronautics: Reston, VA, USA, 2019.
10. Panopoulou, A.; Roulias, D.; Loutas, T.H.; Kostopoulos, V. Health Monitoring of Aerospace Structures Using Fibre Bragg Gratings Combined with Advanced Signal Processing and Pattern Recognition Techniques. *Strain* **2012**, *48*, 267–277. [CrossRef]
11. Chuang, K.; Lin, S.; Ma, C.; Wu, R. Application of a Fiber Bragg Grating-Based Sensing System on Investigating Dynamic Behaviors of a Cantilever Beam Under Impact or Moving Mass Loadings. *IEEE Sens. J.* **2013**, *13*, 389–399. [CrossRef]
12. Kong, C.; Zhao, D.; Zhang, J.; Liang, B. Real-Time Virtual Sensing for Dynamic Vibration of Flexible Structure via Fiber Bragg Grating Sensors. *IEEE Sens. J.* **2022**, *22*, 21706–21718. [CrossRef]
13. Abramson, H.N. *The Dynamic Behavior of Liquids in Moving Containers With Applications to Space Vehicle Technology*; NASA: Washington, DC, USA, 1966.
14. Su, N.; Bian, J.; Peng, S.; Chen, Z.; Xia, Y. Balancing static and dynamic performances of TMD with negative stiffness. *Int. J. Mech. Sci.* **2023**, *243*, 108068. [CrossRef]
15. Chen, Z.; Chen, Z.; Wei, Y. Quasi-Zero Stiffness-Based Synchronous Vibration Isolation and Energy Harvesting: A Comprehensive Review. *Energies* **2022**, *15*, 7066. [CrossRef]
16. Kapasakalis, K.A.; Antoniadis, I.A.; Sapountzakis, E.J. Performance assessment of the KDamper as a seismic Absorption Base. *Struct. Control. Health Monit.* **2020**, *27*, e2482. [CrossRef]
17. Kapasakalis, K.A.; Antoniadis, I.A.; Sapountzakis, E.J. Constrained optimal design of seismic base absorbers based on an extended KDamper concept. *Eng. Struct.* **2021**, *226*, 111312. [CrossRef]
18. Wang, J.; Liu, J.; Li, Y.; Chen, C.L.P.; Liu, Z.; Li, F. Prescribed Time Fuzzy Adaptive Consensus Control for Multiagent Systems With Dead-Zone Input and Sensor Faults. *IEEE Trans. Autom. Sci. Eng.* **2023**, *21*, 1–12. [CrossRef]
19. Wang, J.; Gong, Q.; Huang, K.; Liu, Z.; Chen, C.L.P.; Liu, J. Event-Triggered Prescribed Settling Time Consensus Compensation Control for a Class of Uncertain Nonlinear Systems With Actuator Failures. *IEEE Trans. Neural Networks Learn. Syst.* **2023**, *34*, 5590–5600. [CrossRef]
20. Wang, J.; Wang, C.; Liu, Z.; Chen, C.L.P.; Zhang, C. Practical Fixed-Time Adaptive ERBFNNs Event-Triggered Control for Uncertain Nonlinear Systems With Dead-Zone Constraint. *IEEE Trans. Syst. Man Cybern. Syst.* **2023**, *54*, 1–10. [CrossRef]
21. Goh, C.J.; Caughey, T.K. On the stability problem caused by finite actuator dynamics in the collocated control of large space structures. *Int. J. Control* **1985**, *41*, 787–802. [CrossRef]
22. Mahmoodi, S.N.; Ahmadian, M. Active Vibration Control With Modified Positive Position Feedback. *J. Dyn. Syst. Meas. Control* **2009**, *131*, 041002. [CrossRef]
23. Lieven, N.A.J.; Ewins, D.J.; Inman, D.J. Active modal control for smart structures. *Philos. Trans. R. Soc. London. Ser. A Math. Phys. Eng. Sci.* **2001**, *359*, 205–219. [CrossRef]
24. Baz, A.; Poh, S. Performance of an active control system with piezoelectric actuators. *J. Sound Vib.* **1988**, *126*, 327–343. [CrossRef]
25. Wang, Z.; Wu, W.; Görges, D.; Lou, X. Sliding mode vibration control of an Euler–Bernoulli beam with unknown external disturbances. *Nonlinear Dyn.* **2022**, *110*, 1393–1404. [CrossRef]
26. He, W.; Ge, S.S. Vibration Control of a Flexible Beam With Output Constraint. *IEEE Trans. Ind. Electron.* **2015**, *62*, 5023–5030. [CrossRef]

27. Ma, Y.; Lou, X.; Miller, T.; Wu, W. Fault-Tolerant Boundary Control of an Euler–Bernoulli Beam Subject to Output Constraint. *IEEE Trans. Syst. Man Cybern. Syst.* **2023**, *53*, 4753–4763. [CrossRef]
28. Feng, Y.; Liu, Z. Adaptive Vibration Iterative Learning Control of an Euler–Bernoulli Beam System With Input Saturation. *IEEE Trans. Syst. Man Cybern. Syst.* **2023**, *53*, 2469–2477. [CrossRef]
29. Mirafzal, S.H.; Khorasani, A.M.; Ghasemi, A.H. Optimizing time delay feedback for active vibration control of a cantilever beam using a genetic algorithm. *J. Vib. Control* **2015**, *22*, 4047–4061. [CrossRef]
30. Lin, J.; Chao, W.S. Vibration Suppression Control of Beam-cart System with Piezoelectric Transducers by Decomposed Parallel Adaptive Neuro-fuzzy Control. *J. Vib. Control* **2009**, *15*, 1885–1906. [CrossRef]
31. He, W.; Gao, H.; Zhou, C.; Yang, C.; Li, Z. Reinforcement Learning Control of a Flexible Two-Link Manipulator: An Experimental Investigation. *IEEE Trans. Syst. Man Cybern. Syst.* **2021**, *51*, 7326–7336. [CrossRef]
32. Qiu, Z.-c.; Yang, Y.; Zhang, X.-m. Reinforcement learning vibration control of a multi-flexible beam coupling system. *Aerosp. Sci. Technol.* **2022**, *129*, 107801. [CrossRef]
33. Wang, F.Y.; Zhang, H.; Liu, D. Adaptive Dynamic Programming: An Introduction. *IEEE Comput. Intell. Mag.* **2009**, *4*, 39–47. [CrossRef]
34. Jiang, H. *Real Time Mode Sensing and Attitude Control of Flexible Launch Vehicle with Fiber Bragg Grating Sensor Array*; Florida Institute of Technology: Melbourne, FL, USA, 2011.
35. Orr, J.S. *Robust Autopilot Design for Lunar Spacecraft Powered Descent Using High Order Sliding Mode Control*; The University of Alabama in Huntsville: Huntsville, AL, USA, 2009.
36. Kwon, Y.W.; Bang, H. *The Finite Element Method Using MATLAB*; London CRC Press: Boca Raton, FL, USA, 2000.
37. Ibrahim, R.A. *Liquid Sloshing Dynamics: Theory and Applications*; Cambridge University Press: New York, NY, USA, 2005.
38. Dodge, F.T. *Analytical Representation of Lateral Sloshing by Equivalent Mechanical Models*; NASA Special Publication: Washington, DC, USA, 1966; p. 199.
39. Song, X.; Liang, D. Dynamic displacement prediction of beam structures using fiber bragg grating sensors. *Optik* **2018**, *158*, 1410–1416. [CrossRef]
40. Wu, S.Q.; Zhou, J.X.; Rui, S.; Fei, Q.G. Reformulation of elemental modal strain energy method based on strain modes for structural damage detection. *Adv. Struct. Eng.* **2017**, *20*, 896–905. [CrossRef]
41. Zhao, D.-J.; Wang, Y.-J.; Liu, L.; Wang, Z.-S. Robust Fault-Tolerant Control of Launch Vehicle Via GPI Observer and Integral Sliding Mode Control. *Asian J. Control* **2013**, *15*, 614–623. [CrossRef]
42. Zhong, X.; He, H. An Event-Triggered ADP Control Approach for Continuous-Time System With Unknown Internal States. *IEEE Trans. Cybern.* **2017**, *47*, 683–694. [CrossRef]
43. Igelnik, B.; Yoh-Han, P. Stochastic choice of basis functions in adaptive function approximation and the functional-link net. *IEEE Trans. Neural Netw.* **1995**, *6*, 1320–1329. [CrossRef]
44. Liu, F.; Sun, J.; Si, J.; Guo, W.; Mei, S. A boundedness result for the direct heuristic dynamic programming. *Neural Netw.* **2012**, *32*, 229–235. [CrossRef]
45. Khalil, H.K. *Nonlinear Systems*; Prentice Hall: Upper Saddle River, NJ, USA, 2002.

Disclaimer/Publisher's Note: The statements, opinions and data contained in all publications are solely those of the individual author(s) and contributor(s) and not of MDPI and/or the editor(s). MDPI and/or the editor(s) disclaim responsibility for any injury to people or property resulting from any ideas, methods, instructions or products referred to in the content.

Article

A Sliding Mode Control-Based Guidance Law for a Two-Dimensional Orbit Transfer with Bounded Disturbances

Marco Bassetto, Giovanni Mengali, Karim Abu Salem *, Giuseppe Palaia and Alessandro A. Quarta

Department of Civil and Industrial Engineering, University of Pisa, I-56122 Pisa, Italy; marco.bassetto@ing.unipi.it (M.B.); giovanni.mengali@ing.unipi.it (G.M.); giuseppe.palaia@phd.unipi.it (G.P.); alessandro.antonio.quarta@unipi.it (A.A.Q.)
* Correspondence: karim.abusalem@ing.unipi.it

Abstract: The aim of this paper is to analyze the performance of a state-feedback guidance law, which is obtained through a classical sliding mode control approach, in a two-dimensional circle-to-circle orbit transfer of a spacecraft equipped with a continuous-thrust propulsion system. The paper shows that such an inherently robust control technique can be effectively used to obtain possible transfer trajectories even when the spacecraft equations of motion are affected by perturbations. The problem of the guidance law design is first addressed in the simplified case of an unperturbed system, where it is shown how the state-feedback control may be effectively used to obtain simple mathematical relationships and graphs that allow the designer to determine possible transfer trajectories that depend on a few control parameters. It is also shown that a suitable combination of the controller parameters may be exploited to obtain trade-off solutions between the flight time and the transfer velocity change. The simplified control strategy is then used to investigate a typical heliocentric orbit raising/lowering in the presence of bounded disturbances and measurement errors.

Keywords: sliding mode control; continuous-thrust propulsion system; two-dimensional orbit raising/lowering; spacecraft guidance law

Citation: Bassetto, M.; Mengali, G.; Abu Salem, K.; Palaia, G.; Quarta, A.A. A Sliding Mode Control-Based Guidance Law for a Two-Dimensional Orbit Transfer with Bounded Disturbances. *Actuators* **2023**, *12*, 444. https://doi.org/10.3390/act12120444

Academic Editors: Ti Chen, Junjie Kang, Shidong Xu and Shuo Zhang

Received: 6 November 2023
Revised: 27 November 2023
Accepted: 28 November 2023
Published: 29 November 2023

Copyright: © 2023 by the authors. Licensee MDPI, Basel, Switzerland. This article is an open access article distributed under the terms and conditions of the Creative Commons Attribution (CC BY) license (https://creativecommons.org/licenses/by/4.0/).

1. Introduction

In a preliminary phase of mission design, the use of a state-feedback control law represents a viable option to obtain possible transfer trajectories that may be used as an initial starting point for succeeding (and more refined) analyses. In this context, an interesting approach is based on the use of a rather classical sliding mode control, which is a variable structure (control) method that alters the dynamical behaviour of a nonlinear system through the application of a suitable control signal [1]. In particular, sliding mode control is a basic robust technique, which allows the system trajectory to converge towards the desired target even in the presence of significant perturbations and measurement errors [2,3]. An interesting discussion about the potentialities of a sliding mode control law can be found in the review paper by Hung et al. [4] and in the work by Utkin [5].

The literature about the control of satellites by means of sliding mode techniques is rich, although essentially concentrated on attitude control and terminal guidance maneuvers [6]. In this scenario, Wu et al. [7] investigated the attitude synchronization and tracking problem including model uncertainties, external disturbances, actuator failures, and control torque saturation. By proposing two decentralized sliding mode control laws, that work [7] proved that the control laws guarantee each spacecraft to approach the desired time-varying attitude and angular velocity while maintaining attitude synchronization among the other elements in a typical formation structure. Another example is offered by the work by Massey and Shtessel [8], who adopted a traditional, continuous, high-order sliding mode strategy to control a satellite formation in a robust manner (i.e., compensating for model uncertainties and external disturbances). An adaptive sliding mode tension control method was successfully proposed by Ma et al. [9] for the deployment of tethered satellites, when input tension limitations are taken into account. On the other hand, a terminal sliding mode control law was designed by Liu and Huo [10] for spacecraft rendezvous and docking while considering both model uncertainties and external disturbances, proving that the closed-loop tracking error converges to zero in a finite time. The same problem was also successfully addressed by Dong et al. [11], who constructed a nonsingular terminal sliding surface by introducing a continuous sinusoidal function to solve the inherent singularity problem. More recently, Capello et al. [12] designed two controllers, that is, a first-order sliding mode control for position tracking and a supertwisting second-order sliding mode control for attitude stability, in which the mutual influence was taken into account by the introduction of additional disturbances. Kasaeian et al. [13] presented a robust guidance algorithm to perform a rendezvous between a chaser and a target spacecraft orbiting around the Earth, revealing that sliding mode control guarantees the tracking of the required states and minimum final errors even in the presence of uncertainties and disturbances. Li et al. [14] developed a novel sliding mode control strategy to address the relative position tracking and attitude synchronization problem of spacecraft rendezvous with the requirement of collision avoidance, proving the convergence of relative position and attitude errors even in the presence of external disturbances. Finally, Bassetto et al. [15] discussed how solar sail attitude maneuvers may be designed in a collinear, artificial, equilibrium point by implementing a sliding mode control strategy that uses electrochromic devices as actuators [16–18]. Anyway, there are many other potential feedback control techniques [19–21], to which the interest reader is invited to refer.

In the context of spacecraft trajectory design, a robust state-feedback control law can be used to obtain a possible transfer trajectory that is useful as an initial guess during the subsequent refinement phase [22]. In that case, potential transfer trajectories can be obtained by taking into account the orbit perturbations and the model uncertainties with a reduced computational cost [23]. The aim of this paper is to investigate the potentialities of a sliding mode control strategy in detecting possible trajectories in a typical circle-to-circle orbit transfer scenario, in which the spacecraft propulsion system provides a continuously adjustable and freely steerable propulsive acceleration vector. Among actuators capable of generating variable propulsive acceleration, there are variable thrust ion engines (such as NASA's Evolutionary Xenon Thruster (NEXT)), in which continuous thrust variation can be replaced by a succession of discrete thrust levels that, on average, provide the required propulsive acceleration. For example, NEXT has a total of 40 operating points, with available thrust ranging from 25.5 mN and 236 mN [24]. The proposed approach uses a standard implementation of the sliding mode procedure [1] to obtain a set of preliminary results. In this way, the discussed procedure allows the designer to make a trade-off between the flight time and the required velocity change by selecting the design parameters of the controller. In particular, the simplified control strategy involves three independent parameters of the spacecraft dynamics (on which the resulting propulsive acceleration profile and the characteristics of the transfer trajectory depend), which represent tuning quantities to be selected by the designer. The main limitation of the proposed approach lies in the use of an ideal propulsion system to control the nonlinear dynamics of the spacecraft center of mass. In fact, the time-variation of the thrust vector magnitude, which is an output of the design procedure, can be used a posteriori to check whether the obtained transfer trajectory is compatible with the physical constraints of the thruster, such as the maximum thrust level.

Starting from the simplified scenario, in which the spacecraft orbital dynamics is unaffected by external disturbances or model uncertainties, we firstly discuss how the controller may be tuned by considering the flight time and total velocity change. The procedure is then used to investigate an orbit raising/lowering in the presence of bounded disturbances and measurement errors. The paper is organized as follows. Section 2 presents the mathematical model, i.e., the nonlinear differential equations describing the coplanar orbital motion of a spacecraft around an assigned primary body. Section 3 introduces the sliding mode control technique in its general form, where bounded disturbances are included in the model. Section 4 addresses the design of the sliding mode control law in the simplified case of an unperturbed system. In particular, Section 4 illustrates the time-variation of tracking errors (which can be analytically determined when no disturbance is considered in the mathematical model) and the definition of the control law parameters. The numerical simulations are described in Section 5, while the concluding remarks are drawn in Section 6.

2. Problem Description and Mathematical Model

Consider a spacecraft S that covers a circular parking orbit of radius r_0 around a primary body with center of mass P and gravitational parameter μ. The mission purpose is to transfer the spacecraft to a circular and coplanar target orbit of assigned radius $r_f \neq r_0$ by means of a continuously adjustable (and freely steerable) propulsion system, which gives both a radial (a_r) and a transverse (a_t) component of propulsive acceleration. In this context, the spacecraft two-dimensional dynamics may be described by the classical polar equations of motion [25]:

$$\dot{r} = v_r \tag{1}$$

$$\dot{\theta} = \frac{v_t}{r} \tag{2}$$

$$\dot{v}_r = -\frac{\mu}{r^2} + \frac{v_t^2}{r} + d_r + a_r \tag{3}$$

$$\dot{v}_t = -\frac{v_r v_t}{r} + d_t + a_t \tag{4}$$

where r is the P-S distance and θ is the spacecraft polar angle measured counterclockwise from the P-S line at the initial time $t_0 \triangleq 0$, while v_r (or v_t) is the radial (or transverse) component of the spacecraft velocity vector; see Figure 1. In Equations (3) and (4), the terms $\{d_r, d_t\}$ represent possible unknown bounded disturbance accelerations acting along the radial and transverse directions, with

$$|d_r| \leq D_r \quad , \quad |d_t| \leq D_t \tag{5}$$

where $D_r \geq 0$ and $D_t \geq 0$ are two constant parameters.

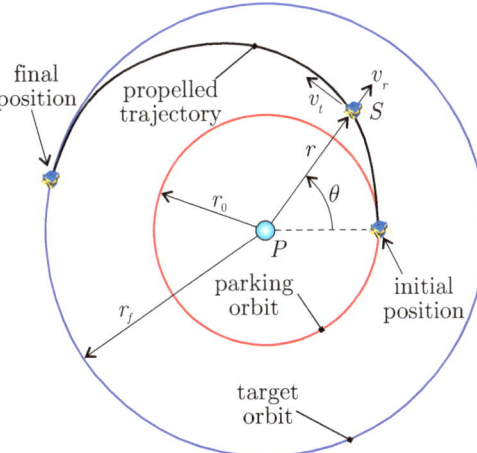

Figure 1. Reference frame and conceptual scheme of the two-dimensional mission scenario.

Bearing in mind that the parking orbit is circular, Equations (1)–(4) are completed by the initial conditions

$$r(t_0) = r_0 \quad , \quad \theta(t_0) = 0 \quad , \quad v_r(t_0) = 0 \quad , \quad v_t(t_0) = \sqrt{\mu/r_0} \tag{6}$$

while the time-variation of the propulsive acceleration components $\{a_r, a_t\}$ are to be found so as to bring the spacecraft states to the desired final values

$$r(t_f) = r_f \quad , \quad v_r(t_f) = 0 \quad , \quad v_t(t_f) = \sqrt{\mu/r_f} \tag{7}$$

within a time interval t_f. Note that the final polar angle $\theta(t_f)$, that is, the angle swept out by the spacecraft during the transfer, is left free.

The spacecraft dynamics is more conveniently rewritten by introducing the dimensionless tracking errors in radial distance (x_1), radial velocity component (x_2), and transverse velocity component (x_3), defined as

$$x_1 \triangleq \frac{r - r_f}{r_0} \equiv \frac{r}{r_0} - \rho \tag{8}$$

$$x_2 \triangleq \frac{v_r}{\sqrt{\mu/r_0}} \tag{9}$$

$$x_3 \triangleq \frac{v_t - \sqrt{\mu/r_f}}{\sqrt{\mu/r_0}} \equiv \frac{v_t}{\sqrt{\mu/r_0}} - \frac{1}{\sqrt{\rho}} \tag{10}$$

where $\rho \triangleq r_f/r_0 \neq 1$ is the dimensionless radius of the target circular orbit. In particular, $\rho \in (0, 1)$ in case of an orbit lowering, whereas $\rho > 1$ in case of an orbit raising. Substituting Equations (8)–(10) into Equations (1), (3), and (4) yields

$$x_1' = x_2 \tag{11}$$

$$x_2' = -\frac{1}{(x_1 + \rho)^2} + \frac{(x_3 + 1/\sqrt{\rho})^2}{x_1 + \rho} + z_r + u_r \tag{12}$$

$$x_3' = -\frac{x_2(x_3 + 1/\sqrt{\rho})}{x_1 + \rho} + z_t + u_t \tag{13}$$

where the prime symbol denotes a derivative taken with respect to the dimensionless time τ, defined as

$$\tau \triangleq \frac{t}{\sqrt{r_0^3/\mu}} \tag{14}$$

with $\tau(t_0) = \tau_0 \triangleq 0$, while

$$u_r \triangleq \frac{a_r}{\mu/r_0^2} \quad , \quad u_t \triangleq \frac{a_t}{\mu/r_0^2} \tag{15}$$

are the two dimensionless control variables, defined as the ratio of the propulsive acceleration components $\{a_r, a_t\}$ to the primary body gravitational acceleration at $r = r_0$. Finally, the two terms $\{z_r, z_t\}$ in Equations (12) and (13) are the dimensionless forms of the disturbance acceleration components, defined as

$$z_r \triangleq \frac{d_r}{\mu/r_0^2} \quad , \quad z_t \triangleq \frac{d_t}{\mu/r_0^2} \tag{16}$$

which, by assumption, satisfy the inequalities

$$|z_r| \leq Z_r \triangleq \frac{D_r}{\mu/r_0^2} \quad , \quad |z_t| \leq Z_t \triangleq \frac{D_t}{\mu/r_0^2} \tag{17}$$

Equations (11)–(13) are integrated with the three initial conditions

$$x_1(\tau_0) = x_{1_0} \triangleq 1 - \rho \quad , \quad x_2(\tau_0) = x_{2_0} \triangleq 0 \quad , \quad x_3(\tau_0) = x_{3_0} \triangleq 1 - 1/\sqrt{\rho} \tag{18}$$

from which it follows that $x_{1_0} < 0$ and $x_{3_0} > 0$ when $\rho > 1$, while $x_{1_0} > 0$ and $x_{3_0} < 0$ when $\rho \in (0, 1)$. Finally, the target states (i.e., the conditions on the target circular orbit) are expressed in a dimensionless form as

$$x_1(\tau_f) = x_{1_f} \triangleq 0 \quad , \quad x_2(\tau_f) = x_{2_f} \triangleq 0 \quad , \quad x_3(\tau_f) = x_{3_f} \triangleq 0 \tag{19}$$

where

$$\tau_f \triangleq \frac{t_f}{\sqrt{r_0^3/\mu}} \tag{20}$$

is the dimensionless flight time. Note that the τ-variation of θ can be obtained by numerically integrating the differential equation

$$\theta' = \frac{x_3 + 1/\sqrt{\rho}}{x_1 + \rho} \tag{21}$$

which is not included in the dynamical system because the final polar angle is left free. However, solving Equation (21) is necessary to obtain the polar trajectory of the spacecraft.

3. State-Feedback Control Design

In this section, a classical sliding mode control law is used to determine the circle-to-circle orbit transfer trajectory. The spacecraft states are brought and maintained on two sliding surfaces, where the system exhibits the desired dynamics of reduced order or one of the states is at its final equilibrium point. More specifically, the first sliding surface is described by the equation

$$s \triangleq x_2 + \lambda x_1 = 0 \tag{22}$$

where $\lambda > 0$ is a dimensionless design parameter, so that the states $\{x_1, x_2\}$ exhibit a first-order dynamics when the system is on that sliding surface. In fact, bearing in mind Equation (11), the condition $s = 0$ implies

$$x_2 = x_1' = -\lambda x_1 \tag{23}$$

from which the τ-variations of the tracking errors $\{x_1, x_2\}$ turn out to be proportional to $e^{-\lambda \tau}$, viz.

$$x_1(\tau) \propto e^{-\lambda \tau} \quad , \quad x_2(\tau) \propto \lambda e^{-\lambda \tau} \tag{24}$$

In other terms, when the system is brought to the sliding surface $s = 0$, both x_1 and x_2 converge exponentially to zero with a convergence rate equal to λ.

Now, in order to bring the system on the sliding surface $s = 0$, it is required that $s' < 0$ when $s > 0$, and $s' > 0$ when $s < 0$. To this end, differentiating s with respect to τ yields

$$s' = x_2' + \lambda x_1' \equiv -\frac{1}{(x_1 + \rho)^2} + \frac{(x_3 + 1/\sqrt{\rho})^2}{x_1 + \rho} + z_r + u_r + \lambda x_2 \tag{25}$$

from which selecting u_r according to the law

$$u_r = \frac{1}{(x_1+\rho)^2} - \frac{(x_3+1/\sqrt{\rho})^2}{x_1+\rho} - \lambda x_2 - \delta \operatorname{sign}(s) \quad (26)$$

where $\operatorname{sign}(\square)$ is the signum function and δ is given by

$$\delta \triangleq Z_r + K > 0 \quad (27)$$

in which $K > 0$ is a dimensionless design parameter, one obtains

$$s' = z_r - (Z_r + K) \operatorname{sign}(s) \quad (28)$$

In this case, $s < 0$ implies $s' = z_r + Z_r + K > 0$ and $s > 0$ implies $s' = z_r - Z_r - K < 0$, while $s = 0$ implies $s' = z_r$, that is, the perturbative term z_r forces the system to leave the sliding surface $s = 0$ once it has been reached.

The second sliding surface is the plane $x_3 = 0$; see Equation (10). Note that the system is driven to the sliding surface $x_3 = 0$ if $x_3' < 0$ when $x_3 > 0$, and if $x_3' > 0$ when $x_3 < 0$. In this context, if the control parameter u_t is selected as

$$u_t = \frac{x_2 \left(x_3 + 1/\sqrt{\rho}\right)}{x_1 + \rho} - \gamma \operatorname{sign}(x_3) \quad (29)$$

with

$$\gamma \triangleq Z_t + c > 0 \quad (30)$$

where $c > 0$ is a dimensionless design parameter, then

$$x_3' = z_t - (Z_t + c) \operatorname{sign}(x_3) \quad (31)$$

In this case, $x_3 < 0$ implies $x_3' = z_t + Z_t + c > 0$, $x_3 > 0$ implies $x_3' = z_t - Z_t - c < 0$, while $x_3 = 0$ implies $x_3' = z_t$, that is, the perturbative term z_t moves the system away from the sliding surface $x_3 = 0$ once it has been reached.

Disturbance Modeling

This section gives a brief description of the source of disturbances (or uncertainties) that will be included in the numerical simulations. The first one is related to the state measurement. In fact, when applying a state-feedback control law, it is necessary to verify whether and how measurement errors or low-frequency sampling affect the control effectiveness. The measured states, denoted as $\{\tilde{x}_1, \tilde{x}_2, \tilde{x}_3\}$, are the sum of a true value x_i and a measurement error X_i, that is,

$$\tilde{x}_1 = x_1 + X_1 \quad (32)$$

$$\tilde{x}_2 = x_2 + X_2 \quad (33)$$

$$\tilde{x}_3 = x_3 + X_3 \quad (34)$$

where X_i is a zero-mean random variable with normal distribution and standard deviation σ_i.

The second source of disturbance is related to the approximation of the signum function in Equations (26) and (29) with a sigmoid-like function. Note, in fact, that the change in sign of u_r (or u_t) each time the system crosses the sliding surface $s = 0$ (or $x_3 = 0$) gives rise to a chattering behaviour, which is typical of the sliding mode control. Such a phenomenon must be mitigated to prevent the switching frequency of the control signals from being too high and, therefore, not applicable. A viable option is to implement a pseudo-sliding mode control [26], which consists of smoothing the discontinuity in the signum function to obtain an arbitrarily close but continuous approximation. One possibility is to approximate the signum function with the sigmoid-like function [26]

$$\mathcal{S} = \mathcal{S}(x) \triangleq \frac{x}{|x| + \kappa} \tag{35}$$

where κ is an arbitrarily small positive scalar. Note that $\mathcal{S}(x) \to \text{sign}(x)$ as $\kappa \to 0$. Using such a pseudo-sliding mode control, however, causes sliding to no longer take place because the (continuous) control only drives the states to a neighbourhood of the switching surfaces [26].

Accordingly, introducing the measured states in Equations (26) and (29) and substituting $\text{sign}(x)$ with $\mathcal{S}(x)$, the control variables become

$$u_r = \frac{1}{(\tilde{x}_1 + \rho)^2} - \frac{(\tilde{x}_3 + 1/\sqrt{\rho})^2}{\tilde{x}_1 + \rho} - \lambda \tilde{x}_2 - \delta \mathcal{S}(\tilde{s}) \tag{36}$$

$$u_t = \frac{\tilde{x}_2 (\tilde{x}_3 + 1/\sqrt{\rho})}{\tilde{x}_1 + \rho} - \gamma \mathcal{S}(\tilde{x}_3) \tag{37}$$

where $\tilde{s} \triangleq \tilde{x}_2 + \lambda \tilde{x}_1$. Measurement errors, low-frequency sampling, and the approximation of the signum function with the sigmoid-like function $\mathcal{S}$ are all treated as disturbance sources. In practice, this situation is equivalent to using ideal sensors and actuators (i.e., sensors capable of measuring the actual states with continuity and actuators capable of adjusting their control signals with continuity) and to perturbing the system with the following (bounded) disturbance accelerations:

$$z_r = \frac{1}{(x_1 + \rho)^2} - \frac{1}{(\tilde{x}_1 + \rho)^2} - \frac{(x_3 + 1/\sqrt{\rho})^2}{x_1 + \rho} + \frac{(\tilde{x}_3 + 1/\sqrt{\rho})^2}{\tilde{x}_1 + \rho} +$$
$$- \lambda (x_2 - \tilde{x}_2) - \delta \left[\text{sign}(s) - \mathcal{S}(\tilde{s}) \right] \tag{38}$$

$$z_t = \frac{x_2 (x_3 + 1/\sqrt{\rho})}{x_1 + \rho} - \frac{\tilde{x}_2 (\tilde{x}_3 + 1/\sqrt{\rho})}{\tilde{x}_1 + \rho} - \gamma \left[\text{sign}(x_3) - \mathcal{S}(\tilde{x}_3) \right] \tag{39}$$

4. Case of an Unperturbed System

The control law described by Equations (26) and (29) takes a simpler form in the case of an unperturbed system, which allows some useful analytical relationships to be found in such a simplified mission scenario. Accordingly, in this section, we analyze the evolution of the tracking errors and address the control law design problem with the significant assumption that $Z_r = Z_t = 0$. In this simplified case, Equations (28) and (31) become

$$s' = -K \, \text{sign}(s) \tag{40}$$

$$x_3' = -c \, \text{sign}(x_3) \tag{41}$$

This means that in the absence of perturbative terms, the value of s (or x_3) approaches zero linearly with respect to τ, and once the sliding surface $s = 0$ (or $x_3 = 0$) is reached for the first time, the term s (or x_3) remains stationary at zero. When Equations (40) and (41) are integrated with respect to the dimensionless time τ, one obtains the τ-variations of s and x_3 before reaching the sliding surfaces $s = 0$ and $x_3 = 0$, respectively. The result is

$$s(\tau) = s_0 - \text{sign}(s_0) \, K \, \tau \tag{42}$$

$$x_3(\tau) = x_{3_0} - \text{sign}(x_{3_0}) \, c \, \tau \tag{43}$$

where $s_0 \triangleq s(\tau_0)$ can be written, according to Equations (18) and (22), as

$$s_0 = x_{2_0} + \lambda\, x_{1_0} \equiv \lambda\, (1 - \rho) \tag{44}$$

while x_{3_0} is given by the last of Equation (18) as a function of ρ. Note that $\{K, c\}$ represent a sort of approach speed to the two sliding surfaces.

The value of τ at which the system reaches the sliding surface $s = 0$ (i.e., $\tau = \tau_s$) or the sliding surface $x_3 = 0$ (i.e., $\tau = \tau_{x_3}$) can be expressed in a compact form using Equations (42) and (43). In fact, enforcing the condition $s = 0$ in Equation (42) gives

$$s_0 - \text{sign}(s_0)\, K\, \tau_s \triangleq 0 \tag{45}$$

from which

$$\tau_s \triangleq \frac{\lambda\, |1 - \rho|}{K} \tag{46}$$

while the condition $x_3 = 0$ in Equation (43) gives

$$x_{3_0} - \text{sign}(x_{3_0})\, c\, \tau_{x_3} \triangleq 0 \tag{47}$$

from which

$$\tau_{x_3} \triangleq \frac{|1 - 1/\sqrt{\rho}|}{c} \tag{48}$$

The value of τ_{x_3} may be written as a function of τ_s in a more convenient way by introducing the dimensionless parameter $\beta > 0$ such that

$$\tau_{x_3} = \beta\, \tau_s \tag{49}$$

Observing that β is a redundant parameter, it may be used in place of c, which can be expressed as a function of $\{K, \lambda, \beta, \rho\}$ as

$$c \triangleq \frac{K\left(1 - 1/\sqrt{\rho}\right)}{\lambda\, \beta\, (\rho - 1)} \tag{50}$$

4.1. The τ-Variation of Tracking Errors and Controls

The τ-variation of the tracking errors is now calculated, thus allowing the expressions of u_r and u_t to be determined through Equations (26) and (29) by simply setting $\delta = K$ and $\gamma = c$. To that end, the differential equation governing the τ-evolution of x_1 is found by substituting Equations (11) and (22) into Equation (42) and bearing in mind Equations (44)–(46), viz.

$$x_1' + \lambda\, x_1 = \begin{cases} \lambda\, (1 - \rho) - \text{sign}(1 - \rho)\, K\, \tau & \text{if } \tau < \tau_s \\ 0 & \text{otherwise} \end{cases} \tag{51}$$

Integrating Equation (51) with respect to τ with the initial condition $x_1(\tau_0) = 1 - \rho$ (see Equation (18)) gives the τ-variation of x_1, that is,

$$\frac{x_1(\tau)}{\text{sign}(1 - \rho)} = \begin{cases} \dfrac{K}{\lambda^2}\left(1 - e^{-\lambda \tau} - \lambda\, \tau\right) + 1 - \rho & \text{if } \tau < \tau_s \\ \dfrac{K}{\lambda^2}\left(1 - e^{-\lambda\, \tau_s}\right) e^{-\lambda\, (\tau - \tau_s)} & \text{otherwise} \end{cases} \tag{52}$$

The τ-variation of x_2 is instead obtained by deriving Equation (52) with respect to τ (see Equation (11)), that is,

$$\frac{x_2(\tau)}{\text{sign}(1-\rho)} = \begin{cases} \frac{K}{\lambda}\left(e^{-\lambda\tau}-1\right) & \text{if } \tau < \tau_s \\ \frac{K}{\lambda}\left(e^{-\lambda\tau_s}-1\right)e^{-\lambda(\tau-\tau_s)} & \text{otherwise} \end{cases} \tag{53}$$

Finally, the τ-variation of x_3 is governed by the differential equation

$$x_3' = \begin{cases} -c\,\text{sign}(x_3) & \text{if } \tau < \beta\tau_s \\ 0 & \text{otherwise} \end{cases} \tag{54}$$

which must be solved recalling the initial condition $x_3(\tau_0) = 1 - 1/\sqrt{\rho}$ (see Equation (18)), and the result is

$$\frac{x_3(\tau)}{\text{sign}(1-1/\sqrt{\rho})} = \begin{cases} |1 - 1/\sqrt{\rho}| - c\tau & \text{if } \tau < \beta\tau_s \\ 0 & \text{otherwise} \end{cases} \tag{55}$$

Note that in the absence of perturbative terms, the maximum values of $|x_1|$ and $|x_3|$ occur when $\tau = \tau_0$, that is,

$$\max(|x_1|) = |x_{1_0}| \equiv |1-\rho| \tag{56}$$

$$\max(|x_3|) = |x_{3_0}| \equiv |1 - 1/\sqrt{\rho}| \tag{57}$$

while the maximum value of $|x_2|$ (which corresponds to the maximum of $|v_r|$) is reached when $\tau = \tau_s$, viz.

$$\max(|x_2|) = \frac{K}{\lambda}|e^{-\lambda\tau_s}-1| \tag{58}$$

The dimensionless propulsive acceleration components $\{u_r, u_t\}$ in absence of perturbative terms are simply obtained by substituting Equations (52), (53), and (55) into Equations (26) and (29) and setting $\delta = K$ and $\gamma = c$. Those expressions, which are here omitted for the sake of conciseness, change according to whether $\beta < 1$, $\beta = 1$, or $\beta > 1$. In particular, u_r exhibits a discontinuity equal to $K\,\text{sign}(1-\rho)$ when $\tau = \tau_s$, whereas u_t exhibits a discontinuity equal to $c\,\text{sign}(1-\sqrt{\rho})$ when $\tau = \beta\tau_s$. Accordingly, if $\beta \neq 1$, the profile of the magnitude $u \triangleq \sqrt{u_r^2 + u_t^2}$ has two discontinuities (one at $\tau = \tau_s$, the other at $\tau = \beta\tau_s$). Otherwise (i.e., when $\beta = 1$), the profile of u presents a single discontinuity at $\tau = \tau_s$.

4.2. Control Parameter Selection

For a given value of ρ, the design of the sliding mode control law amounts to selecting the values of the triplet $\{K, \lambda, \beta\}$. The previous expressions allow the flight time and the total velocity change to be determined and the dimensionless parameters in the control law to be established, according to arbitrary criteria. More precisely, when the spacecraft orbital dynamics is unaffected by external disturbances or model uncertainties, the dimensionless flight time τ_f and the total velocity change Δv can be calculated with analytical expressions or graphic plots that only depend on the design parameters $\{K, \lambda, \beta\}$.

For example, the value of τ_f can be obtained by assuming that the orbit transfer terminates when the tracking errors x_1 and x_2 are sufficiently close to zero. To that end, the value of τ_f is defined as the instant at which the exponent $\lambda(\tau - \tau_s)$ in Equations (52) and (53) satisfies the equality $\lambda(\tau_f - \tau_s) = n$, for an assigned value of $n \in \mathbb{R}^+$. In this context, using Equation (46), one obtains

$$\tau_f = \tau_s + \frac{n}{\lambda} \equiv \frac{\lambda|1-\rho|}{K} + \frac{n}{\lambda} \qquad (59)$$

Note that τ_f can be minimized with respect to λ by enforcing the necessary condition

$$\frac{\partial \tau_f}{\partial \lambda} = 0 \qquad (60)$$

in Equation (59), from which

$$\lambda = \lambda^\star \triangleq \sqrt{\frac{nK}{|1-\rho|}} \qquad (61)$$

so that, by assuming $\lambda = \lambda^\star$, the expression of the dimensionless flight time becomes

$$\tau_f = 2\sqrt{\frac{n|1-\rho|}{K}} \equiv 2\tau_s|_{\lambda=\lambda^\star} \qquad (62)$$

A suitable value of n may be chosen by evaluating the tracking error x_1 at the final time $\tau = \tau_f$, that is,

$$x_1(\tau_f) = \frac{e^{-n}(1-e^{-n})}{n} x_{1_0} \qquad (63)$$

Figure 2, which describes the variation of $x_1(\tau_f)/x_{1_0}$ with n when $\lambda = \lambda^\star$, shows that a value of $n = 4$ (when $x_1(\tau_f)/x_{1_0} \simeq 0.0045$) is reasonable from a practical point of view. In fact, the percentage error in orbital radius, that is, the function

$$\epsilon_r \triangleq \frac{|r(t_f) - r_f|}{r_f} \times 100 \simeq \frac{0.45|1-\rho|}{\rho} \qquad (64)$$

is less than 1% when $n = 4$ and $\rho > 0.310$; see Figure 3. Therefore, it is assumed that $n = 4$ in the rest of the paper.

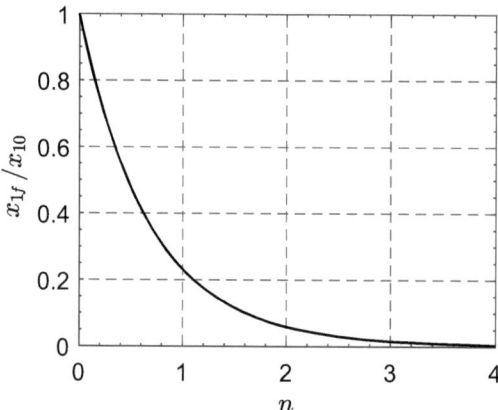

Figure 2. Variation in x_{1_f}/x_{1_0} with n when $\lambda = \lambda^\star$.

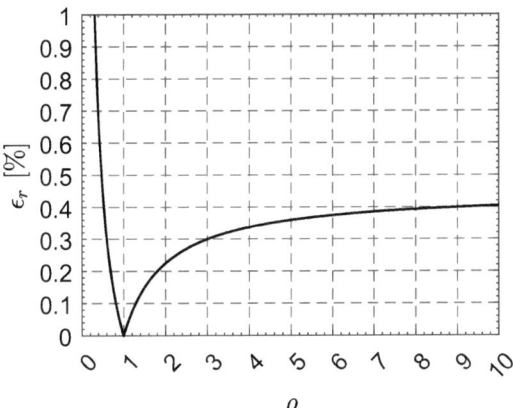

Figure 3. Percentage error in (final) orbital radius as a function of ρ when $n = 4$.

According to Equations (61) and (62), the expressions of $\lambda^\star$ and τ_f when $n = 4$ become

$$\lambda^\star = 2\sqrt{\frac{K}{|1-\rho|}} \tag{65}$$

$$\tau_f = 4\sqrt{\frac{|1-\rho|}{K}} \equiv \frac{8}{\lambda^\star} \tag{66}$$

Note that Equation (66) relates the flight time τ_f to the value of λ necessary to minimize the flight time for fixed values of K and ρ. Such a value of λ is a function of K and ρ, as described by Equation (65). This means that for given values of ρ and τ_f, λ can be chosen by reversing Equation (66), that is, by setting $\lambda = \lambda^\star \equiv 8/\tau_f$. In this case, the value of K is related to ρ and λ (or to ρ and τ_f) through Equation (65), and Figure 4 shows the variation in τ_f with K and ρ.

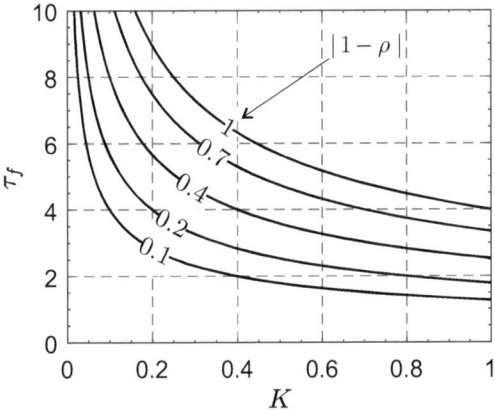

Figure 4. Variation in τ_f with $\{K, \rho\}$ when $\lambda = \lambda^\star$ and $n = 4$.

The value of β can be chosen with the aid of another parameter that usually determines the transfer performance. More precisely, β may be related to the total velocity change Δv of the transfer, defined as

$$\Delta v \triangleq \int_0^{\tau_f} u \, d\tau \tag{67}$$

Note that $\beta \in (0, 2]$, since $\beta \to 0$ (or $\beta = 2$) means that the sliding surface $x_3 = 0$ is reached at the beginning (or at the end) of the transfer; see Equations (49) and (62). Figure 5 shows the values of β (referred to as $\beta^\star$) that minimize the total velocity change when $\lambda = \lambda^\star$, $n = 4$, and $\rho = \{0.723, 1.524\}$ (the same values of ρ that will be used for some numerical applications of the proposed control law) as a function of $K \in (0, 1]$. In fact, K corresponds to the magnitude of the discontinuity of u_r when $\tau = \tau_s$ (see Section 4.1), and a value of K greater than 1 would imply a discontinuity of $|a_r|$ greater than the gravitational acceleration on the parking orbit. Figure 6, instead, shows the variation in Δv with K when $\lambda = \lambda^\star$, $n = 4$, and $\beta = \beta^\star$. Note that the function $\Delta v|_{\beta=\beta^\star}(K)$ exhibits a global minimum. When $\rho = 0.723$, such a minimum is reached when $K \simeq 0.097$ and the corresponding values of Δv and $\beta^\star$ are $\Delta v \simeq 0.357$ and $\beta^\star \simeq 1.368$, respectively. If, instead, $\rho = 1.524$, such a minimum is reached when $K \simeq 0.032$ and the corresponding values of Δv and $\beta^\star$ are $\Delta v \simeq 0.324$ and $\beta^\star \simeq 1.242$, respectively.

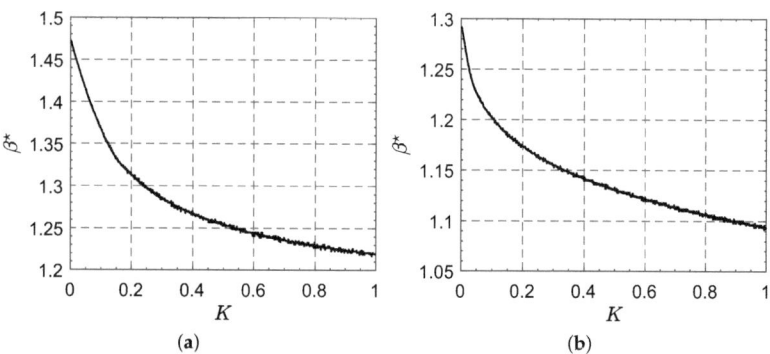

Figure 5. Variation in $\beta^\star$ with K when $\lambda = \lambda^\star$ and $n = 4$. (**a**) $\rho = 0.723$; (**b**) $\rho = 1.524$.

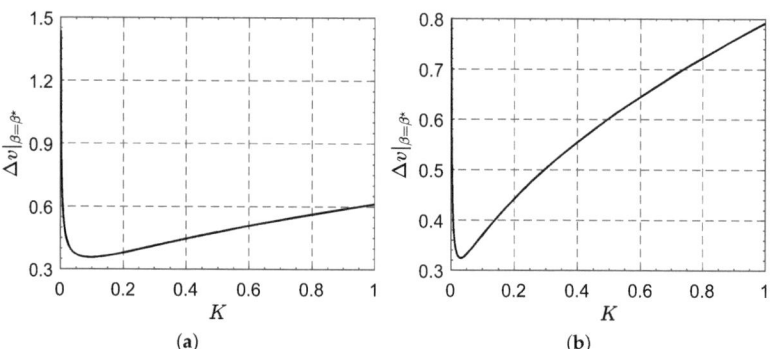

Figure 6. Variation in $\Delta v|_{\beta=\beta^\star}$ with K when $\lambda = \lambda^\star$ and $n = 4$. (**a**) $\rho = 0.723$; (**b**) $\rho = 1.524$.

In essence, the design of the control law only requires the choice of the single parameter K. For example, we have already seen that a value of K may be determined using Equations (65) and (66) by fixing the total flight time τ_f, or using Figure 6 by looking for the value of K that minimizes the function $\Delta v|_{\beta=\beta^\star}(K)$, so that K may be thought of as sort of trade-off parameter, as discussed in the next section.

5. Numerical Simulations and Mission Application

The proposed control strategy is now used to analyze two classical circle-to-circle interplanetary transfers. In particular, the radius of the circular parking orbit is $r_0 = r_\oplus \triangleq 1$ au, which is consistent with a spacecraft that leaves the Earth's sphere of influence using a parabolic escape trajectory, with the simplifying assumption that the Earth's heliocentric orbit is circular. The radii of the target orbits are $r_f = \{0.723, 1.524\}$ au, so the analyzed mission scenarios describe simplified ephemeris-free Earth–Venus and Earth–Mars orbit transfers.

Bearing in mind Equations (32)–(34), it is assumed that (i) the sensors measure the states once per day; (ii) $\sigma_1 = \sigma_2 = \sigma_3 = 10^{-4}$, which means that the measurement error in $\{x_1, x_2, x_3\}$ is less than 0.01% with a probability of 68.3%; and (iii) the sigmoid-like function $\mathcal{S} = \mathcal{S}(x)$ of Equation (35) is obtained with $\kappa = 10^{-2}$. Although the numerical simulations consider measurement errors, low-frequency sampling, and the approximation of the signum function with the sigmoid-like function, the parameters used in the control law can be those found in Section 4.2 thanks to the robustness of the proposed approach.

For example, assume that $\lambda = \lambda^\star$, $n = 4$, $\beta = \beta^\star$, and select K such that $\Delta v|_{\beta=\beta^\star}$ is minimized (the corresponding value of K will be referred to as K_v), so that according to Figure 6, one has $K_v \simeq 0.0969$ (or $K_v \simeq 0.0320$) when $\rho = 0.723$ (or $\rho = 1.524$). The numerical simulations give a flight time of about 394 days (or 949 days) in the Earth–Venus (or Earth–Mars) mission scenario. Moreover, Figure 7 shows the corresponding (two-dimensional) heliocentric trajectories, while Figure 8 shows the time-variations of the propulsive acceleration components $\{a_r, a_t\}$. In particular, each black dot in Figure 8 corresponds to one day (i.e., to the sampling period of the states), while the red lines show the propulsive acceleration components in case of ideal sensors and actuators (that is, when only the approximation of the signum function with the sigmoid-like function is taken into account).

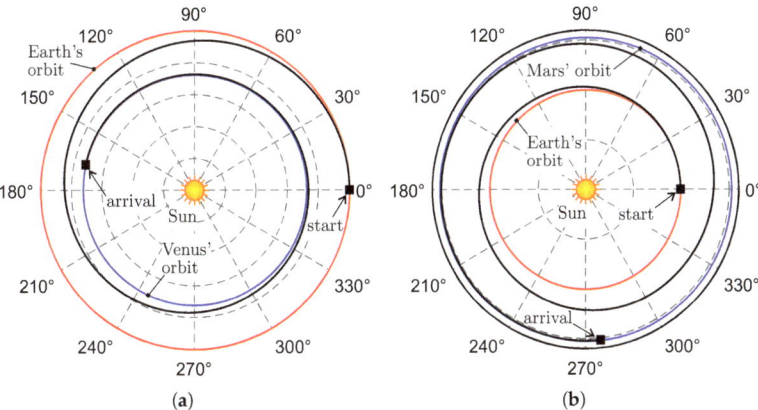

Figure 7. Transfer trajectories when $K = K_v$ in the two interplanetary mission scenarios. (**a**) Earth–Venus case; (**b**) Earth–Mars case.

A second case considered in the simulations is when the flight time is assigned, $\lambda = \lambda^\star \equiv 8/\tau_f$, $n = 4$, and $\beta = \beta^\star$. For example, by assuming that the transfer time coincides with the Hohmann transfer one, the values of τ_f become

$$\tau_f = \tau_H \triangleq \pi \sqrt{\frac{(1+\rho)^3}{8}} \simeq \begin{cases} 2.512 & \text{if } \rho = 0.723 \\ 4.454 & \text{if } \rho = 1.524 \end{cases} \quad (68)$$

which correspond to a flight time of 146 days in the Earth–Venus scenario and to 259 days in Earth–Mars case. In these cases, by using Equation (66), the values of $\lambda^\star$ and K, respectively referred to as $\lambda_H^\star$ and K_H, are given by

$$\lambda_H^\star = \frac{8}{\tau_H} \simeq \begin{cases} 3.185 & \text{if } \rho = 0.723 \\ 1.796 & \text{if } \rho = 1.524 \end{cases} \quad (69)$$

$$K_H = |1 - \rho| \left(\frac{\lambda_H^\star}{2}\right)^2 \simeq \begin{cases} 0.702 & \text{if} \quad \rho = 0.723 \\ 0.423 & \text{if} \quad \rho = 1.524 \end{cases} \qquad (70)$$

while the values of $\beta^\star$ are chosen by using Figure 5 to minimize the total velocity change, viz.

$$\beta^\star \simeq \begin{cases} 1.234 & \text{if} \quad \rho = 0.723 \\ 1.138 & \text{if} \quad \rho = 1.524 \end{cases} \qquad (71)$$

In this context, Figure 9 shows the interplanetary transfer trajectories, while Figure 10 collects the time-variations of the propulsive acceleration components $\{a_r, a_t\}$ for the two mission scenarios.

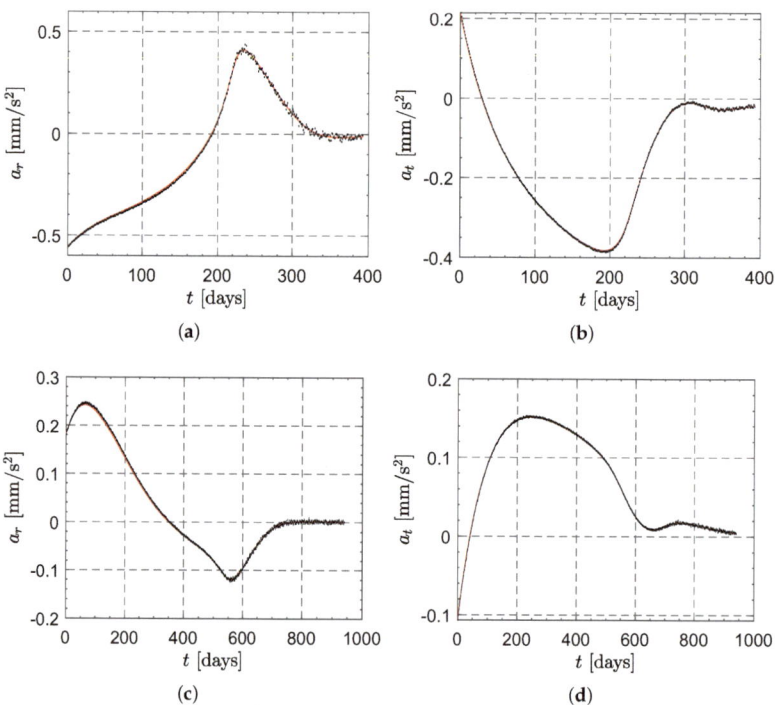

Figure 8. Time-variations of a_r and a_t when $K = K_v$ in two typical interplanetary mission scenarios. (**a**) Radial component, Earth–Venus case; (**b**) transverse component, Earth–Venus case; (**c**) radial component, Earth–Mars case; (**d**) transverse component, Earth–Mars case.

Once the control parameters are selected and the transfer trajectory is obtained, it is possible to evaluate the time-variation of the magnitude of the propulsive acceleration vector $a \triangleq \sqrt{a_r^2 + a_t^2}$ during the transfer. Figure 11 shows the values of a as a function of time in the four cases previously described. The curves depicted in that figure can be used to evaluate, a posteriori, the feasibility of the obtained transfer trajectory according to the actual thruster installed on board. In particular, Figure 11a (or Figure 11b) indicates that the maximum value of a during an Earth-Venus (or Earth-Mars) transfer with $K = K_v$ is about $0.6 \, \text{mm/s}^2$ (or $0.25 \, \text{mm/s}^2$), while Figure 11c (or Figure 11d) shows that the maximum value of a is roughly $4.2 \, \text{mm/s}^2$ (or $2.5 \, \text{mm/s}^2$) for an Earth–Venus (or Earth–Mars) case when $\tau_f = \tau_H$. Therefore, if, for example, the installed thruster gives a maximum propulsive acceleration of $0.3 \, \text{mm/s}^2$, when $K = K_v$, one concludes that the transfer trajectory obtained in the Earth–Mars case can be theoretically flown, while the result in the Earth–Venus scenario gives a trajectory that violates the propulsive constraint. In the latter case (that is, in the Earth–Venus scenario with $K = K_v$), the designer could suitably change the control law parameters in order to reduce the maximum value of a reached during the transfer. For example, when $K = 0.03$ and $\beta = 1.48$, the maximum value of a reduces to about 0.29 mm/s^2, while the flight time rises to roughly 1413 days.

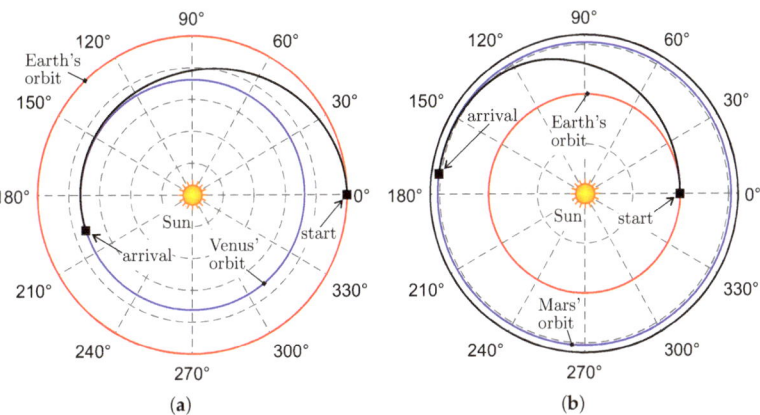

Figure 9. Transfer trajectories when $\tau_f = \tau_H$ in two typical interplanetary mission scenarios. (**a**) Earth–Venus case; (**b**) Earth–Mars case.

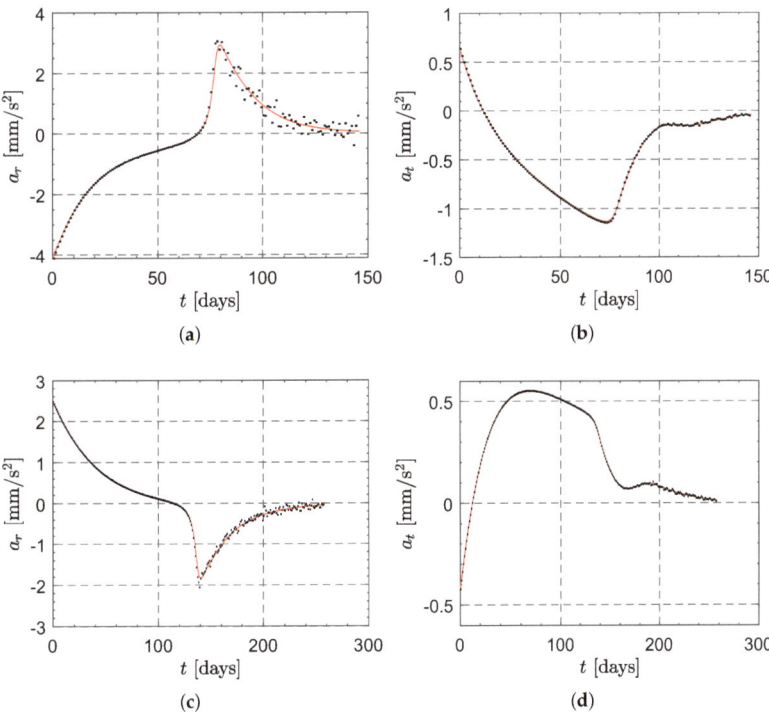

Figure 10. Time-variations of a_r and a_t when $\tau_f = \tau_H$ in two typical interplanetary mission scenarios. (**a**) Radial component, Earth–Venus case; (**b**) transverse component, Earth–Venus case; (**c**) radial component, Earth–Mars case; (**d**) transverse component, Earth–Mars case.

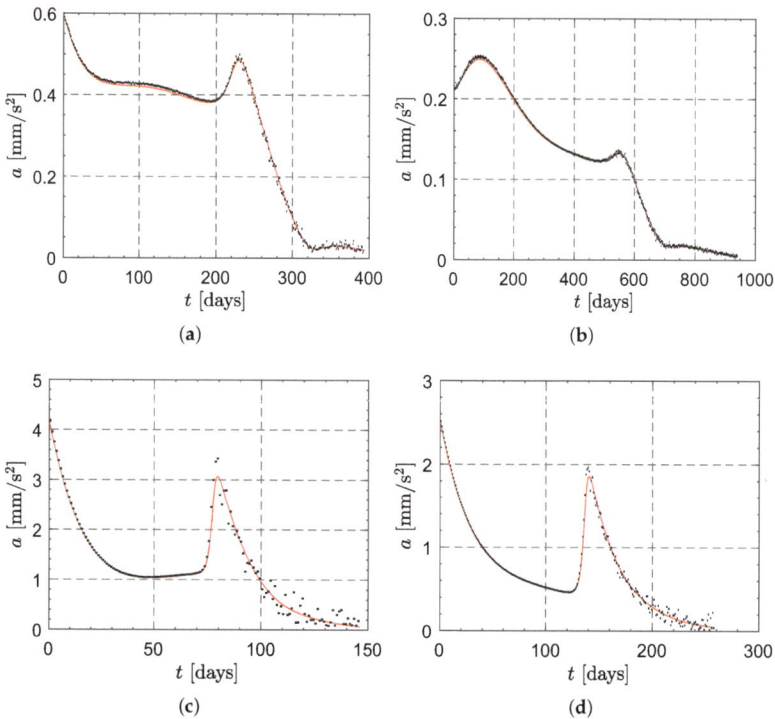

Figure 11. Time-variations of a in the four test mission scenarios. (**a**) Earth–Venus case, $K = K_v$; (**b**) Earth–Mars case, $K = K_v$; (**c**) Earth–Venus case, $\tau_f = \tau_H$; (**d**) Earth–Mars case, $\tau_f = \tau_H$.

The previous results may be easily extended to trade-off solutions between the flight time and the total velocity change necessary to complete the transfer. Recall in fact that, for a given value of ρ, the flight time is a function of K according to Equation (66), while the total velocity change depends on K as Figure 6 shows. Therefore, it is possible to plot the value of $\Delta v|_{\beta=\beta^*}$ as a function of τ_f. The results are shown in Figure 12, where K ranges within the interval $[K_v, 1]$ and the black squares correspond to the cases in which the flight time equals the Hohmann transfer time. Figure 12 represents a simple and effective means to identify reasonable compromise solutions, which are useful in a preliminary analysis of the trajectory design.

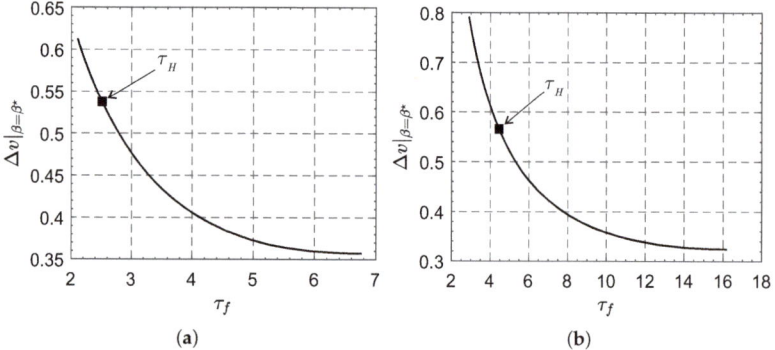

Figure 12. Trade-off solution between flight time and total velocity change in two typical interplanetary mission scenarios. (**a**) Earth–Venus case; (**b**) Earth–Mars case.

6. Conclusions

In this paper, we have investigated the capabilities of a sliding mode control technique of rapidly generating a possible trajectory for a spacecraft in a typical circle-to-circle orbit transfer mission scenario. Such a control strategy, which is a robust technique usually used for controlling nonlinear systems affected by disturbances, has been employed here to obtain simple mathematical relations and graphs that allow the designer to estimate the (possible) transfer trajectory characteristics as a function of few tuning parameters.

The proposed approach has shown to be effective even in the presence of bounded disturbances due to measurement errors and low-frequency sampling. It may be effectively employed in an early phase of trajectory planning, that is, just before the usual refinement phase that provides the nominal spacecraft trajectory to be tracked during the transfer. In particular, the discussed approach has the scope of reducing the complexity of the mathematical model and the computational cost required to obtain a possible solution to the transfer problem. However, rather strong simplifying assumptions have been adopted, such as the use of a sort of ideal thruster with a freely steerable thrust vector, or the definition of the error in terms of the desired states. A more accurate estimate of the actual propulsive acceleration profile can be obtained by relaxing some of those assumption. This aspect represents the natural extension of this work.

Author Contributions: Conceptualization, M.B.; methodology, M.B. and G.M.; software, M.B.; writing—original draft preparation, M.B., A.A.Q. and G.M.; writing—review and editing, all. All authors have read and agreed to the published version of the manuscript.

Funding: This research received no external funding.

Institutional Review Board Statement: Not applicable.

Informed Consent Statement: Not applicable.

Data Availability Statement: Data are contained within the article.

Conflicts of Interest: The authors declare no conflicts of interest.

Notation

a	propulsive acceleration magnitude (mm/s^2)
a_r	radial component of the propulsive acceleration (mm/s^2)
a_t	transverse component of the propulsive acceleration (mm/s^2)
c	speed of approach to $x_3 = 0$ when $Z_t = 0$
d_r	radial component of the disturbance acceleration (mm/s^2)
D_r	maximum of $|d_r|$, (mm/s^2)
d_t	transverse disturbance acceleration, (mm/s^2)
D_t	maximum of $|d_t|$ (mm/s^2)
K	speed of approach to $s = 0$ when $Z_r = 0$
K_H	value of K corresponding to $\tau_f = \tau_H$
K_v	value of K that minimizes the total velocity change
n	dimensionless positive parameter; see Equation (59)
P	primary body center of mass
r	orbital radius (au)
s	linear combination of $\{x_1, x_2\}$; see Equation (22)
S	spacecraft center of mass
$\mathcal{S}$	sigmoid-like function; see Equation (35)
t	time (days)
u	magnitude of command signal
u_r	dimensionless value of a_r
u_t	dimensionless value of a_t
v_r	radial velocity component (km/s)
v_t	transverse velocity component (km/s)
X	normally distributed random number
$\{x_1, x_2, x_3\}$	dimensionless tracking errors along $\{r, v_r, v_t\}$

z_r	dimensionless radial component of the disturbance acceleration
Z_r	maximum magnitude of z_r
z_t	dimensionless transverse component of the disturbance acceleration
Z_t	maximum magnitude of z_t
α_t	thrust angle (rad)
β	ratio of τ_{x_3} to τ_s
δ	auxiliary parameter; see Equation (27)
Δv	dimensionless velocity change
γ	auxiliary parameter; see Equation (30)
ϵ_r	percentage error in orbital radius
θ	polar angle (rad)
λ	convergence rate of x_1 and x_2
$\lambda_H^\star$	value of $\lambda^\star$ corresponding to $\tau_f = \tau_H$
μ	primary body gravitational parameter (km^3/s^2)
ρ	ratio of r_f to r_0
σ	specific standard deviation
τ	dimensionless time
τ_H	dimensionless Hohmann transfer time
τ_s	time to reach the condition $s = 0$
τ_{x_3}	time to reach the condition $x_3 = 0$
Subscripts	
0	initial
f	final
Superscripts	
$\cdot$	derivative with respect to t
$\prime$	derivative with respect to τ
$\star$	design value
$\sim$	measured

References

1. Slotine, J.J.E.; Li, W. *Applied Nonlinear Control*; Prentice-Hall: Englewood Cliffs, NJ, USA, 1991; Chapter 7, pp. 276–310.
2. Gambhire, S.J.; Kishore, D.R.; Londhe, P.S.; Pawar, S.N. Review of sliding mode based control techniques for control system applications. *Int. J. Dyn. Control* **2020**, *9*, 363–378. [CrossRef]
3. Qureshi, M.S.; Das, S.; Swarnkar, P.; Gupta, S. Design and Implementation of Sliding Mode Control for Uncertain Systems. *Mater. Today Proc.* **2018**, *5*, 4299–4308. [CrossRef]
4. Hung, J.; Gao, W.; Hung, J. Variable structure control: A survey. *IEEE Trans. Ind. Electron.* **1993**, *40*, 2–22. [CrossRef]
5. Utkin, V. Sliding mode control design principles and applications to electric drives. *IEEE Trans. Ind. Electron.* **1993**, *40*, 23–36. [CrossRef]
6. Chen, T.; Shan, J.; Wen, H.; Xu, S. Review of attitude consensus of multiple spacecraft. *Astrodynamics* **2022**, *6*, 329–356. [CrossRef]
7. Wu, B.; Wang, D.; Poh, E.K. Decentralized sliding-mode control for spacecraft attitude synchronization under actuator failures. *Acta Astronaut.* **2014**, *105*, 333–343. [CrossRef]
8. Massey, T.; Shtessel, Y. Continuous Traditional and High-Order Sliding Modes for Satellite Formation Control. *J. Guid. Control Dyn.* **2005**, *28*, 826–831. [CrossRef]
9. Ma, Z.; Sun, G. Adaptive sliding mode control of tethered satellite deployment with input limitation. *Acta Astronaut.* **2016**, *127*, 67–75. [CrossRef]
10. Liu, S.; Huo, W. Terminal sliding mode control for space rendezvous and docking. In Proceedings of the International Conference on Computational Intelligence and Communication Networks (CICN), Jabalpur, India, 12–14 December 2015. [CrossRef]
11. Dong, J.; Li, C.; Jiang, B.; Sun, Y. Fixed-time nonsingular terminal sliding mode control for spacecraft rendezvous. In Proceedings of the 29th Chinese Control and Decision Conference (CCDC), Chongqing, China, 28–30 May 2017. [CrossRef]
12. Capello, E.; Punta, E.; Dabbene, F.; Guglieri, G.; Tempo, R. Sliding-mode control strategies for rendezvous and docking maneuvers. *J. Guid. Control Dyn.* **2017**, *40*, 1481–1488. [CrossRef]
13. Kasaeian, S.A.; Assadian, N.; Ebrahimi, M. Sliding mode predictive guidance for terminal rendezvous in eccentric orbits. *Acta Astronaut.* **2017**, *140*, 142–155. [CrossRef]
14. Li, Q.; Yuan, J.; Wang, H. Sliding mode control for autonomous spacecraft rendezvous with collision avoidance. *Acta Astronaut.* **2018**, *151*, 743–751. [CrossRef]
15. Bassetto, M.; Niccolai, L.; Boni, L.; Mengali, G.; Quarta, A.A.; Circi, C.; Pizzurro, S.; Pizzarelli, M.; Pellegrini, R.C.; Cavallini, E. Sliding mode control for attitude maneuvers of Helianthus solar sail. *Acta Astronaut.* **2022**, *198*, 100–110. [CrossRef]
16. Aliasi, G.; Mengali, G.; Quarta, A.A. Artificial Lagrange points for solar sail with electrochromic material panels. *J. Guid. Control Dyn.* **2013**, *36*, 1544–1550. [CrossRef]

17. Boni, L.; Bassetto, M.; Niccolai, L.; Mengali, G.; Quarta, A.A.; Circi, C.; Pellegrini, R.C.; Cavallini, E. Structural response of Helianthus solar sail during attitude maneuvers. *Aerosp. Sci. Technol.* **2023**, *133*, 108152. [CrossRef]
18. Quarta, A.A.; Mengali, G. Solar sail orbit raising with electro-optically controlled diffractive film. *Appl. Sci.* **2023**, *13*, 7078. [CrossRef]
19. Wang, X.; Roy, S.; Farì, S.; Baldi, S. Adaptive Vector Field Guidance Without a Priori Knowledge of Course Dynamics and Wind. *IEEE/ASME Trans. Mechatron.* **2022**, *27*, 4597–4607. [CrossRef]
20. Bai, Y.; Yan, T.; Fu, W.; Li, T.; Huang, J. Robust Adaptive Composite Learning Integrated Guidance and Control for Skid-to-Turn Interceptors Subjected to Multiple Uncertainties and Constraints. *Actuators* **2023**, *12*, 243. [CrossRef]
21. Feng, C.; Chen, W.; Shao, M.; Ni, S. Trajectory Tracking and Adaptive Fuzzy Vibration Control of Multilink Space Manipulators with Experimental Validation. *Actuators* **2023**, *12*, 138. [CrossRef]
22. Xie, R.; Dempster, A.G. An on-line deep learning framework for low-thrust trajectory optimisation. *Aerosp. Sci. Technol.* **2021**, *118*, 107002. [CrossRef]
23. Peloni, A.; Rao, A.V.; Ceriotti, M. Automated Trajectory Optimizer for Solar Sailing (ATOSS). *Aerosp. Sci. Technol.* **2018**, *72*, 465–475. [CrossRef]
24. Patterson, M.J.; Benson, S.W. NEXT ion propulsion system development status and performance. In Proceedings of the 43rd AIAA/ASME/SAE/ASEE Joint Propulsion Conference and Exhibit, Cincinnati, OH, USA, 8–11 July 2007; Paper AIAA-2007-5199.
25. Battin, R. *An Introduction to the Mathematics and Methods of Astrodynamics*; AIAA education series; American Institute of Aeronautics and Astronautics: Reston, VA, USA, 1999; Chapter 8, pp. 408–418. [CrossRef]
26. Edwards, C.; Spurgeon, S.K. *Sliding Mode Control: Theory and Applications*; CRC Press: Boca Raton, FL, USA, 1998; Chapter 1, pp. 15–17.

Disclaimer/Publisher's Note: The statements, opinions and data contained in all publications are solely those of the individual author(s) and contributor(s) and not of MDPI and/or the editor(s). MDPI and/or the editor(s) disclaim responsibility for any injury to people or property resulting from any ideas, methods, instructions or products referred to in the content.

Article

Active Vibration Control Using Loudspeaker-Based Inertial Actuator with Integrated Piezoelectric Sensor

Minghao Chen, Qibo Mao *, Lihua Peng and Qi Li

School of Aircraft Engineering, Nanchang HangKong University, 696 South Fenghe Avenue, Nanchang 330063, China; 2206082500003@stu.nchu.edu.cn (M.C.); penglihua2021@163.com (L.P.); l921123187@163.com (Q.L.)
* Correspondence: qbmao@nchu.edu.cn; Tel.: +86-150-7913-1754

Abstract: With the evolution of the aerospace industry, structures have become larger and more complex. These structures exhibit significant characteristics such as extensive flexibility, low natural frequencies, numerous modes, and minimal structural damping. Without implementing vibration control measures, the risk of premature structural fatigue failure becomes imminent. In present times, the installation of inertial actuators and control signal acquisition units typically requires independent setups, which can be cumbersome for practical engineering purposes. To address this issue, this study introduces a novel approach: an independent control unit combining a loudspeaker-based inertial actuator (LBIA) with an integrated piezoelectric ceramic sensor. This unit enables autonomous vibration control, offering the advantages of ease of use, low cost, and lightweight construction. Experimental verification was performed to assess the mechanical properties of the LBIA. Additionally, a mathematical model for the LBIA with an integrated piezoelectric ceramic sensor was developed, and its efficacy as a control unit for thin plate structure vibration control was experimentally validated, showing close agreement with numerical results. Furthermore, the LBIA's benefits as an actuator for low-frequency mode control were verified through experiments using external sensors. To further enhance control effectiveness, a mathematical model of the strain differential feedback controller based on multi-bandpass filtering velocity improvement was established and validated through experiments on the clamp–clamp thin plate structure. The experimental results demonstrate that the designed LBIA effectively reduces vibration in low-frequency bands, achieving vibration energy suppression of up to 12.3 dB and 23.6 dB for the first and second modes, respectively. Moreover, the LBIA completely suppresses the vibration of the fourth mode. Additionally, the improved control algorithm, employing bandpass filtering, enhances the effectiveness of the LBIA-integrated sensor, enabling accurate multimodal damping control of the structure's vibrations for specified modes.

Keywords: active vibration control; loudspeaker-based inertial actuator; feedback control

Citation: Chen, M.; Mao, Q.; Peng, L.; Li, Q. Active Vibration Control Using Loudspeaker-Based Inertial Actuator with Integrated Piezoelectric Sensor. *Actuators* **2023**, *12*, 390. https://doi.org/10.3390/act12100390

Academic Editors: Ti Chen, Junjie Kang, Shidong Xu and Shuo Zhang

Received: 12 September 2023
Revised: 13 October 2023
Accepted: 13 October 2023
Published: 17 October 2023

Copyright: © 2023 by the authors. Licensee MDPI, Basel, Switzerland. This article is an open access article distributed under the terms and conditions of the Creative Commons Attribution (CC BY) license (https://creativecommons.org/licenses/by/4.0/).

1. Introduction

Vibration presents substantial challenges in aerospace engineering, as it embodies unwanted energy that can trigger aeroelastic instability, particularly flutter. Additionally, aircraft vibration can induce fatigue, generate excessive noise, and cause discomfort for passengers and crew. Consequently, it becomes imperative to implement efficient vibration control methods to ensure both structural integrity and overall vehicle safety. Among the various vibration control strategies, passive control usually uses dynamic vibration absorbers (DVAs), which are extensively employed for controlling structural vibrations [1–5]. However, one drawback of DVAs is their limited effectiveness in attenuating vibrations at multiple resonance frequencies. Significant advancements have been achieved by implementing inertial actuators (or proof mass actuators) [6,7], resulting in improved reduction of vibrations across multiple resonance frequencies [8]. These actuators exhibit enhanced adaptability to changes in controlled structural parameters, making them suitable for a wide range of operating conditions including low-frequency scenarios where precise vibration control is required. Over the past two decades, inertial actuators have found extensive use as vibration and structural sound suppression devices, demonstrating their high effectiveness in reducing vibrations during bridge construction [9] and minimizing noise levels in aircraft cabins, thereby enhancing safety and passenger comfort [10,11].

To date, many active vibration control (AVC) strategies have utilized a centralized architecture based on negative velocity feedback, which allows for precise damping of structural vibrations [12–15]. However, as the control system expands, potential issues may arise, particularly in cases of sensor failure, which can result in a complete system breakdown. In contrast, a totally decentralized control strategy allows each control unit to operate independently, also known as self-sufficient controllers [16,17]. In this strategy, inertial actuators and accelerometers are employed, with their signals processed by a time integrator to serve as control units in active control systems, enabling precise control over structural vibrations. To ensure system stability, it is crucial to align the sensors and actuators perfectly during application. However, this method often presents imperfect alignment control during installation [18]. If the inertial actuator and sensor are designed as a monolithic structure, perfect alignment of the system can be ensured and the stability of the system is theoretically guaranteed [19].

On the other hand, the distributed control strategy in vibration control systems often necessitates complex alignment of each control unit [20,21]. A massive number of sensors and actuators need to be installed and the operation is very complex [22]. Meanwhile, commercially available velocity sensors currently suffer from drawbacks such as their large size, high cost, and inconvenient usage [23]. Additionally, displacement sensors or acceleration sensors require corresponding circuitry to convert their outputs into velocity signals. To overcome the drawback of complex alignment, a novel solution has been developed: a self-sufficient control unit [17]. This innovative device enables precise alignment control and integrates a compact, portable velocity sensor directly into the control unit, eliminating the need for external circuit modules. It combines the basic elements of a control unit, containing a velocity sensor for vibration detection and an actuator driven by a control signal.

Apart from the design of the inertial actuators, selecting an appropriate active control algorithm is crucial to improve control efficiency. One rather appealing solution for the active control of broadband vibration is using self-contained control units with direct velocity feedback (DVBF) laws implemented through velocity sensors and collocated inertial actuator pairs. These control units are achieved by employing a velocity sensor or an accelerometer processed by an external integrated circuit and a parallel inertial actuator pair, eliminating the need for an external support to react off; thus, the control unit is simple to mount on the structure to be controlled [24].

Earlier work focused on implementing a decentralized array of DVBF control units with the ability to adjust the feedback gain to reduce the structure's vibration significantly [25]. Additionally, Zilletti et al. [26] demonstrated that maximizing power absorption is equivalent to minimizing the kinetic energy if the primary structure is an SDOF structure. By increasing the feedback gain, the peak resonance frequency of the SDOF system decreases due to the active damping effect. However, the dynamic behavior of the inertial actuator leads to overflow at its resonant frequency, making the system unstable as the feedback gain approaches the maximum steady-state gain. To solve this problem, the optimal feedback gain of the vibration control board can be achieved by maximizing the power absorption [27,28]. Moreover, the margin of feedback gain can be further improved by phase compensation of inertial actuators or other control algorithms, thus enhancing the stability of the system.

Among these control algorithms, positive position feedback (PPF) stands out as a second-order low-pass filter widely used in structural vibration control due to its simplicity, robustness, and effectiveness in addressing high-frequency signal saturation [29–32]. Notably, Fanson and Caughey first proposed the PPF control technique; PPF is insensitive to the uncertainty of the structure's natural damping ratio, providing stable and reliable vibration control [33–35]. The term "positive position" refers to feeding position measurements into the compensator in a positive manner and positively feeding position signals from the compensator back to the structure [3]. This characteristic makes this algorithm well-suited for collocated actuator/sensor pairs [36–38]. Previous research by Friswell and Inman proposed PPF control as an output feedback controller, using optimal control technology to address the instability problem caused by not considering mode overflow in the SDOF system. The effectiveness of this method was verified using centralized and distributed control architectures [39]. Similarly, Sim and Lee used an accelerated feedback (AFC) strategy employing second-order filters, also known as resonant controllers, to target desired modes of multi-degree-of-freedom systems [40,41]. Zhao et al. studied the application of a nonlinear positive position feedback controller with Duffing oscillator and derived the closed expression of optimal control parameters [42]. As a result, the low-pass filter as a PPF controller has been shown to work well for signal or multimodal control [43].

However, using multiple low-pass filters as PPF controllers to control multimodal vibrations may lead to phase overlap and potential multimodal control problems. The combined action of multiple PPF controllers affects the overall phase response of the system, making it challenging to control multiple modes simultaneously. Furthermore, incorrect placement of the second PPF controller may interfere with the operation of the first one, rendering it ineffective [44]. To enhance the control stability of the inertial actuator under the DVBF method, this study adopts a bandpass filter based on the second-order compensator principle proposed by Rohlfing [14]. The objective is to address the problem and improve the traditional DVBF stability. By adjusting the center frequency and damping of each bandpass filter, precise control of different modes can be achieved; then, overall system stability can be improved. This method offers a promising solution for improving phase-related issues and enhancing the effectiveness of multimodal control in vibration reduction.

The rest of this study is structured as follows. In Section 2, the designed loudspeaker-based inertial actuator with an integrated piezoelectric sensor will be described and experimentally verified. Section 3 presents the control performances of the proposed LBIA by using the DVBF approach for thin plate structures. In addition, the piezoelectric sensor model of the integrated control unit is established. In Section 4, a parallel connection of multiple bandpass filters is used as a controller to improve the control performance based on the proposed LBIA; then, the corresponding experimental results are presented. Section 5 concludes with a discussion and summary of the relevant research results and experimental conclusions.

2. LBIA with Integrated Piezoelectric Sensor Configuration and Performance

2.1. LBIA with Integrated Piezoelectric Sensor Design

The inertial actuator consists of several essential components, such as an electromagnetic coil, permanent magnet, elastic element, and base. Its operation relies on supplying power to the electromagnetic induction coil, which induces specific and regular changes in the surrounding magnetic field. As a result, the Lorentz force is generated, propelling the reciprocal motion of the actuator and permanent magnet. With these fundamental principles and components in mind, it becomes apparent that the loudspeaker principle seamlessly aligns with the inertial actuator. Therefore, it is possible to modify a loudspeaker as an inertial actuator, as depicted in Figure 1.

The loudspeaker employed in this design is a moving-coil loudspeaker, commonly referred to as an electric loudspeaker. Upon inputting the control signal to the loudspeaker, the resulting Lorentz force propels the paper cone to reciprocate, thereby generating the desired control force. To optimize the effectiveness of the control force, the surface of the paper cone is connected to a natural rubber base using neutral silicone rubber. This rubber support base enhances the overall actuator performance and improves passive control effects. Neutral silicone rubber possesses favorable characteristics such as lower damping, reduced energy absorption, and efficient transfer of the control force, making it an ideal material choice for the base.

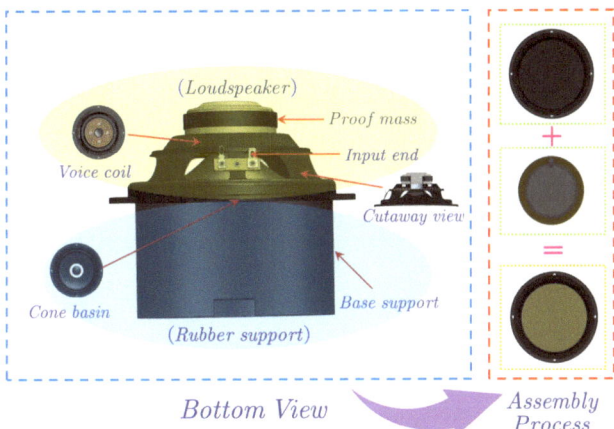

Figure 1. Configuration design of loudspeaker–sensor integrated actuator.

Furthermore, it is essential to highlight that Figure 1 illustrates the utilization of a thin-film piezoelectric ceramic sensor with a small volume and electrodes positioned on its surface. To ensure accurate measurement and prevent contact with the sensor's electrode surface, a ring-shaped support base is employed, facilitating the integrated design of the sensor and actuator. This configuration allows the applied force to be transmitted to the vibrating structure's surface through the ring shape, approximating the force as a point force based on findings from [40]. This feature makes it well-suited for active vibration control.

A photograph of the proposed LBIA is presented in Figure 2. The mechanical and electrical parameters of the LBIA are listed in Table 1. Notice that the proposed LBIA also integrates a piezoelectric sensor, resulting in an integrated control unit. This integration greatly improves installation efficiency and reduces the occurrence of poor stability due to alignment errors.

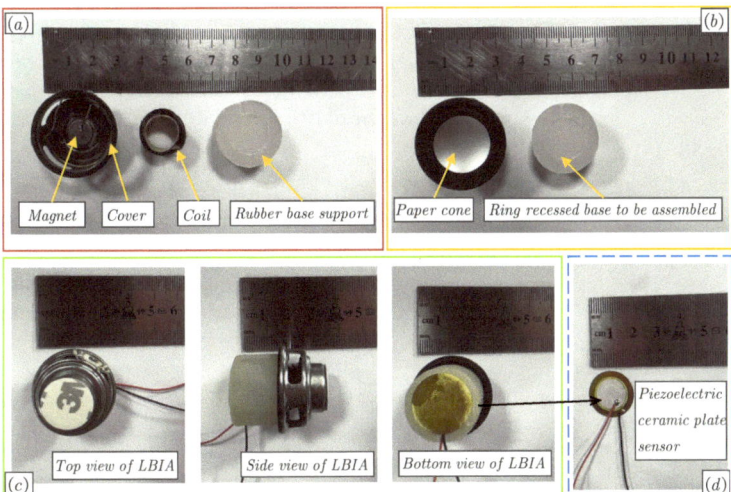

Figure 2. LBIA with integrated piezoelectric sensor: (**a**) before assembly; (**b**) partial assembly; (**c**) after assembly; (**d**) sensor.

Table 1. Performance of the inertial actuator (LBIA).

Parameter	Notation	Value	Unit
Proof mass	M_a	0.0236	kg
Base support mass	M_b	0.0088	kg
Suspension stiffness	K_a	2800	N/m
Suspension damping coefficient	C_a	0.2	Ns/m
Natural frequency	ω_n	55	Hz
Damping ratio	ζ	5.2252	
Voice coil coefficient	Bl	0.018	N/A
Coil resistance	R_e	4	Ω
Coil inductance	L_e	88.1×10^{-5}	H

2.2. Dynamic Model for LBIA

This subsection focuses on describing the dynamic characteristics of the LBIA itself. Since the loudspeaker is an electromechanical coupling system, a corresponding dynamic model can be established based on this, and the specific schematic diagram is shown in Figure 3.

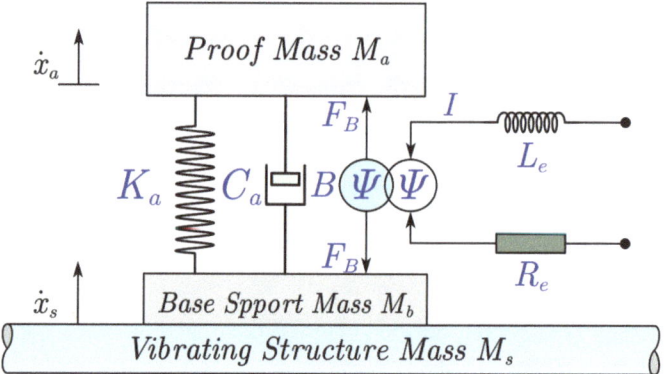

Figure 3. The electromechanical schematic for LBIA with integrated piezoelectric sensor.

The loudspeaker itself outputs volume velocity, which excites the mass block of the supporting base. Additionally, considering the mechanical characteristics of the actuator on the main structure, a coupling dynamic model between the inertial actuator and the main structure is established. This model can be considered a mass-spring-damping system, as shown in Figure 3. The dynamic model can be expressed as follows:

$$M_a \ddot{x}_a + C_a(\dot{x}_a - \dot{x}_s) + K_a(x_a - x_s) = Bl \cdot I = F_B \tag{1}$$

$$U_{in} = B(\dot{x}_a - \dot{x}_s) + R_e I + L_e \dot{I} \tag{2}$$

where U_{in} is the input voltage of the inertial actuator, and x_a and x_s are the displacement of the proof mass and base support of the inertial actuator, respectively.

The controlling force (inertial force) output of the inertial actuator is

$$F_c = Bl \cdot I - C_a \dot{x}_a - K_a x_a \tag{3}$$

In the state-space domain way, the actuator dynamics from Equations (1)–(3) can be rearranged as

$$\begin{cases} \dot{\mathbf{x}} = \mathbf{A}\mathbf{x} + \mathbf{B}U_{in} \\ F_c = \mathbf{C}\mathbf{x} + \mathbf{D}U_{in} \end{cases} \tag{4}$$

where the output F_c is the control force vector generated by the actuators. In the voltage-driven configuration, the state of the system is completely defined by both differential equations in Equations (1) and (2). Therefore, the state vector includes the current as $\mathbf{x}^T = \begin{bmatrix} x & \dot{x} & I \end{bmatrix}$.

The state matrices are, respectively,

$$\mathbf{A} = \begin{bmatrix} 0 & 1 & 0 \\ -\dfrac{K_a}{M_a} & -\dfrac{C_a}{M_a} & \dfrac{B}{M_a} \\ 0 & -\dfrac{B}{L_e} & \dfrac{R_e}{L_e} \end{bmatrix}, \quad \mathbf{B} = \begin{bmatrix} 0 \\ 0 \\ \dfrac{1}{L_e} \end{bmatrix}, \quad \mathbf{C} = \begin{bmatrix} -K_a \\ -C_a \\ B \end{bmatrix}^T, \quad \mathbf{D} = 0 \quad (5)$$

According to the state space method, the transfer function of the input voltage to the output force of the inertial actuator can be obtained as follows:

$$T_a(\omega) = \frac{F_c(\omega)}{U_{in}(\omega)} = \mathbf{C}(j\omega \mathbf{E} - \mathbf{A})^{-1}\mathbf{B} \quad (6)$$

Equation (6) reflects the relationship between the input signal and the output force of the actuator. Based on Equation (6), the mechanical characteristics of the actuator can be solved, which provides a reference for the selection of the active control algorithm.

2.3. LBIA Performance Test

To obtain the mechanical performance of the proposed LBIA, the actuator's output characteristics were tested to study the variation in output force and the effective working frequency range. Based on this, a test platform was constructed, as depicted in Figure 4.

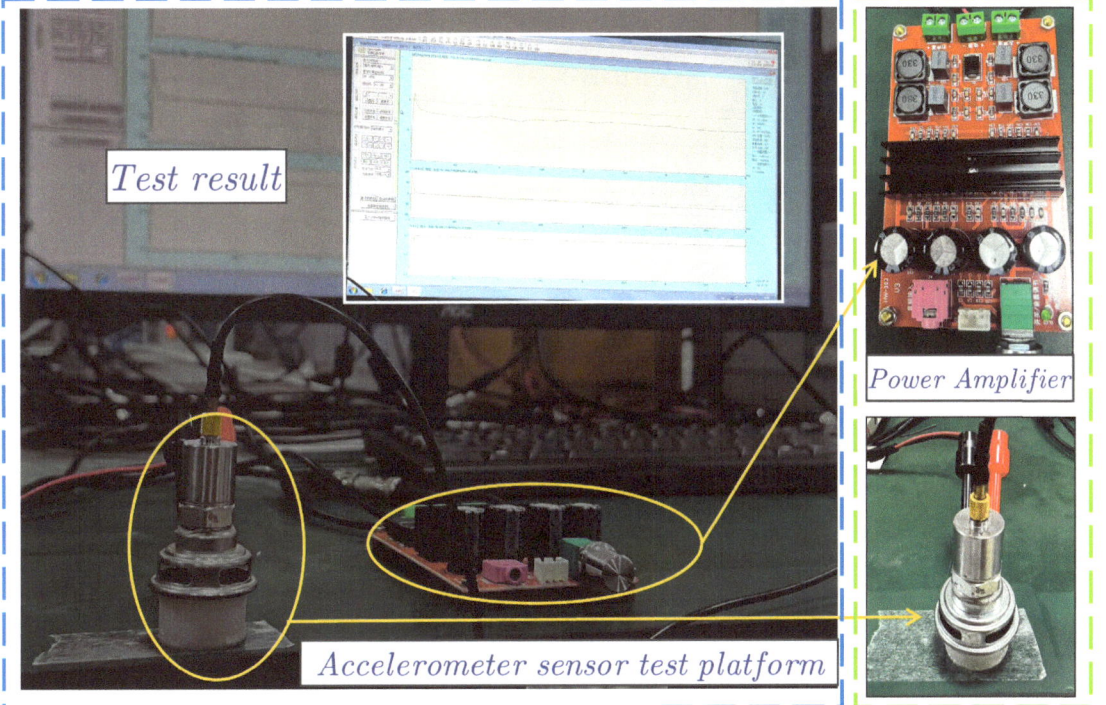

Figure 4. Picture of the experimental setup for measuring mechanical performance of LBIA with integrated piezoelectric sensor.

Figure 4 shows the experimental setup for measuring the mechanical performance of the inertial actuator. The experimental measurements focused on establishing the relationship between the amplitude of the LBIA with the integrated piezoelectric sensor input signal and the input force. To evaluate the mechanical performance, a YD-186 accelerometer from Sinocera Piezotronics Inc. (Yangzhou, China) (with a weight of 30 g and sensitivity of $10.2\,\mathrm{mv}/(\mathrm{ms}^{-2})$) was installed on the proof mass and base support of the LBIA with an integrated piezoelectric sensor to measure its acceleration. The COINV dynamic signal analyzer (Beijing, China) (which has 24 channels but only uses the first two channels) was used to obtain its frequency response function. The mechanical properties of the inertial actuator can then be determined.

Importantly, the negligible mass (approximately 0.1 g) of the piezoelectric ceramic sensor allows us to disregard it; thus, it was not installed during the tests, as shown in Figure 4. However, the mass of the accelerometer cannot be disregarded for the LBIA. The output characteristics can be expressed as

$$T_a = \frac{F_c}{U_{in}} = \frac{M_a a_a}{U_{in}} \quad s.t. v_s = 0 \qquad (7)$$

During the variable amplitude output force response test, a range of sinusoidal signals with frequencies of 155 Hz and varying amplitudes were employed. The selected test signal had an input voltage from 50 mV to 350 mV, with test points recorded at 50 mV intervals. By utilizing Newton's second law, the acceleration signals accurately portrayed the mechanical performance of the actuator. The experimental results are presented in Figure 5.

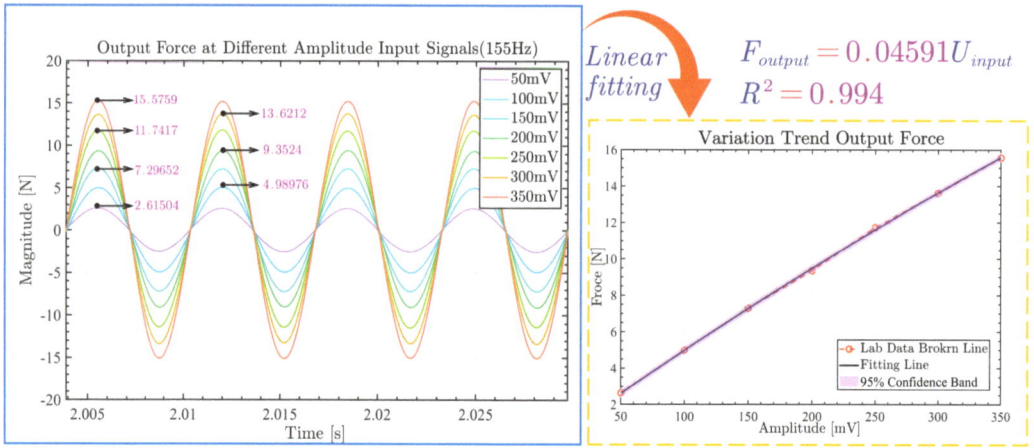

Figure 5. Output force test results at different input amplitude signals.

The test results show that at 155 Hz, the output force of the inertial actuator is accurate and free from distortion, aligning with the first mode of the controlled structure shown in Figure 5. Additionally, the output force of the inertial actuator shows a proportional increase in relation to the input amplitude, illustrating an approximate linear relationship with a goodness of fit value of 0.94.

For the variable frequency output force response test, a sinusoidal signal ranging from 10 Hz to 1000 Hz with an amplitude of 100 mV was utilized. Figure 6 illustrates that the natural frequency of LBIA is 30 Hz due to the accelerometer mass. Notably, the inertial actuator's output control force remains relatively stable in both magnitude and phase when the frequency exceeds the natural frequency. In this frequency band, the inertial actuator effectively functions as an exciter, producing an ideal point force output. Additionally, the frequency response test results of the actuator align closely with the transfer function of the theoretical model. The bearing shape and output force characteristics were also verified in [45].

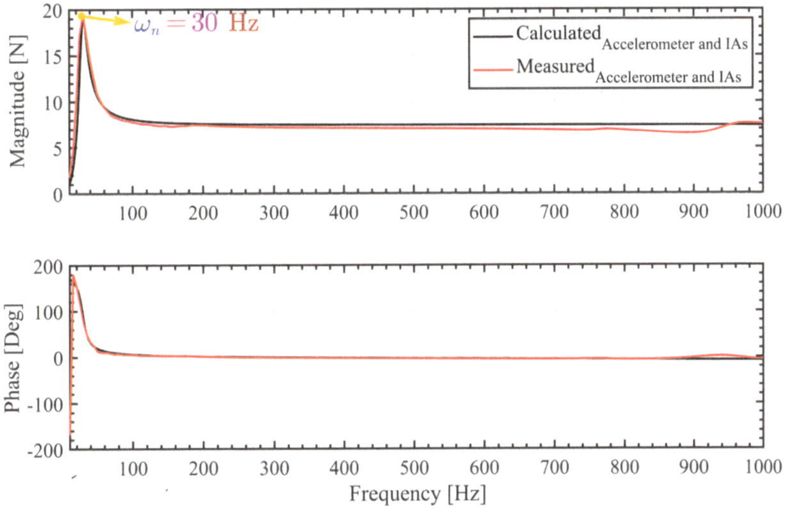

Figure 6. Output force test results at different frequencies.

From Figures 5 and 6, it can be found that the proposed LBIA offers several advantages:

- The whole configuration adopts an integrated design that is easy to disassemble, and because of the alignment design, the control system is stable;
- The speaker, serving as the actuator, produces sound vibrations characterized by low distortion and high fidelity. Its flat frequency response enables clear resolution and accurate reproduction of audio signals across a wide frequency range, from low to high;
- The rubber structure is used as the support structure, the structural damping is small, and the output control force can be effective.

This section demonstrates the establishment of an accurate dynamic model for the LBIA, offering valuable insights into its mechanical performance. Moreover, it serves as a foundational basis for further research on control algorithms.

3. Control Performance of LBIA with Integrated Piezoelectric Sensor by Using Strain Differential Feedback (SDF) Approach

In this section, our focus is to discuss the effectiveness of the integrated design as a control unit for thin plate structures. To demonstrate this, we utilize the strain differential feedback (SDF) approach based on the integrated piezoelectric ceramic sensor.

Figure 7 illustrates the behavior of the piezoelectric ceramic sensor under the influence of a stress field. When the sensor senses structural vibrations, it generates the corresponding induced electricity. According to [46], the induced charge generated by piezoelectric ceramic sensors can be described by Equation (8) as follows:

$$Q_e = \frac{h}{2} e \int_0^{2\pi} \int_0^R \left(\frac{\partial^2 w(r,\theta,t)}{\partial r^2} + \frac{1}{r} \frac{\partial w(r,\theta,t)}{\partial r} + \frac{1}{r^2} \frac{\partial^2 w(r,\theta,t)}{\partial \theta^2} \right) r \, dr \, d\theta \quad (8)$$

$$for \begin{cases} r = \sqrt{(x-x_0)^2 + (y-y_0)^2} \\ \theta = \text{atan2}(y - y_0, x - x_0) \end{cases}$$

where $w(r,\theta,t)$ is the polar coordinate form of displacement of the thin plate, e is the piezoelectric strain constant of the sensor, R is the radius of the selected circular piezoelectric ceramic sensor, and x_0 and y_0 represent the central position of the piezoelectric ceramic sensor.

As depicted in Figure 7, the sensor generates a charge signal in response. This signal can be perceived as a voltage source connected in series with the capacitor. Once connected to a signal conditioning circuit—specifically, a current amplifier—the output signal undergoes conversion into a current signal. Reference [47] provides insights into the understanding of this process from both physical and mathematical perspectives, assuming the piezoelectric strain constant and alignment with the Y-axis. Essentially, it involves differentiating Equation (8) once. The current output of piezoelectric ceramics is expressed as follows:

$$I_e = \frac{dQ_e}{dt} = \frac{h}{2}e \int_0^{2\pi} \int_0^R \left(\frac{\partial^2 v(r,\theta)}{\partial r^2} + \frac{1}{r}\frac{\partial v(r,\theta)}{\partial r} + \frac{1}{r^2}\frac{\partial^2 v(r,\theta)}{\partial \theta^2} \right) r\, dr\, d\theta \quad (9)$$

Equation (9) can be expressed as

$$\begin{cases} v(r,\theta) = \sum_{k=1}^m \phi_k \Phi_k(r,\theta) = \mathbf{\Phi}^T \boldsymbol{\phi} \\ I_e = \frac{h}{2}e \int_0^{2\pi} \int_0^R \left(\sum_{k=1}^m \phi_k \frac{\partial^2 \Phi_k(r,\theta)}{\partial r^2} + \sum_{k=1}^m \frac{1}{r}\phi_k \frac{\partial \Phi_k(r,\theta)}{\partial r} + \right. \\ \left. \sum_{k=1}^m \frac{1}{r^2}\phi_k \frac{\partial^2 \Phi_k(r,\theta)}{\partial \theta^2} \right) r\, dr\, d\theta \end{cases} \quad (10)$$

where $\Phi_k(r,\theta)$ and ϕ_k represent the mth structural modal shape and modal velocity, respectively. The modal index of the structure on the x and y axes is represented by $k=(m_x, m_y)$, respectively.

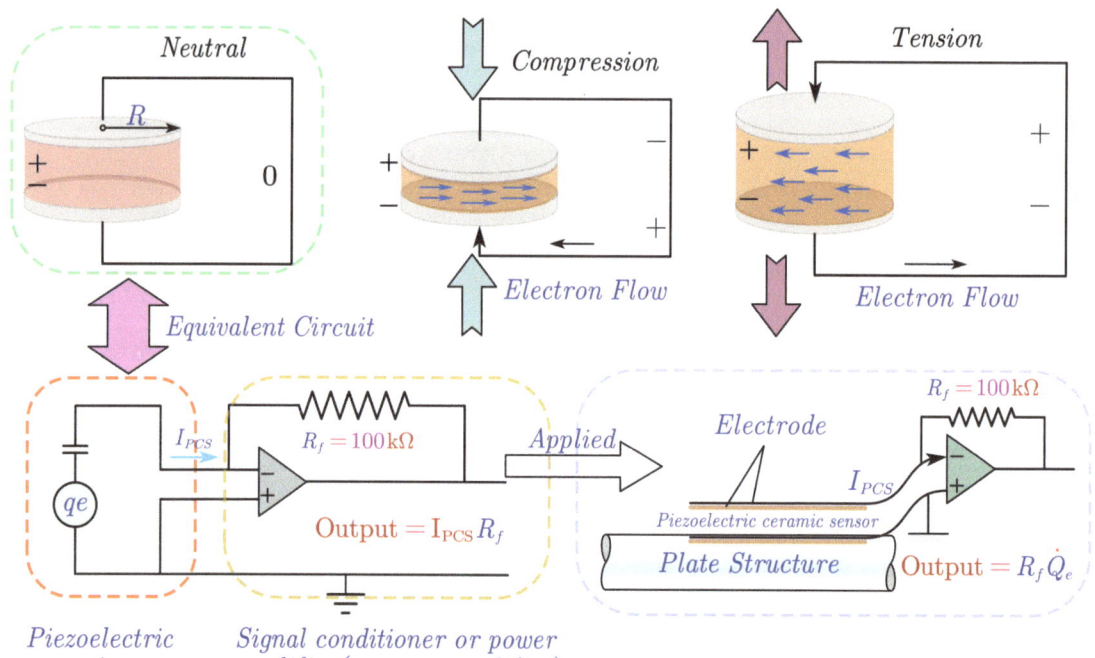

Figure 7. Schematic diagram of piezoelectric ceramic sensor.

Equation (10) can be further simplified to matrix form:

$$I_e = P\phi \tag{11}$$

$$P_k = \frac{h}{2}e \int_0^{2\pi} \int_0^R \left(\sum_{k=1}^m \frac{\partial^2 \Phi_k(r,\theta)}{\partial r^2} + \sum_{k=1}^m \frac{1}{r} \frac{\partial \Phi_k(r,\theta)}{\partial r} + \sum_{k=1}^m \frac{1}{r^2} \frac{\partial^2 \Phi_k(r,\theta)}{\partial \theta^2} \right) r \, dr \, d\theta \tag{12}$$

The output voltage of the sensor is determined under the assumption that the current amplifier operates ideally with infinite internal resistance:

$$U_{in} = R_f \dot{Q}_e = R_f I_e = R_f P\phi \tag{13}$$

where R_f is the amplifier constant.

The combination model of the piezoelectric ceramic sensor and the inertial actuator can be obtained by utilizing the mathematical model of the piezoelectric ceramic sensor and the dynamic model of the inertial actuator:

$$\left[M_a \left(j\omega R_e - \omega^2 L_e \right) + C_a(j\omega L_e + R_e) + K_a \left(L_e + \frac{R_e}{j\omega} \right) + (Bl)^2 \right] \cdot \dot{x}_a$$
$$= U_{in} \cdot \frac{\left(Bl \left(R_f \Gamma + Bl \right) + C_a(j\omega L_e + R_e) + K_a \left(L_e + \frac{R_e}{j\omega} \right) \right)}{R_f \Gamma} \tag{14}$$

where Γ is a direct multiple relationship between the current I and the velocity $\dot{x}_s$, which is expressed as $\Gamma = P\phi/\dot{x}_s$.

Finally, the control situation of the actuator and piezoelectric ceramic sensor is expressed as a transfer function:

$$T_{SDF}(\omega) = \frac{\dot{x}_a}{U_{in}} = \frac{\left(Bl \left(R_f \Gamma + Bl \right) + C_a(j\omega L_e + R_e) + K_a \left(L_e + \frac{R_e}{j\omega} \right) \right)}{R_f \Gamma \left[M_a(j\omega R_e - \omega^2 L_e) + C_a(j\omega L_e + R_e) + K_a \left(L_e + \frac{R_e}{j\omega} \right) + (Bl)^2 \right]} \tag{15}$$

Meanwhile, Figure 8b experimentally illustrates the Nyquist diagram of the open-loop transfer function testing before and after signal conditioning of the sensor in the LBIA. The diagram demonstrates the sensor's ability to convert its charge (displacement signal) into current (velocity signal) after signal conditioning. Notably, it can be seen from Figure 8a that effective vibration control is achieved when the phase detection indicates that the first and second modes are close to 0 degrees.

Additionally, the fourth mode also exhibits a discernible control effect, as observed in the phase analysis. Moreover, conditioning the signal is accompanied by a 90-degree phase change, resulting from the charge signal transforming into a current signal through the conditioning circuit. It can be described as a differential process, and its specific derivation process is given by Equation (9). Based on this process, the active vibration control based on it can be referred to as strain differential feedback (SDF) control. In conclusion, this analysis provides valuable support for subsequent experiments.

Figure 9 depicts the experimental setup, comprising a thin aluminum plate as the subject of investigation. The geometrical and physical properties of the plate are summarized in Table 2. To stimulate the structure, a sinusoidal scanning signal, generated by the commercial inertial actuator (DAEX58FP electrodynamic exciter, Nikon, Tokyo, Japan), is used for excitation in a frequency range of 10 Hz to 800 Hz. The vibration signal is collected by a piezoelectric ceramic sensor integrated with LBIA and then passed through the power amplifier to form a closed-loop control system. To ensure a more realistic case study, the primary excitation positions on the board were deliberately chosen to avoid the node lines associated with the previous modes of the structure. The selected position (x_p, y_p) for the primary source was set at (125 mm, 65 mm). By adopting this approach, the lower-order modes are excited, allowing us to observe the effect of active control on their vibration modes. Similarly, to avoid the node line position, an LBIA with an integrated piezoelectric ceramic sensor is installed at (x_c, y_c) = (120 mm, 160 mm).

The experiment described in this paper involved collecting vibration energy signals at the same location as LBIA. Specifically, the vibration energy point of the measuring point is on the opposite side of the same-positioned thin plate as LBIA. These signals were obtained using a high-precision laser vibrometer (Polytec IVS-500, Karlsruhe, Germany). In addition to this, an accelerometer sensor was included in Figure 9 to observe vibrations at different positions. This served as a backup measure in case of laser vibrometer failure, ensuring the test's success when integrated with the LBIA design that incorporates piezoelectric sensors.

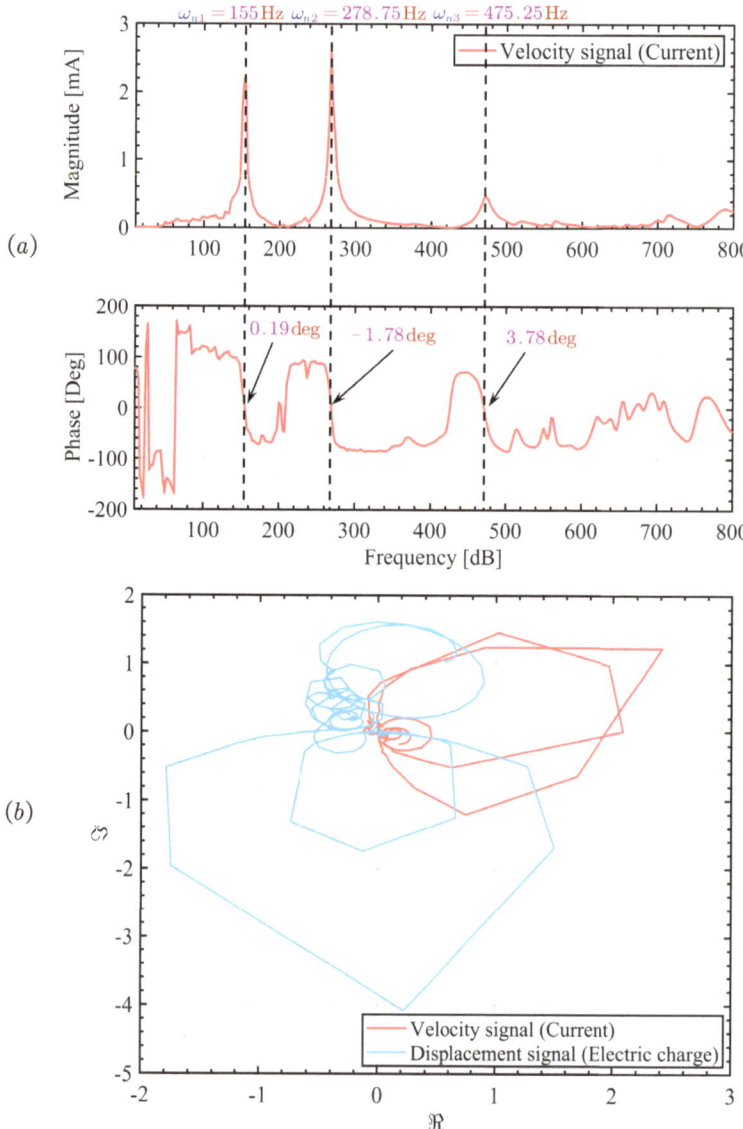

Figure 8. Open loop experiment diagram of integrated sensor in LBIA: (**a**) Bode diagram; (**b**) Nyquist diagram.

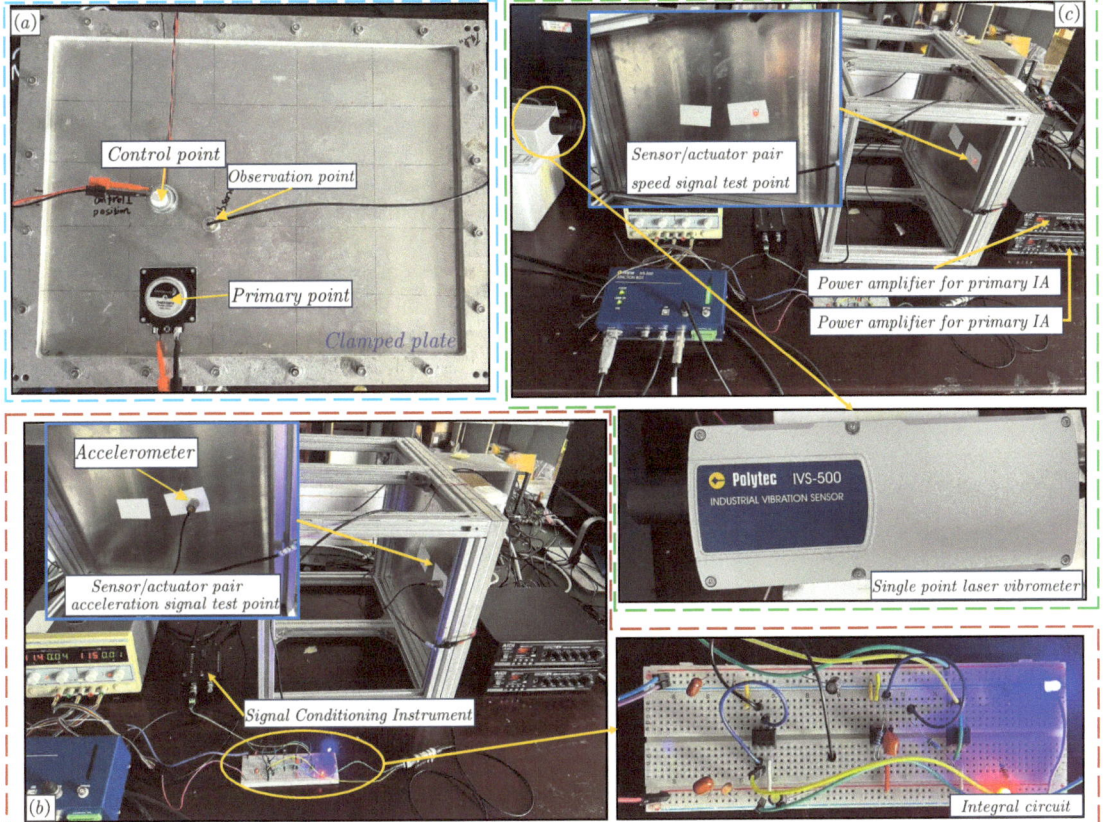

Figure 9. DVBF experiment of LBIA: (**a**) DVBF based on LBIA with integrated sensor; (**b**) DVBF based on accelerometer; (**c**) DVBF based on accelerometer.

Table 2. Geometry and physical parameters of the plate.

Parameter	Numerical Value	Unit
Planar dimensions	$l_x \times l_y = 440 \times 340$	mm
Thickness	$h = 3$	mm
Young's modulus	$E = 7 \times 10^{10}$	N/m^2
Poisson ratio	$\nu = 0.33$	
Density	$\rho = 2700$	kg/m^3

The LBIA, as shown in Figure 2, is equipped with integrated sensors that facilitate autonomous signal acquisition and output control. Moreover, the open-loop transfer function for LBIA with integrated piezoelectric sensors has been tested, as demonstrated in Figure 8.

From Figure 10a, it can be observed that the integrated design of LBIA proves effective in actively controlling the thin plate structure. The vibration energy in the first and second modes reduces by 8.1 dB and 7.3 dB, respectively. There is also some suppression in the fourth mode.

To provide a more accurate assessment of LBIA's control performance, the control performances for LBIA using DVBF with different external sensors (such as a collocated accelerometer and non-collocated laser vibrometer) are also presented in this study. Figure 9b illustrates the placement of the accelerometer sensor on the sheet's back, aligned with the

LBIA. Notably, the signal from the accelerometer sensor requires filtering through the signal conditioning instrument before being processed by the time integrator, which converts it into a velocity signal. This step completes the experimental setup for direct velocity feedback. Additionally, as shown in Figure 9c, the laser vibrometer's test point is located at the same position as the accelerometer sensor. The laser vibrometer directly collects the velocity signal of the vibration structure as the control signal, offering high acquisition accuracy without requiring any additional signal processing circuit. As a result, the LBIA's control effect on the structure can be more precisely evaluated.

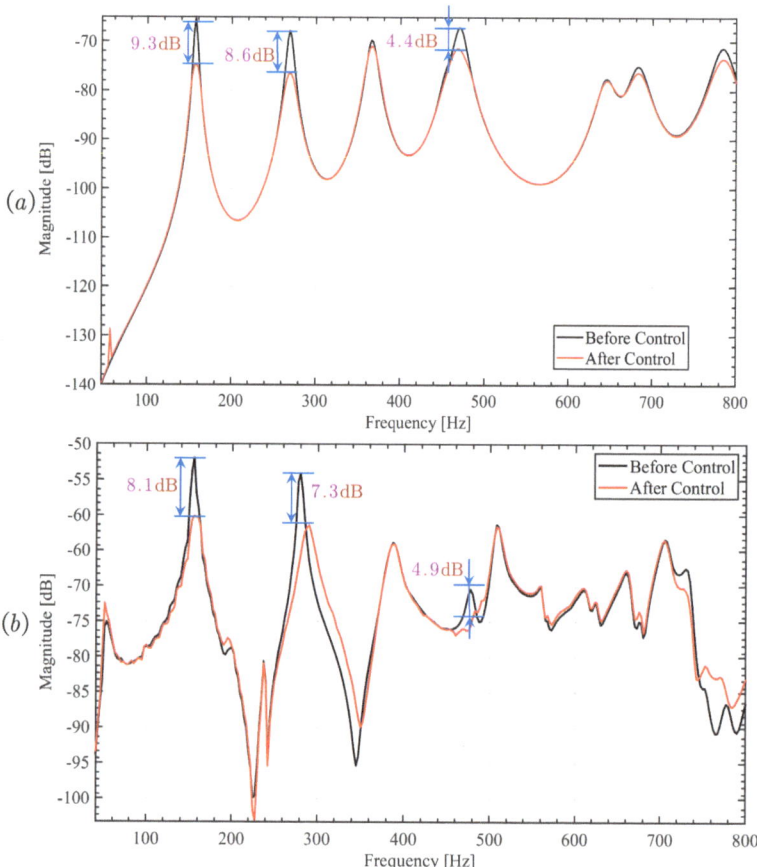

Figure 10. Control performance of LBIA with integrated sensor by using SDF approach: (**a**) calculation result; (**b**) experiment result.

Figure 11 demonstrates the control effect of the LBIA by illustrating the vibration energy of the system before and after implementing the control measures. Both the contact sensor and the non-contact sensor, utilizing negative velocity feedback, are employed in these measures. The LBIA significantly reduces vibration energy in the first and second modes, with reductions of 10.1 dB and 18.6 dB (contact sensor), and 12.3 dB and 23.6 dB (non-contact sensor), respectively. The use of high-precision non-contact sensors, such as a laser vibrometer, has demonstrated a positive control effect, which aligns with the numerical results. However, it is important to note that the integrated sensor design of the LBIA proves effective in active control but falls short in comparison to external sensors. The disparity arises mainly due to the piezoelectric ceramic sensor's sensitivity to the external environment, leading to interference signals that adversely affect control stability.

Despite this limitation, the low cost of the sensor (about USD 0.153) makes it a promising candidate for further research and suitable for large-scale vibration control in the industry. Consequently, the following section aims to enhance the control unit's effectiveness through multiple parallel bandpass filters.

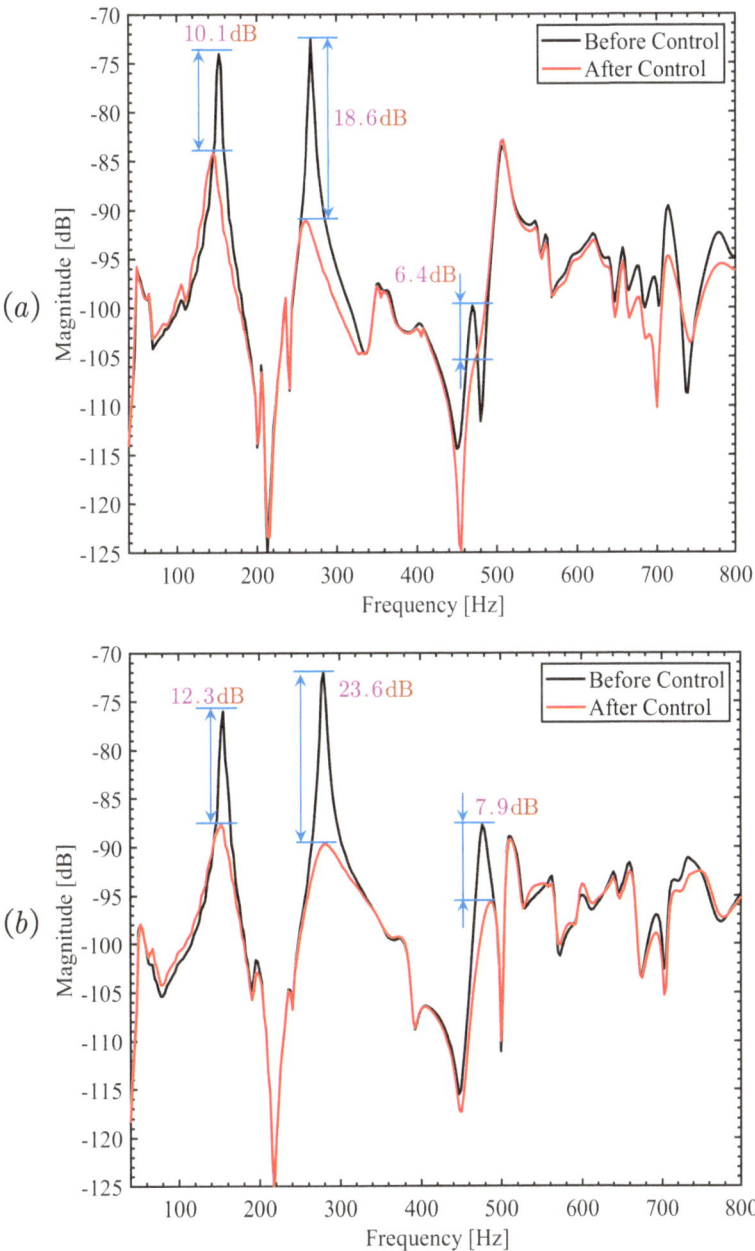

Figure 11. The effect diagram of control by DVBF: (**a**) accelerometer (contact sensor); (**b**) laser vibrometer (non-contact sensor).

4. Enhancing the Strain Differential Feedback Algorithm through Bandpass Filtering

4.1. Control Principle and Control Effect Simulation

To further improve the control performance of the proposed LBIA with an integrated piezoelectric sensor, the feedback control algorithm of the bandpass filter utilizes the strain differential signal (velocity signal) obtained from the LBIA with the integrated piezoelectric sensor as input. It enhances the damping of the controlled modes by employing multiple parallel bandpass filters, thereby facilitating active control of the multiple modes. Figure 12 depicts the control principal diagram of this algorithm.

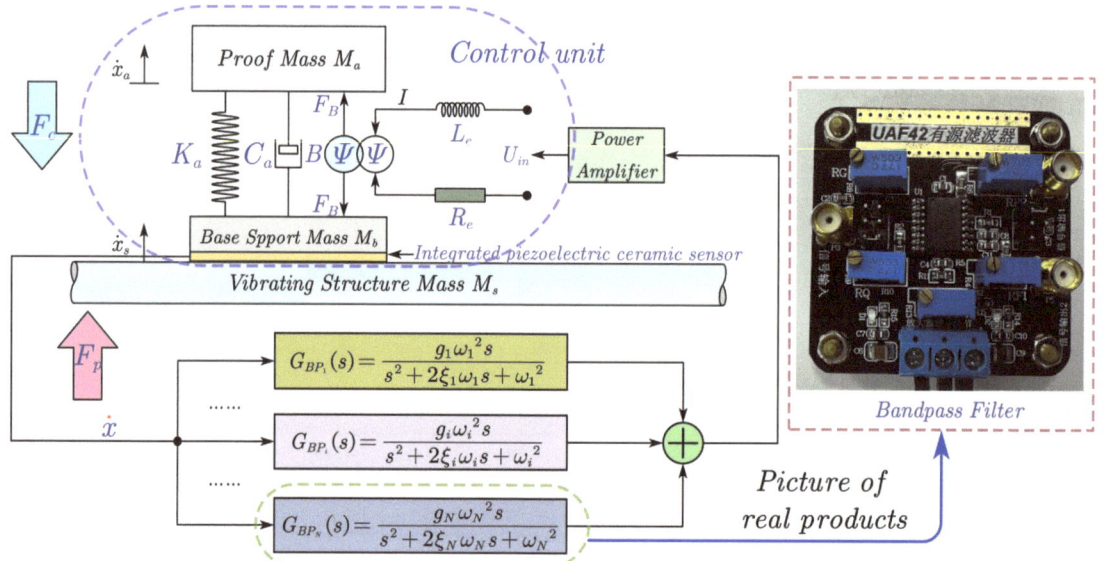

Figure 12. Schematic diagram of DVBF algorithm improved by bandpass filtering.

The LBIA employs an integrated piezoelectric ceramic sensor as the feedback mechanism. When the controlled structure is stimulated by F_p, the feedback sensor collects its vibration velocity signal. Subsequently, this signal is fed into an N-branch parallel bandpass filter, where it undergoes integration through an adder. Finally, the signal is sent to the LBIA via a power amplifier to generate the corresponding controlling force F_c, thus achieving closed-loop active control. Importantly, each bandpass filter's natural frequency, damping ratio, and gain can be independently adjusted to suit specific requirements. The controller described is a parallel N-branch bandpass filter, and its output signal represents the summation of the basic velocity output from each branch bandpass filter, which can be expressed using the following Equation (16):

$$U_{in} = Gain \cdot \sum_{i=1}^{N} G_{BP_i}(s)\dot{x} = Gain \cdot \dot{x} \sum_{i=1}^{N} \frac{g_i \omega_i s}{s^2 + 2\xi_i \omega_i s + \omega_i^2}$$
$$= Gain \cdot \dot{x} \sum_{i=1}^{N} \frac{g_i \omega_i}{-\omega^2 + 2\xi_i \omega_i s + \omega_i^2} \qquad (16)$$

where g_i, ξ_i, and ω_i represent the velocity feedback gain, damping ratio, and natural frequency of the nth branch bandpass filters, respectively. *Gain* refers to the amplification gain of a power amplifier.

According to Equations (1) and (2), we can derive the velocity transfer function of the inertial actuator after passing through N-branches:

$$T_{BP}(\omega) = \frac{\dot{x}_a}{\dot{x}_s} = \left[Gain \cdot \omega^2 \frac{Bl}{j\omega L_e + R_e} \cdot \sum_{i=1}^{N} \frac{g_i \omega_i}{\omega^2 - 2\xi_i \omega_i s - \omega_i^2} + j\omega \left(\frac{(Bl)^2}{j\omega L_e + R_e} + C_a \right) + K_a \right] \Big/ \left[-\omega^2 M_a + j\omega \left(\frac{(Bl)^2}{j\omega L_e + R_e} + C_a \right) + K_a \right] \quad (17)$$

The velocity transfer in Equation (17) will be utilized to assess the multimodal control effect of the proposed controller combined with LBIA with integrated piezoelectric sensor, thereby providing theoretical calculation support for subsequent experiments.

4.2. Experiment Setup and Results

The primary objective of this experiment is to enhance the control effect of the LBIA with an integrated piezoelectric sensor by incorporating a low-cost, second-order bandpass filter, specifically designed using UAF42 (by Burr-Brown Inc., Tucson, AZ, USA). Figure 13 illustrates the schematic diagram of this specific bandpass filter.

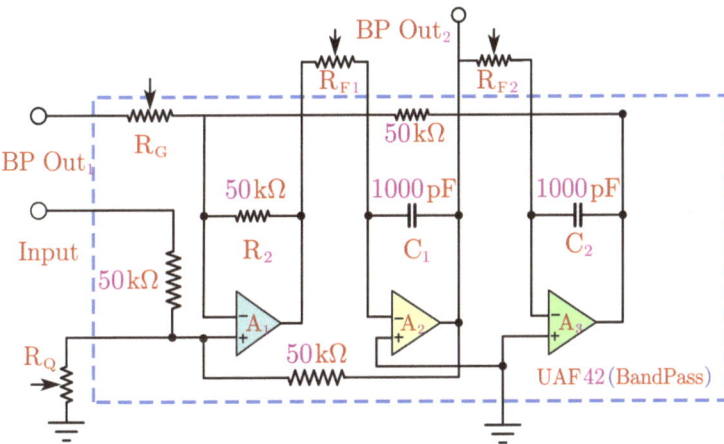

Figure 13. Diagram of a bandpass filter with reverse amplification function.

According to Figure 13, to facilitate the parameter adjustment in the experimental setup, R_G is set to a fixed value of 50 kΩ. The equation for adjusting the natural frequency, gain, and damping ratio of the bandpass filter designed based on UAF42 is derived as follows:

$$g_{BP} = \frac{25}{R_Q + 25} \quad (18)$$

$$\omega_n = \frac{10^9}{2\pi} \sqrt{\frac{1}{(R_{F1} \cdot R_{F2})}} \quad (19)$$

$$\xi_n = \frac{1}{2 \cdot Q} = \sqrt{\frac{R_{F2}}{R_{F1}}} \cdot \frac{R_Q}{2R_Q + R_G} \quad (20)$$

where Q is the quality factor, which should be noted as inversely proportional to the passband gain. It is important to acknowledge that the Q value and the passband gain g_{BP} cannot be simultaneously maximized. When adjusting the setting, a certain balance must be achieved between the two.

After establishing the controller model, it is necessary to conduct further testing to evaluate the actual control effect of the LBIA with an integrated piezoelectric sensor combined with a bandpass filter. To accomplish this, the focus of the study is placed on a thin plate as the research object. A dedicated experimental platform is built, as illustrated in Figure 14. This experiment serves as an extension of the previous experiments. The position arrangement is the same as in the previous experiment but different control algorithms are used.

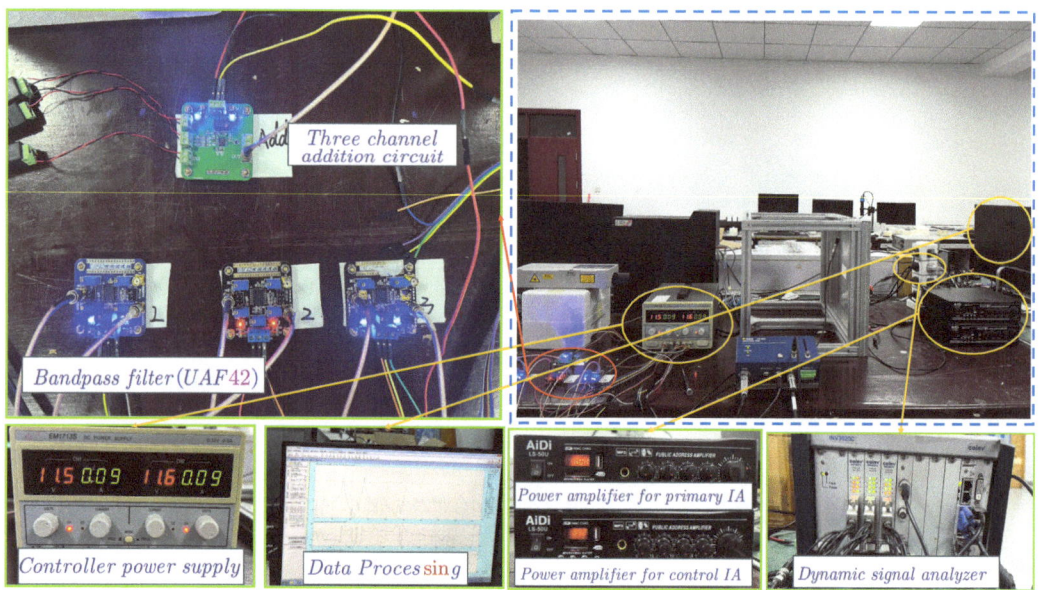

Figure 14. Experimental setup for bandpass filter velocity feedback control for a clamped–clamped thin plate using the LBIA.

To observe changes in the vibration energy of the controlled system, accelerometer sensors and a laser vibrometer are installed. Additionally, data acquisition and analysis software are configured for analyzing the experimental data. In this system, an integrated piezoelectric ceramic sensor is used to detect the vibration signal of the thin plate. The detected signal is then transmitted to three parallel bandpass filters. The output signals of these filters are combined using a three-way adder, and the resulting control signal is sent to the power amplifier to drive the inertial actuator for vibration suppression. This configuration forms a closed-loop control system.

Before conducting the experiment, it is essential to assess the characteristics of the multimodal controller based on multi-branch parallel features. This evaluation will pave the way for subsequent experiments. During the test, a sinusoidal signal sweeping from 0 to 800 Hz is passed through a dynamic analyzer equipped with three parallel bandpass filters.

Importantly, based on the theory and prior experiments mentioned above, the LBIA is capable of effectively controlling the first two and fourth modes from this control position. As a result, we can accurately predict the natural frequency of the bandpass filter using Equation (19). The calculated natural frequencies for the three bandpass filters are set to 156.25 Hz, 278.75 Hz, and 476.25 Hz, corresponding to the first two modes and the fourth mode of the thin plate structure. Additionally, the damping ratios for the filters are set to 0.12, 0.09, and 0.0625, respectively. Figure 15 shows the theoretical and experimental results of the velocity transfer functions of three parallel bandpass filters.

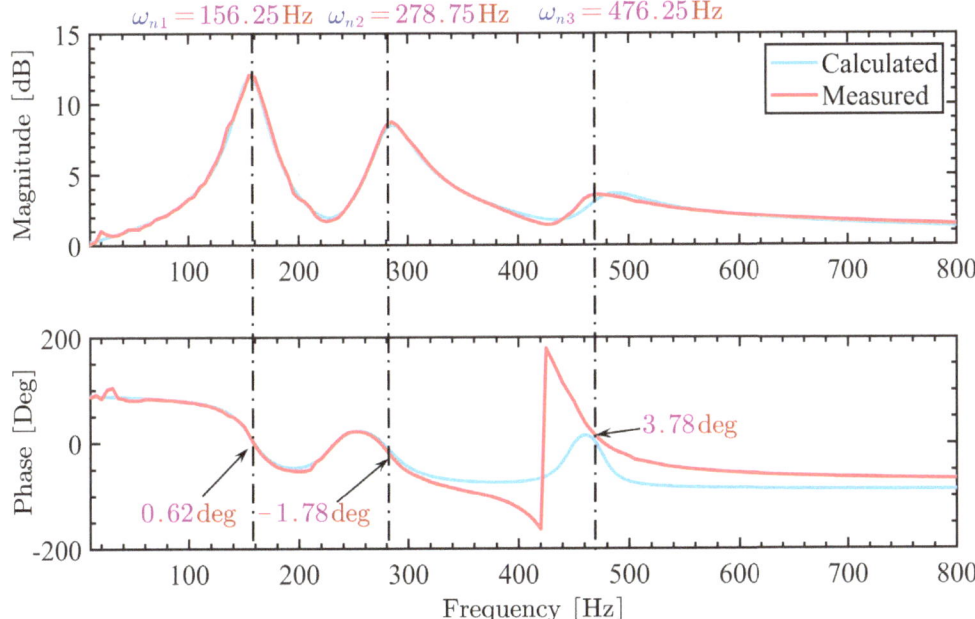

Figure 15. The velocity transfer function test diagram of bandpass filter is also combined.

According to Figure 15, the theoretical calculation of the bandpass filter is basically consistent with the experimental results. However, an interesting observation was made regarding a phase jump in the bandpass filter before reaching the third natural frequency of the control. Specifically, the phase first decays to −180 degrees and then rapidly jumps to approximately +180 degrees. This phase jump is a result of the relative positions of the poles and zeros of the filter. As the signal approaches the third-order operating frequency, the phase of the poles and zeros undergoes changes, leading to the occurrence of this phase jump. Additionally, at the operating frequency, the contributions of the poles and zeros to the phase cancel each other out, resulting in a phase of 0 degrees.

It is important to note that the phase jump is a normal phenomenon and does not adversely affect the performance of the filter. Despite the jump, the phase remains close to 0 degrees at the operating frequency and does not occur within the control frequency.

Figure 16 illustrates the calculated and experimental control performances of the LBIA with integrated piezoelectric sensor combined with three parallel bandpass filters. The experimental and calculation results demonstrate a high level of consistency. Notably, Figure 16b presents the control effect curve of the experimental test, revealing a reduction of approximately 10.6 dB in the peak value of the first mode and around 13.7 dB in the peak value of the second mode. Furthermore, the fourth mode is entirely suppressed. Intriguingly, the first and second modes exhibit the emergence of two new modes characterized by minimal damping. This damping effect primarily arises from the utilization of the bandpass filter, which efficiently absorbs vibration energy and enhances the control effect. The experimental curve aligns closely with the numerical calculation results, effectively introducing active damping for controlling structural vibration.

Nevertheless, disparities between the experimental and numerical approaches persist, attributable to several factors. Firstly, the measurement of structural vibration incurs inherent noise. Secondly, the experimental setup exhibits a resonance shift caused by non-ideal boundary conditions, in contrast to the simulations where full clamping is assumed. Finally, the imperfect match between the filter and sensor also impacts the performance of the control system.

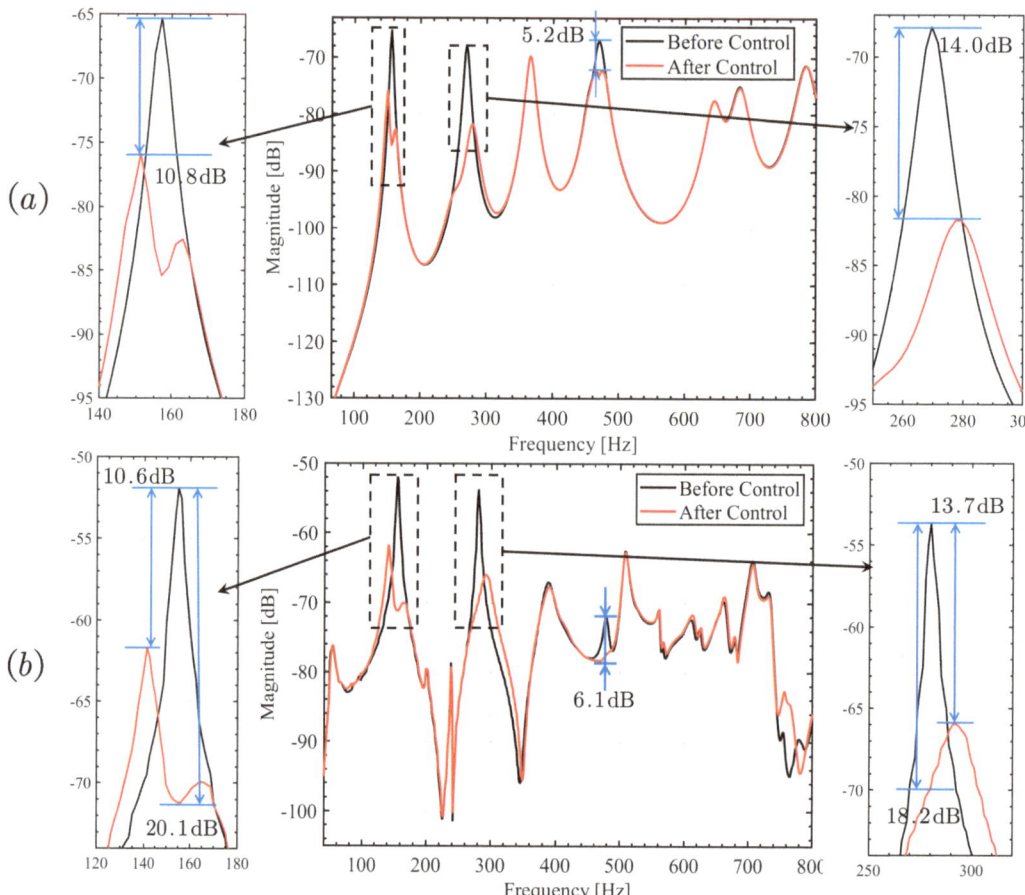

Figure 16. Bandpass filtering algorithm multimodal control effect. (**a**) Calculation result. (**b**) Experimental result.

Despite these variations, the model accurately captures the trends in the experimental results and serves as a suitable basis for comparison, thereby showcasing its efficacy.

Figure 17 displays time domain plots of acceleration before and after applying three different control solutions, all measured using the same positions of the velocity sensor and laser vibrometer. By examining the time domain diagram obtained from the velocity sensor, a comparison of the structural vibration state can be made between various control schemes under both uncontrolled and controlled conditions. Furthermore, Table 3 presents the percentage reduction in vibration for three control approaches.

Table 3. Comparison of FFT peak reduction rate with three control algorithms.

	Mode	Direct Velocity Feedback (Laser Vibrometer)	Strain Differential Feedback	Bandpass Filter
Reduction (%)	I	79.55%	49.53%	66.08%
	II	88.16%	77.55%	93.18%
	IV	61.05%	49.53%	66.08%

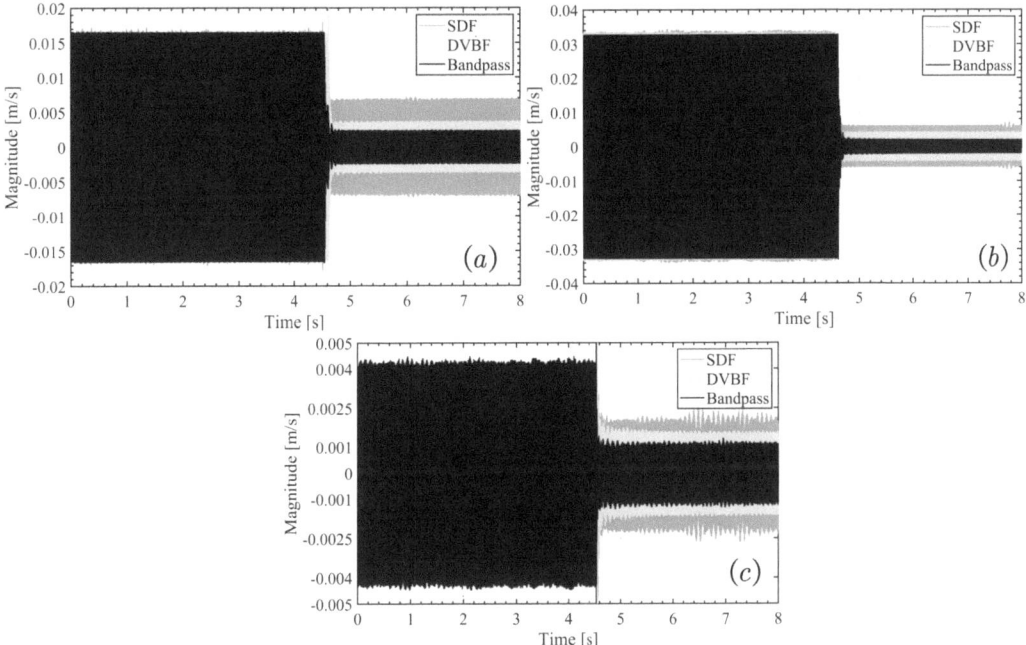

Figure 17. The experimental time domain results with different control approaches for (**a**) the first mode, (**b**) the second mode, and (**c**) the fourth mode.

From the observations of the percentage reduction, it can be concluded that mode II exhibits less damping than mode I, and mode I has less damping than mode IV. This pattern arises because, at this specific frequency, the inertial actuator efficiently transmits its force, leading to reduced vibration amplitudes. This mainly depends on the choice of control position and the design of the inertial actuator. Furthermore, the bandpass filter demonstrates the most effective control effect. It is capable of significantly reducing the vibration of the mode and further enhancing the control effect of the LBIA.

5. Discussion and Conclusions

This paper introduces an LBIA-integrated piezoelectric ceramic sensor control unit that serves as an inertial actuator for active vibration control. To achieve multimodal vibration control and enhance the control unit's effectiveness, a velocity feedback controller with multiple second-order bandpass filters in parallel is also used. The main findings of this study are outlined below:

Firstly, we completed the structural design of the LBIA as an inertial actuator and established its dynamic model and impedance model. The actuator's output characteristics and frequency response are tested on the experimental platform to verify the linear relationship between input and output, leading to the determination of the effective working frequency band.

Secondly, the integration of the piezoelectric ceramic sensor into LBIA as a separate control unit allows for establishing a mathematical model of the LBIA with piezoelectric ceramic sensor. The piezoelectric ceramic sensor converts its charge signal into a current signal through a current amplifier, resembling a differential process, which has been theoretically and experimentally proven. The resulting output current signal from the integrated sensor can be directly employed for actively controlling the vibrations of the structure. Additionally, the effectiveness of the control unit utilizing the SDF control algorithm independently for vibration control of the thin plate structure is verified through numerical calculations and experiments.

Moreover, the effectiveness of LBIA for active control of thin plate structures is experimentally verified through its traditional direct velocity negative feedback algorithm. Experimental verification is conducted using the velocity sensor and acceleration sensor on the thin plate structure. The results demonstrate the homemade inertial actuator's effectiveness in controlling the first two modes of vibration of the thin plate structure and the fourth mode as well. The control effect is shown in Figure 10, with the observed reductions of 12.3 dB, 23.6 dB, and 7.9 dB being consistent with the numerical calculation results, thus confirming the superiority and reliability of the control design.

It is noteworthy that the integrated design of the control unit eliminates the need for additional cumbersome alignment operations to address the control problem. The sensors are adjusted to become velocity signals that are directly usable in the control loop, avoiding the need for additional external circuitry. Furthermore, the control unit adopts a lightweight design (32.4 g) with a low cost (about USD 2.5), enabling simple operation, cost-effectiveness, compact size, and effective control of the low-frequency-band mode of structural vibration.

Moreover, to enhance the control effect of the LBIA with an integrated piezoelectric ceramic thin plate, a well-designed bandpass filter velocity negative feedback control algorithm is developed. The main advantage of the bandpass filter lies in its ability to attenuate the vibration of multiple modes simultaneously. By setting the natural frequency of the bandpass filter equal to the control frequency of the target mode, the control effect of the target mode is significantly improved. Taking the thin plate structure as an example, the performance of the proposed bandpass filter's DVBF control is experimentally and numerically evaluated. The results show that, compared with the control unit, the control effect of the second mode is improved from 7.3 dB to 13.7 dB, with some improvement also observed in the first and second modes. The homemade inertial actuator effectively responds to the control signal processed by the filter and suppresses the vibration of the thin plate. The fourth-order natural frequency, as defined in this paper, may result from external noise; it can be conceptualized as a "non-resonant" frequency. Notably, it can be effectively regulated using a specifically designed bandpass filter to attenuate vibrations stemming from local responses. Consequently, this technique can be extended to address structural vibrations in diverse multiphysical environments, making it a promising direction for future research.

In summary, the LBIA with an integrated piezoelectric ceramic sensor, combined with the bandpass filter DVBF control algorithm proposed in this study, demonstrates remarkable results in multimodal vibration control. The lightweight design and low cost of the control unit provide valuable insights for its industrial application. Future research directions will include parameter optimization of the current method and its application to more complex structures, enabling wider applications of bandpass filters in multimodal control.

Author Contributions: M.C., original draft preparation, software, and experiment. Q.M., conceptualization, resources, project administration, funding acquisition, writing—review and editing, and supervision. L.P., data curation. Q.L., investigation. All authors have read and approved the final manuscript.

Funding: This work was sponsored by the National Natural Science Foundation of China (No. 12364058 and No. 51975266), the Aeronautical Science Foundation of China (No. 2020Z073056001), and the Graduate Innovative Special Fund Projects of Jiangxi Province (No. YC2023-S698).

Data Availability Statement: The data that support the findings of this study are available from the corresponding author upon reasonable request

Conflicts of Interest: The authors declare no conflict of interest.

References

1. Fuller, C.C.; Elliott, S.; Nelson, P.A. *Active Control of Vibration*; Academic Press: London, UK, 1996; pp. 1–2.
2. Hansen, C.; Snyder, S.; Qiu, X.; Brooks, L.; Moreau, D. *Active Control of Noise and Vibration*; CRC Press: Boca Raton, FL, USA, 2012; pp. 1–13.
3. Mao, Q.; Pietrzko, S. *Control of Noise and Structural Vibration*; Springer: London, UK, 2013; pp. 213–258.
4. Peláez-Rodríguez, C.; Magdaleno, A.; Iglesias-Pordomingo, Á.; Pérez-Aracil, J. Evolutionary Computation-Based Active Mass Damper Implementation for Vibration Mitigation in Slender Structures Using a Low-Cost Processor. *Actuators* 2023, *12*, 254. [CrossRef]
5. Mao, Q.; Yuan, W.; Wu, J. Design and Experimental Study of a Multiple-Degree-of-Freedom Vibration Absorber by Using an Inertial Actuator. *Int. J. Struct. Stab. Dyn.* 2023, in press. [CrossRef]
6. Elliott, S.J.; Rohlfing, J.; Gardonio, P. Multifunctional design of inertially-actuated velocity feedback controllers. *J. Acoust. Soc. Am.* 2012, *131*, 1150–1157. [CrossRef]
7. Huyanan, S.; Sims, N.D. Vibration control strategies for proof-mass actuators. *J. Vib. Control* 2007, *13*, 1785–1806. [CrossRef]
8. Mao, Q.; Li, S.; Huang, S. Inertial actuator with virtual mass for active vibration control. *Int. J. Acoust. Vib.* 2020, *25*, 445–452. [CrossRef]
9. Díaz, I.M.; Reynolds, P. Robust saturated control of human-induced floor vibrations via a proof-mass actuator. *Smart Mater. Struct.* 2009, *18*, 125024. [CrossRef]
10. Caresta, M.; Kessissoglou, N. Active control of sound radiated by a submarine hull in axisymmetric vibration using inertial actuators. *J. Vib. Acoust.* 2012, *134*, 011002. [CrossRef]
11. Haase, T.; Unruh, O.; Algermissen, S.; Pohl, M. Active control of counter-rotating open rotor interior noise in a Dornier 728 experimental aircraft. *J. Sound Vib.* 2016, *376*, 18–32. [CrossRef]
12. Díaz, C.G.; Gardonio, P. Feedback control laws for proof-mass electrodynamic actuators. *Smart Mater. Struct.* 2007, *16*, 1766. [CrossRef]
13. Zilletti, M.; Elliott, S.J.; Gardonio, P. Self-tuning control systems of decentralised velocity feedback. *J. Sound Vib.* 2010, *329*, 2738–2750. [CrossRef]
14. Rohlfing, J.; Gardonio, P.; Thompson, D.J. Comparison of decentralized velocity feedback control for thin homogeneous and stiff sandwich panels using electrodynamic proof-mass actuators. *J. Sound Vib.* 2011, *330*, 843–867. [CrossRef]
15. Camperi, S.; Tehrani, M.G.; Elliott, S.J. Local tuning and power requirements of a multi-input multi-output decentralised velocity feedback with inertial actuators. *Mech. Syst. Signal Process.* 2019, *117*, 689–708. [CrossRef]
16. Elliott, S.J.; Gardonio, P.; Sors, T.C.; Brennan, M.J. Active vibroacoustic control with multiple local feedback loops. *J. Acoust. Soc. Am.* 2002, *111*, 908–915. [CrossRef] [PubMed]
17. Debattisti, N.; Bacci, M.L.; Cinquemani, S. Distributed wireless-based control strategy through Selective Negative Derivative Feedback algorithm. *Mech. Syst. Signal Process.* 2020, *142*, 106742. [CrossRef]
18. Baudry, M.; Micheau, P.; Berry, A. Decentralized harmonic active vibration control of a flexible thin plate using piezoelectric actuator-sensor pairs. *J. Acoust. Soc. Am.* 2006, *119*, 262–277. [CrossRef]
19. Balas, M. Feedback control of flexible systems. *IEEE Trans. Automat. Contr.* 1978, *23*, 673–679. [CrossRef]
20. Gardonio, P.; Bianchi, E.; Elliott, S.J. Smart panel with multiple decentralized units for the control of sound transmission. Part I: Theoretical predictions. *J. Sound Vib.* 2004, *274*, 163–192. [CrossRef]
21. Gardonio, P.; Bianchi, E.; Elliott, S.J. Smart panel with multiple decentralized units for the control of sound transmission. Part II: Design of the decentralized control units. *J. Sound Vib.* 2004, *274*, 193–213. [CrossRef]
22. Gardonio, P.; Bianchi, E.; Elliott, S.J. Smart panel with multiple decentralized units for the control of sound transmission. Part III: Control system implementation. *J. Sound Vib.* 2004, *274*, 215–232.
23. Silva, T.M.P.; Hameury, C.; Ferrari, G.; Balasubramanian, P.; Franchini, G.; Amabili, M. Particle swarm optimization of a non-collocated MIMO PPF active vibration control of a composite sandwich thin plate. *J. Sound Vib.* 2023, *555*, 117723. [CrossRef]
24. Engels, W.P.; Baumann, O.N.; Elliott, S.J.; Fraanje, R. Centralized and decentralized control of structural vibration and sound radiation. *J. Acoust. Soc. Am.* 2006, *119*, 025009. [CrossRef]
25. Camperi, S.; Tehrani, M.G.; Elliott, S.J. Parametric study on the optimal tuning of an inertial actuator for vibration control of a thin plate: Theory and experiments. *J. Sound Vib.* 2018, *435*, 1–22. [CrossRef]
26. Zilletti, M.; Gardonio, P.; Elliott, S.J. Optimisation of a velocity feedback controller to minimise kinetic energy and maximise power dissipation. *J. Sound Vib.* 2014, *333*, 4405–4414. [CrossRef]
27. Elliott, S.J.; Zilletti, M.; Gardonio, P. Self-tuning of local velocity feedback controllers to maximize power absorption. In *Recent Advances Structural Dynamics: Proceedings of the X International Conference, Southampton, UK, 12–14 July 2010*; Plymouth University: Plymouth, UK, 2010.
28. Gardonio, P.; Miani, S.; Blanchini, F.; Casagrande, D.; Elliott, S.J. Thin plate with decentralised velocity feedback loops: Power absorption and kinetic energy considerations. *J. Sound Vib.* 2012, *331*, 1722–1741. [CrossRef]
29. Goh, C.J. Analysis and Control of Quasi-Distributed Parameter Systems. Ph.D. Dissertation, California Institute of Technology, Pasadena, CA, USA, 1983.
30. Fanson, J.L. An Experimental Investigation of Vibration Suppression in Large Space Structures Using Positive Position Feedback. Ph.D. Dissertation, California Institute of Technology, Pasadena, CA, USA, 1978.

31. Dosch, J.J.; Inman, D.J.; Garcia, E.A. A self-sensing piezoelectric actuator for collocated control. *J. Intell. Mater. Syst. Struct.* **1992**, *3*, 166–185. [CrossRef]
32. Dosch, J.J.; Leo, D.J.; Inman, D.J. Comparison of vibration control schemes for a smart antenna. In Proceedings of the 31st IEEE Conference on Decision and Control, Tucson, AZ, USA, 16–18 December 1992.
33. Jha, A.K.; Inman, D.J. Optimal sizes and placements of piezoelectric actuators and sensors for an inflated torus. *J. Intell. Mater. Syst. Struct.* **2003**, *14*, 563–576. [CrossRef]
34. El-Khoury, O.; Adeli, H. Recent advances on vibration control of structures under dynamic loading. *Arch. Comput. Methods Eng.* **2013**, *20*, 353–360. [CrossRef]
35. Fanson, J.L.; Caughey, T.K. Positive position feedback control for large space structures. *AIAA J.* **1990**, *28*, 717–724. [CrossRef]
36. Haase, F.; Kauba, M.; Mayer, D.; Van der Auweraer, H.; Gajdatsy, P.; de Oliveira, L.; da Silva, M.; Sas, P.; Deraemaeker, A. Active vibration control of an automotive firewall for interior noise reduction. In Proceedings of the Adaptronic Congress 2008, Berlin, Germany, 1 May 2008.
37. Omidi, E.; Mahmoodi, N. Hybrid positive feedback control for active vibration attenuation of flexible structures. *IEEE/ASME Trans. Mechatron.* **2014**, *20*, 1790–1797. [CrossRef]
38. Preumont, A. *Vibration Control of Active Structures: An Introduction*; Springer: London, UK, 2018; pp. 47–66.
39. Friswell, M.I.; Inman, D.J. The relationship between positive position feedback and output feedback controllers. *Smart Mater. Struct.* **1999**, *8*, 285–291. [CrossRef]
40. Sim, E.; Lee, S.W. Active vibration control of flexible structures with acceleration feedback. *J. Guid. Control Dyn.* **1993**, *16*, 413–415. [CrossRef]
41. Zhao, G.; Paknejad, A.; Raze, G.; Deraemaeker, A.; Kerschen, G.; Collette, C. Nonlinear positive position feedback control for mitigation of nonlinear vibrations. *Mech. Syst. Signal Process.* **2019**, *132*, 457–470. [CrossRef]
42. Paknejad, A.; Zhao, G.; Osée, M.; Deraemaeker, A.; Robert, F.; Collette, C. A novel design of positive position feedback controller based on maximum damping and H2 optimization. *J. Vib. Control.* **2020**, *26*, 1155–1164. [CrossRef]
43. Balasubramanian, P.; Ferrari, G.; Hameury, C.; Silva, T.M.; Buabdulla, A.; Amabili, M. An experimental method to estimate the electro-mechanical coupling for active vibration control of a non-collocated free-edge sandwich thin plate. *Mech. Syst. Signal Process.* **2023**, *188*, 110043. [CrossRef]
44. Omidi, E.; Mahmoodi, S.N.; Shepard, W.S., Jr. Multi positive feedback control method for active vibration suppression in flexible structures. *Mechatronics* **2016**, *33*, 23–33. [CrossRef]
45. Kournoutos, N.; Cheer, J. A system for controlling the directivity of sound radiated from a structure. *J. Acoust. Soc. Am.* **2020**, *147*, 231–241. [CrossRef]
46. Zhong, H.; Wu, J.; Bao, B.; Mao, Q. A composite beam integrating an in-situ FPCB sensor membrane with PVDF arrays for modal curvature measurement. *Measurement* **2020**, *166*, 108241. [CrossRef]
47. Mao, Q.; Zhong, H. Sound power estimation for beam and thin plate structures using polyvinylidene fluoride films as sensors. *Sensors* **2017**, *17*, 1111. [CrossRef]

Disclaimer/Publisher's Note: The statements, opinions and data contained in all publications are solely those of the individual author(s) and contributor(s) and not of MDPI and/or the editor(s). MDPI and/or the editor(s) disclaim responsibility for any injury to people or property resulting from any ideas, methods, instructions or products referred to in the content.

Article

A Linear Iterative Controller for Software Defined Control Systems of Aero-Engines Based on LMI

Xiaoxiang Ji, Jiao Ren, Jianghong Li * and Yafeng Wu

School of Power and Energy, Northwestern Polytechnical University, Xi'an 710072, China; jixiaoxiang@mail.nwpu.edu.cn (X.J.); renjiao@mail.nwpu.edu.cn (J.R.); yfwu@nwpu.edu.cn (Y.W.)
* Correspondence: jhli@nwpu.edu.cn

Abstract: Currently, most control systems of the aero-engines possess a central controller. The core tasks for the control system, such as control law calculations, are executed in this central controller, and its performance and reliability greatly impact the entire control system. This paper introduces a control system design named Software Defined Control Systems (SDCS), which features a controller-decentralized architecture. In SDCS, a network composed of a set of nodes serves as the controller, so there is no central controller in the system, and computations are distributed throughout the entire network. Since the controller is decentralized, there is a need for decentralized control tasks. To address this, this paper introduces a method for designing decentralized control tasks using periodic linear iteration. Each node in the network periodically broadcasts its own state and updates its next-step state as a weighted sum of its current state and the received current states of other nodes in the network. Each node in the network acts as a linear dynamic controller and maintains an internal state through information exchange with other nodes. We modeled the decentralized controller and obtained the model of the entire control system, and the workload of each obtained decentralized control task is balanced. Then, we obtained a parameter tuning method for each decentralized controller node based on Linear Matrix Inequalities (LMI) to stabilize the closed-loop system. Finally, the effectiveness of the proposed method was verified through digital simulation.

Keywords: decentralized controller; linear iterative; aero-engines; dynamic controller; Linear Matrix Inequalities (LMI)

Citation: Ji, X.; Ren, J.; Li, J.; Wu, Y. A Linear Iterative Controller for Software Defined Control Systems of Aero-Engines Based on LMI. *Actuators* **2023**, *12*, 259. https://doi.org/10.3390/act12070259

Academic Editors: Ti Chen, Junjie Kang, Shidong Xu and Shuo Zhang

Received: 28 May 2023
Revised: 17 June 2023
Accepted: 20 June 2023
Published: 23 June 2023

Copyright: © 2023 by the authors. Licensee MDPI, Basel, Switzerland. This article is an open access article distributed under the terms and conditions of the Creative Commons Attribution (CC BY) license (https:// creativecommons.org/licenses/by/ 4.0/).

1. Introduction

The traditional aero-engine control system uses a centralized control scheme, and the controller is connected to the analog sensors and actuator through cables and connectors. The controller executes most of the tasks of the control system, such as control law calculation, digital-to-analog conversion (D/A), analog-to-digital conversion (A/D), and so on, and the workload of the controller is large. Meanwhile, the centralized control scheme increases the difficulty and cost of upgrading and maintaining aero-engines throughout their entire life cycle. In addition, the presence of rotating components results in a large amount of cables within the system, which reduces the thrust-to-weight ratio of the aero-engines. To overcome the drawbacks of centralized control schemes, researchers have introduced the design concept of distributed control into the design of aero-engine control systems [1–5].

The Distributed Control System (DCS) first appeared in industrial process control [6,7], and it has been widely adopted in many fields [8–10]. The core design concept of the distributed control system for aero-engines is functional decentralization, which means that some functions of a centralized control system, such as signal conditioning and state monitoring, are executed by intelligent sensors and actuators, while the central controller focuses on the core tasks of the control system, such as control law calculation [11]. In a DCS, the controller, intelligent sensors, and intelligent actuators are connected through a common and standardized digital bus [12,13]. Compared to centralized control, DCS has the following advantages: (1) digital intelligent sensors and actuators reduce the control tasks of the central controller. (2) The digital bus reduces the weight of the aero-engine control system. (3) Modular design solutions lower the difficulty and cost of system upgrades and maintenance processes.

In addition to the above advantages, DCS also has some disadvantages:

1. As the performance of aircraft and aero-engines gradually improve, the functionality and complexity of central control tasks also rapidly increase, which requires high-performance, multi-core microprocessors as controllers, and it places high demands on the thermal management system of the aviation engine.
2. Due to the increasing functionality and complexity of core control tasks, the amount of software code in the control system rapidly increases, reducing the software reliability of the control system.
3. Control tasks are centralized on the central controller. The central controller determines the performance of the aero-engine's control system, and its damage or failure have a significant impact on the control system.

The analysis of the DCS for aero-engines mentioned above shows that the drawbacks in the system are caused by the central controller present in the system. The drawbacks of DCS suggest that a feasible scheme is to make the controllers in the control system decentralized. This involves using a network of decentralized nodes as the system controller, where multiple microprocessors work together to execute the control tasks of the central controller of DCS. As a result, the workload of each decentralized node is significantly reduced. With the decrease in workload, the software code volume for each node also decreases, thereby improving the software reliability of the control system. Additionally, several low-performance, but highly reliable microprocessors, can be used as decentralized controllers, which enhances the reliability of individual nodes.

The origin of the decentralized control design scheme dates back to the 1970s, and its concept was mainly proposed to solve the control problem of large-scale systems. Large-scale systems are characterized by extensive spatial distribution, numerous external signals, or dynamic changes in control structure. The core problem is that the control system is too large, and the control problem is too complex [14]. It is generally difficult to achieve control objectives by using more powerful microprocessors and larger storage space for large-scale systems. Due to the characteristics of large-scale systems, it is usually necessary to analyze and process the controlled plants, divide the system control problems into independent sub-problems, and handle signal delays caused by large spaces [15]. Distributed control in the industrial process control field adopts this design concept. This includes the later network control systems (NCS) and wireless networked control systems (WNCS). In WNCS, wireless communication is used to transmit data between system components. Compared with wired communication, wireless communication has its advantages and disadvantages. Its advantages include flexibility in installation, easy configuration, and strong adaptability; disadvantages include communication delays, data packet loss, etc. Ref. [16] conducted a study on the decentralized control system for the autonomous guided vehicles (AGVs) path planning problem, proposing two decentralized control methods to solve the AGV control problem: task allocation for AGVs through consensus method and path planning coordination through decentralized control strategy. In this design scheme, the microprocessors of several vehicles form the controller for the AGV problem, specifically targeting path optimization in a large spatial range. However, for each vehicle, there still exists a central controller in its control system. Ref. [17] researched the decentralized control of automation systems, such as smart factories and smart cities. In addition to data related to control tasks, there is also data interaction, such as images and voices, resulting in high communication requirements for the system. Inspired by concepts, such as software-defined networking (SDN), a decentralized data interaction control method was designed for large space and large data flow systems through extending Lyapunov drift-plus-penalty (LDP) control to a new queuing system to adjust the data flow within the system. Ref. [18] studied the decentralized control problem of two coupled power systems, including wind turbines and diesel generators. In this paper, the system was not decoupled and treated as a whole, and a decentralized controller was designed consisting of two PI-lead controllers. This scheme mainly focused on the research of PI controllers and parameter tuning, but the application of this scheme is limited for cases where the control effect of some PI controllers is poor. Ref. [19] proposed a decentralized control scheme for aero-engines, which replaced the central controller in traditional aero-engine control systems with multiple controllers to control different subsystems of the aero-engine. However, in this scheme, the controllers are fixedly mapped to the subsystems, and this means that there exists a central controller for each subsystem, and the control system structure is relatively inflexible. In the aforementioned studies, each decentralized controller can independently complete specific control tasks.

Ref. [20] proposed a controller decentralized design method, namely, the Software-Defined Control System (SDCS). In SDCS, the network composed of decentralized nodes serves as the controller to execute control tasks, such as control law execution. Therefore, the control tasks executed by the decentralized nodes are also decentralized, and each decentralized control task is a part of the core control tasks. The core control tasks are the sum of these decentralized control tasks in a specific way, no individual decentralized node can complete the system control task alone. The aero-engine is a safety-critical system that requires high reliability of the control system. When applying SDCS to the aero-engine control system, highly reliable but low-performance microprocessors should be used as decentralized nodes. This requires that the workload and software code volume of each decentralized control task are low, and complex control functions need to be executed.

The contributions of this paper are as follows. A linear iterative-based decentralized design scheme for control tasks was introduced. In this scheme, a network composed of several decentralized nodes act as the controller to undertake the system control task. Each decentralized node broadcasts its own state information to other decentralized nodes in each cycle period. Other nodes receive the broadcasted state information, and when all decentralized nodes have completed broadcasting their state information, all nodes update their own states as a weighted sum of their current state and the received states from other nodes. Therefore, each task node can act as a small dynamic controller. Through the linear iterative process, a model of the decentralized controller was constructed, which obtained the task executed on each decentralized node and the model of the entire control system. Due to the controller being located in the forward path and the linear iterative process introducing new internal state variables, conventional state feedback or output feedback design schemes cannot be used for decentralized controller node parameter tuning. Additionally, since the parameters with respect to the Lyapunov condition are nonlinear, conventional Lyapunov methods and LMI methods cannot be directly applied to parameter tuning of decentralized controller nodes. A parameter tuning method for decentralized controller nodes based on LMI was presented for the designed decentralized control task to ensure that the control system is Schur stable.

The remainder of this paper is organized as follows.

Section 2 describes a linear model for aero-engines. Section 3 presents the structure of the Software Defined Control System for aero-engines and a periodic linear iterative-based decentralized controller implementation scheme, along with control system modeling. Section 4 introduces a parameter tuning method for the controller based on LMI. Section 5 conducts simulation verification of the designed scheme. Section 6 is the final chapter of this paper, which summarizes and discusses the future outlook.

2. Aero-Engine Model

The mathematical model of aero-engines is crucial for the design of aero-engines control systems. As aero-engines are complex, time-varying, and strongly nonlinear systems, their mathematical models should be nonlinear [21–24]. The current controllers are designed based on linear systems for aero-engines [25,26]. The decentralized controller proposed later in this paper is designed based on a periodic linear iterative scheme and is linear, hence requiring a linear model of the aero-engines.

The modeling methods for aero-engines generally include two types: identification method and analytical method. The identification method requires corresponding experimental conditions, has high costs, and some of its algorithms have large computational consumption. Moreover, the resulting model may lose certain characteristics of the engine. The analytical method is based on the principles of aero-engines to build models. Firstly, the components of the aero-engines are modeled; then, a series of nonlinear equations that describe the working process of the aero-engines by following the aerodynamic and thermodynamic principles during their operation are utilized to obtain the mathematical model of the aero-engines [27,28].

In this section, we analyze the case of a turbofan engine using the method described in Refs. [27,28]. This section first introduces the nonlinear model of turbofan engines and then describes the linearization method for the nonlinear model of turbofan engines.

2.1. Nonlinear Model of Turbofan Engines

In this sub-section, the models of various components of turbofan engines are presented, including the inlet, fan, compressor, combustor, turbine, bypass duct, mixer, and exhaust nozzle.

(1) Intake

When air enters the aero-engines, it first flows through the intake. The intake's inlet parameters are:

When $H \leq 11$ km,

$$
\begin{aligned}
T_1 &= 288.15 - 0.0065H \\
P_1 &= 101,325 \times \left(1 - \frac{H}{44,331}\right)^{5.25588}
\end{aligned}
\tag{1}
$$

When $H > 11$ km,

$$
\begin{aligned}
T_1 &= 216.5 \\
P_1 &= 22,632 \times e^{\frac{11,000-H}{6342}}
\end{aligned}
\tag{2}
$$

where; T_1 and P_1 are the total temperature and the total pressure at the inlet of the intake, respectively.

(2) Fan

The inlet parameters of aero-engines fan are:

$$
\begin{aligned}
T_2 &= T_1\left(1 + \frac{k_1-1}{2}M_a^2\right) \\
P_2 &= \sigma_1 P_1\left(1 + \frac{k_1-1}{2}M_a^2\right)^{\frac{k_1}{k_1-1}}
\end{aligned}
\tag{3}
$$

where; T_2 and P_2 are the total temperature and the total pressure at the inlet of the fan, respectively, M_a is Mach number, σ_1 is the total pressure recovery coefficient of the intake, and k_1 is the air adiabatic index.

The outlet parameters of aero-engines fan are:

$$
\begin{aligned}
T_{21} &= T_2\left(1 + \frac{\pi_F^{\frac{k_F-1}{k_F}} - 1}{\eta_F}\right) \\
P_{21} &= \pi_F P_2
\end{aligned}
\tag{4}
$$

where; T_{21} and P_{21} are the total temperature and the total pressure at the outlet of the fan, respectively, k_F is the air adiabatic index, π_F is the fan pressure ratio, and η_F is the efficiency of the fan.

(3) Compressor

The inlet parameters of aero-engines compressor are:

$$T_{22} = T_{21}$$
$$P_{22} = \sigma_2 P_{21} \tag{5}$$

where; T_{22} and P_{22} are the total temperature and the total pressure at the inlet of the compressor, respectively, σ_2 is the total pressure recovery coefficient of the fan.

The outlet parameters of aero-engines compressor are:

$$T_3 = T_{22}\left(1 + \frac{\pi_C^{\frac{k_C-1}{k_C}} - 1}{\eta_C}\right)$$
$$P_3 = \pi_C P_{22} \tag{6}$$

where; T_3 and P_3 are the total temperature and the total pressure at the outlet of the compressor, respectively, π_C is the pressure ratio of the compressor, η_C is the efficiency of the compressor, and k_C is the air adiabatic index.

(4) Combustion chamber

According to the energy conservation law, the simplified energy balance equation for the combustion chamber is:

$$W_{mf} H_u \eta_r + W_{ma} c_p T_3 = W_{ma} c_p T_4 \tag{7}$$

where; T_4 is the temperature at the outlet of the combustion chamber, W_{mf} is the fuel flow, H_u is the calorific value of the fuel, η_r is the combustion efficiency, W_{ma} is the air flow at the inlet of the combustion chamber, and c_p is the specific heat capacity of air at constant pressure. The outlet pressure P_4 of the combustion chamber can be calculated using the above equation.

$$P_4 = \sigma_3 P_3 \tag{8}$$

where; P_4 is the pressure at the outlet of the combustion chamber, and σ_3 is the total pressure recovery coefficient of the combustion chamber.

(5) High-pressure turbine

The outlet parameters of aero-engines high-pressure turbine are:

$$T_{41} = \frac{q_{mg,TH} T_4 + C_{HPTCool} q_{maC,totalTcool}}{q_{mg,TH,total}} \left(1 - \left(1 - \pi_{TH}^{\frac{1-k_{TH}}{k_{TH}}}\right)\eta_{TH}\right)$$
$$P_{41} = \frac{P_4}{\pi_{TH}} \tag{9}$$

where; T_{41} and P_{41} are the total temperature and the total pressure at the outlet of the high-pressure turbine, respectively, $q_{mg,TH}$ is the gas flow of the high-pressure turbine, $C_{HPTCool}$ is the proportion coefficient of the high-pressure compressor bleed air used to cool the high-pressure turbine, $q_{maC,totalTcool}$ is the total flow of added air, $q_{mg,TH,total}$ is the outlet air flow of the high-pressure turbine, π_{TH} is the high-pressure turbine pressure ratio, η_{TH} is the efficiency of the high-pressure turbine, and k_{TH} is the gas adiabatic index.

(6) Low-pressure turbine

The outlet parameters of aero-engines low-pressure turbine are:

$$T_5 = \frac{q_{mg,TL} T_{42} + C_{LPTCool} q_{maC,totalTcool}}{q_{mg,TL,total}} \left(1 - \left(1 - \pi_{TL}^{\frac{1-k_{TL}}{k_{TL}}}\right)\eta_{TL}\right)$$
$$P_5 = \frac{P_{42}}{\pi_{TL}} \tag{10}$$

where; T_5 and P_5 are the total temperature and the total pressure at the outlet of the low-pressure turbine, respectively, $q_{mg,TL}$ is the gas flow converted by the high-pressure turbine, and T_{42} is the inlet temperature of the low-pressure turbine, which is approximately equals to the outlet temperature of the high-pressure turbine T_{41}. P_{42} is the inlet pressure of the low-pressure turbine, which is approximately equals to the outlet pressure of the high-pressure turbine P_{41}. $C_{LPTCool}$ is the proportion coefficient of the high-pressure compressor bleed air used to cool the low-pressure turbine, $q_{mg,TL,total}$ is the outlet air flow of the low-pressure turbine, π_{TL} is the low-pressure turbine pressure ratio, η_{TL} is the efficiency of the low-pressure turbine, and k_{TL} is the gas adiabatic index.

(7) Bypass duct

The outlet parameters of aero-engines bypass duct are:

$$\begin{aligned} T_6 &= T_{21} \\ P_6 &= \sigma_4 P_{21} \end{aligned} \quad (11)$$

where; T_6 and P_6 are the total temperature and the total pressure at the outlet of the bypass duct, respectively, and σ_4 is the total pressure recovery coefficient of the bypass duct.

(8) Mixer

The gas flow $q_{mg,7}$ at the outlet of the mixer is the sum of the air flow $q_{ma,6}$ at the outlet of the bypass duct and the gas flow $q_{mg,5}$ at the outlet of the low-pressure turbine, let σ_5 denotes the total pressure recovery coefficient of the mixer, then the physical parameters at the outlet of the mixer are:

$$\begin{aligned} q_{mg,7} &= q_{mg,5} + q_{ma,6} \\ h_7 &= \frac{h_5 q_{mg,5} + h_6 q_{ma,6}}{q_{mg,7}} \\ P_7 &= \sigma_5 \cdot \frac{P_5 q_{mg,5} + P_6 q_{ma,6}}{q_{mg,7}} \end{aligned} \quad (12)$$

where: h_5, h_6, and h_7 are the specific enthalpy of the gas at the outlet of the low-pressure turbine, air at the outlet of the bypass duct, and gas at the outlet of the mixer, respectively. From h_7, it is easy to obtain the temperature T_7 at the outlet of the mixer.

(9) Nozzle

The outlet parameters of aero-engines nozzle are:

$$\begin{aligned} \frac{P_8}{P_S} &= \pi_{NZ} \\ q_{mg,8} &= K_q \frac{P_8 A_8 q(\lambda_8)}{\sqrt{T_7}} \end{aligned} \quad (13)$$

where; P_8 are the pressure at the outlet of nozzle, A_8 is the sectional area of the nozzle, K_q is the state coefficient of the nozzle, $q(\lambda_8)$ refers to the function related to the characteristics of the bypass duct and the core duct, π_{NZ} is the available pressure drop of the nozzle, and P_{S0} denotes the standard atmospheric pressure.

2.2. Common Working Equations

In this paper, the high-pressure turbine and high-pressure compressor form the high-pressure rotor in the turbofan engine. The high-pressure compressor is driven by the high-pressure turbine. The low-pressure turbine and fan form the low-pressure rotor. After passing through the inlet duct and fan, the gas flow is divided into two parts, with one entering the bypass duct and the other entering the core flow. Based on the flow rate, pressure and power balance between each engine component and the principle of constant rotational speed, the following common working equation can be obtained.

(1) Power balance equation of the high-pressure rotor:

$$P_H = P_{CH} + P_{ex,H} + D_H \left(\frac{dn_H}{dt} \right) \quad (14)$$

where; P_H is the power of high-pressure turbine, P_{CH} is the power of high-pressure compressor, $P_{ex,H}$ is the power lost by transmission friction force, $D_H\left(\frac{dn_H}{dt}\right)$ is the acceleration power of high-pressure rotor, and $D_H = (\pi/30)^2 J_H n_H$, J_H is the moment of inertia of the high-pressure rotor, and n_H iss the speed of the high-pressure rotor.

When the aero-engine is in stable state, $dn_H/dt = 0$, and ignore the power lost by transmission friction force $P_{ex,H}$, that Equation (14) is simplified into:

$$P_H = P_{CH} \tag{15}$$

(2) Power balance equation of the low-pressure rotor:

When the engine is in a stable state, $dn_L/dt = 0$. Ignoring the power loss $P_{ex,L}$ caused by transmission friction, a simplified power balance equation for the low-pressure rotor can be obtained, similar to that of the high-pressure rotor.

$$P_L = P_{CL} \tag{16}$$

where; P_L is the power of low-pressure turbine, and P_{CL} is the power of low-pressure compressor.

(3) Flow balance equation of the fan:

After passing through the fan, the gas is divided into two parts: one enters the bypass duct and the other enters the high-pressure compressor. Ignoring loss of gas flow, we have:

$$q_{maF} = q_{maC} + q_{ma6} \tag{17}$$

where; q_{maF} denotes the air flow from the fan, q_{maC} denotes the air flow into the high-pressure compressor, and q_{ma6} denotes the air flow into bypass duct.

(4) Flow balance equation of the high-pressure turbine:

The gas flow of the high-pressure turbine satisfies the following equation.

$$q_{mg,4} = q_{maC} + q_{mf} \tag{18}$$

where; $q_{mg,4}$ denotes the air flow into the high-pressure turbine, q_{maC} denotes the air flow from the high-pressure compressor, and q_{mf} denotes the fuel flow.

(5) Flow balance equation of the low-pressure turbine:

The air flow at the inlet of the low-pressure turbine is equal to the air flow at the outlet of the high-pressure gas turbine $q_{mg,42}$, then:

$$q_{mg,42} = q_{mg,4} + C_{HPTCool} q_{maC,totalTcool} \tag{19}$$

where; $C_{HPTCool}$ is the proportion coefficient of the high-pressure compressor bleed air used to cool the high-pressure turbine, and $q_{maC,totalTcool}$ is the total air flow into the high-pressure turbine.

(6) Flow balance equation of the nozzle:

The gas flow of the nozzle $q_{mg,8}$ satisfies the following equation.

$$q_{mg,8} = q_{mg,5} + q_{ma,16} \tag{20}$$

where; $q_{mg,5}$ is the gas flow at the low-pressure turbine outlet, and $q_{ma,16}$ is the gas flow into the bypass duct.

2.3. Linearization of Nonlinear Models

The current design of control systems is mainly linear, so it is necessary to obtain a linear model of aero-engines. Aero-engines have strong non-linear characteristics. The common method is to select several operating points within the flight envelope and establish a linear model at these points, thus obtaining a segmented linear model of the aero-engine [29].

The partial derivative method is a common method for obtaining the linear model of aero-engines [30]. This method first perturbs a given state variable with a small disturbance, while keeping all other control variables and state variables constant. The partial derivatives of the state variables with the perturbation are then calculated to obtain the state matrix and output matrix of the state space model of the aero-engine. Then, a small perturbation is given to a given control variable, while keeping all other control variables and all state variables constant. The input matrix of the state space model is obtained by taking the partial derivatives of the control variable with the perturbation. This method only has theoretical significance because most state variables and control variables are coupled, and it is difficult to change only a single state variable or control variable without affecting other state variables or control variables. This leads to large modeling errors in this method.

This section briefly introduces the fitting method of linearization for aero-engines [31]. Firstly, a small perturbation state variable model of the aero-engine is obtained based on the selected state variables, control variables, and output variables. The linear dynamic response is calculated by giving a small step input to each control variable. Similarly, the nonlinear dynamic response is obtained by giving the same step input to the control variables of the nonlinear model of the aero-engine. Using the nonlinear dynamic response data as a reference, the matrices in the linear model are fitted to make the linear dynamic response data as close as possible to the nonlinear dynamic response data, thus obtaining the linear model of the aero-engine.

Here, taking the bivariate state-space model of aero-engines as an example, the state-space model is given below:

$$\begin{cases} \dot{x} = Ax + Bu \\ y = Cx \end{cases} \quad (21)$$

where; the state variables are selected as the low-pressure rotor speed n_L and high-pressure rotor speed n_H, the output variables are selected as the low-pressure rotor speed n_L and the turbine pressure ratio $P_iT = \pi_{TL}\pi_{TH}$, the control variables are selected as the fuel flow W_{mf}, and the area of the nozzle A_8. That is, $x = \begin{bmatrix} n_L & n_H \end{bmatrix}^T$, $u = \begin{bmatrix} W_{mf} & A_8 \end{bmatrix}^T$, and $y = \begin{bmatrix} n_L & P_iT \end{bmatrix}^T$.

Let,

$$A = \begin{bmatrix} a_{11} & a_{12} \\ a_{21} & a_{22} \end{bmatrix}, B = \begin{bmatrix} b_{11} & b_{12} \\ b_{21} & b_{22} \end{bmatrix}, C = \begin{bmatrix} c_{11} & c_{12} \\ c_{21} & c_{22} \end{bmatrix}$$

From the selected output variables and state variables, $c_{11} = 1$, $c_{12} = 0$.

Giving a small step to the fuel flow while keeping the nozzle area constant, then:

$$\frac{\Delta n_L}{\Delta W_{mf}} \frac{(W_{mf})_0}{(n_L)_0} = \frac{a_{12}b_{21} - a_{22}b_{11}}{a_{11}a_{22} - a_{12}a_{21}} + \frac{(\lambda_1 - a_{22})b_{11} + a_{12}b_{21}}{\lambda_1(\lambda_1 - \lambda_2)}e^{\lambda_1-t} - \frac{(\lambda_2 - a_{22})b_{11} + a_{12}b_{21}}{\lambda_2(\lambda_1 - \lambda_2)}e^{\lambda_2-t}$$

$$\frac{\Delta n_H}{\Delta W_{mf}} \frac{(W_{mf})_0}{(n_H)_0} = \frac{a_{21}b_{11} - a_{11}b_{21}}{a_{11}a_{22} - a_{12}a_{21}} + \frac{(\lambda_1 - a_{11})b_{21} + a_{21}b_{11}}{\lambda_1(\lambda_1 - \lambda_2)}e^{\lambda_1-t} - \frac{(\lambda_2 - a_{11})b_{21} + a_{21}b_{11}}{\lambda_2(\lambda_1 - \lambda_2)}e^{\lambda_2-t}$$

$$c_{21}\frac{\Delta n_L}{\Delta W_{mf}}\frac{(W_{mf})_0}{(n_L)_0} = -c_{22}\frac{\Delta n_H}{\Delta W_{mf}}\frac{(W_{mf})_0}{(n_H)_0}$$

where; λ_1 and λ_2 are the eigenvalues of the state matrix A.

At the same steady-state point, by keeping the nozzle area and fuel flow constant, respectively, small steps are applied to the nozzle area and fuel flow to obtain the component-level nonlinear model's small step response data for aero-engines.

Linear and nonlinear models are given the same amplitude of control variable step at the same steady-state point; the target is to make the linear and nonlinear dynamic responses as close as possible. By establishing and fitting the equation group, the state–space model of aero-engines at that steady-state point can be obtained.

3. Aero-Engines' Linear Iterative Controller

3.1. Software Defined Control Systems of the Aero-Engines

Ref. [32] proposed the Software Defined Control Systems (SDCS) for aero-engines. The SDCS for aero-engines is based on the DCS for aero-engines, and it has been proposed alongside advancements in computer technology, communication technology, sensor technology, data storage technology, and other related support technologies. First, a brief introduction to the DCS for aero-engines will be provided. The DCS for aero-engines was developed based on a centralized control system. In the centralized control system for aero-engines, the controller is connected to the engine's sensors, actuators, and other devices via cables. These devices do not have signal processing or control capabilities and interact with the controller through cables. The controller undertakes all control functions of the control system, including signal acquisition and processing, control law calculation, and control signal output. The structure of the DCS for aero-engines is shown in Figure 1 [33]. In the aero-engine DCS, signal conditioning, A/D, D/A, and other functions are delegated to the sensors and actuators, while the controller only performs core control tasks, such as control law calculations [34,35]. The sensors and actuators integrate microprocessors internally, making them intelligent sensors and intelligent actuators. Communication between the controller, intelligent sensors, and intelligent actuators is performed through a digital bus. Intelligent sensors convert measurement signals into digital signals and provide them to the controller. The functions performed by intelligent sensors include A/D, redundancy management, and interface with the digital bus. Intelligent actuators receive control commands from the controller and control the actuator. The functions performed by intelligent actuators include D/A, closed-loop feedback, redundancy management, and interface with the digital bus. The controller sends control commands to intelligent actuators at a fixed rate through the digital bus. The bus structure of the aero-engine DCS is generally divided into circle structure and linear structure. The circle bus connects each node through a circular structure and has a simple structure. The linear bus uses fewer cables but has a more complex structure. Commonly used buses in aero-engine DCS research include MIL-STD-1553, Field Bus, CAN, and others. Currently, the tasks of controllers are becoming increasingly complex and require new, high-performance microprocessors as controllers. Therefore, this has a certain impact on the reliability of the control system.

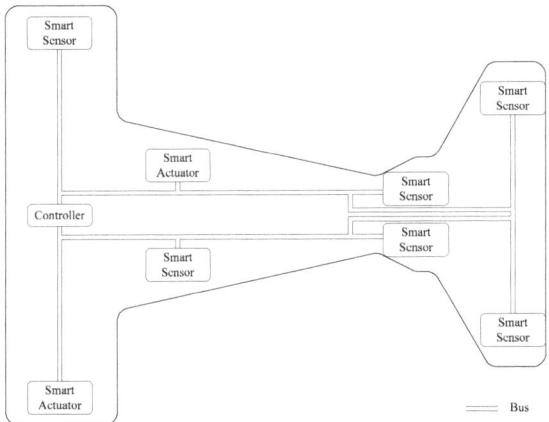

Figure 1. Overview of a traditional aero-engine DCS.

The feature of SDCS is that the controllers are decentralized. In an SDCS, a decentralized network is formed by a set of low-cost, highly reliable, but low-performance, microprocessor nodes. As shown in Figure 2, each microprocessor node has computing, storage, and communication functions, and data can be transmitted between nodes. Aero-engines SDCS replaces the central controller in aero-engines' DCS with this decentralized network, and the entire decentralized network acts as the controller of the aero-engine control system. Some microprocessors in the network possess bus communication capabilities, allowing the entire decentralized network to communicate directly with intelligent sensors and actuators in the aero-engine control system. In Figure 2, green circles represent conventional nodes, and blue circles represent nodes with bus communication capabilities.

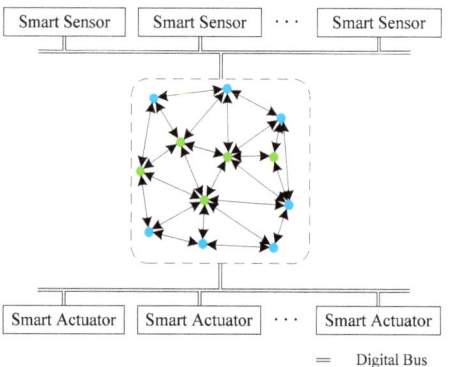

Figure 2. Overview of an aero-engine SDCS.

With the structure of aero-engines, the SDCS requires decentralized control tasks to be performed on each decentralized node. The process of obtaining decentralized control tasks from core control tasks is named control task virtualization, and each decentralized task is called a virtual control task (VCT). It can be seen that the core control tasks of the control system are the sum of these VCTs, and each VCT is mapped to a decentralized node in the network. Since each VCT is a subtask of the core control tasks of the control system, the task load on each decentralized node is effectively reduced, which allows for the use of low-cost, high-reliability, and low-performance microprocessors as decentralized nodes of the control system. At the same time, the amount of software code on each node is effectively reduced, which can improve the software reliability of each decentralized node.

3.2. Linear Iterative Controller Design Scheme

For the structure of a SDCS, nodes in the network act as controllers to execute system control tasks. Sensors, actuators, and nodes in the network can exchange information with each other, and each node maintains its own internal state. Let $\gamma = \{v_1, v_2, \cdots, v_n\}$ denote the set of all nodes, $\gamma_D = \{vd_1, vd_2, \cdots, vd_d\}$ denote the set of all error tracking nodes, $\gamma_A = \{va_1, va_2, \cdots, va_a\}$ denote the set of nodes that send data to the actuators, $\gamma_C = \{vc_1, vc_2, \cdots, vc_c\}$ denote the set of nodes that execute VCTs, and γ_{Ci} represent the neighboring node set of node vc_i in γ_C.

The linear iterative working mechanism of the nodes in the network is that each VCT node in the network broadcasts its own state information to other nodes in the network at each cycle. Other task nodes receive the broadcasted state information. After all, the VCT nodes have completed their state broadcasting, and each VCT node updates its own state as a weighted sum of its previous cycle state and the received state information of other VCT nodes. Thus, each task node can act as a small dynamic controller.

Consider the example of the linear iterative network shown in Figure 3, where each node maintains a scalar state. Nodes broadcast their own states to other nodes in the network in a certain order (it is known from the analysis process below that the broadcast order has no effect on the iteration solution model). In each cycle, node vc_4 broadcasts its state first, then node vc_5 broadcasts its state, and so on, with node vc_2 being the last node to broadcast its state in each cycle. Any node $vc_i \in \gamma_C$ receives the states of other nodes in γ_C, and updates its own state as a weighted sum of its previous cycle state and the received states of all other nodes. Let $z_i[k]$ denote the state of node $vc_i \in \gamma_C$ in the kth cycle. Taking the example of placing a decentralized controller in the forward path to complete unity feedback control, where u_1 is the tracking error between the output and input, and u_2 is the output of the decentralized controller, based on the linear iterative process of the decentralized controller, we have:

$$z_i[k+1] = w_{ii}z_i[k] + \sum_{v_j \in \gamma_{Ci}} \left(w_{ij}z_j[k] \right) + h_i u_1[k] \quad (22)$$

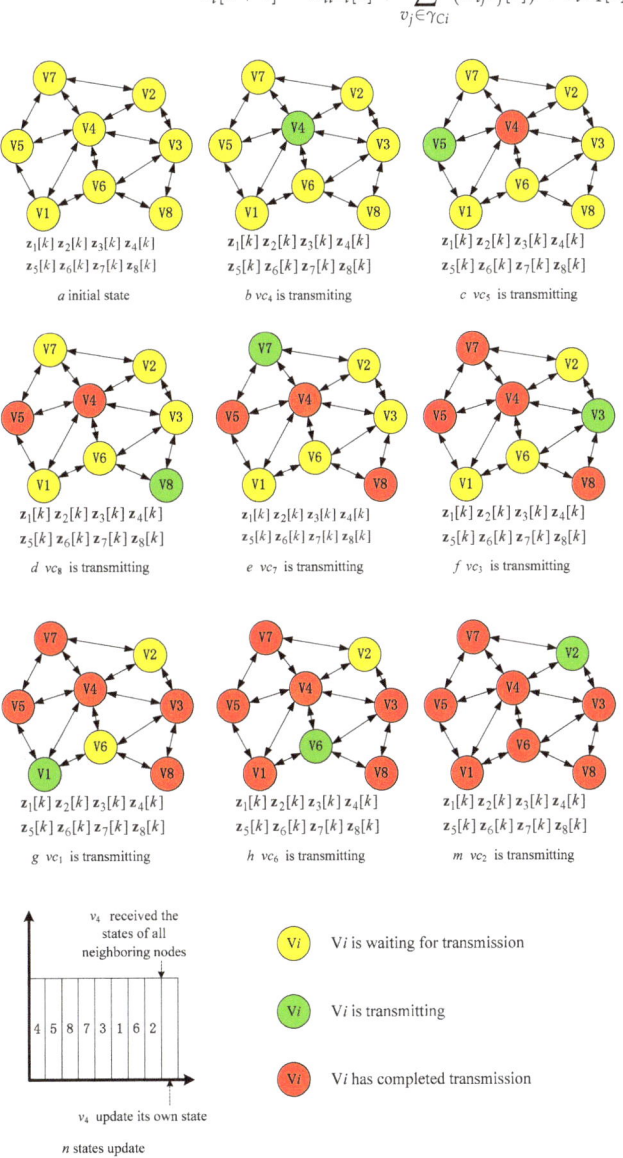

Figure 3. Example of a linear iteration scheme.

The actuator input u_2 is obtained as a linear combination of the tracking error value u_1 and the node states z_i in γ_C:

$$u_2[k] = \sum_{v_j \in \gamma_C} (g_j z_j[k]) \tag{23}$$

The vector $z[k] = [z_1[k] \quad z_2[k] \quad \cdots \quad z_N[k]]^T$, which is a concatenation of all node states in γ_C (where N is the number of internal state variables maintained by the node in γ_C), we obtain the iteration process of the entire network as:

$$z[k+1] = \begin{bmatrix} w_{11} & w_{12} & \cdots & w_{1N} \\ w_{21} & w_{22} & \cdots & w_{2N} \\ \vdots & \vdots & \ddots & \vdots \\ w_{N1} & w_{N2} & \cdots & w_{NN} \end{bmatrix} z[k] + \begin{bmatrix} h_1 \\ h_2 \\ \vdots \\ h_N \end{bmatrix} u_1[k] \tag{24}$$

$$u_2[k] = \begin{bmatrix} g_1 & g_2 & \cdots & g_N \end{bmatrix} z[k] \tag{25}$$

Let,

$$\begin{bmatrix} w_{11} & w_{12} & \cdots & w_{1N} \\ w_{21} & w_{22} & \cdots & w_{2N} \\ \vdots & \vdots & \ddots & \vdots \\ w_{N1} & w_{N2} & \cdots & w_{NN} \end{bmatrix} = W$$

$$\begin{bmatrix} h_1 \\ h_2 \\ \vdots \\ h_N \end{bmatrix} = H$$

$$\begin{bmatrix} g_1 & g_2 & \cdots & g_N \end{bmatrix} = G$$

Then,

$$\begin{aligned} z[k+1] &= Wz[k] + Hu_1[k] \\ u_2[k] &= Gz[k] \end{aligned} \tag{26}$$

This sub-section describes the working process of the linear iteration strategy for nodes in the network and obtains the system model of the linear iteration strategy. From the above analysis, it can be concluded that the broadcasting order of nodes in Figure 3 has no impact on modeling the linear iteration strategy of the network, and the tasks of each node are balanced. The key to designing the linear iteration strategy is to determine the link weights (i.e., W, H, and G in Equation (26)). A simple method for obtaining link weight parameters will be introduced in the following sub-sections of this paper.

3.3. Control System Modeling

As shown in Figure 4, this is a schematic diagram of the control system structure for aero-engines, with the controller placed in the forward path. In Figure 4, u represents the system input, y represents the system output, u_1 represents the tracking error between the output and input, and u_2 represents the output of the controller. The control system adopts unity negative feedback design. The controller calculates the control signal using the difference between the system reference input and output and sends the control signal to the actuator to complete the system control task. The system control task is centralized at a specific controller node.

For SDCS, the system feature is that the controller is decentralized, that is, a controller is composed of several decentralized nodes in the network with computing, storage, and communication capabilities. The corresponding SDCS control system structure diagram is shown in Figure 5. In the figure, u is the system input, y is the system output, u_1 is the tracking error between the output and input, u_2 is the output of the decentralized node based controller, the dashed box represents the decentralized nodes, the blue lines represent communication between nodes, and for simplicity, not all data transmission relationships between decentralized nodes are shown. Each decentralized node has computing, storage, and communication capabilities. The node used to calculate the tracking error between the output and input is called the error tracking node, which is isomorphic to other nodes in the network.

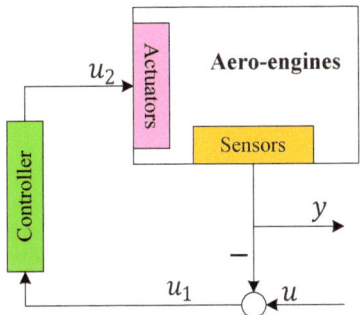

Figure 4. Structure of a unit feedback control system.

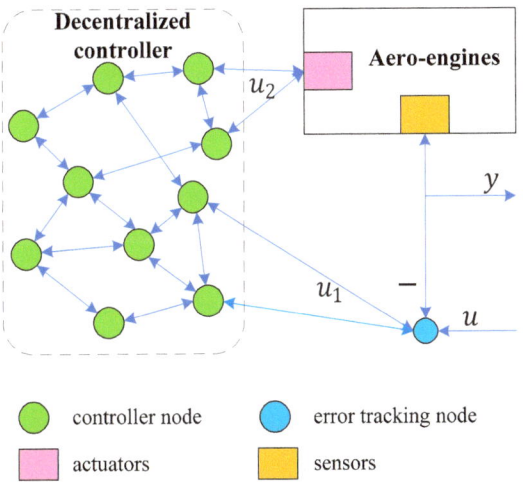

Figure 5. Decentralized controller structure with unit negative feedback.

The general time-discrete linear time-invariant model of the aero-engine in Figure 5 is:

$$\begin{aligned} x[k+1] &= Ax[k] + Bu_2[k] \\ y[k] &= Cx[k] \end{aligned} \quad (27)$$

From the control system structure in Figure 5,

$$u_1[k] = u[k] - y[k] \quad (28)$$

Based on the periodic linear iterative strategy derived from Equations (24) and (25), the state–space model of the control system can be represented as $\tilde{x}[k] = \begin{bmatrix} x[k]^T & z[k]^T \end{bmatrix}^T$.

In Figure 5, when the nodes in the network adopt the linear iterative strategy shown in Figure 3, the state equation of the closed-loop system is:

$$\tilde{x}[k+1] = \begin{bmatrix} A & BG \\ -HC & W \end{bmatrix} \tilde{x}[k] - \begin{bmatrix} 0 \\ H \end{bmatrix} u[k] \qquad (29)$$

Let,

$$\begin{bmatrix} A & BG \\ -HC & W \end{bmatrix} = \tilde{A}$$

$$\begin{bmatrix} 0 \\ H \end{bmatrix} = \tilde{B}$$

Then,

$$\tilde{x}[k+1] = \tilde{A}\tilde{x}[k] - \tilde{B}u[k] \qquad (30)$$

From the above analysis, it can be concluded that the control-related computing tasks performed by the decentralized controller nodes responsible for system control are described by Equation (22), which is a linear weighted sum process. The linear weighted sum of a single node is equivalent to a VCT. It can be seen that, for each VCT, only simple addition and multiplication operations are required, with minimal computational requirements and low utilization of node computing resources. Each VCT actually uses its own state and the state of other task nodes to perform linear iterations in order to maintain its own state.

Based on the running process and mathematical description of the linear iterative SDCS closed-loop control system presented in this section, the following advantages of this method can be identified:

1. Low resource consumption

   In the linear iteration scheme proposed in this section, the task of each VCT is a linear weighted sum of the node's own state and the states of other task nodes. The computation is simple and requires minimal resources, resulting in low computational overhead for the nodes. As aero-engines are safety-critical systems that often use mature and stable but low-performance electronic devices, the SDCS control system design based on the linear iteration scheme is suitable for aero-engine control systems. It can use limited computing resources to perform simple periodic tasks and complete system control tasks.

2. Balanced resource consumption

   As mentioned earlier, the task of each VCT is a linear weighted sum of multiple node states, resulting in equal task loads for each task node and balanced resource consumption across the nodes.

3. Good software performance

   Since the tasks of each task node are the same, with only different link weights, there is no need to design the software system for each VCT separately. In addition, the simple weighted summation task of each VCT results in less code and simpler tasks, making it less likely to affect the control system due to software system failures during operation.

4. Easy to develop

   This solution does not require homogeneous hardware nodes, as long as the nodes meet the basic requirements of computing, communication, and storage performance, they can be easily added to the control system for system expansion.

   In this section, a SDCS model based on the linear iterative strategy was obtained, and a fully distributed decentralized controller was implemented, where VCT represents the linear weighted combination process executed on a single task node. Not every node needs to directly exchange data with sensors and actuators, as data exchange mainly occurs between decentralized nodes in the network. The tuning of link weights in the linear iteration process of this scheme will be described in detail in the next section.

4. Controller Parameter Tuning Method Based on LMI

To ensure the normal running of the control system, it is essential to ensure that the control system is stable. For the design scheme in this paper, since the linear iterative process of the decentralized controller is discrete, it is required that the system be Schur stable.

The most intuitive method is to use the Lyapunov method [36], which involves finding a matrix P that satisfies the following equation.

$$\begin{aligned} & P > 0 \\ & \tilde{A}^T P \tilde{A} - P < 0 \end{aligned} \tag{31}$$

However, since the linear iterative method of the decentralized controller in this paper does not impose any restrictions on the network topology of the nodes, and the parameters P and $\tilde{A}(W, H, G)$ in Equation (31) are nonlinear with respect to the Lyapunov condition, it is difficult to solve using general Linear Matrix Inequality (LMI) methods. Here, the following lemma is introduced,

Lemma 1. *System $\tilde{A}$ is Schur stable if and only if there exist matrices P and Q, such that the following equation holds:*

$$\begin{aligned} & P > 0 \\ & Q > 0 \\ & Q = P^{-1} \\ & \tilde{A}^T Q^{-1} \tilde{A} - P < 0 \end{aligned} \tag{32}$$

Proof. From Equation (31), according to the Schur complement theorem, we can conclude that,

$$\begin{bmatrix} P & \tilde{A}^T \\ \tilde{A} & P^{-1} \end{bmatrix} > 0$$

Let, $Q = P^{-1}$, then,

$$\begin{bmatrix} P & \tilde{A}^T \\ \tilde{A} & Q \end{bmatrix} > 0$$

According to the Schur complement theorem, we can obtain that,

$$\tilde{A}^T Q^{-1} \tilde{A} - P < 0$$

Additionally, P and Q are positive definite matrices. □

From Refs. [37,38], we can obtain the following lemma.

Lemma 2. *For positive definite matrices P and Q, P and Q are the optimal solutions to the following optimization problem, if and only if $Q = P^{-1}$:*

$$\min \; trace(PQ)$$
$$s.t. \; \begin{cases} P > 0 \\ Q > 0 \\ P > Q^{-1} \end{cases}$$

By Lemma 1 and Lemma 2, we can easily derive the following conclusion:

System $\tilde{A}$ is Schur stable, if and only if the following optimization problem has a solution.

$$\min \ trace(PQ)$$
$$s.t. \begin{cases} P > 0 \\ Q > 0 \\ P > Q^{-1} \\ \tilde{A}^T Q^{-1} \tilde{A} - P < 0 \end{cases} \quad (33)$$

In the above optimization condition, the four constraint conditions are linear, but the objective function $\min \ trace(PQ)$ is nonlinear. In order to facilitate solving the problem using LMI method, the objective function can be linearized at any P_0 and Q_0 [38], that is:

$$trace(PQ)|_{P_0,Q_0} = trace(P_0Q + PQ_0) + c \quad (34)$$

In Equation (34), c is a constant. Therefore, the optimization objective can take $trace(P_0Q + PQ_0)$ as the objective function. Since $trace(P_0Q + PQ_0)$ is linear, the LMI method can be used to solve the above optimization problem. Based on this conclusion, the following $\tilde{A}(W, H, G)$ parameter tuning algorithm is given. First, the LMI problem is presented:

$$\min \ trace(P_k Q_{k+1} + P_{k+1} Q_k)$$
$$s.t. \begin{cases} P_{k+1} > 0 \\ Q_{k+1} > 0 \\ P_{k+1} > Q_{k+1}^{-1} \\ \tilde{A}_{k+1}^T Q_{k+1}^{-1} \tilde{A}_{k+1} - P_{k+1} < 0 \end{cases} \quad (35)$$

The flowchart for the parameter tuning algorithm is shown in Figure 6.

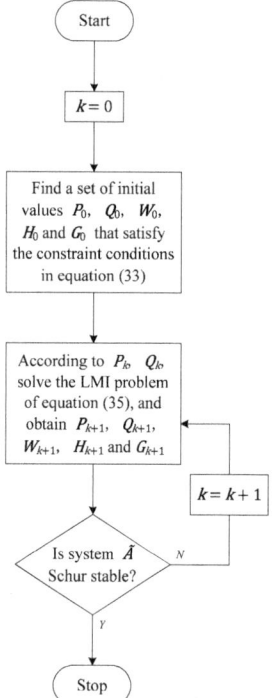

Figure 6. The flowchart for the parameter tuning algorithm.

In the above algorithm, during each iteration, P_k and Q_k are equivalent to P_0 and Q_0 in Equation (34), and the matrix variables P_{k+1} and Q_{k+1} to be solved are equivalent to P and Q in Equation (34). Therefore, the optimization objective function in Equation (35), $\min\ trace(P_k Q_{k+1} + P_{k+1} Q_k)$, is linear. The parameters W_{k+1}, H_{k+1}, and G_{k+1} obtained from the above algorithm are the required parameters W, H, and G for the decentralized controller. Under this control parameterization, the control system $\widetilde{A}$ is Schur stable.

5. Simulation Analysis

The previous sections introduced a decentralized controller design based on linear iteration, provided the mathematical model of the control system, and presented a controller parameter tuning method based on LMI. The purpose of this simulation is to verify the feasibility of the proposed decentralized controller design of the aero-engines SDCS and parameter tuning algorithm in this paper.

By using the modeling method in Section 2, we obtain the discrete-time state space model of a turbofan engine under the conditions of 0 altitude and 0 Mach number, which is given by:

$$\begin{bmatrix} n_L[k+1] \\ n_H[k+1] \end{bmatrix} = A \begin{bmatrix} n_L[k] \\ n_H[k] \end{bmatrix} + B \begin{bmatrix} W_{mf}[k] \\ A_8[k] \end{bmatrix}$$

$$\begin{bmatrix} n_L[k] \\ P_i T[k] \end{bmatrix} = C \begin{bmatrix} n_L[k] \\ n_H[k] \end{bmatrix}$$

where,

$$A = \begin{bmatrix} 0.9561 & -0.0139 \\ -0.0168 & 0.9721 \end{bmatrix}$$

$$B = \begin{bmatrix} 0.0152 & 0.0117 \\ 0.0104 & 0.0081 \end{bmatrix}$$

$$C = \begin{bmatrix} 1 & 0 \\ -3.7430 & 6.8260 \end{bmatrix}$$

n_L and n_H are the low-pressure and high-pressure rotor speeds of the turbofan engine, respectively, $P_i T = \pi_{TL} \pi_{TH}$ is the turbine pressure ratio, W_{mf} is the fuel flow, and A_8 is the area of the nozzle.

Using the LMI based algorithm designed in Section 4 and the MATLAB YALMIP toolbox, with "sdpt3" solver, the parameters W, H, and G of the decentralized controller are calculated as follows:

$$W = \begin{bmatrix} 2.0579 & 1.6296 \\ 1.6139 & 1.2523 \end{bmatrix} \times 10^{-5}$$

$$H = \begin{bmatrix} 4.0930 \times 10^{-4} & -5.9683 \times 10^{-5} \\ -5.9683 \times 10^{-5} & -7.7824 \times 10^{-5} \end{bmatrix}$$

$$G = \begin{bmatrix} -0.1846 & 0.2545 \\ 0.2545 & -0.3239 \end{bmatrix}$$

It can be verified that $\widetilde{A}(W, H, G)$ is Schur stable. Therefore, it can be seen that only two nodes are required to form a decentralized controller of the aero-engine's control system.

Figure 7 shows the schematic diagram of the low-pressure rotor speed response of an aero-engine under corresponding inputs. Figure 8 shows the schematic diagram of the turbine pressure ratio response of an aero-engine under corresponding inputs.

From the simulation results of Figures 7 and 8, it can be seen that the response of the low-pressure rotor speed and turbine pressure ratio under the corresponding input is stable. In the design scheme proposed in this paper, a decentralized controller design scheme is first presented to obtain the VCT of the SDCS. The model of the control system is obtained through the model of the decentralized controller. A control parameter tuning scheme is given using LMI, which obtains the parameters of each VCT. For the parameter tuning scheme, the objective is to ensure the stability of the control system. By analyzing the result of the parameter tuning scheme, the system $\widetilde{A}(W, H, G)$ is Schur stable, meanwhile, the simulation results show that the system is stable, demonstrating the effectiveness of the proposed design.

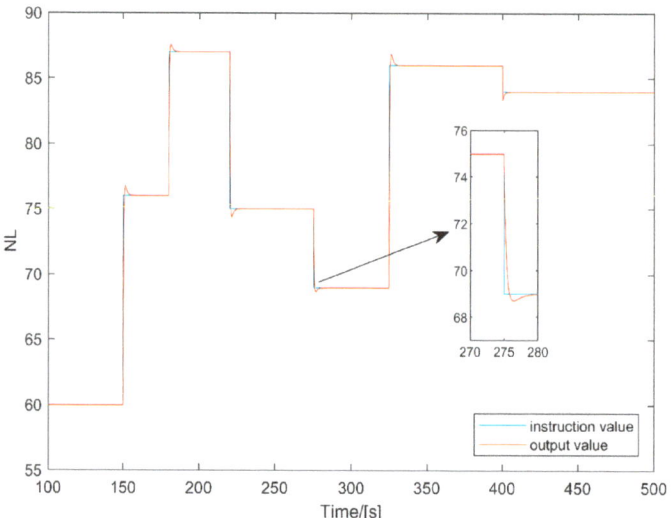

Figure 7. Schematic diagram of the low-pressure rotor speed response.

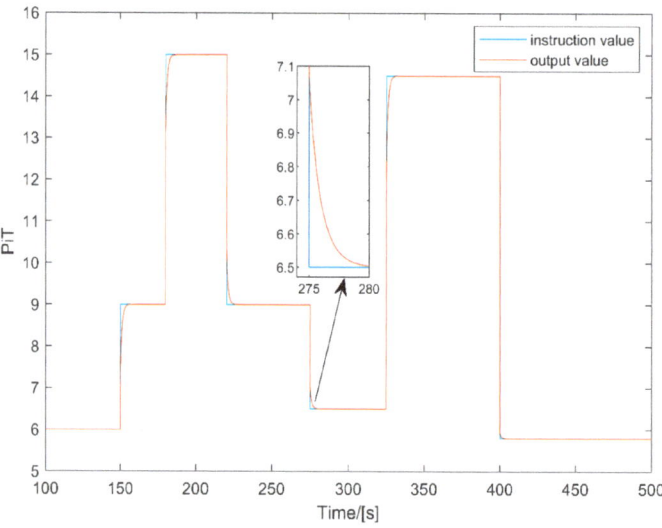

Figure 8. Schematic diagram of the turbine pressure ratio response.

In the design process of this paper, a decentralized controller can be composed using only a small number of nodes, making the design of the control system simple. Furthermore, because the internal nodes of the controller use periodic broadcasting for linear iteration, the communication resources consumed by the controller increase as the number of nodes executing the VCT increases. From the simulation results, it can be seen that using a decentralized node with the same number as the system order can complete the control task well.

The proposed design scheme makes the application of SDCS in aero-engine control systems simple. From the perspective of control system design, existing aero-engine models can be used to obtain the control tasks of decentralized controller nodes through the algorithm introduced in this paper, simplifying the system design. From the perspective of software design, the control tasks executed by each decentralized node in the controller are simply linearly weighted sums of the node's own state and the states of other nodes. Different VCTs executed on different nodes only have different weights on the state, and the software structure is the same, making software code writing and development easy and short. From the perspective of hardware development, since the consumption of computing, storage, and communication resources between different VCTs is indistinguishable, and homogeneous nodes can be used for hardware system design, making hardware design easy without requiring specific designs for different VCTs, effectively reducing the difficulty and cycle of the hardware design process.

6. Conclusions

This paper briefly introduces the architecture of SDCS for aero-engines. The SDCS is composed of a network of low-cost, low-performance, and highly reliable nodes that serve as the controller for the aero-engine control system. The characteristic of SDCS is that the controller is decentralized, and there is no central controller node in the control system. In order to meet the decentralized characteristic of the SDCS controller, this paper introduces a virtualization scheme for control tasks based on node linear iteration, obtaining the virtual control tasks executed on each node. The linear iterative scheme for decentralized controllers is modeled, and the mathematical model of the aero-engine control system is obtained. In the proposed scheme, the control tasks executed on each node have balanced workloads, and the workloads of the decentralized nodes are small, making it easy to use low-cost, low-performance, and highly reliable microprocessors as decentralized controllers of the control system. To stabilize the control system with the designed controller, a controller parameter tuning scheme based on the LMI method is introduced. Finally, the feasibility of the proposed design scheme is verified through digital simulation.

Author Contributions: Conceptualization, X.J. and J.L.; data curation, J.R.; investigation, X.J. and J.R.; methodology, X.J., J.L. and Y.W.; software, X.J. and J.R.; supervision, J.L. and Y.W.; validation, X.J. and J.R.; writing—original draft, X.J.; writing—review and editing, X.J., J.L., J.R. and Y.W. All authors have read and agreed to the published version of the manuscript.

Funding: This research received no external funding.

Data Availability Statement: The data used to support the findings of this study are included within the article.

Conflicts of Interest: The authors declare that they have no competing financial and non-financial interests.

References

1. Wang, X.; Du, X.; Wang, X.; Sun, X.; Li, Y. Controller Design of Aero-engines under the Distributed Architecture with Time Delays. In Proceedings of the 2020 39th Chinese Control Conference (CCC), Shenyang, China, 27–29 July 2020.
2. De Giorgi, M.G.; Strafella, L.; Ficarella, A. Neural Nonlinear Autoregressive Model with Exogenous Input (NARX) for Turboshaft Aeroengine Fuel Control Unit Model. *Aerospace* **2021**, *8*, 206. [CrossRef]
3. Zhang, L.; Xie, S.; Zhang, Y.; Ren, L.; Zhou, B.; Wang, H.; Peng, J.; Wang, L.; Li, Y. Aero-Engine DCS Fault-Tolerant Control with Markov Time Delay Based on Augmented Adaptive Sliding Mode Observer. *Asian J. Control.* **2020**, *22*, 788–802. [CrossRef]
4. Lv, C.; Chang, J.; Bao, W.; Yu, D. Recent Research Progress on Airbreathing Aero-Engine Control Algorithm. *Propuls. Power Res.* **2022**, *11*, 1–57. [CrossRef]
5. Culley, D.; Thomas, R.; Saus, J. Concepts for Distributed Engine Control. In Proceedings of the 43rd AIAA/ASME/SAE/ASEE Joint Propulsion Conference & Exhibit, Cincinnati, OH, USA, 8–11 July 2007.

6. Gianluca, A. Interconnected Dynamic Systems: An Overview on Distributed Control. *IEEE Contr. Syst. Mag.* **2013**, *33*, 76–88.
7. Xu, S.; Bao, J. Distributed Control of Plantwide Chemical Processes. *J. Process. Control.* **2009**, *19*, 1671–1687. [CrossRef]
8. Chen, T.; Shan, J. Distributed Tracking of Multiple Under-actuated Lagrangian Systems with Uncertain Parameters and Actuator Faults. In Proceedings of the 2019 American Control Conference (ACC), Philadelphia, PA, USA, 10–12 July 2019.
9. Chen, T.; Shan, J. Distributed Adaptive Attitude Control for Multiple Underactuated Flexible Spacecraft. In Proceedings of the 2018 Annual American Control Conference (ACC), Milwaukee, WI, USA, 27–29 June 2018.
10. Chen, T.; Shan, J. Distributed Tracking of a Class of Underactuated Lagrangian Systems with Uncertain Parameters and Actuator Faults. *IEEE Trans. Ind. Electron.* **2020**, *67*, 4244–4253. [CrossRef]
11. Zhao, Y.; Qiu, H.; Song, H. Distributed Measurement and Control System of the Test-rig of Ram-compressed Rotor Aero-engine Based on PLC and PXI bus. In Proceedings of the 2012 IEEE International Conference on Control Applications, Dubrovnik, Croatia, 3–5 October 2012.
12. Li, G.; Wang, X.; Ren, X. Multi-package Transmission Aero-engine DCS Neural Network Sliding Mode Control Based on Multi-Kernel LS-SVM Packet Dropout Online Compensation. *PLoS ONE* **2020**, *15*, e0234356.
13. Zhang, S.; Zhang, D.; Zhang, Z.; Zhang, Q.; Lu, B. Dynamic Control Allocation Fault-Tolerant Method for a Class of Distributed Control Systems. In Proceedings of the 2018 37th Chinese Control Conference (CCC), Wuhan, China, 25–27 July 2018.
14. Lubomir, B. Decentralized Control: An Overview. *Annu. Rev. Control* **2008**, *32*, 87–98.
15. Lubomir, B. Decentralized Control: Status and Outlook. *Annu. Rev. Control* **2014**, *38*, 1367–5788.
16. Maria, F.; Agostino, M.; Giovanni, P.; Walter, U. A Decentralized Control Strategy for the Coordination of AGV Systems. *Control Eng. Pract.* **2018**, *70*, 86–97.
17. Yang, C.; Jaime, L.; Antonia, T.; Andreas, M. Decentralized Control of Distributed Cloud Networks with Generalized Network Flows. *IEEE. Trans. Commun.* **2023**, *71*, 256–268.
18. Shojaee, M.; Azizi, M. Optimal Decentralized Control of a Wind Turbine and Diesel Generator System. *Optim. Control Appl. Meth.* **2023**, *44*, 677–698. [CrossRef]
19. Pan, M.; Cao, L.; Zhou, W.; Huang, J.; Chen, Y. Robust Decentralized Control Design for Aircraft Engines: A Fractional Type. *Chin. J. Aeronaut.* **2019**, *32*, 347–360. [CrossRef]
20. Ji, X.; Li, J.; Ren, J.; Wu, Y.; Wang, K. Fully Connected Clustering Based Software Defined Control System and Node Failure Analysis. *J. Northwest. Polytech. Univ.* **2019**, *37*, 1238–1247. [CrossRef]
21. Stöhr, M.; Geigle, K.; Hadef, R.; Boxx, I.; Carter, C.; Grader, M.; Gerlinger, P. Time-Resolved Study of Transient Soot Formation in an Aero-engine Model Combustor at Elevated Pressure. *Proc. Combust. Inst.* **2019**, *37*, 5421–5428. [CrossRef]
22. Zheng, Q.; Fang, J.; Hu, Z.; Zhang, H. Aero-Engine On-Board Model Based on Batch Normalize Deep Neural Network. *IEEE Access* **2019**, *7*, 54855–54862. [CrossRef]
23. Pang, S.; Li, Q.; Feng, H. A Hybrid Onboard Adaptive Model for Aero-engine Parameter Prediction. *Aerosp. Sci. Technol.* **2020**, *105*, 105951. [CrossRef]
24. Ren, L.; Ye, Z.; Zhao, Y. A Modeling Method for Aero-engine by Combining Stochastic Gradient Descent with Support Vector Regression. *Aerosp. Sci. Technol.* **2020**, *99*, 105775. [CrossRef]
25. Zhao, Y.; Zhao, J.; Fu, Y.; Shi, Y.; Chen, C. Rate Bumpless Transfer Control for Switched Linear Systems with Stability and Its Application to Aero-Engine Control Design. *IEEE Trans. Ind. Electron.* **2020**, *67*, 4900–4910. [CrossRef]
26. Shi, Y.; Sun, X. Bumpless Transfer Control for Switched Linear Systems and Its Application to Aero-Engines. *IEEE Trans. Circuits I.* **2021**, *68*, 2171–2182. [CrossRef]
27. Morteza, M.; Ali, R.; Ali, J.; Milad, E. Design and Implementation of MPC for Turbofan Engine Control System. *Aerosp. Sci. Technol.* **2019**, *92*, 99–113.
28. Morteza, M.; Ali, R. Analyzing Different Numerical Linearization Methods for the Dynamic Model of a Turbofan Engine. *Mech. Ind.* **2019**, *20*, 303.
29. Gou, L.; Zeng, X.; Wang, Z.; Han, G.; Lin, C.; Cheng, X. A Linearization Model of Turbofan Engine for Intelligent Analysis Towards Industrial Internet of Things. *IEEE Access.* **2019**, *7*, 145313–145323. [CrossRef]
30. Chen, Q.; Huang, J.; Pan, M.; Lu, F. A Novel Real-Time Mechanism Modeling Approach for Turbofan Engine. *Energies* **2019**, *12*, 3791. [CrossRef]
31. Ling, Y.; Zhou, W.; Zhu, P.; Zeng, J. Modeling of a Large Envelope System for Turbofan Engine. *J. Nanjing Univ. Aeronaut. Astronaut.* **2021**, *53*, 529–536.
32. Ji, X.; Li, J.; Ren, J.; Wu, Y. A Decentralized LQR Output Feedback Control for Aero-Engines. *Actuators* **2023**, *12*, 164. [CrossRef]
33. Belapurkar, R. Stability and Performance of Proplusion Control Systems with Distributed Control Architectures and Failure. Ph.D. Thesis, The Ohio State University, Columbus, OH, USA, 2012.
34. Thompson, H.; Fleming, P. Distributed Aero-Engine Control Systems Architecture Selection Using Multi-Objective Optimisation. In Proceedings of the 5th IFAC Workshop on Algorithm & Architecture for Real Time Control (AARTC' 98), Cancun, Mexico, 15–17 April 1998.
35. Skira, C.; Agnello, M. Control Systems for the Next Century's Fighter Engines. *J. Eng. Gas Turbines Power.* **1992**, *114*, 749–754. [CrossRef]
36. Ricardo, C.; Pedro, L. LMI Conditions for Robust Stability Analysis Based on Polynomially Parameter-dependent Lyapunov Functions. *Syst. Control Lett.* **2006**, *55*, 52–61.

37. Bruno, M.; Joao, Y.; Eduardo, S. LMI-based Consensus of Linear Multi-agent Systems by reduced-order Dynamic Output Feedback. *ISA Trans.* **2020**, *129*, 121–129.
38. Ghaoui, L.; Oustry, F.; AitRami, M. A cone complementarity linearization algorithm for static output-feedback and related problems. *IEEE Trans. Automat. Contr.* **1997**, *42*, 1171–1176. [CrossRef]

Disclaimer/Publisher's Note: The statements, opinions and data contained in all publications are solely those of the individual author(s) and contributor(s) and not of MDPI and/or the editor(s). MDPI and/or the editor(s) disclaim responsibility for any injury to people or property resulting from any ideas, methods, instructions or products referred to in the content.

Article

Robust Adaptive Composite Learning Integrated Guidance and Control for Skid-to-Turn Interceptors Subjected to Multiple Uncertainties and Constraints

Yu Bai [1,2], Tian Yan [2,*], Wenxing Fu [2], Tong Li [2] and Junhua Huang [3]

1. Northwest Institute of Nuclear Technology, Xi'an 710024, China
2. Unmanned Systems Research Institute, Northwestern Polytechnical University, Xi'an 710072, China
3. Xi'an Modern Control Technology Research Institute, Xi'an 710076, China
* Correspondence: tianyan@nwpu.edu.cn

Abstract: This paper investigates a novel robust adaptive dynamic surface control scheme based on the barrier Lyapunov function (BLF), online composite learning, disturbance observer, and improved saturation function. It is mainly designed for a class of skid-to-turn (STT) interceptor integrated guidance and control (IGC) design problems under multi-source uncertainties, state constraints, and input saturation. The serial-parallel estimation model used in this study estimates the system states and provides "critic" information for the neural network and disturbance observer; then, these three are combined to realize online composite learning of the multiple uncertainties of the system and improve the interception accuracy. In addition, the state and input constraints are resolved by adopting the BLF and the improved saturation function, while the design of the auxiliary system ensures stability. Finally, a series of simulation results show that the proposed IGC scheme with a direct-hit intercept strategy achieves a satisfactory effect, demonstrating the validity and robustness of the scheme.

Keywords: integrated guidance and control; online composite learning; dynamic surface control; multiple uncertainties; multi-constraints

1. Introduction

As proposed in [1], the conventional design framework of separating the guidance subsystem and control subsystem possesses many advantages, such as being beneficial to stability analysis and engineering implementation, which makes it extremely widely used. However, under the background of a separate study of guidance and control loops, this framework also gives rise to several serious drawbacks that reduce the adjustability and robustness of missiles [2]. What follows is a failure to take full advantage of the missile's overall effectiveness. Therefore, it is necessary to put forward an integrated guidance and control (IGC) scheme that is different from the above design framework and can overcome the preceding defects [3].

As a type of control framework with better overall performance than the conventional one, the guidance loop and the control loop is designed as a whole, making the guidance and control circuits operate synergistically. It is worth noting that the guidance and control synthesis of interceptors, as a challenging task, has become a hotspot research area of current technology [4]. In recent decades, numerous advanced control algorithms for IGC have been proposed, including model predictive control [5], sliding mode control [6,7], trajectory linearization control [8], and active disturbance rejection control [9]. To be more specific, the unique strict feedback form of the IGC model makes it suitable to control the system using the classic backstep method. To address the issue that the explosion terms inevitably appear in the classic backstep method, a novel design framework based on dynamic surface control (DSC) is provided [10]. The feasible non-singular terminal dynamic surface control for the IGC system is investigated in [11], which considers the terminal impact angular constraint and multiple disturbances comprehensively. In [12], a novel IGC method of the fuzzy adaptive dynamic surface is proposed for different maneuvering targets by introducing the fuzzy adaptive technique into the dynamic surface-based control framework. Meanwhile, different constraints and types of uncertain perturbations are considered in this study. In brief, various technologies based on DSC have been widely utilized in national defense and military industries.

There are multiple uncertainties and all types of constraints in actual systems, which need to be further investigated [13–15]. Plenty of achievements have been achieved in the research on systems with uncertain disturbances. Although the future course of action of the target in [16], as a type of uncertainty, cannot be predicted, the effect of the target maneuver can be counteracted by utilizing adaptive control techniques. Meanwhile, many control algorithms have been reported based on adaptive control [17], DSC [18,19], command filtered control [20], sliding mode control [21,22], barrier Lyapunov function (BLF) [23,24], and a fixed-time differentiator [25] for the constrained variables of IGC system. In [26], a three-dimensional integrated guidance and control law is developed, which relies on the advantage of dynamic surface control and extended state observer techniques to address input saturation and actuator failure. Both the studies in [26] and [27] are based on dynamic surface and extended state observer techniques, but the hyperbolic tangent function and auxiliary system in [27] are introduced to sort out the problem caused by input saturation and impact angle constrained. In addition, by combining numerous methods including backstepping, command filter, sliding mode control (SMC), and super twisting extended state observer (STESO), [28] proposes a control scheme with superior interception performance to solve the IGC problem of a 6-DOFs interceptor. By exploiting the relationships between the virtual commands of the constrained states and the tracking errors, the solution investigated in [29] can be employed in an IGC system with multi-constraints.

Although a lot of achievements have been achieved, the problems of multi-uncertainties and state constraints in the design of IGC have not been systematically solved. Motivated by the above research, a novel controller is investigated in this paper for interceptor IGC with multi-source uncertainties, state constraint, and input saturation. Our study mainly performs the following valuable work.

1. As elaborated in the paper, this study proposes a novel adaptive dynamic surface control framework based on online composite learning and the BLF principle for the IGC by considering multiple uncertainties and the overload constraint;
2. Different from previous studies in which the learning law is designed only by tracking errors, the serial-parallel estimation model established in this study estimates the system state and provides "critic" information for the neural network (NN) and disturbance observer that approximates the system uncertainties. Then, these three are combined to realize online composite learning of the multiple uncertainties of the system and improve interception accuracy;
3. Aiming at the state constraint of the interceptor, the application of the improved saturation function and BLF can restrict the specific state to a certain range, while the design of the auxiliary system guarantees the system stability.

The rest of this paper is organized as follows. Section 2.1 presents the IGC model of the skid-to-turn (STT) missiles with state constraint, input saturation, and multi-source uncertainties. Section 2.2 introduces the intelligent approximation scheme. Section 3 designs a novel adaptive dynamic surface online composite learning IGC algorithm. Section 4 analyzes the stability of the proposed algorithm. Section 5 verifies the effectiveness and robustness of the proposed algorithm through nonlinear simulation. Finally, the conclusion of this study is given in Section 6.

2. Preliminaries

2.1. IGC Model

The longitudinal dynamic model of STT interceptor ignoring gravity is generalized below [30]:

$$
\begin{aligned}
\dot{\alpha} &= q + \frac{(-F_x \sin\alpha + F_z \cos\alpha)}{mV} \\
\dot{V} &= \frac{(F_x \cos\alpha + F_z \sin\alpha)}{m}, \\
\dot{q} &= \frac{M}{I_{yy}}, \\
\dot{\theta} &= q, \\
\dot{n}_L &= \frac{-n_L + Vq}{T_\alpha}, \\
\gamma_M &= \theta - \alpha,
\end{aligned}
\quad (1)
$$

where q denotes the pitch angle rate of the missile, θ, α, γ_M, m, n_L, and V represent the pitch angle, attack angle, flight-path angle, mass, normal acceleration, and velocity of the missile respectively. M and I_{yy} are the pitching moment and moment of inertia around the pitch axis. In addition, aerodynamic forces F_x and F_z can be expressed as [2]:

$$
\begin{aligned}
F_x &= 0.5\rho V^2 S C_x(\alpha) \\
F_z &= 0.5\rho V^2 S C_z(\alpha, M_m)
\end{aligned}
\quad (2)
$$

where $C_z(\alpha, M_m) = C_{z_0}(M_a) + C_z^\alpha(M_a)\alpha + C_z^{\delta_e}(M_a)\delta_e$.

The pitch moment M can be obtained as:

$$
\begin{aligned}
M &= 0.5\rho V^2 S l C_m(\alpha, M_m, \delta_e) \\
C_m(\alpha, M_m, \delta_e) &= C_{m_0}(\alpha, M_m) + C_m^{\delta_e}\delta_e \\
C_{m_0}(\alpha, M_m) &= C_{m1}(\alpha) + C_{m2}(\alpha)M_m
\end{aligned}
\quad (3)
$$

where C_x, C_z, C_z^α, $C_z^{\delta_e}$, C_m, C_{m_0}, $C_m^{\delta_e}$, C_{m1}, and C_{m2} are the aerodynamic coefficients. It is important for us to note that $T_\alpha = \frac{\alpha}{\gamma_M}$ can be treated as a time constant in our study according to [31].

As shown in Figure 1, M and T denote the missile and the target, respectively, and the remaining definitions are presented in Table 1.

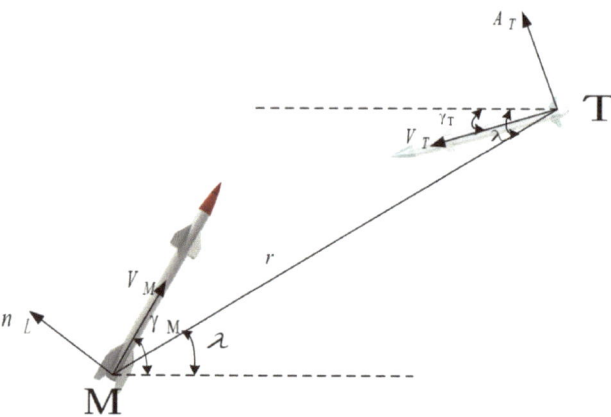

Figure 1. Missile–target planar intercept geometry diagram.

Table 1. Nomenclature.

Variable	Implication
r	The relative distance along the line of sight (LOS)
V_r	The projections of relative velocity along the LOS
V_λ	The projections of relative velocity perpendicular to the LOS
A_{Tr}	The projections of target acceleration along the LOS
$A_{T\lambda}$	The projections of target acceleration perpendicular to the LOS
λ	LOS angle
ρ	Atmosphere density
S	Reference area
l	Reference length

Thus, the missile–target plane interception kinematics can be obtained as:

$$\begin{aligned}\dot{r} &= V_r \\ \dot{V}_r &= \frac{V_\lambda^2}{r} + A_{Tr} - \sin(\lambda - \gamma_M)n_L \\ \dot{\lambda} &= \frac{V_\lambda}{r} \\ \dot{V}_\lambda &= -\frac{V_\lambda V_r}{r} + A_{T\lambda} - \cos(\lambda - \gamma_M)n_L\end{aligned} \quad (4)$$

To ensure a direct hit, the following intercept strategy is selected for our study [31].

$$V_\lambda \to c_0\sqrt{r} \quad (5)$$

where $c_0 > 0$ is a designed scalar, and the interception strategy ξ is defined as follows:

$$\xi = V_\lambda - c_0\sqrt{r} \quad (6)$$

Then a direct-hit intercept for the missile can be achieved as $\xi \to 0$, and the proof of effectiveness for Equation (5) can be found in [32].

By substituting (1), (2), (3), and (4) into (6), the IGC model for missiles can be described as follows [2]:

$$\begin{cases}\dot{\xi} = -\frac{V_\lambda V_r}{r} - \frac{c_0 V_r}{2\sqrt{r}} - \cos(\lambda - r_M)n_L + A_{T\lambda} \\ \dot{n}_L = -\frac{n_L}{T_a} + \frac{V}{T_a}q \\ \dot{q} = \frac{\rho V^2 Sl(c_{m1}(\alpha)+c_{m2}(\alpha)M_m)}{2I_{yy}} + \frac{\rho V^2 Sl c_m^{\delta_e}}{2I_{yy}}\delta_e\end{cases} \quad (7)$$

By defining $\zeta_1 = \xi, \zeta_2 = n_L, \zeta_3 = q$ and

$$\begin{aligned}f_1(\overline{\zeta}_1) &= -\frac{V_\lambda V_r}{r} - \frac{c_0 V_r}{2\sqrt{r}}, g_1(\overline{\zeta}_1) = -\cos(\lambda - \gamma_M), d_1(t) = A_{T\lambda} \\ f_2(\overline{\zeta}_2) &= -\frac{n_L}{T_a}, g_2(\overline{\zeta}_2) = \frac{V}{T_a} \\ f_3(\overline{\zeta}_3) &= \frac{\rho V^2 Sl(c_{m1}(\alpha)+c_{m2}(\alpha)M_m)}{2I_{yy}}, g_3(\overline{\zeta}_3) = \frac{\rho V^2 Sl c_m^{\delta_e}}{2I_{yy}}\end{aligned} \quad (8)$$

The above model (7) can be converted into a strict-feedback form as follows:

$$\begin{cases}\dot{\zeta}_1 = f_1(\overline{\zeta}_1) + g_1(\overline{\zeta}_1)\zeta_2 + d_1 \\ \dot{\zeta}_2 = f_2(\overline{\zeta}_2) + g_2(\overline{\zeta}_2)\zeta_3 \\ \dot{\zeta}_3 = f_3(\overline{\zeta}_3) + g_3(\overline{\zeta}_3)u \\ y = \zeta_1\end{cases} \quad (9)$$

where $u \in R$ and $y \in R$ are the input and output of the system, respectively. $\zeta(t) = [\zeta_1, \zeta_2, \zeta_3]^T$ denotes the state vector, $\overline{\zeta}_i = [\zeta_1, \cdots, \zeta_i]^T, i = 1, 2, 3$. In engineering practice, there are two challenging problems. On the one hand, due to the limitation of the missile's maneuvering ability, its normal acceleration will be restricted, i.e., the normal acceleration n_L satisfies $|n_L| \leq n_{L\max}$. Considering $\zeta_2 = n_L$, there exists

$$|\zeta_2| \leq \zeta_{2\max} \tag{10}$$

where $\zeta_{2\max} > 0$ is a known positive constant.

On the other hand, due to the physical constraints of the actuator, input saturation is inevitable in the control input of the system. Thus, the control input saturation can be described via the improved saturation function as:

$$u = \text{SAT}(u_0) = \begin{cases} \text{sign}(u_0)u_{\max}, & |u_0| > \Gamma_{p2} \\ -\frac{\text{sign}(u_0)u_0^2}{4\lambda_2} + \frac{(u_{\max}+\lambda_2)u_0}{2\lambda_2} - \frac{\text{sign}(u_0)(u_{\max}-\lambda_2)^2}{4\lambda_2}, & ||u_0| - u_{\max}| \leq \lambda_2 \\ u_0, & |u_0| < \Gamma_{q2} \end{cases} \tag{11}$$

where $\Gamma_{p2} = u_{\max} + \lambda_2$, $\Gamma_{q2} = u_{\max} - \lambda_2$. $u_{\max}$ represents the maximum value of u. λ_2 is a positive constant to be designed. Obviously, the improved saturation function in our study is both continuous and differentiable when $|u_0| = u_{\max}$.

Invoking (11) into (8), the IGC system can be rewritten as:

$$\begin{cases} \dot{\zeta}_1 = f_1(\overline{\zeta}_1) + g_1(\overline{\zeta}_1)\zeta_2 + d_1 \\ \dot{\zeta}_2 = f_2(\overline{\zeta}_2) + g_2(\overline{\zeta}_2)\zeta_3 \\ \dot{\zeta}_3 = f_3(\overline{\zeta}_3) + g_3(\overline{\zeta}_3)\text{SAT}(u_0) \end{cases} \tag{12}$$

where $f_i(\overline{\zeta}_i), i = 1, 2, 3$, $g_i(\overline{\zeta}_i), i = 1, 2, 3$.

For the IGC system (12), some reasonable assumptions are made as follows.

Assumption 1. *The uncertainty in the integrated design and its differentiation are bounded, i.e., $|d_1| \leq \phi_0$, $|\dot{d}_1| \leq \phi_{10}$, where ϕ_0 and ϕ_1 are positive scalars.*

Assumption 2. *Due to the measurement errors of the missile instrument, the continuous functions $f_1(\overline{\zeta}_1), f_2(\overline{\zeta}_2)$, and $f_3(\overline{\zeta}_3)$ can be considered unknown nonlinear uncertainty terms.*

Lemma 1 [33]. *For any $|p| < q_0$, the following inequality always holds.*

$$\ln\left(\frac{q_0^2}{q_0^2 - p^2}\right) < \frac{p^2}{q_0^2 - p^2}$$

Remark 1. *Assumption 1 is often seen in the design of disturbance observer-based controllers [34].*

Remark 2. *Technically, a direct-hit intercept for the missile can be achieved if the relative distance between the missile and the target along the line of sight reaches some sufficiently small value r_0. Therefore, the range r satisfies $r \geq r_0$ during the guidance process.*

The control objective of this study is to construct a novel IGC scheme to ensure that the interception strategy ζ can converge to the neighborhood of zero and achieve direct hit intercept under the conditions of multiple uncertainties and constraints.

2.2. Intelligent Approximation Scheme

The following expression is employed to approximate the nonlinear uncertainties in the system.

$$\hat{f}_i(\overline{\zeta}_i) = \hat{w}_i^T \varphi_i(\overline{\zeta}_i)$$

where U_B is the variable space, and $\overline{\zeta}_i \in U_B$ is the input. Denotes $\hat{w}_i^T$ as a weight vector that can be updated online. There exists an optimal weight vector satisfying

$f_i(\overline{\zeta}_i) = \hat{w}_i^{*T} \varphi_i(\overline{\zeta}_i) + v_i$, and $\sup\limits_{\overline{\zeta}_i \in \Omega_{\overline{\zeta}_i}} |v_i| < v_m$. In other words, $v_m > 0$ denotes the supremum value of the approximation error v_i, and the former is an arbitrarily small positive constant. As an element of the radial basis function NN, $\varphi_i(\overline{\zeta}_i)$ is defined as:

$$\varphi_i(\overline{\zeta}_i) = \exp\left(-\frac{\left(\zeta_j - \mu_j^l\right)^2}{2\left(\sigma_j^l\right)^2}\right)$$

where μ_j^l and σ_j^l are the center and the variance of the ith basis function.

3. The Design of Adaptive Composite Learning IGC Scheme

In this section, we design the control input so that the interception strategy ξ converges to the neighborhood of zero in the presence of unknown target maneuver, missile normal acceleration constraint, and missile actuator saturation. Combined with Equations (7) and (12), it is more appropriate to adopt a three-loop control structure to complete the integrated controller design. The control structure diagram of the integrated guidance and control design is shown in Figure 2. The outer loop is designed to drive ξ to the neighborhood of zero using the normal acceleration command n_{LC} as a virtual control input. The intermediate loop is used to make the actual normal acceleration n_L track the normal acceleration command n_{LC}, using the pitch angle rate command q_C as a virtual control input. Similarly, the inner loop is used to make the actual pitch rate q track pitch rate command q_C, using the elevator deflection δ_e as the control input.

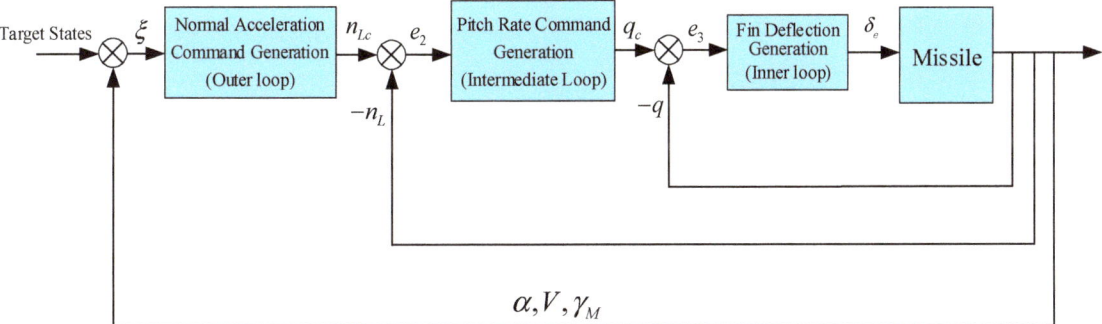

Figure 2. The control structure diagram of IGC design.

A. Outer Loop

Define $F_1(\overline{\zeta}_1) = H_{F_1} f_1(\overline{\zeta}_1)$, where $H_{F_1} > 0$ is a designable parameter. As mentioned above, for the dynamics equation of ζ_1 in the IGC system (12), the estimation of $F_1(\overline{\zeta}_1)$ is completed through NN.

$$\begin{aligned}\dot{\zeta}_1 &= f_1(\overline{\zeta}_1) + g_1(\overline{\zeta}_1)\zeta_2 + d_1(t) \\ &= H_{F_1}^{-1} w_1^{*T} \varphi(\overline{\zeta}_1) + g_1(\overline{\zeta}_1)\zeta_2 + D_1(t)\end{aligned} \quad (13)$$

where $D_1 = H_{F_1}^{-1} v_1 + d_1$. w_1^* denotes the optical weight vector. v_1, as the approximation error of NN, satisfies $|v_1| \le v_m$.

From Assumption 2, D_1 meets the following requirements:

$$|D_1| \le \mu_1, \left|\dot{D}_1\right| \le v_1 \quad (14)$$

where μ_1 and v_1 are unknown positive constants.

The first error surface is defined as $s_1 = \zeta_1$, and then the nominal virtual control ζ_{2d} is constructed as:

$$\zeta_{2d} = -\frac{\left(k_1 s_1 + H_{F_1}^{-1} \hat{w}_1^T \varphi_1(\overline{\zeta}_1) + \hat{D}_1 + k_{\chi_1} \chi_1\right)}{g_1(t, \overline{\zeta}_1)} \tag{15}$$

where $k_1 > 0$ and $k_{\chi_1} > 0$ are the user-defined parameters, and χ_1 is the auxiliary system to be designed. $\hat{D}_1$ and $\hat{w}_1$ represent the estimated values of D_1 and w_1^*, respectively.

At this point, a first-order filter is introduced into the system, which meets the following conditions.

$$\varepsilon_2 \dot{\zeta}_{2f} + \zeta_{2f} = \zeta_{2d}, \zeta_{2f}(0) = \zeta_{2d}(0) \tag{16}$$

where ε_2 is the filter parameter, and the error signal of the filter (16) is presented as:

$$z_2 = \zeta_{2f} - \zeta_{2d} \tag{17}$$

Differentiating z_2 with respect to time,

$$\dot{z}_2 = -\varepsilon_2^{-1} z_2 - \dot{\zeta}_{2d} \tag{18}$$

The actual virtual control ζ_{2s} can be generated through the following saturation function.

$$\zeta_{2s} = \begin{cases} \text{sign}(\zeta_{2f}) \overline{\zeta}_2, & |\zeta_{2f}| > \Gamma_{pi} \\ -\frac{\text{sign}(\zeta_{2f}) \zeta_{2f}^2}{4\lambda_1} + \frac{(\overline{\zeta}_2 + \lambda_1) \zeta_{2f}}{2\lambda_1} - \frac{\text{sign}(\zeta_{2f})(\overline{\zeta}_2 - \lambda_1)^2}{4\lambda_1}, & \left||\zeta_{2f}| - \overline{\zeta}_2\right| \leq \lambda_1 \\ \zeta_{2f}, & |\zeta_{2f}| < \Gamma_{qi} \end{cases} \tag{19}$$

where $\overline{\zeta}_2$ denotes the maximum magnitude of $\overline{\zeta}_{2s}$, $\lambda_1 > 0$ is a positive constant to be designed, and $\Gamma_{pi} = \overline{\zeta}_2 + \lambda_1$, $\Gamma_{qi} = \overline{\zeta}_2 - \lambda_1$.

Additionally, some variables are defined as:

$$\begin{aligned} T_b &= \zeta_{2\max} - \overline{\zeta}_2, \Xi_2 = \frac{1}{T_b^2 - s_2^2} \\ s_2 &= \zeta_2 - \zeta_{2s}, \Phi_1 = \left(\zeta_{2s} - \zeta_{2f}\right) \end{aligned} \tag{20}$$

Differentiating s_1 with respect to time and invoking (13), (15), (17), and (20), then the derivative of s_1 is obtained as:

$$\begin{aligned} \dot{s}_1 &= H_{F_1}^{-1} w_1^{*T} \varphi(\overline{\zeta}_1) + D_1(t) + g_1(\overline{\zeta}_1) \left[\zeta_2 - \zeta_{2s} + \zeta_{2s} - \zeta_{2f} + \zeta_{2f} - \zeta_{2d}\right] \\ &\quad - \left[k_1 s_1 + H_{F_1}^{-1} \hat{w}_1^T \varphi_1(\overline{\zeta}_1) + \hat{D}_1(t) + k_{\chi_1} \chi_1\right] \\ &= -k_1 s_1 + H_{F_1}^{-1} \tilde{w}_1^T \varphi(\overline{\zeta}_1) + \tilde{D}_1(t) + g_1(\overline{\zeta}_1) s_2 + g_1(\overline{\zeta}_1) z_2 + g_1(\overline{\zeta}_1) \Phi_1 - k_{\chi_1} \chi_1 \end{aligned} \tag{21}$$

where $\tilde{w}_1 = w_1^* - \hat{w}_1$, $\tilde{D}_1(t) = D_1(t) - \hat{D}_1(t)$.

To analyze and compensate for the effects caused by the introduced saturation function, the auxiliary system is constructed as follows:

$$\dot{\chi}_1 = \begin{cases} -\frac{2g_1(\overline{\zeta}_1) \Phi_1 s_1 - 2k_{\chi_1} \chi_1 s_1 + g_1^2(\overline{\zeta}_1) \Phi_1^2}{2\chi_1} - k_{\chi_1} \chi_1 + g_1(\overline{\zeta}_1) \Phi_1 & |\chi_1| > \omega_1 \\ 0 & |\chi_1| \leq \omega_1 \end{cases} \tag{22}$$

where $\omega_1 > 0$ is a scalar.

The prediction error h_{1p} is designed as:

$$h_{1p} = \zeta_1 - \hat{\zeta}_1 \tag{23}$$

Inspired by the serial-parallel estimation model (SPEM), $\hat{\zeta}_1$, as the predicted value of ζ_1, is constructed as:

$$\dot{\hat{\zeta}}_1 = H_{F_1}^{-1}\hat{w}_1^T\varphi_1(\overline{\zeta}_1) + g_1(\overline{\zeta}_1)\zeta_2 + \hat{D}_1(t) + \Gamma_{h_{1p}}h_{1p} \tag{24}$$

where $\Gamma_{h_{1p}}$ is a design scalar, and $\hat{\zeta}_1(0) = \zeta_1(0)$.

The first composite intelligent learning law of NN is updated in the following form:

$$\dot{\hat{w}}_1 = \tau_1\left[H_{F_1}^{-1}(p_{h_1}h_{1p} + s_1)\varphi_1(\overline{\zeta}_1) - \Gamma_1\hat{w}_1\right] \tag{25}$$

As shown below, the disturbance observer is used to cope with the uncertainty of $\hat{D}_1$.

$$\hat{D}_1 = H_1(\zeta_1 - \eta_1) \tag{26}$$

$$\dot{\eta}_1 = H_{F_1}^{-1}\hat{w}_1^T\varphi(\overline{\zeta}_1) + g_1(\overline{\zeta}_1)\zeta_2 + \hat{D}_1(t) - H_1^{-1}(p_{h_1}h_{1p} + s_1) \tag{27}$$

where H_1 is a positive parameter to be designed, and η_1 is an auxiliary variable.

Substituting (27) into (26), then the derivative of $\tilde{D}_1$ is obtained as:

$$\begin{aligned}\dot{\tilde{D}}_1 &= \dot{D}_1 - H_1\left[\begin{array}{c}H_{F_1}^{-1}w_1^{*T}\varphi(\overline{\zeta}_1) + D_1(t) + g_1(\overline{\zeta}_1)\zeta_2 - \\ H_{F_1}^{-1}\hat{w}_1^T\varphi(\overline{\zeta}_1) - g_1(\overline{\zeta}_1)\zeta_2 - \hat{D}_1(t) + H_1^{-1}(p_{h_1}h_{1p} + s_1)\end{array}\right] \\ &= \dot{D}_1 - H_1\left[H_{F_1}^{-1}\tilde{w}_1^T\varphi(\overline{\zeta}_1) + \tilde{D}_1(t)\right] - (p_{h_1}h_{1p} + s_1)\end{aligned} \tag{28}$$

Considering the Lyapunov function as:

$$V_1 = \frac{1}{2}s_1^2 + \frac{1}{2\tau_1}\tilde{w}_1^T\tilde{w}_1 + \frac{1}{2}\tilde{D}_1^2 + \frac{1}{2}p_{h_1}h_{1p}^2 + \frac{1}{2}\chi_1^2 + \frac{1}{2}z_2^2 \tag{29}$$

Substituting (24) into (23) yields, then the derivative of prediction error h_{1p} is constructed below:

$$\dot{h}_{1p} = H_{F_1}^{-1}\tilde{w}_1^T\varphi(\overline{\zeta}_1) + \tilde{D}_1(t) - \Gamma_{h_{1p}}h_{1p} \tag{30}$$

Define $p_1 = \tilde{w}_1^T\varphi(\overline{\zeta}_1)$, and then:

$$h_{1p}\dot{h}_{1p} = h_{1p}\left[H_{F_1}^{-1}p_1 + \tilde{D}_1(t)\right] - \Gamma_{h_{1p}}h_{1p}^2 \tag{31}$$

Differentiating V_1 with respect to time and invoking (21), (22), (25), (28), and (31), $\dot{V}_1$ is calculated as:

$$\begin{aligned}\dot{V}_1 &= s_1\dot{s}_1 - \frac{1}{\tau_1}\tilde{w}_1^T\dot{\hat{w}}_1 + \tilde{D}_1\dot{\tilde{D}}_1 + p_{h_1}h_{1p}\dot{h}_{1p} + \chi_1\dot{\chi}_1 + z_2\dot{z}_2 \\ &= -k_1s_1^2 + g_1(t,\overline{\zeta}_1)s_1s_2 + g_1(t,\overline{\zeta}_1)s_1z_2 + \Gamma_1\tilde{w}_1^T\hat{w}_1 + \tilde{D}_1(t)\dot{D}_1(t) - H_1H_{F_1}^{-1}\tilde{D}_1(t)p_1 \\ &\quad -H_1\tilde{D}_1^2(t) - \Gamma_{h_{1p}}p_{h_1}h_{1p}^2 - k_{\chi_1}\chi_1^2 + g_1(\overline{\zeta}_1)\Phi_1\chi_1 - \frac{g_1^2(\overline{\zeta}_1)\Phi_1^2}{2} - \frac{1}{\varepsilon_2}z_2^2 - \zeta_{2d}z_2\end{aligned} \tag{32}$$

The following facts should be considered:

$$g_1(t,\overline{\zeta}_1)s_1z_2 \leq \frac{g_1^2(t,\overline{\zeta}_1)}{2l_1^2}s_1^2 + \frac{l_1^2}{2}z_2^2 \tag{33}$$

$$\Gamma_1\tilde{w}_1^T\hat{w}_1 \leq \frac{\Gamma_1}{2}w_1^{*2} - \frac{\Gamma_1}{2}\tilde{w}_1^T\tilde{w}_1 \tag{34}$$

$$\tilde{D}_1\dot{D}_1 \leq \frac{1}{2}\tilde{D}_1^2 + \frac{1}{2}v_1^2 \tag{35}$$

$$-\tilde{D}_1(t)p_1 \leq \frac{1}{2}\vartheta_1\tilde{D}_1^2(t)\beta_1^2 + \frac{1}{2\vartheta_1}\tilde{w}_1^T\tilde{w}_1 \tag{36}$$

$$g_1(t,\bar{\zeta}_1)\Phi_1\chi_1 \leq \frac{g_1^2(t,\bar{\zeta}_1)\Phi_1^2}{2} + \frac{1}{2}\chi_1^2 \tag{37}$$

$$-\dot{\zeta}_{2d}z_2 \leq \frac{\left|\dot{\zeta}_{2d}\right|^2}{2}z_2^2 + \frac{1}{2} \tag{38}$$

where $\|\varphi_1(\bar{\zeta}_1)\| \leq \beta_1$, $\left|\dot{D}_1\right| \leq v_1$.

According to the above inequalities, the derivative of V_1 can be given by:

$$\begin{aligned}\dot{V}_1 \leq &-\left[k_1 - \frac{g_1^2(t,\bar{\zeta}_1)}{2t_1^2}\right]s_1^2 - \left(\frac{\Gamma_1}{2} - \frac{H_1 H_{F_1}^{-1}}{2\theta_1}\right)\widetilde{\omega}_1^T\widetilde{\omega}_1 - \left(H_1 - \frac{H_1 H_{F_1}^{-1}\theta_1\beta_1^2+1}{2}\right)\widetilde{D}_1^2 \\ &-\Gamma_{h_{1p}}p_{h_1}h_{1p}^2 - \left(k_{\chi_1} - \frac{1}{2}\right)\chi_1^2 - \left(\frac{1}{\varepsilon_2} - \frac{\left|\dot{\zeta}_{2d}\right|^2 + r_1^2}{2}\right)z_2^2 + C_1 + g_1(t,\bar{\zeta}_1)s_1s_2\end{aligned} \tag{39}$$

where $C_1 = \frac{\Gamma_1}{2}w_1^{*2} + \frac{1}{2}v_1^2 + \frac{1}{2}$.

B. Intermediate Loop

Define $F_2(t,\bar{\zeta}_2) = H_{F_2}f_2(t,\bar{\zeta}_2)$, where $H_{F_2} > 0$ is a designable parameter. As mentioned above, for the dynamics equation of ζ_2 in IGC system (12), $F_2(t,\bar{\zeta}_2)$ is estimated through NN.

$$\dot{\zeta}_2 = H_{F_2}^{-1}w_2^{*T}\varphi_2(\bar{\zeta}_2) + H_{F_2}^{-1}v_2 + g_2(t,\bar{\zeta}_2)\zeta_3 \tag{40}$$

where w_2^* denotes the optical weight vector. v_2, as the approximation error of NN, satisfies $|v_2| \leq v_m$.

The second error surface $s_2 = \zeta_2 - \zeta_{2s}$ can be obtained from (20), and then the nominal virtual control ζ_{3d} is constructed as:

$$\zeta_{3d} = -\left(k_2s_2 + H_{F_2}^{-1}\hat{w}_2^T\varphi_2(\bar{\zeta}_2) + \Xi_2^{-1}g_1(t,\bar{\zeta}_1)s_1 - \dot{\zeta}_{2s}\right)/g_2(t,\bar{\zeta}_1) \tag{41}$$

where $k_2 > 0$ is a constant to be designed, and $\hat{w}_2$ represents the estimate value of w_2^*.

At this point, a first-order filter is introduced into the system, which meets the following conditions.

$$\varepsilon_3\dot{\zeta}_{3f} + \zeta_{3f} = \zeta_{3d}, \zeta_{3f}(0) = \zeta_{3d}(0) \tag{42}$$

where ε_3 is the filter parameter and the error signal of the filter (42) is presented as:

$$z_3 = \zeta_{3f} - \zeta_{3d} \tag{43}$$

Its dynamic satisfies:

$$\dot{z}_3 = -\varepsilon_3^{-1}z_3 - \dot{\zeta}_{3d} \tag{44}$$

The third error surface is designed as $s_3 = \zeta_3 - \zeta_{3f}$. Invoking (41), (43), and (44), the derivative of s_2 can be given by:

$$\begin{aligned}\dot{s}_2 &= H_{F_2}^{-1}w_2^{*T}\varphi(\bar{\zeta}_2) + g_2(t,\bar{\zeta}_2)(s_3+z_3) + H_{F_2}^{-1}v_2 - \dot{\zeta}_{2s} \\ &\quad - \left[k_2s_2 + H_{F_2}^{-1}\hat{w}_2^T\varphi(\bar{\zeta}_2) + \Xi_2^{-1}g_1(t,\bar{\zeta}_1)s_1 - \dot{\zeta}_{2s}\right] \\ &= -k_2s_2 + H_{F_2}^{-1}\widetilde{w}_2^T\varphi(\bar{\zeta}_2) + H_{F_2}^{-1}v_2 - \Xi_2^{-1}g_1(t,\bar{\zeta}_1)s_1 + g_2(t,\bar{\zeta}_2)(s_3+z_3)\end{aligned} \tag{45}$$

where $\widetilde{w}_2 = w_2^* - \hat{w}_2$.

The prediction error h_{2p} is defined as:

$$h_{2p} = \zeta_2 - \hat{\zeta}_2 \tag{46}$$

Inspired by SPEM, $\hat{\zeta}_2$, as the predicted value of ζ_2, is constructed as:

$$\dot{\hat{\zeta}}_2 = H_{F_2}^{-1}\hat{w}_2^T \varphi_2(\overline{\zeta}_2) + g_2(t,\overline{\zeta}_2)\zeta_3 + \Gamma_{h_{2p}} h_{2p} \tag{47}$$

where $\Gamma_{h_{2p}}$ is a positive scalar to be designed, and $\hat{\zeta}_2(0) = \zeta_2(0)$.

The second composite intelligent learning law of NN is updated in a novel form, which fuses tracking error with prediction error.

$$\dot{\hat{w}}_2 = \tau_2 \left[H_{F_2}^{-1}\left(p_{h_2} h_{2p} + \Xi_2 s_2\right)\varphi(\overline{\zeta}_2) - \Gamma_2 \hat{w}_2 \right] \tag{48}$$

Considering the barrier Lyapunov function as:

$$V_2 = \frac{1}{2} \ln \frac{T_b^2}{T_b^2 - s_2^2} + \frac{1}{2\tau_2} \tilde{w}_2^T \tilde{w}_2 + \frac{1}{2} p_{h_2} h_{2p}^2 + \frac{1}{2} z_3^2 \tag{49}$$

A specified compact set is defined as:

$$Z_1 = \{s_2 | |s_2| < T_b\} \tag{50}$$

Substituting (47) into (46) yields, then the derivative of prediction error h_{2p} is constructed as:

$$\dot{h}_{2p} = H_{F_2}^{-1} \tilde{w}_2^T \varphi(\overline{\zeta}_2) + H_{F_2}^{-1} \nu_2 - \Gamma_{h_{2p}} h_{2p} \tag{51}$$

Define $p_2 = \tilde{w}_2^T \varphi(\overline{\zeta}_2)$, and then:

$$h_{2p}\dot{h}_{2p} = h_{2p}\left[H_{F_2}^{-1} p_2 + H_{F_2}^{-1} \nu_2\right] - \Gamma_{h_{1p}} h_{1p}^2 \tag{52}$$

Differentiating V_2 with respect to time when $s_2 \in Z_1$ and invoking (43), (44), (45), (48), and (52), the derivative of V_2 is calculated by:

$$\begin{aligned}\dot{V}_2 &= \frac{s_2 \dot{s}_2}{T_b^2 - s_2^2} - \frac{1}{\tau_2}\tilde{w}_2^T \dot{\hat{w}}_2 + p_{h_2} h_{2p} \dot{h}_{2p} + z_3 \dot{z}_3 \\ &= -k_2 \Xi_2 s_2^2 + H_{F_2}^{-1} \nu_2 \Xi_2 s_2 + g_2(t,\overline{\zeta}_2)\Xi_2 s_2 s_3 + \Gamma_2 \tilde{w}_2^T \hat{w}_2 + g_2(t,\overline{\zeta}_2)\Xi_2 s_2 z_3 \\ &\quad -g_1(t,\overline{\zeta}_1) s_1 s_2 + H_{F_2}^{-1} p_{h_2} h_{2p} \nu_2 - \Gamma_{h_{2p}} p_{h_2} h_{2p}^2 - \frac{1}{\varepsilon_3} z_3^2 - \dot{\zeta}_{3d} z_3\end{aligned} \tag{53}$$

The following facts should be considered:

$$H_{F_2}^{-1} \nu_2 \Xi_2 s_2 \leq \frac{H_{F_2}^{-2} \Xi_2^2}{2} s_2^2 + \frac{\nu_m^2}{2} \tag{54}$$

$$g_2(t,\overline{\zeta}_2)\Xi_2 s_2 z_3 \leq \frac{g_2^2(t,\overline{\zeta}_2)\Xi_2}{2 l_2^2} s_2^2 + \frac{l_2^2 \Xi_2}{2} z_3^2 \tag{55}$$

$$\Gamma_2 \tilde{w}_2^T \hat{w}_2 \leq \frac{\Gamma_2}{2} w_2^{*2} - \frac{\Gamma_2}{2} \tilde{w}_2^T \tilde{w}_2 \tag{56}$$

$$H_{F_2}^{-1} p_{h_2} h_{2p} \nu_2 \leq \frac{H_{F_2}^{-2} p_{h_2}^2}{2} h_{2p}^2 + \frac{\nu_m^2}{2} \tag{57}$$

$$-\dot{\zeta}_{3d} z_3 \leq \frac{\left|\dot{\zeta}_{3d}\right|^2}{2} z_3^2 + \frac{1}{2} \tag{58}$$

According to the above inequalities, the derivative of V_2 can be given by:

$$\dot{V}_2 \leq -\left[k_2\Xi_2 - \frac{H_{F_2}^{-2}\Xi_2^2}{2} - \frac{g_2^2(t,\bar{\zeta}_2)\Xi_2^2}{2l_2^2}\right]s_2^2 - \frac{\Gamma_2}{2}\tilde{w}_2^T\tilde{w}_2 - \left(\Gamma_{h_{1p}}p_{h_2} - \frac{H_{F_2}^{-2}p_{h_2}^2}{2}\right)h_{2p}^2 \\ -\left(\frac{1}{\varepsilon_3} - \frac{|\dot{\zeta}_{3d}|^2}{2} - \frac{l_3^2\Xi_2}{2}\right)z_3^2 - g_1(t,\bar{\zeta}_1)s_1s_2 + g_2(t,\bar{\zeta}_2)\Xi_2s_2s_3 + C_2 \quad (59)$$

where $C_2 = v_m^2 + \frac{\Gamma_2}{2}w_2^{*2} + \frac{1}{2}$.

C. Inner Loop

Define $F_3(t,\bar{\zeta}_3) = H_{F_3}f_3(t,\bar{\zeta}_3)$, where $H_{F_3} > 0$ is a designable parameter. As mentioned above, for the dynamics equation of ζ_3 in the IGC system (12), $F_3(t,\bar{\zeta}_3)$ is estimated through NN.

$$\dot{\zeta}_3 = H_{F_3}^{-1}w_3^{*T}\varphi(\bar{\zeta}_3) + H_{F_3}^{-1}v_3 + g_3(t,\bar{\zeta}_3)u \quad (60)$$

where w_3^* denotes the optical weight vector. v_3, as the approximation error of NN, satisfies $|v_3| \leq v_m$.

The virtual control u_0 is constructed as:

$$u_0 = -\left(k_3s_3 + H_{F_3}^{-1}\hat{w}_3^T\varphi(\bar{\zeta}_3) + \Xi_2g_2(t,\bar{\zeta}_2)s_2 - \dot{\zeta}_{3s} + k_{\chi_2}\chi_2\right)/g_3(t,\bar{\zeta}_1) \quad (61)$$

where $k_3 > 0$ and $k_{\chi_2} > 0$ are the user-defined parameters, and χ_2 is the auxiliary system to be designed. $\hat{w}_3$ represents the estimate value of w_3^*.

Invoking (60) and (61), the derivative of s_3 can be further obtained as:

$$\dot{s}_3 = -k_3s_3 + H_{F_3}^{-1}\tilde{w}_3^T\varphi(\bar{\zeta}_3) + H_{F_3}^{-1}v_3 - \Xi_2g_2(t,\bar{\zeta}_2)s_2 + g_3(t,\bar{\zeta}_3)\Phi_2 - k_{\chi_2}\chi_2 \quad (62)$$

where $\tilde{w}_3 = w_3^* - \hat{w}_3$, $\Phi_2 = (u - u_0)$.

To analyze and compensate for the effects caused by the introduced saturation function, the auxiliary system is constructed as follows:

$$\dot{\chi}_2 = \begin{cases} -\frac{2g_3(t,\bar{\zeta}_3)\Phi_2s_3 - 2k_{\chi_2}\chi_2s_3 + g_3^2(t,\bar{\zeta}_3)\Phi_2^2}{2\chi_2} - k_{\chi_2}\chi_2 + g_3(t,\bar{\zeta}_3)\Phi_2 & |\chi_2| > \omega_2 \\ 0 & |\chi_2| \leq \omega_2 \end{cases} \quad (63)$$

where $\omega_2 > 0$ is a scalar.

The prediction error h_{3p} is designed as:

$$h_{3p} = \zeta_3 - \hat{\zeta}_3 \quad (64)$$

Inspired by the SPEM, $\hat{\zeta}_3$, as the predicted value of ζ_3, is constructed as:

$$\dot{\hat{\zeta}}_3 = H_{F_3}^{-1}\hat{w}_3^T\varphi(\bar{\zeta}_3) + g_3(t,\bar{\zeta}_1)u + \Gamma_{h_{3p}}h_{3p} \quad (65)$$

where $\Gamma_{h_{3p}}$ is a positive scalar to be designed, and $\hat{\zeta}_3(0) = \zeta_3(0)$.

The third learning law of NN is updated in the following form:

$$\dot{\hat{w}}_3 = \tau_3\left[H_{F_3}^{-1}(p_{h_3}h_{3p} + s_3)\varphi_3(\bar{\zeta}_3) - \Gamma_3\hat{w}_3\right] \quad (66)$$

Considering the Lyapunov function as:

$$V_3 = \frac{1}{2}s_3^2 + \frac{1}{2\tau_3}\tilde{w}_3^T\tilde{w}_3 + \frac{1}{2}p_{h_3}h_{3p}^2 + \frac{1}{2}\chi_2^2 \quad (67)$$

Substituting (65) into (64) yields, the derivative of h_{3p} is computed as:

$$\dot{h}_{3p} = H_{F_3}^{-1} \tilde{w}_3^T \varphi(\bar{\zeta}_3) + H_{F_3}^{-1} v_3 - \Gamma_{h_{3p}} h_{3p} \tag{68}$$

Define $p_3 = \tilde{w}_3^T \varphi(\bar{\zeta}_3)$, and then:

$$h_{3p} \dot{h}_{3p} = h_{3p} \left[H_{F_3}^{-1} p_3 + H_{F_3}^{-1} v_3 \right] - \Gamma_{h_{3p}} h_{3p}^2 \tag{69}$$

Differentiating V_3 with respect to time and invoking (62), (63), (66), and (69), $\dot{V}_3$ is calculated as:

$$\begin{aligned}
\dot{V}_3 &= s_3 \dot{s}_3 - \tfrac{1}{\tau_3} \tilde{w}_3^T \dot{\hat{w}}_3 + p_{h_3} h_{3p} \dot{h}_{3p} + \chi_3 \dot{\chi}_3 \\
&= s_3 \Big[-k_3 s_3 + H_{F_3}^{-1} \tilde{w}_3^T \varphi(\bar{\zeta}_3) + H_{F_3}^{-1} v_3 - \Xi_2 g_2(t, \bar{\zeta}_2) s_2 + g_3(t, \bar{\zeta}_3) \Phi_2 - k_{\chi_2} \chi_2 \Big] \\
&\quad + p_{h_3} \Big[h_{3p} \big(H_{F_3}^{-1} p_3 + H_{F_3}^{-1} v_3 \big) - \Gamma_{h_{3p}} h_{3p}^2 \Big] - k_{\chi_2} \chi_2^2 + g_2(t, \bar{\zeta}_2) \Phi_2 \chi_2 - \tfrac{1}{2} \Psi_{\chi_2} \\
&\quad - \tfrac{1}{\tau_3} \tilde{w}_3^T \tau_3 \Big[H_{F_3}^{-1} (p_{h_3} h_{3p} + s_3) \varphi_3(\bar{\zeta}_3) - \Gamma_3 \hat{w}_3 \Big] \\
&= -k_3 s_3^2 + H_{F_3}^{-1} v_3 s_3 + \Gamma_3 \tilde{w}_1^T \hat{w}_3 + H_{F_3}^{-1} p_{h_3} h_{3p} v_3 - \Gamma_{h_{3p}} p_{h_3} h_{3p}^2 - k_{\chi_2} \chi_2^2 \\
&\quad + g_3(t, \bar{\zeta}_3) \Phi_2 \chi_2 - g_3^2(t, \bar{\zeta}_3) \Phi_2^2 - g_2(t, \bar{\zeta}_2) \Xi_2 s_2 s_3
\end{aligned} \tag{70}$$

The following facts should be considered:

$$H_{F_3}^{-1} v_3 s_3 \leq \frac{H_{F_3}^{-2}}{2} s_3^2 + \frac{v_m^2}{2} \tag{71}$$

$$\Gamma_3 \tilde{w}_1^T \hat{w}_3 \leq \frac{\Gamma_3}{2} w_3^{*2} - \frac{\Gamma_3}{2} \tilde{w}_3^T \tilde{w}_3 \tag{72}$$

$$H_{F_3}^{-1} p_{h_3} h_{3p} v_3 \leq \frac{H_{F_3}^{-2} p_{h_3}^2}{2} h_{3p}^2 + \frac{v_m^2}{2} \tag{73}$$

$$g_3(t, \bar{\zeta}_3) \Phi_2 \chi_2 \leq \frac{g_3^2(t, \bar{\zeta}_3)}{2} \Phi_2^2 + \frac{1}{2} \chi_2^2 \tag{74}$$

According to the above inequalities, the derivative of V_3 can be given by:

$$\begin{aligned}
\dot{V}_3 &\leq -k_3 s_3^2 + \frac{H_{F_3}^{-2}}{2} s_3^2 + \frac{v_m^2}{2} + \frac{\Gamma_3}{2} w_3^{*2} - \frac{\Gamma_3}{2} \tilde{w}_3^T \tilde{w}_3 + \frac{H_{F_3}^{-2} p_{h_3}^2}{2} h_{3p}^2 + \frac{v_m^2}{2} - \Gamma_{h_{3p}} p_{h_3} h_{3p}^2 \\
&\quad - k_{\chi_2} \chi_2^2 + \frac{g_3^2(t, \bar{\zeta}_3)}{2} \Phi_2^2 + \tfrac{1}{2} \chi_2^2 - \frac{g_3^2(t, \bar{\zeta}_3)}{2} \Phi_2^2 - g_2(t, \bar{\zeta}_2) \Xi_2 s_2 s_3 \\
&\leq -\left(k_3 - \frac{H_{F_3}^{-2}}{2} \right) s_3^2 - \frac{\Gamma_3}{2} \tilde{w}_3^T \tilde{w}_3 - \left(\Gamma_{h_{3p}} p_{h_3} - \frac{H_{F_3}^{-2} p_{h_3}^2}{2} \right) h_{3p}^2 - \left(k_{\chi_2} - \tfrac{1}{2} \right) \chi_2^2 - g_2(t, \bar{\zeta}_2) \Xi_2 s_2 s_3 + C_3
\end{aligned} \tag{75}$$

where $C_3 = \frac{v_m^2}{2} + \frac{\Gamma_3}{2} w_3^{*2} + \frac{v_m^2}{2}$.

As you can see, a novel robust adaptive dynamic surface IGC scheme based on the serial-parallel estimation model, neural networks, and the disturbance observer is proposed in Section 3. The recursive design details of the scheme can be summarized in Figure 3. To be specific, the serial-parallel estimation model estimates the system states and provides "critic" information for the neural network and disturbance observer; then, these three are combined to realize online composite learning of the multiple uncertainties of the system and improve the interception accuracy. In addition, the state and input constraints are resolved by adopting the BLF and the improved saturation function.

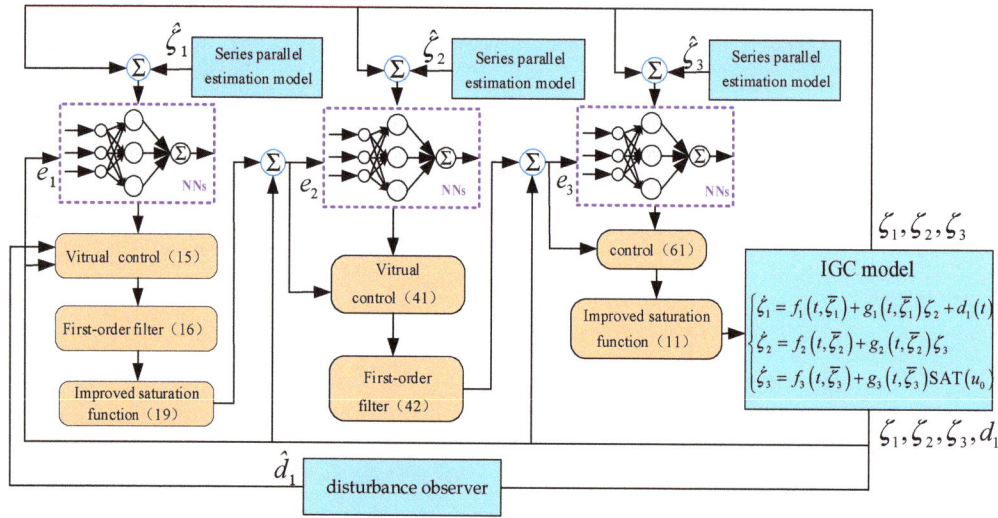

Figure 3. The recursive design details of the scheme proposed in Section 3.

4. Stability Analysis

The Lyapunov function is chosen as:

$$
\begin{aligned}
V &= V_1 + V_2 + V_3 \\
&= \tfrac{1}{2}s_1^2 + \tfrac{1}{2}\ln\tfrac{T_b^2}{T_b^2 - s_2^2} + \tfrac{1}{2}s_3^2 + \tfrac{1}{2\tau_1}\widetilde{w}_1^T\widetilde{w}_1 + \tfrac{1}{2\tau_2}\widetilde{w}_2^T\widetilde{w}_2 + \tfrac{1}{2\tau_3}\widetilde{w}_3^T\widetilde{w}_3 + \tfrac{1}{2}\widetilde{D}_1^2 \\
&\quad + \tfrac{1}{2}p_{h_1}h_{1p}^2 + \tfrac{1}{2}p_{h_2}h_{2p}^2 + \tfrac{1}{2}p_{h_3}h_{3p}^2 + \tfrac{1}{2}\chi_1^2 + \tfrac{1}{2}\chi_2^2 + \tfrac{1}{2}z_2^2 + \tfrac{1}{2}z_3^2
\end{aligned}
\tag{76}
$$

Differentiating V with respect to time and invoking (39), (59), and (75), $\dot{V}$ can be represented as:

$$
\begin{aligned}
\dot{V} &\leq -\left[k_1 - \tfrac{g_1^2(t,\bar{\zeta}_1)}{2l_1^2}\right]s_1^2 - \left[k_2\Xi_2 - \tfrac{H_{F_2}^{-2}\Xi_2^2}{2} - \tfrac{g_2^2(t,\bar{\zeta}_2)\Xi_2}{2l_2^2}\right]s_2^2 - \left(k_3 - \tfrac{H_{F_3}^{-2}}{2}\right)s_3^2 - \tfrac{\Gamma_2}{2}\widetilde{w}_2^T\widetilde{w}_{21} \\
&\quad - \left(\tfrac{\Gamma_1}{2} - \tfrac{H_1 H_{F_1}^{-1}}{2\theta_1}\right)\widetilde{w}_1^T\widetilde{w}_1 - \tfrac{\Gamma_3}{2}\widetilde{w}_3^T\widetilde{w}_3 - \left(H_1 - \tfrac{H_1 H_{F_1}^{-1}\vartheta_1\beta_1^2 + 1}{2}\right)\widetilde{D}_1^2 - \Gamma_{h_{1p}}p_{h_1}h_{1p}^2 \\
&\quad - \left(\Gamma_{h_{1p}}p_{h_2} - \tfrac{H_{F_2}^{-2}p_{h_2}^2}{2}\right)h_{2p}^2 - \left(\Gamma_{h_{3p}}p_{h_3} - \tfrac{H_{F_3}^{-2}p_{h_3}^2}{2}\right)h_{3p}^2 - \left(k_{\chi_1} - \tfrac{1}{2}\right)\chi_1^2 \\
&\quad - \left(k_{\chi_2} - \tfrac{1}{2}\right)\chi_2^2 - \left(\tfrac{1}{\varepsilon_2} - \tfrac{|\dot{\zeta}_{2d}|^2 + \iota_1^2}{2}\right)z_2^2 - \left(\tfrac{1}{\varepsilon_3} - \tfrac{|\dot{\zeta}_{3d}|^2}{2} - \tfrac{\iota_2^2 \Xi_2}{2}\right)z_3^2 + C
\end{aligned}
\tag{77}
$$

where $C = C_1 + C_2 + C_3$.

Obviously, differentiating $|\dot{\zeta}_{2d}|$ and $|\dot{\zeta}_{3d}|$ with respect to time, there exists as:

$$
\begin{aligned}
|\dot{\zeta}_{2d}| &\leq Q_1\left(s_1, s_2, z_2, \chi_1, \widetilde{D}_1, y_r, \dot{y}_r, \ddot{y}_r\right) \\
|\dot{\zeta}_{3d}| &\leq Q_2\left(s_1, s_2, s_3, z_2, z_3, \chi_1, \chi_2, \widetilde{D}_1, y_r, \dot{y}_r, \ddot{y}_r\right)
\end{aligned}
\tag{78}
$$

where Q_1 and Q_2 are positive continuous functions.

A specified compact set is designed as:

$$
Z = \left\{Q\left(s_1, s_2, s_3, z_2, z_3, \chi_1, \widetilde{D}_1, y_r, \dot{y}_r, \ddot{y}_r\right) : V \leq I_1\right\}
\tag{79}
$$

where $I_1 > 0$. Then, it can be concluded from (78) and (79) that the continuous functions Q_1 and Q_2 have maximum values within the scope of set Z, i.e.,

$$|Q_1| \leq N_1, |Q_2| \leq N_2 \tag{80}$$

The appropriate parameters are selected as follows:

$$\begin{cases} k_1 \geq \frac{g_1^2(t,\bar{\zeta}_1)}{2l_1^2} + \frac{K}{2}, k_2 \geq \frac{H_{F_2}^{-2}\Xi_2}{2} + \frac{g_2^2(t,\bar{\zeta}_2)}{2l_2^2} + \frac{K}{2}, k_3 \geq \frac{H_{F_3}^{-2}}{2} + \frac{K}{2} \\ \Gamma_1 \geq \frac{H_1 H_{F_1}^{-1}}{\vartheta_1} + \frac{K}{\tau_1}, \Gamma_2 \geq \frac{K}{\tau_2}, \Gamma_3 \geq \frac{K}{\tau_3}, H_1 \geq \frac{H_1 H_{F_1}^{-1}\vartheta_1\beta_1^2+K+1}{2} \\ k_{\chi_1} \geq \frac{K}{2} + \frac{1}{2}, k_{\chi_2} \geq \frac{K}{2} + \frac{1}{2}, \Gamma_{h_{1p}} \geq \frac{K}{2}, \Gamma_{h_{2p}} \geq \frac{H_{F_2}^{-2}p_{h_2}+K}{2} \\ \Gamma_{h_{3p}} \geq \frac{H_{F_3}^{-2}p_{h_3}+K}{2}, \varepsilon_2^{-1} \geq \frac{N_1^2+l_1^2+K}{2}, \varepsilon_3^{-1} \geq \frac{N_2^2+l_2^2\Xi_2+K}{2} \end{cases} \tag{81}$$

where $K > 0$ represents a constant.

Theorem 1. *Consider the IGC system (12) under Assumption 1 with multiple uncertainties and actuator saturation, the controller (61), and NN learning laws (25), (48), (66) with parameters satisfying (81). If $s_2(0) \in Z_1 := \{s_2||s_2| < T_b\}$, the following properties hold:*

1. *The output of the system ζ_1 can converge to a neighborhood of zero;*
2. *All signals such as $s_1, s_2, s_3, z_2, z_3, \chi_1, \chi_2, h_{1p}, h_{2p}, h_{3p}$, and $\widetilde{D}_1$ are uniformly ultimately bounded;*
3. *The constraint of the state variable ζ_2 will not be violated.*

When $s_2 \in Z$, inequality (77) can be rewritten in the following form if the design parameters are selected as inequation (81).

$$\begin{aligned}\dot{V} \leq &-\frac{K}{2}s_1^2 - \frac{K\Xi_2}{2}s_2^2 - \frac{K}{2}s_3^2 - \frac{K}{2\tau_1}\widetilde{w}_1^T\widetilde{w}_1 - \frac{K}{2\tau_2}\widetilde{w}_2^T\widetilde{w}_2 - \frac{K}{2\tau_3}\widetilde{w}_3^T\widetilde{w}_3 \\ &- \frac{K}{2}\widetilde{D}_1^2 - \frac{K}{2}p_{h_1}h_{1p}^2 - \frac{K}{2}p_{h_2}h_{2p}^2 - \frac{K}{2}p_{h_3}h_{3p}^2 - \frac{K}{2}\chi_1^2 - \frac{K}{2}\chi_2^2 - \frac{K}{2}z_2^2 - \frac{K}{2}z_3^2 + C \end{aligned} \tag{82}$$

Invoking (82) and Lemma 1, $\dot{V}$ is calculated as:

$$\begin{aligned}\dot{V} \leq &-\frac{K}{2}s_1^2 - \frac{K}{2}\ln\frac{T_b^2}{T_b^2-s_2^2} - \frac{K}{2}s_3^2 - \frac{K}{2\tau_1}\widetilde{w}_1^T\widetilde{w}_1 - \frac{K}{2\tau_2}\widetilde{w}_2^T\widetilde{w}_2 - \frac{K}{2\tau_3}\widetilde{w}_3^T\widetilde{w}_3 \\ &- \frac{K}{2}\widetilde{D}_1^2 - \frac{K}{2}p_{h_1}h_{1p}^2 - \frac{K}{2}p_{h_2}h_{2p}^2 - \frac{K}{2}p_{h_3}h_{3p}^2 - \frac{K}{2}\chi_1^2 - \frac{K}{2}\chi_2^2 - \frac{K}{2}z_2^2 - \frac{K}{2}z_3^2 + C \end{aligned} \tag{83}$$

That is:

$$\dot{V} \leq -KV + C \tag{84}$$

$\dot{V} < 0$ can be received from (84) when $V = I_1$ and $K > \frac{C}{I_1}$, i.e., if $V(0) \leq I_1$, then $V(t) \leq I_1$ always holds for $\forall t \geq 0$. In other words, $V(0) \leq I_1$ is an invariant set.
By solving the inequality (84), we have:

$$0 \leq V(t) \leq \left(V(0) - \frac{C}{K}\right)e^{-Kt} + \frac{C}{K} \tag{85}$$

Substituting (29) into (85) yields, the following expression can be obtained.

$$|s_1| \leq \sqrt{2\left(V(0) - \frac{C}{K}\right)e^{-Kt} + \frac{2C}{K}}, \forall t > 0 \tag{86}$$

According to (86), $|s_1| \leq \sqrt{\frac{2C}{K}}$ can be acquired when $t \to \infty$, i.e., s_1 is uniformly ultimately bounded. Therefore, by choosing appropriate design parameters, s_1 can be arbitrarily small and ζ_1 can converge to a certain neighborhood of zero.

Invoking (49) and (85), we have:

$$\frac{1}{2}\ln\frac{T_b^2}{T_b^2 - s_2^2} \leq \left(V(0) - \frac{C}{K}\right)e^{-Kt} + \frac{C}{K} \quad (87)$$

$$|s_2| \leq T_b\sqrt{1 - e^{-2((V(0) - \frac{C}{K})e^{-Kt} + \frac{C}{K})}} \quad (88)$$

Similarly, it can be obtained from Equation (88) that s_2 is uniformly ultimately bounded. In addition, it can be concluded that the rest of signals such as s_3, z_2, z_3, χ_1, χ_2, h_{1p}, h_{2p}, h_{3p} and $\tilde{D}_1$ in the closed-loop system are uniformly ultimately bounded.

According to inequation (88), $|s_2| \leq T_b$ always holds, and $\zeta_{2s} \leq \bar{\zeta}_2$ can be obtained from the definition of saturation function (19). Then,

$$|\zeta_2| < T_b + \bar{\zeta}_2 = \zeta_{2\max} \quad (89)$$

According to Equation (89), for any $\forall t > 0$, there is always $|\zeta_2| < \zeta_{2\max}$. Thus, the state constraint of ζ_2 will not be violated.

The proof is completed.

5. Simulation Study

In this section, two simulation situations for the terminal guidance phase of a surface-to-air missile are taken into account to verify the effectiveness and robustness of the proposed control scheme. The simulations are aimed at the missile intercepts heading-on target.

5.1. Simulation Parameters

The setting of the missile's inherent parameters includes $\rho = 0.2641$ kg/m^3, $S_{ref} = 0.0286$ m^2, $L_{ref} = 0.1888$ m, $m = 144$ kg, and $I_{yy} = 136$ kg·m^2. The velocity of the missile is assumed to be constant during the terminal guidance phase, so it is defined as $V_M = 900$ m/s. The remaining initial trim conditions of the missile and the target can be found in Table 2. Similarly, the constraints of normal acceleration and elevator deflection of the missile are listed in Table 3. Motivated by [31], the aerodynamic coefficients of the missile are represented as:

$$C_x = 0.0083\alpha - 0.57 + 0.004\delta_e,$$
$$C_z = -0.1796\alpha - 0.0077 - 0.09\delta_e,$$
$$C_m = -0.435\alpha - 0.1078 - 0.675\delta_e$$

Table 2. The initial conditions of the missile and the target.

Variable	Variable
Pitch angle ϑ	0.315 rad
Angle of attack α	0.1 rad
Target velocity V_T	300 m/s
elevator deflection δ_e	0°

Table 3. The constraints of acceleration and input saturation.

Variable	Maximum Value	Minimum Value
Normal acceleration n_L, m/s^2	40	−40
Elevator deflection δ_e, deg	30	−30

The initial range along the LOS between the missile and the target is $r(0) = 8900$ m; the LOS angle is $\lambda(0) = 0.1648$ rad. The initial position of the missile and the target are given as: $x_M(0) = 0, y_M(0) = 0, x_T(0) = 8900$ m, and $y_T(0) = 1480$ m.

5.2. Simulation Results

5.2.1. Effectiveness Verification

To evaluate the effectiveness and superiority of the adaptive composite learning integrated guidance and control (ACLIGC) scheme proposed in our study, it is compared with the conventional backstepping integrated guidance and control (CBIGC) algorithm in [34]. For convenience, the two methods are denoted as ACLIGC and CBIGC respectively.

In this section, the target maneuver acceleration is given as $A_T = 30 \, \text{m/s}^2$ in scenario 1. The parameters of the controller and filters are chosen as: $k_1 = 1, k_5 = 5, k_3 = 50, p_{h_1} = 1.5$, $p_{h_2} = p_{h_3} = 15, \Gamma_{h_{1p}} = \Gamma_{h_{2p}} = \Gamma_{h_{3p}} = 5, H_{F_1} = H_{F_2} = H_{F_3} = 1$, and $\varepsilon_2 = \varepsilon_3 = 0.5$. The parameters associated with online composite learning are set as: $\tau_1 = \tau_2 = 20, \tau_3 = 30$, $\Gamma_1 = 0.01$, and $\Gamma_2 = \Gamma_3 = 0.001$.

Figures 4–8 show the simulation results when the maneuvering acceleration of the target is constant. It can be seen that the trajectories of the missile and target and the x and y coordinates of the missile and target are presented in Figure 4.

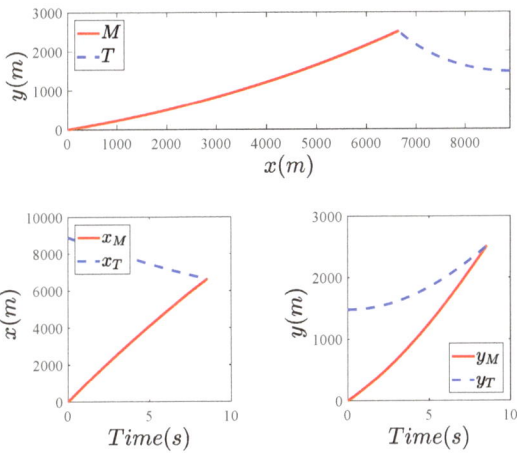

Figure 4. The trajectories of the missile and target and x and y coordinates of the missile and target in scenario 1.

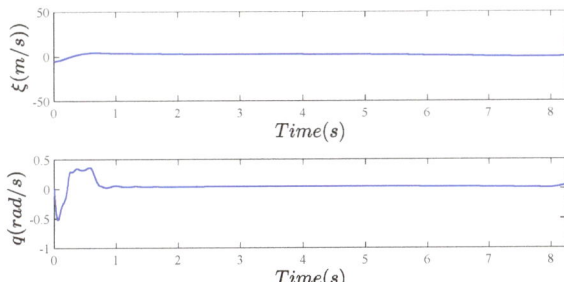

Figure 5. The time responses of the intercept strategy ξ and pitch angle rate q in scenario 1.

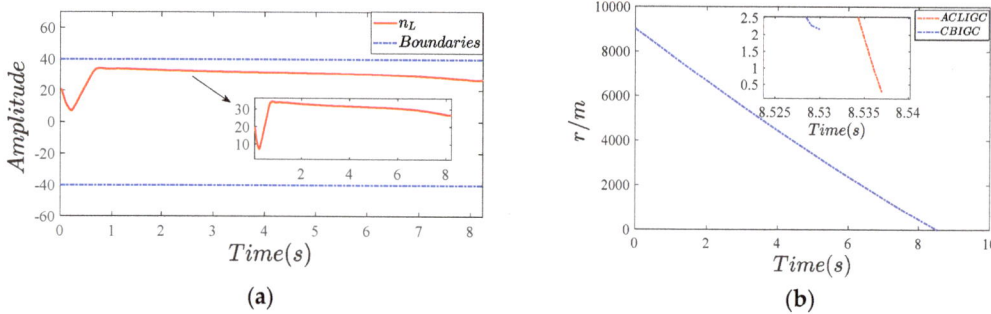

Figure 6. (**a**) The time responses of normal acceleration n_L in scenario 1; (**b**) the time response of range along the LOS in scenario 1.

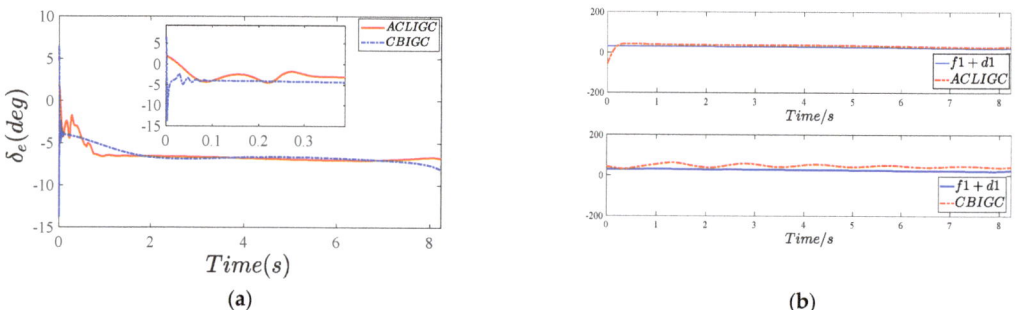

Figure 7. (**a**) The time responses of elevator deflection δ_e in scenario 1; (**b**) ACLIGC and CBIGC to estimate $f_1 + d_1$ in scenario 1.

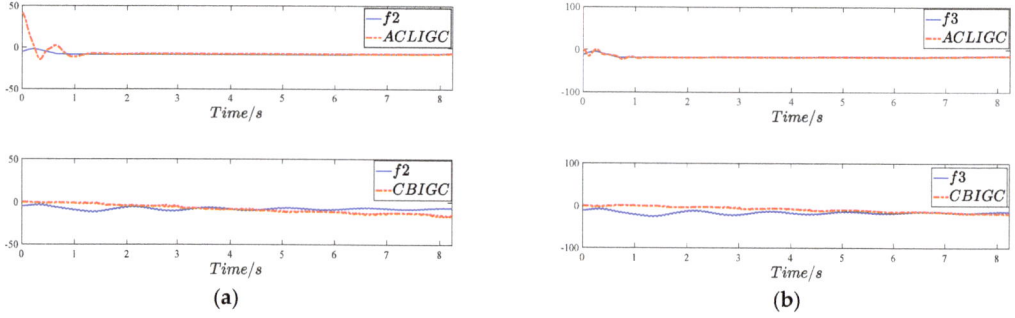

Figure 8. (**a**) ACLIGC and CBIGC to estimate f_2 in scenario 1; (**b**) ACLIGC and CBIGC to estimate f_3 in scenario 1.

The trajectories of state variables for the IGC system such as intercept strategy ξ and pitch angle rate q are illustrated in Figure 5. It can be seen from Table 3 and Figure 6a that n_L always maintains in the constrained area. The final miss distances of different methods are shown in Table 4, and the time response of range along the LOS is presented in Figure 6b. Table 4 and Figure 6b indicate that the ACLIGC proposed in our study has a smaller final miss distance compared with another value.

Table 4. The final miss distance in various schemes.

Scheme	Miss Distance, m	
	Scenario 1	Scenario 2
ACLIGC	0.267	0.332
CBIGC	2.161	2.648

As shown in Figure 7a, the actuator saturation problem can be solved via both CBIGC and ACLIGC, while the proposed ACLIGC can achieve better performance, with a smaller elevator deflection amplitude and smoother time responses. As can be seen from Figures 7b and 8, compared with the traditional backstepping-based adaptive learning approach in [35], the online composite learning algorithm proposed in this paper achieves higher accuracy in estimating the system multi-source uncertainties.

5.2.2. Robustness Verification

In scenario 2, the target maneuver acceleration is changed to a time-varying form $A_T = 10 + 10\sin 0.1\pi t$ m/s^2 to verify the robustness of the proposed ACLIGC.

A series of satisfactory results are shown in Figures 9–13 when the maneuvering acceleration of the target is in a time-varying form. From the trajectories of the missile and target, x and y coordinates of the missile and target shown in Figure 9, it can be seen that the state variables for the IGC system such as intercept strategy $\tilde{\zeta}$ and pitch angle rate q converge to the neighborhood of some value from Figure 10 respectively. In addition, Figure 10 indicates that the intercept strategy $\tilde{\zeta}$ converges to the neighborhood of zero at 0.9 s, achieving a direct hit of the target.

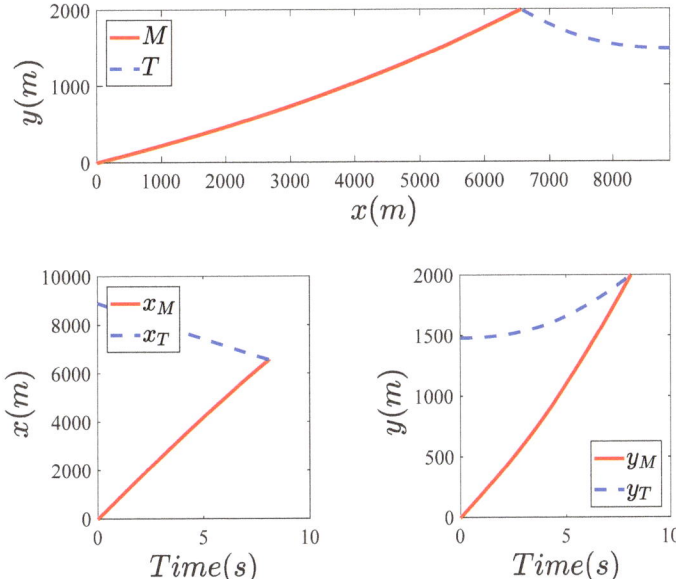

Figure 9. The trajectories of the missile and target and x and y coordinates of the missile and target in scenario 2.

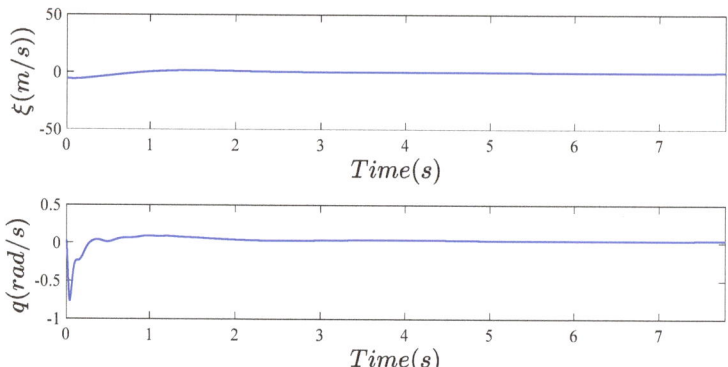

Figure 10. Time responses of intercept strategy ξ and pitch angle rate q in scenario 2.

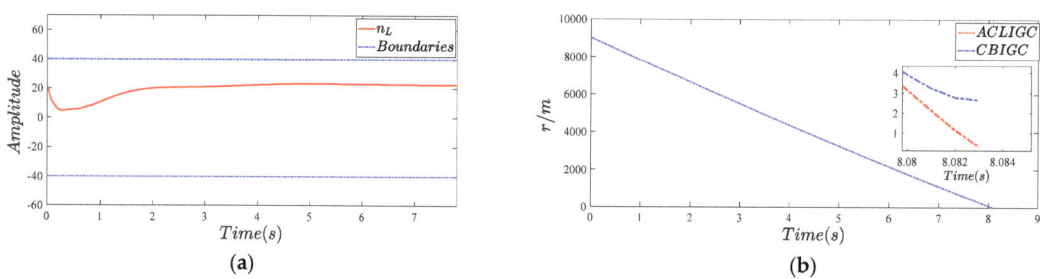

Figure 11. (**a**) Time responses of normal acceleration n_L in scenario 2; (**b**) time response of range along the LOS in scenario 2.

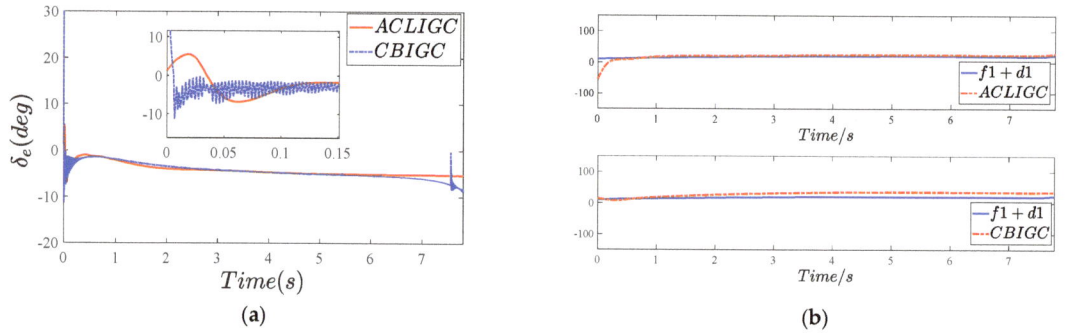

Figure 12. (**a**) The time responses of elevator deflection δ_e in scenario 2; (**b**) ACLIGC and CBIGC to estimate $f_1 + d_1$ in scenario 2.

Moreover, according to Table 3 and Figure 11a, the constraint of n_L will not be violated. The final miss distances for ACLIGC and CBIGC are shown in Table 4, and the curves of r are given in Figure 11b. It is clear that the ACLIGC proposed in our study obtains a smaller value of 0.33 m to realize direct hit interception. The contrast curve in Figure 12a shows that the elevator deflection amplitude of ACLIGC is 8°, while that of CBIGC is 30°. Meanwhile, the time response curve of the elevator deflection corresponding to the ACLIGC is smoother. Similar to scenario 1, ACLIGC achieves higher estimation accuracy for the uncertainties and time-varying disturbance, as indicated by Figures 12b and 13.

Remark 3. In a nutshell, because of adopting improved saturation functions, disturbance observer, prediction error with SPEM, and online composite learning, the proposed ACLIGC is capable of achieving good performance despite the target performing various forms of maneuver in different scenarios, as illustrated by the simulation results in Figures 4–13.

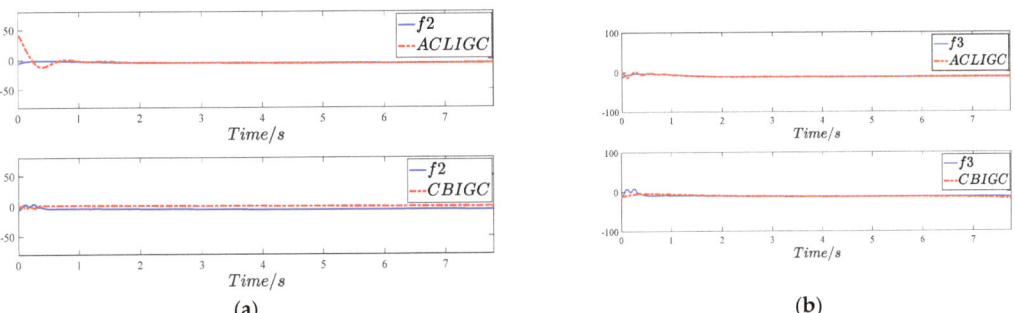

Figure 13. (a) ACLIGC and CBIGC to estimate f_2 in scenario 2; (b) ACLIGC and CBIGC to estimate f_3 in scenario 2.

6. Conclusions

This study considers the problems in the IGC design for a class of STT interceptors, such as input saturation, state constraint, and unknown nonlinear uncertainties. The online composite learning based on disturbance observer is exploited to compensate for multi-source uncertainties. Meanwhile, the state constraint problem can be solved by introducing BLF and improved saturation function. This limits the virtual control and the corresponding tracking error to a certain region, and the missile acceleration constraint will not be violated. Moreover, the design of the auxiliary system ensures the stability of the system. In addition, the saturation function and auxiliary system are designed to handle in case of actuator saturation sufficiently. Finally, stability analysis shows that all signals in the closed-loop system are uniformly ultimately bounded in the presence of input saturation, state constraint, and uncertainties. The effectiveness and robustness of the proposed scheme are illustrated via numerical simulation, and the results indicate that it can achieve direct hit under constant and time-varying maneuvering of the target.

Author Contributions: Conceptualization, Y.B. and T.Y.; methodology, Y.B.; software, Y.B.; formal analysis, T.Y.; writing—original draft preparation, T.Y.; writing—review and editing, W.F. and T.L.; supervision, J.H. All authors have read and agreed to the published version of the manuscript.

Funding: This research was funded by the National Natural Science Foundation of China (No. 62176214, No. 62101590) and Fundamental Research Funds for the Central Universities.

Data Availability Statement: Not applicable.

Conflicts of Interest: The authors declare no conflict of interest.

References

1. Chen, Z.; Shima, T. Nonlinear optimal guidance for intercepting a stationary target. *J. Guid. Control Dyn.* **2019**, *42*, 2418–2431. [CrossRef]
2. Wang, Z.; Yuan, J.P. Fuzzy adaptive fault tolerant IGC method for STT missiles with time-varying actuator faults and multisource uncertainties. *J. Franklin. Inst.* **2020**, *357*, 59–81. [CrossRef]
3. Liu, W.K.; Wei, Y.Y.; Duan, G.R. Barrier Lyapunov function-based integrated guidance and control with input saturation and state constraints. *Aerosp. Sci. Technol.* **2019**, *84*, 845–855. [CrossRef]
4. Liang, X.L.; Hou, M.Z.; Duan, G.R. Adaptive dynamic surface control for integrated missile guidance and autopilot in the presence of input saturation. *J. Aerosp. Eng.* **2015**, *28*, 04014121. [CrossRef]
5. Park, J.; Kim, Y.; Kim, J.H. Integrated guidance and control using model predictive contrSol with flight path angle prediction against pull-up maneuvering target. *Sensors* **2020**, *20*, 3143. [CrossRef] [PubMed]

6. Guo, J.G.; Xiong, Y.; Zhou, J. A new sliding mode control design for integrated missile guidance and control system. *Aerosp. Sci. Technol.* **2018**, *78*, 54–61. [CrossRef]
7. Zhou, X.H.; Wang, W.H.; Liu, Z.H.; Liang, C.; Lai, C. Impact angle constrained three-dimensional integrated guidance and control based on fractional integral terminal sliding mode control. *IEEE Access* **2019**, *7*, 126857–126870. [CrossRef]
8. Zhou, H.; Zhao, H.; Huang, H.Q.; Zhao, X. Integrated guidance and control design of the suicide UCAV for terminal attack. *J. Syst. Eng. Electron.* **2017**, *28*, 546–555.
9. Shao, X.L.; Wang, H.L. Back-stepping active disturbance rejection control design for integrated missile guidance and control system via reduced-order ESO. *ISA Trans.* **2015**, *57*, 10–22.
10. Liu, W.K.; Wei, Y.Y.; Duan, G.R.; Hou, M.Z. Integrated guidance and control with input saturation and disturbance observer. *J. Control Decis.* **2018**, *5*, 277–299. [CrossRef]
11. Zhang, C.; Wu, Y.J. Non-singular terminal dynamic surface control based integrated guidance and control design and simulation. *ISA Trans.* **2016**, *63*, 112–120.
12. Jiang, S.; Tian, F.Q.; Sun, S.Y.; Liang, W.G. Integrated guidance and control of guided projectile with multiple constraints based on fuzzy adaptive and dynamic surface. *Def. Technol.* **2020**, *16*, 1130–1141. [CrossRef]
13. Yan, X.H.; Shao, G.W.; Yang, Q.Y.; Yu, L.; Yao, Y.W.; Tu, S.X. Adaptive Robust Tracking Control for Near Space Vehicles with Multi-Source Disturbances and Input–Output Constraints. *Actuators* **2022**, *11*, 273. [CrossRef]
14. Wu, B.; Wu, J.L.; Zhang, J.; Tang, G.J.; Zhao, Z.J. Adaptive Neural Control of a 2DOF Helicopter with Input Saturation and Time-Varying Output Constraint. *Actuators* **2022**, *11*, 336. [CrossRef]
15. Panchal, B.; Mate, N.; Talole, S.E. Continuous-time predictive control-based integrated guidance and control. *J. Guid. Control Dyn.* **2017**, *40*, 1579–1595. [CrossRef]
16. He, S.M.; Song, T.; Lin, D.F. Impact angle constrained integrated guidance and control for maneuvering target interception. *J. Guid. Control Dyn.* **2017**, *40*, 2653–2661. [CrossRef]
17. Zhou, D.; Xu, B. Adaptive dynamic surface guidance law with input saturation constraint and autopilot dynamics. *J. Guid. Control Dyn.* **2016**, *39*, 1155–1162. [CrossRef]
18. Guo, J.; Zhou, J.; Zhao, B. Three-dimensional integrated guidance and control for strap-down missiles considering seeker's field-of-view angle constraint. *Trans. Inst. Meas. Control* **2020**, *42*, 1097–1109. [CrossRef]
19. Liu, X.D.; Huang, W.W.; Du, L.F. An integrated guidance and control approach in three-dimensional space for hypersonic missile constrained by impact angles. *ISA Trans.* **2017**, *66*, 164–175. [CrossRef]
20. Wang, L.; Zhang, W.H.; Wang, D.H.; Peng, K.; Yang, H.B. Command filtered back-stepping missile integrated guidance and autopilot based on extended state observer. *Adv. Mech. Eng.* **2017**, *9*, 1687814017733251. [CrossRef]
21. Li, Z.B.; Dong, Q.L.; Zhang, X.Y.; Gao, Y.F. Impact angle-constrained integrated guidance and control for supersonic skid-to-turn missiles using backstepping with global fast terminal sliding mode control. *Aerosp. Sci. Technol.* **2022**, *122*, 107386. [CrossRef]
22. Chao, M.; Wang, X.M.; Sun, R.S. A novel non-singular terminal sliding mode control-based integrated missile guidance and control with impact angle constraint. *Aerosp. Sci. Technol.* **2019**, *94*, 105368.
23. Tian, J.Y.; Chen, H.F.; Liu, X.C.; Yang, H.B.; Zhang, S.F. Integrated strapdown missile guidance and control with field-of-view constraint and actuator saturation. *IEEE Access* **2020**, *8*, 123623–123638. [CrossRef]
24. Peng, Q.; Guo, J.G.; Zhou, J. Integrated guidance and control system design for laser beam riding missiles with relative position constraints. *Aerosp. Sci. Technol.* **2020**, *98*, 105693. [CrossRef]
25. Ai, X.L.; Shen, Y.C.; Wang, L.L. Adaptive integrated guidance and control for impact angle constrained interception with actuator saturation. *Aeronaut. J.* **2019**, *123*, 1437–1453. [CrossRef]
26. Wang, W.H.; Xiong, S.F.; Wang, S.; Song, S.Y.; Lai, C. Three dimensional impact angle constrained integrated guidance and control for missiles with input saturation and actuator failure. *Aerosp. Sci. Technol.* **2016**, *53*, 169–187. [CrossRef]
27. Wang, S.; Wang, W.H.; Xiong, S.F. Impact angle constrained three-dimensional integrated guidance and control for STT missile in the presence of input saturation. *ISA Trans.* **2016**, *64*, 151–160. [CrossRef]
28. Khankalantary, S.; Sheikholeslam, F. Robust extended state observer-based three dimensional integrated guidance and control design for interceptors with impact angle and input saturation constraints. *ISA Trans.* **2020**, *104*, 299–309. [CrossRef] [PubMed]
29. Zhang, D.H.; Ma, P.; Wang, S.Y.; Chao, T. Multi-constraints adaptive finite-time integrated guidance and control design. *Aerosp. Sci. Technol.* **2020**, *107*, 106334. [CrossRef]
30. Huang, J.; Lin, C.F. Application of sliding mode control to bank-to-turn missile systems. In Proceedings of the First IEEE Regional Conference on Aerospace Control Systems, Westlake Village, CA, USA, 25–27 May 1993.
31. Zhou, J.L.; Yang, J.Y. Smooth sliding mode control for missile interception with finite-time convergence. *J. Guid. Control Dyn.* **2015**, *38*, 1311–1318. [CrossRef]
32. Shtessel, Y.B.; Shkolnikov, I.A.; Levant, A. Guidance and control of missile interceptor using second-order sliding modes. *IEEE Trans. Aerosp. Electron. Syst.* **2009**, *45*, 110–124. [CrossRef]
33. Liu, W.K.; Wei, Y.Y.; Hou, M.Z.; Duan, G.R. Integrated guidance and control with partial state constraints and actuator faults. *J. Frankl. Inst.-Eng. Appl. Math.* **2019**, *356*, 4785–4810. [CrossRef]

34. Chen, M.; Ge, S.S. Adaptive neural output feedback control of uncertain nonlinear systems with unknown hysteresis using disturbance observer. *IEEE Trans. Ind. Electron.* **2015**, *62*, 7706–7716. [CrossRef]
35. Liu, W.K.; Wei, Y.Y.; Duan, G.R. Integrated guidance and control with input saturation. In Proceedings of the 2017 36th Chinese Control Conference (CCC), Dalian, China, 26–28 July 2017.

Disclaimer/Publisher's Note: The statements, opinions and data contained in all publications are solely those of the individual author(s) and contributor(s) and not of MDPI and/or the editor(s). MDPI and/or the editor(s) disclaim responsibility for any injury to people or property resulting from any ideas, methods, instructions or products referred to in the content.

Article

A Multi-Scale Attention Mechanism Based Domain Adversarial Neural Network Strategy for Bearing Fault Diagnosis

Quanling Zhang [1], Ningze Tang [1], Xing Fu [1], Hao Peng [2], Cuimei Bo [3] and Cunsong Wang [1,*]

1. Institute of Intelligent Manufacturing, Nanjing Tech University, Nanjing 210009, China
2. College of Mechanical and Power Engineering, Nanjing Tech University, Nanjing 211816, China
3. College of Electrical Engineering and Control Science, Nanjing Tech University, Nanjing 211816, China
* Correspondence: wangcunsong@njtech.edu.cn

Abstract: There are a large number of bearings in aircraft engines that are subjected to extreme operating conditions, such as high temperature, high speed, and heavy load, and their fatigue, wear, and other failure problems seriously affect the reliability of the engine. The complex and variable bearing operating conditions can lead to differences in the distribution of data between the source and target operating conditions, as well as insufficient labels. To solve the above challenges, a multi-scale attention mechanism-based domain adversarial neural network strategy for bearing fault diagnosis (MADANN) is proposed and verified using Case Western Reserve University bearing data and PT500mini mechanical bearing data in this paper. First, a multi-scale feature extractor with an attention mechanism is proposed to extract more discriminative multi-scale features of the input signal. Subsequently, the maximum mean discrepancy (MMD) is introduced to measure the difference between the distribution of the target domain and the source domain. Finally, the fault diagnosis process of the rolling is realized by minimizing the loss of the feature classifier, the loss of the MMD distance, and maximizing the loss of the domain discriminator. The verification results indicate that the proposed strategy has stronger learning ability and better diagnosis performance than shallow network, deep network, and commonly used domain adaptive models.

Keywords: bearing; multi-scale feature extractor; attention mechanism; domain adversarial; fault diagnosis

Citation: Zhang, Q.; Tang, N.; Fu, X.; Peng, H.; Bo, C.; Wang, C. A Multi-Scale Attention Mechanism Based Domain Adversarial Neural Network Strategy for Bearing Fault Diagnosis. *Actuators* **2023**, *12*, 188. https://doi.org/10.3390/act12050188

Academic Editor: Zongli Lin

Received: 27 March 2023
Revised: 22 April 2023
Accepted: 26 April 2023
Published: 27 April 2023

Copyright: © 2023 by the authors. Licensee MDPI, Basel, Switzerland. This article is an open access article distributed under the terms and conditions of the Creative Commons Attribution (CC BY) license (https://creativecommons.org/licenses/by/4.0/).

1. Introduction

Rotating bearing is some of the core components of the most important machinery equipment, such as the aero-engine, the high-speed axle box, etc. Under harsh environments, such as high temperature and high pressure for a long time, the performance of the rolling bearing will inevitably deteriorate, even leading to the failure of the aero-engine, the high-speed axle box, and other equipment [1–3]. Furthermore, due to the closed-loop regulation of the system, external environmental interference, especially the change in working conditions, the fault characteristics of the system are easily covered up [4]. If the fault cannot be identified timely and effectively, it will cause great economic losses and even cause great accidents. Therefore, bearing fault diagnosis is very important in aerospace, automobile, and railway industries [5,6].

Driven by this motivation, various fault diagnosis methods have been fully developed in recent years. Especially with the rapid development of signal processing, data mining and artificial intelligence technology, data-driven fault diagnosis methods have been applied to the field of bearing fault diagnosis [7]. Some machine learning based methods have been successfully applied. The machine learning-based bearing fault diagnosis method generally includes signal feature extraction [8] and fault classification. Common feature extraction methods include Fourier transform [9], wavelet transform [10], variational mode decomposition [11], etc. Fault classification methods commonly include artificial neural network [12–14] and support vector machine [15–17]. Although these fault diagnosis methods can realize automatic fault identification and improve the efficiency of fault diagnosis, these machine learning-based methods have a shallow structure and rely on manual experience. Their diagnosis accuracy is closely related to feature extraction. Facing the above challenges, the deep learning-based diagnosis methods have made great progress because deep learning has stronger feature capture, better big data processing capabilities, and superior performance in multi-layer nonlinear mapping and processing large-scale mechanical data than the shallow network [18]. What is more, the use of a multi-layer structure can eliminate the dependence on human and expert knowledge. Among many deep learning methods, the convolutional neural network (CNN) has been successfully applied in the field of intelligent fault identification due to its weight sharing, local perception, and strong anti-noise ability [19–29].

The above fault diagnosis methods are all based on constant working conditions. However, in practical engineering, operational conditions of the equipment are not constant due to the continuous change in the production environment and working conditions. The neural network-based fault diagnosis method under constant working conditions is not enough to effectively identify all fault types. The changing working conditions will cause vibration signal amplitude changes, pulse interval changes, and other problems. Deep learning models, such as CNN, cannot solve the problem of data distribution difference under variable working conditions because it is expensive to collect a large number of labeled data. Therefore, domain adaptive technology, combined with CNN, is proposed to solve the problem of difficulty to obtain labeled data under current working conditions. For instance, Wang et al. [30] used a domain adversarial neural network (DANN) with a domain discriminator to mine domain invariant features under different devices. Li et al. [31] proposed a migration learning network based on DANN to identify shared fault types in two domains and to learn new fault types. Lu et al. [32] proposed a depth domain adaptive structure. This structure can adapt both the conditional distribution and the edge distribution in the multi-layer neural network and use maximum mean discrepancy (MMD) to measure the distribution difference. Wu et al. [33] proposed a novel intelligent recognition method based on an adversarial domain adaptation convolutional neural network (ADACNN). The ADACNN introduced MMD in the prediction label space for domain adaptation to alleviate the problem of algorithm performance degradation, which is caused by the distribution deviation between the test data and the training data. Wu et al. [34] adopted a cost-sensitive depth classifier to solve the problem of class imbalance, and they used the domain counter subnet with MMD to simultaneously minimize the marginal and conditional distribution differences between the source domain and the target domain. Liu et al. [35] proposed a migration learning fault diagnosis model based on a deep full convolution conditional Wasserstein adversarial network (FCWAN), which uses the conditional countermeasure mechanism to enhance the effect of migration domain adaptation and further improve the accuracy of diagnosis. Zou et al. [36] proposed a deep convolution Wasserstein adversarial network (DCWAN)-based fault transfer diagnosis model. This model solved the problem of inadequate self-adaptive measurement of feature distribution differences under different working conditions, increased variance constraints to improve the aggregation of extracted features, and expanded the margins between different types of features in the source domain. Wu et al. [37] proposed a Gaussian-guided adversarial adaption transfer network (GAATN) for bearing fault diagnosis. GAATN introduced a Gaussian-guided distribution alignment strategy to make the data distribution of two domains close to the Gaussian distribution to reduce data distribution discrepancies.

In summary, most scholars have studied various deep learning methods from different angles to improve their performance in bearing fault diagnosis. However, the importance of the features extracted by the feature extractors is different. The existing domain adaptive methods seldom pay attention to the more discriminative features and use a single scale extraction when extracting features, and the model performance will be poor due to the lack of information. Therefore, a multi-scale attention mechanism domain adversarial neural network for bearing fault diagnosis (MADANN) will be discussed in this article. Specifically, the main contributions are as follows:

(1) A feature extractor based on a multi-scale convolution structure and attention mechanism is designed. It is adopted to broaden the network width, fuse feature information of different scales, focus on the key features with identification ability to suppress irrelevant features, and improve the accuracy of fault identification.

(2) A class domain adaptation based on the maximum mean difference is designed. MMD is introduced into the predictive label space for domain adaptation to measure the distribution difference between the target and source domains.

(3) Experimental results on a public bearing dataset and data collected by the test bench confirm that the proposed methodology has higher recognition accuracy.

The rest of this paper is arranged as follows. Section 1 introduces the relevant theories of domain adversarial network, maximum mean discrepancy, and attention mechanism. Section 2 introduces the proposed rolling bearing fault diagnosis model of domain adversarial migration based on multi-scale and attention mechanism. Section 3 uses two different data sets to verify the effectiveness of the proposed method. Finally, this is all summarized in Section 4.

2. Theoretical Background

2.1. Domain Adversarial Neural Network

The DANN network is composed of three parts: feature extractor G_f, label classifier G_y, and domain discriminator G_d. A gradient reverse layer (GRL) is added between the feature extractor and the domain discriminator.

The structure of DANN is as shown in Figure 1. First, the source domain data $X_s = \{x_s^i, y_s^i\}_{i=1}^{n_s}$ and the target domain data $X_t = \{x_t^i\}_{i=1}^{n_t}$ are input to the feature extractor G_f to extract the source domain feature $G(x_i^s, \theta_f)$ and target domain feature $G(x_i^t, \theta_f)$, as well as to input the extracted source domain feature $G(x_i^s, \theta_f)$ to the label classifier for classification. The label L_y loss operation is:

$$L_y^i(\theta_f, \theta_y) = L_y^i(G_y(G_f(x_i^s)), y_i^s) = P_i^s \log \frac{1}{G_y(G_f(x_i^s)), y_i^s}, \tag{1}$$

$$L_y(\theta_f, \theta_y) = \frac{1}{n_s}\sum_{i=1}^{n_s} L_y^i(G_y(G_f(x_i^s)), y_i^s), \tag{2}$$

where P_i^s represents 0 or 1. If the true category of sample i is equal to s, take 1, otherwise take 0. θ_f represents parameters in the feature extraction module. θ_y represents parameters in the fault diagnosis classification module. y_i is the label of the bearing. $G_f(x_i^s)$ is the output of the ith source domain sample mapped by the feature extractor, and n_s is the number of samples.

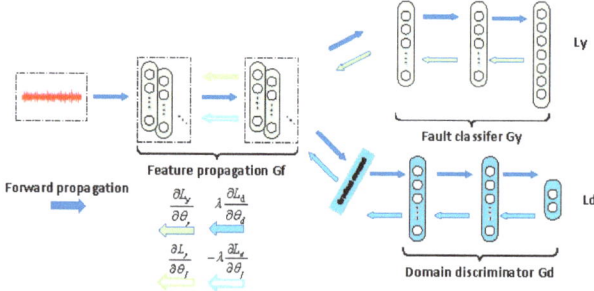

Figure 1. DANN network.

At the same time, input the source domain feature $G(x_i^s, \theta_f)$ and the target domain feature $G(x_i^s, \theta_f)$ to the domain discriminator to determine whether the extracted feature is from the target domain or the source domain. Since adding a gradient reversal layer between the domain discriminator and the feature extractor, the gradient of the incoming feature extractor G_f during the reverse propagation of L_d is $-\lambda \frac{\partial L_d}{\partial \theta_f}$. At this time, G_f optimization will increase the error of the domain discriminator, and the parameter θ_f is learned by maximizing the loss function L_d of the domain discriminator, while the gradient in the domain discriminator G_d is $\frac{\partial L_d}{\partial \theta_d}$, and the parameter θ_d is learned by minimizing the loss function L_d of the domain discriminator. The domain discriminator loss operation L_d is:

$$L_d = \frac{1}{n_s} \sum_{i=1}^{n_s} L_d^i(\theta_f, \theta_d) + \frac{1}{n_t} \sum_{j=1}^{n_t} L_d^j(\theta_f, \theta_d), \tag{3}$$

$$L_d^i(\theta_f, \theta_d) = L_d(G_d(G_f(x_i)), d_i) = d_i \log \frac{1}{G_d(G_f(x_i))} + (1-d_i) \log \frac{1}{G_d(G_f(x_i))}, \tag{4}$$

where θ_f and θ_d, respectively, represent the parameters of the feature extractor and the domain discriminator, n_s is the number of samples in the source domain, and n_t is the number of samples in the target domain.

The overall objective function is:

$$L(\theta_f, \theta_y, \theta_d) = \frac{1}{n_s} \sum_{i=1}^{n_s} L_y^i(\theta_f, \theta_y) - \lambda (\frac{1}{n_s} \sum_{i=1}^{n_s} L_d^i(\theta_f, \theta_d) + \frac{1}{n_t} \sum_{j=1}^{n_t} L_d^j(\theta_f, \theta_d)), \tag{5}$$

The final optimization result is obtained in $\hat{\theta}_f, \hat{\theta}_d, \hat{\theta}_y$ and the expression is:

$$\hat{\theta}_f, \hat{\theta}_y = \underset{\theta_f, \theta_y}{\operatorname{argmin}} L(\theta_f, \theta_y, \hat{\theta}_d), \tag{6}$$

$$\hat{\theta}_d = \underset{\theta_d}{\operatorname{argmin}} L(\hat{\theta}_f, \hat{\theta}_y, \theta_d), \tag{7}$$

2.2. Maximum Mean Discrepancy

Suppose there are two data sets, source domain data set $X_s = \{x_s^i, y_s^i\}_{i=1}^{n_s}$ with label and target domain data set $X_t = \{x_t^i\}_{i=1}^{n_t}$ without label. Where n_s represents the number of samples of the source domain data, n_t represents the number of samples of the target domain data, and y_s^i represents the data label of the source domain. These two datasets have the same label space $y^s = y^t$ and follow different distributions $P_s(X), P_t(X)$. Therefore, the square of the MMD distance of x_s, x_t can be defined as:

$$MMD^2(X_s, X_t) = \left\| \frac{1}{n_s} \sum_{i=1}^{n_s} \Phi(x_i^s) - \frac{1}{n_t} \sum_{j=1}^{n_t} \Phi(x_j^t) \right\|_H^2, \tag{8}$$

where $\Phi(\cdot)$ represents the nonlinear mapping function of the reproducing kernel Hilbert space (RKHS).

To simplify the above functions, the kernel function is introduced in the formula, and the square of MMD distance is rewritten as:

$$MMD^2(X_s, X_t) = \frac{1}{n_s n_s} \sum_{i=1}^{n_s} \sum_{j=1}^{n_s} k(x_i^s, x_j^s) + \frac{1}{n_t n_t} \sum_{i=1}^{n_t} \sum_{j=1}^{n_t} k(x_i^t, x_j^t) - \frac{2}{n_s n_t} \sum_{i=1}^{n_s} \sum_{j=1}^{n_t} k(x_i^s, x_j^t), \tag{9}$$

where $k(x_i^s, x_j^t) = \langle \Phi(x_i^s), \Phi(x_j^t) \rangle$ represents a kernel function.

Select the Gaussian kernel as the kernel function because it can map data to an infinite dimensional space. The formula of the Gaussian kernel function is as follows:

$$k(x^s, x^t) = e^{-\frac{\|x^s - x^t\|^2}{2\sigma^2}}, \tag{10}$$

where σ is the kernel bandwidth, and, if $\sigma \to 0$, the MMD will be 0. Similarly, if the larger bandwidth is $\sigma \to \infty$, the MMD will also be 0. To solve this problem, the kernel bandwidth σ is selected as the median distance between all sample pairs, that is:

$$\sigma^2 = E\|x^s - x^t\|^2, \tag{11}$$

Different kernel functions will be mapped to different regenerated kernel Hilbert spaces to form different distributions. To reduce the influence of Gaussian kernel functions on the results, multiple Gaussian kernels are used to construct multi kernel functions. The definition of multi kernel functions is as follows:

$$k(x^s, x^t) = \sum_{i=1}^{n} k_i(x^s, x^t), \tag{12}$$

where $k_i(x^s, x^t)$ represents the ith basic kernel function.

2.3. Attention Mechanism

The attention mechanism filters information by adaptively weighting the features of different signal segments, highlights the fault features with important information, and suppresses irrelevant features.

The attention mechanism is shown in Figure 2. C represents the number of characteristic channels, and L represents the number of characteristic channels. $F_{sq}(\cdot)$ is the compression operation, is the excitation operation, and $F_{scale}(\cdot)$ is the product operation. First is the compression operation. Along the direction of the feature channel, use global average pooling to compress features of size $L \times C$ into vectors of size $1 \times C$. There, the characteristics of each channel are compressed into a channel characteristic response value with a global receptive field.

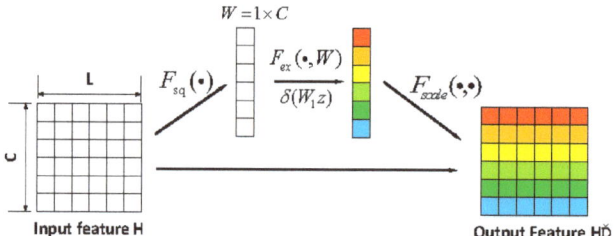

Figure 2. Attention mechanism.

The calculation process is as follows:

$$z = F_{sq}(H) = \frac{1}{L}\sum_{i=1}^{L} u_c(i), \tag{13}$$

where z is the output after compression, $i = 1, 2 \cdots, L$, and $u_c(i)$ is the output value of column i in the characteristic channel c.

The second is the excitation operation. Adding two full connection layers to predict the importance of each channel to obtain the importance of different channels. The specific implementation is as follows:

$$y = F_{ex}(z) = \sigma(g(z, W)) = \sigma(W_2 \delta(W_1 z)), \tag{14}$$

where $\sigma(\cdot)$ is the sigmoid activation function, W_1, W_2 are the weight matrix of the two fully connected layers, and $\delta(\cdot)$ is the Relu activation function.

Finally, the operation is multiplication, and the channel weights obtained by the above operations are weighted to the original features channel by multiplication so as to obtain the feature sequence after attention screening. The specific implementation is as follows:

$$X_c = F_{scale}(U, y) = U \times y, \tag{15}$$

When the rolling bearing has a local fault, the fault position will generate pulse excitation and resonance to other parts, which makes the vibration signal components complex. Therefore, the signal characteristics collected at different times under the same working condition are different. Some characteristics can be used to accurately diagnose the fault information, and some may cause interference, which reduces the generalization ability of the model. To focus on more discriminative features and suppress irrelevant features, this paper uses a one-dimensional attention module to obtain the weight coefficients of different features.

3. A Multi-Scale Attention Mechanism Domain Adversarial Neural Network for Bearing Fault Diagnosis

3.1. Fault Diagnosis Method Framework

The fault diagnosis method framework proposed in this paper firstly uses the multi-scale convolution structure, and this structure is used to widen the width of the network, extract sensitive features of different dimensions, and fuse the information of different scale features. Then, introduce an attention mechanism into the feature extractor to focus more on the key features, and suppress the attention of irrelevant features, thus helping to improve the accuracy of fault identification. introducing MMD into the prediction tag space for domain adaptation, measuring the difference between the distribution of the target domain and the source domain, and improving the ability of the feature extractor to extract domain invariant features. The domain discriminator distinguishes whether the data come from the target domain or the source domain, and it finally inputs the data into the classifier for fault classification.

Figure 3 shows the framework of fault diagnosis method for domain adversarial migration based on multi-scale and attention mechanism, which is mainly composed of four parts: a feature extractor, based on multi-scale and attention mechanism, as well as a domain discriminator, a feature classifier, and a category domain adaptation design, based on the maximum mean discrepancy.

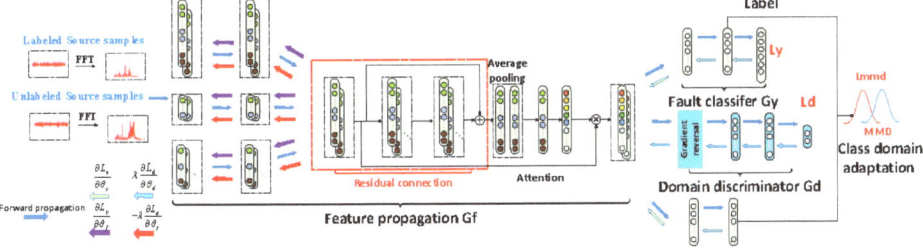

Figure 3. Fault diagnosis method framework.

The feature extractor is composed of three layers of one-dimensional convolutional neural networks with different scales and an attention mechanism embedded in residual blocks. Introduce an attention mechanism into the feature extractor to focus on more useful features and to suppress irrelevant features.

The classification module is composed of the full connection layer. The fault features extracted by the feature extractor are classified by the softmax layer. The domain recognition module is composed of two fully connected neural network layers. The category domain adaptation design uses the MMD distance as the target loss function.

In the process of model training, the function of the feature extractor is to extract the common features of the target domain data and the source domain data. The function of the domain discriminator is to distinguish whether the data are from the target domain or the source domain. The function of the feature classifier is to correctly classify the fault signal. The class domain adaptation design is to reduce the difference in the distribution of the source domain and the target domain data in the prediction tag space and improve the ability of the feature extractor to extract domain invariant features.

3.2. Feature Extraction Method Based on a Multi-Scale Module and an Attention Mechanism

The feature module includes a multi-scale module and an attention mechanism. There are three convolution modules with different scales in the multi-scale module. First, in the convolution module of the first scale, the input data are convoluted as follows:

$$x_{i1}{}^z = \delta_1(x_i{}^{z-1} \times \omega_1{}^z + b_1{}^z), \tag{16}$$

In the convolution module of the second scale, the input data are convoluted as follows:

$$x_{i2}{}^z = \delta_2(x_i{}^{z-1} \times \omega_2{}^z + b_2{}^z), \tag{17}$$

In the convolution module of the third scale, the input data are convoluted as follows:

$$x_{i3}{}^z = \delta_3(x_i{}^{z-1} \times \omega_3{}^z + b_3{}^z), \tag{18}$$

Then, the features extracted from the three scales are fused:

$$x_i{}^z = x_{i1}{}^z + x_{i2}{}^z + x_{i3}{}^z, \tag{19}$$

where $x_i{}^{z-1}$ represents the output of the previous convolution module of the data, $x_i{}^z$ represents the output of the current convolution module of the data, z represents the convolution module, ω^z and represents the parameters in each convolution calculation, and $\delta(\cdot)$ represents the activation function.

Then, $x_i{}^z$ inputs the residual block in the attention module to extract the deep abstract representation of the set features, and the formula is as follows:

$$x_i{}^{z+1} = x_i{}^z + \sum_{j=1}^{L}(F(x_i{}^{z,j}, W_j)), \tag{20}$$

where $x_i{}^{z+1}$ is the output of the residual block, W_j is the weight matrix of each residual block, L is the number of residual blocks, and F is the residual map to be learned. Then, give different weights to the characteristics of different channels. First, perform global average pooling on input $x_i{}^z$, and the results are as follows:

$$v_m{}^z = GAP(x_i{}^z) = \frac{1}{L}\sum_{n=1}^{L} x_{i,m}{}^z(n), \tag{21}$$

where m represents the mth channel in $x_i{}^z$, and the feature vectors obtained through the two fully connected layers are used to adjust $x_i{}^z$, and the adjusted $x_i{}^z$ is:

$$x_i{}^z = x_i{}^z + v_m{}^z \times x_i{}^z, \tag{22}$$

$$G(x_i{}^s, \theta_f) = x_i{}^s, \tag{23}$$

$$G(x_i{}^t, \theta_f) = x_i{}^t, \tag{24}$$

where x_i^s and x_i^t in the above expression represent feature outputs of the source domain data and the target domain data after the feature extractor.

3.3. Design of Feature Classifier

The fault diagnosis classification module is composed of a full connection layer. The source domain features extracted by the feature module are input to the fault diagnosis module. The formula is as follows:

$$x_i^{s,fc} = \delta(x_i^s; \theta^{fc}) = \sigma(\omega_f \times x_i^s + b_f) \tag{25}$$

where $\theta^{fc} = \{\omega_{fc}, b_{fc}\}$ is the parameter of the full connection layer, $\sigma(\cdot)$ is the activation function, and x_i^s is the source domain feature.

The softmax function is selected as the label prediction, and its output is the probability of each type of sample. The formula is as follows:

$$h_i^s = [p(y_i^s = 0 | x_i^{s,fc}) \cdots p(y_i^s = 5 | x_i^{s,fc})], \tag{26}$$

The loss of the fault classifier is:

$$L_y(x_i^s) = \frac{1}{n_s} \sum_{i=1}^{n_s} L_y^i(x_i^{s,fc}, y_i^s), \tag{27}$$

$$L_y^i(x_i^{s,fc}, y_i^s) = P_i^s \log \frac{1}{G_y(G_f(x_i^s)), y_i^s}, \tag{28}$$

where P_i^s represents 0 or 1. If the true category of sample i is equal to s, take 1, otherwise take 0. y_i is the label of the bearing, $G_f(x_i^s)$ is the output of the ith source domain sample mapped by the feature extractor, n_s is the number of samples, and $G_y(\cdot)$ is the output of the classifier.

3.4. Design of Domain Discriminator

In the domain classification, the feature extraction is performed on the target domain data using the Formulas (16)–(22) to obtain the feature output, which is then input to the full connection layer of the domain discriminator. The formula is as follows:

$$x_i^{t,fc} = \delta(x_i^t; \theta^{fc}) = \sigma(\omega_f \times x_i^t + b_f), \tag{29}$$

Obtain $x_i^{t,fc}$. It is a binary classification problem to consider whether the data comes from the source domain or the target domain at the output layer. The formula is as follows:

$$L_d = \frac{1}{n_s} \sum_{i=1}^{n_s} L_d^i(x_i^{s,fc}) + \frac{1}{n_t} \sum_{j=1}^{n_t} L_d^j(x_j^{t,fc}), \tag{30}$$

$$L_d^i(x^i) = L_d(G_d(G_f(x_i)), d_i) = d_i \log \frac{1}{G_d(G_f(x_i))} + (1 - d_i) \log \frac{1}{G_d(G_f(x_i))}, \tag{31}$$

where n_s is the number of samples in the source domain, n_t is the number of samples in the target domain, and $G_d(\cdot)$ is the output of the domain classification module.

3.5. Class Domain Adaptation Design Based on the Maximum Mean Difference

The category domain adaptation design is to reduce the difference between the data distribution of the source domain and the target domain in the predicted tag space, improve the ability of the feature extractor to extract domain invariant features, calculate the MMD distance between the distribution of the source domain and the target domain in the tag space, take it as the objective loss function of the category field adaptation, and use the

MMD distance loss to minimize the difference in the conditional distribution between the source domain and the target domain.

The formula is as follows:

$$L_{MMD} = \frac{1}{n_s n_s}\sum_{i=1}^{n_s}\sum_{j=1}^{n_s} k(x_i^s, x_j^s) + \frac{1}{n_t n_t}\sum_{i=1}^{n_t}\sum_{j=1}^{n_t} k(x_i^t, x_j^t) - \frac{2}{n_s n_t}\sum_{i=1}^{n_s}\sum_{j=1}^{n_t} k(x_i^s, x_j^t), \quad (32)$$

3.6. Total Loss Function Design

Because a gradient reversal layer is added between the domain discriminator and the feature extractor, the gradient that is transmitted to the feature extractor G_f during the backpropagation of L_d is $-\lambda \frac{\partial L_d}{\partial \theta_f}$. At this time, G_f optimization will increase the error of the domain discriminator, and the parameter θ_f is learned by maximizing the loss function L_d of the domain discriminator, while the gradient in the domain discriminator G_d is $\frac{\partial L_d}{\partial \theta_d}$, and the parameter θ_d is learned by minimizing the loss function L_d of the domain discriminator. The overall loss function includes three parts: the feature classification loss function in Formula (27), the domain classification loss function of Formula (30), and the category domain adaptation loss function of Formula (32). So, the overall loss function is:

$$\begin{aligned} L(\theta_f, \theta_y, \theta_d) &= L_y(x_i^s) - \lambda_1 L_d + \lambda_2 L_{MMD} \\ &= \frac{1}{n_s}\sum_{i=1}^{n_s} L_y^i(x_i^s) - \frac{\lambda_1}{n_s}\sum_{i=1}^{n_s} L_d^i(x_i^{s,fc}) - \frac{\lambda_1}{n_t}\sum_{j=1}^{n_t} L_d^j(x_i^{t,fc}) \\ &+ \frac{\lambda_2}{n_s n_s}\sum_{i=1}^{n_s}\sum_{j=1}^{n_s} k(x_i^s, x_j^s) + \frac{\lambda_2}{n_t n_t}\sum_{i=1}^{n_t}\sum_{j=1}^{n_t} k(x_i^t, x_j^t) - \frac{2\lambda_2}{n_s n_t}\sum_{i=1}^{n_s}\sum_{j=1}^{n_t} k(x_i^s, x_j^t) \end{aligned} \quad (33)$$

The optimization parameters are as follows:

$$\hat{\theta}_f, \hat{\theta}_y = \underset{\theta_f, \theta_y}{\arg\min} L(\theta_f, \theta_y, \theta_d), \quad (34)$$

$$\hat{\theta}_d = \underset{\theta_d}{\arg\min} L(\hat{\theta}_f, \hat{\theta}_y, \theta_d), \quad (35)$$

where $\theta_f, \theta_d, \theta_y$, respectively, represent parameters in the feature extraction module, the domain classification module, and the fault diagnosis classification module.

3.7. Algorithm Flow

Figure 4 introduces the process of the fault diagnosis model proposed in this paper, mainly including three parts: data processing, training process, and testing process. The specific steps are as follows:

(1) The bearing vibration data under different working conditions are collected and normalized, and then they are converted into frequency–domain signals using fast Fourier transform as input, which is divided into source domain data $X_s = \{x_s^i, y_s^i\}_{i=1}^{n_s}$ and target domain data $X_t = \{x_t^i\}_{i=1}^{n_t}$. Finally, the source domain data is divided into two parts: the verification set and the training set, and the target domain data is divided into two parts: the test set and the training set.

(2) The training sets of the source domain data and the target domain data are input into the shared multi-scale feature extractor, and the source domain multi-scale features $x_{i1}^s, x_{i2}^s, x_{i3}^s$ and the target domain multi-scale features $x_{i1}^t, x_{i2}^t, x_{i3}^t$ are extracted, respectively, via Equations (16)–(18). Additionally, use Formula (19) to fuse the multi-scale features of the source domain and the target domain to obtain x_i^s, x_i^t. Through the attention mechanism, the source domain feature x_i^s and the target domain feature x_i^t, with more discriminative power, are extracted through Formulas (20)–(22), and the feature x_i^s extracted from the source domain is input to the feature classifier for classification. The classification loss $L_y(x_i^s)$ is calculated by Formulas (25)

and (27), and then the features extracted from the source domain and the target domain are input to the category domain adapter to calculate the MMD loss L_{MMD} by Formula (32), and the domain discriminator is used to calculate the domain discriminator loss L_d by Formulas (29) and (30), and the three loss functions are constructed into a total loss function $L(\theta_f, \theta_y, \theta_d)$. Finally, the model is iteratively trained to minimize the classification loss and MMD loss and maximize the domain discriminator loss.

(3) The model is tested, and the target domain test set is input into the feature extractor and classifier for actual fault diagnosis to test the effectiveness of diagnosis.

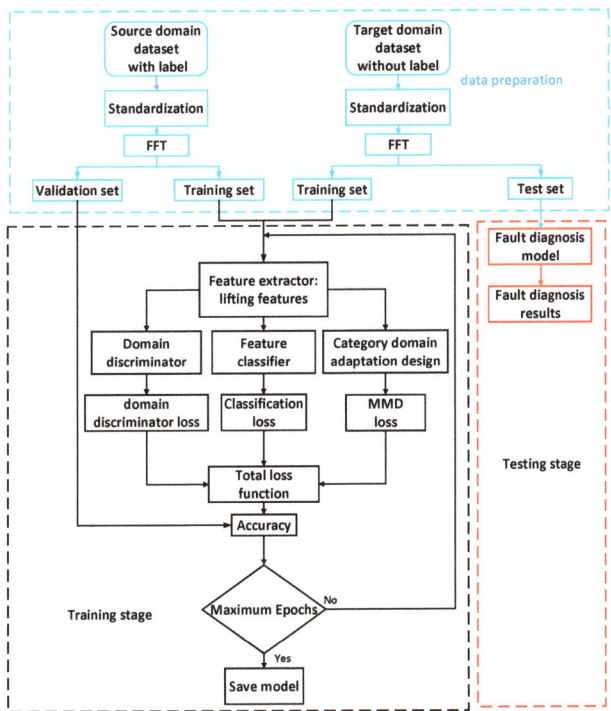

Figure 4. Fault diagnosis model process.

4. Application Results and Analysis

4.1. Case Western Reserve University Bearing Data Analysis

4.1.1. Data Preparation

In this paper, the rolling bearing data set of Case Western Reserve University (CWRU) is used for verification. The download link is http://engineering.case.edu/bearingdatacenter/ (accessed on 10 October 2021). The sampling frequency of the selected data is 12 kHz. The bearings used are divided into a normal state, inner ring fault, outer ring fault, and rolling element fault. As shown in Figure 5, the test bed uses EDM technology to arrange single point faults on the inner ring, rolling element, and outer ring (three o'clock direction) of the bearing. The faults at each position have different fault degrees. The fault diameters are 0.007 inches, 0.014 inches, and 0.021 inches, respectively. Figure 5 is from the bearing data center of the Case School of Engineering.

Three different load states of sample data were selected: 1HP (1772 r/min), 2HP (1750 r/min), and 3HP (1730 r/min), which were divided into three data sets: A, B, and C. An amount of 2048 data points of normal bearing vibration data of Western Reserve University and vibration data of inner ring, rolling element, and outer ring fault are selected as a sample. Table 1 shows the composition of experimental samples. Six transmission tasks are set: A→ B, C, B→A, C, C→A, B. An amount of 300 samples are collected for each faulty bearing state, of which 200 are training samples, and 100 are test samples. Each transmission task is performed five times to take the average value. When the motor load changes, speed will slightly shift. It is a fast process.

Figure 5. Case Western Reserve University bearing testing rig.

Table 1. Composition of experimental samples.

Type	Length	Quantity	Label
Normal	2048	300	9
Inner ring fault (0.007 inch)	2048	300	0
Rolling element failure (0.007 inch)	2048	300	1
Outer ring fault (0.007 inch)	2048	300	2
Inner ring fault (0.014 inch)	2048	300	3
Rolling element failure (0.014 inch)	2048	300	4
Outer ring fault (0.014 inch)	2048	300	5
Inner ring fault (0.021 inch)	2048	300	6
Rolling element failure (0.021 inch)	2048	300	7
Outer ring fault (0.021 inch)	2048	300	8

4.1.2. Performance Comparison and Analysis of Different Algorithms

To confirm the advantages of the proposed fault diagnosis method (Figure 3) under variable operating conditions (loads), the shallow model, the deep model, and the domain adaptive model are selected for comparative experiments, which are SVM, CNN, CNN-LSTM, DACNN, and ADACNN, respectively. (1) SVM extracts ten time–domain features and three frequency–domain features, and then it inputs them into SVM for fault diagnosis under variable conditions. (2) CNN uses a three-layer convolution pooling layer for feature extraction, sends it to the softmax layer for fault diagnosis, and then uses the target domain test set for migration testing of the trained model. The sample size of each operating condition is 3000, and each health state includes 200 training samples and 100 test samples. (3) CNN-LSTM adds an LSTM layer based on CNN to capture the long-term dependence between time series data. The sample size of each operating condition is 3000, and each health state includes 200 training samples and 100 test samples. (4) The DACNN method proposed in document 35 extracts the common features of the source domain and the target domain through a discriminant classifier, uses adversarial learning, and finally inputs the test set of the target domain into the classifier for classification. The sample size of each operating condition is 2000, and each health state includes 100 training samples and 100 test samples. (5) The ADACNN method proposed in document 31 uses MMD distance to measure the difference between the distribution of the target domain and the source domain. The structure of the feature extractor, classifier, and domain discriminator is the same as DACNN. The sample size of each operating condition is 3000, and each health state includes 200 training samples and 100 test samples. Table 2 and Figure 6 show the results obtained by the above method.

Table 2. Average accuracy of different algorithms.

Methods	A-B	A-C	B-A	B-C	C-A	C-B	Average
SVM	70	74	61.6	67.6	65.7	63.3	67.0
CNN	87.3	77.8	91.5	92.7	80.0	79.9	84.9
CNN-LSTM	87.3	81.4	93.1	92.6	82.2	83.4	86.7
DACNN	98.1	95.1	98	98.8	94.6	98.7	97.2
ADACNN	98.6	96.2	98	99.2	96.6	98	97.7
MADANN	99.9	99.7	99.9	100	99.8	100	99.8

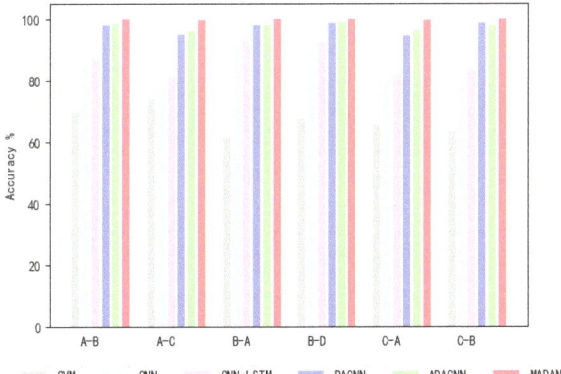

Figure 6. Comparison of accuracy of different algorithms.

It can be concluded, from Table 2 and Figure 6, that: (1) the generalization ability of conventional shallow models, such as SVM, is poor under variable load conditions. (2) For a single depth model, such as CNN and CNN-LSTM, superimposed by two depth models, the average accuracy rate of fault identification is only 84.9% and 86.7%, respectively, when the operating conditions change. Because the change in data distribution has a significant impact on the depth model, the classification effect is poor, which also reveals the importance of reducing the distribution difference between the two fields. (4) Compared with CNN and CNN-LSTM models, the accuracy rate of DACNN is 97.2%, indicating that both feature alignment and domain adversarial learning can mitigate the impact of data distribution deviation caused by variable load conditions. (5) The accuracy rate of the ADACNN algorithm proposed in the document [29] is 97.7%, which is slightly higher than that of DACNN, indicating that introducing MMD domain adaptation into feature space and prediction tag space can alleviate the problem of algorithm performance degradation caused by the distribution deviation between test data and training data. However, the above algorithms use CNN to directly extract features, without considering more discriminative features, so the highest diagnostic accuracy is only 97.7%. In this paper, we use the attention mechanism to consider the weight of each feature extracted from the convolution layer, and then we screen out important features and use the multi-scale convolution structure to broaden the width of the network to achieve the extraction of sensitive features in different dimensions, Finally, the MMD domain is used to adaptively alleviate the problem of algorithm performance degradation caused by the difference of data distribution. The accuracy of this method is greatly improved compared with the above methods.

4.1.3. Feature Visualization and Analysis

To further verify the advantages of the proposed method in fault diagnosis under variable operating conditions, CNN, DACNN, and ADACNN are used as comparisons. Taking B-C as an example, T-SNE visualization is used to analyze the last full connection layer of the classifier. The feature visualization results are shown in Figures 7 and 8. Figure 7 shows the distribution of target domain sample convolution results by different models. Figure 8 shows the distribution of target domain sample features extracted by different models.

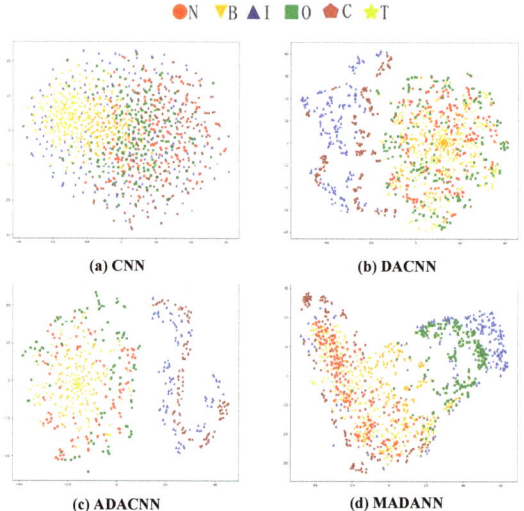

Figure 7. T-SNE visualization of convolution results.

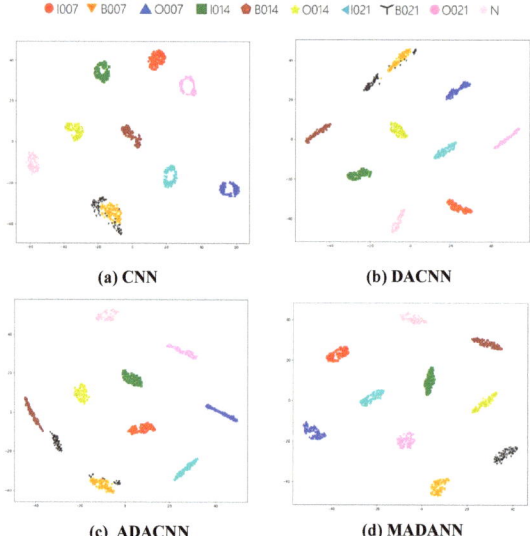

Figure 8. T-SNE visualization of different models.

It can be analyzed from Figure 8a that, for CNN, the fault features of 0.007-inch rolling element and 0.021-inch rolling element are seriously overlapped, and it is impossible to distinguish which type of features are. Other fault features are obvious. (2) It can be analyzed, from Figure 8b,c, that the impact of data distribution shift caused by variable load conditions, forming obvious clusters, is alleviated due to the introduction of feature alignment and domain adversarial learning. Although the fault features of the 0.007-inch rolling element and the 0.021-inch rolling element are still partially overlapped, the situation is improved compared with CNN. (3) It can be seen from Figure 8d that the multi-scale convolution structure broadens the width of the network to achieve the extraction of sensitive features in different dimensions. The channel attention mechanism is introduced into the feature extractor to focus more on the key features with discriminant power, suppress the attention of irrelevant features, and combine feature alignment and domain confrontation learning to extract features more suitable for classification. The fault features of the 0.007-inch rolling element and the 0.021-inch rolling element are clearly separated, and there is no aliasing. This proves, again, that the proposed fault identification method, based on MADANN, has better identification ability under different load conditions.

4.2. Data Analysis of PT500mini Mechanical Bearing Fault Simulation Test Bed

4.2.1. Data Preparation

The PT500mini mechanical bearing gear fault simulation test-bed is used to simulate bearing fault and collect data. The test bed is shown in Figure 9 below. The sampling frequency of selected data is 48 kHz. The bearings used are divided into normal state (N), inner ring fault (I), outer ring fault (O), rolling element fault (B), comprehensive fault (C), and cage fault (T). The inner ring fault is an inner ring crack of 0.3 mm, the outer ring fault is an outer ring crack of 0.3 mm, the rolling element fault is a peeling pit of 3 mm, the comprehensive fault is a crack of 0.3 mm on the inner and outer rings, and the cage fault is a cage fracture.

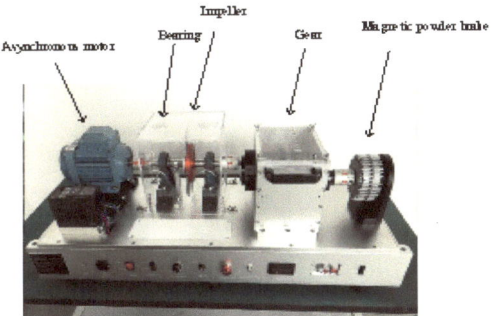

Figure 9. PT500mini mechanical bearing gear fault simulation test bed.

The sample data has three different rotational speeds: 1000 r/min, 1500 r/min, and 2000 r/min, which are divided into three data sets: A, B, and C. An amount of 2048 data points of vibration data are selected as a sample, and 1000 samples are collected in each state. Among them, 700 are test sets and 300 are test sets. Table 3 below shows the composition of bearing test samples. Six transmission tasks are set: A→B, C, B→A, C, C→A, B. Table 4 below gives the details of the experimental data set built under variable operating conditions.

Table 3. Composition of experimental samples.

Sample Type	Sample Length	Number of Samples	Category Tag
N	2048	1000	0
B	2048	1000	1
C	2048	1000	2
I	2048	1000	3
O	2048	1000	4
T	2048	1000	5

Table 4. Transmission tasks.

Domain Adaptation	Source Domain	Target Domain	Accuracy
A-B	1000 r/min	1500 r/min	99.1
A-C	1000 r/min	2000 r/min	98.6
B-A	1500 r/min	1000 r/min	99.1
B-C	1500 r/min	2000 r/min	99.5
C-A	2000 r/min	1000 r/min	99.3
C-B	2000 r/min	1500 r/min	99.9

4.2.2. Experimental Results and Analysis

A variable load condition is a scene with a small difference in signal characteristic distribution between the source condition and the target condition. To verify the accuracy of the proposed method in the case of the large difference in distribution, the variable speed condition is selected for fault diagnosis in this paper. Figure 10 is the accuracy curve and confusion matrix of 500 iterations under each variable working condition. From the accuracy curve under each variable working condition in Figure 10, it can be seen that the accuracy of different tasks is constantly rising. Although it will decline during the iteration, it will eventually stabilize. (1) For task A-B, as shown in Figure 10a,b, it can be analyzed that the accuracy rate can reach 98.6% by the confusion matrix, and a small number of samples are misclassified. For task A-C, as shown in Figure 10c,d, it can be analyzed that the accuracy rate can reach 98.6%, which is slightly lower than that of task A-B. Because the large change in rotational speed of A-C results in a large difference in the characteristic distribution between the two working conditions, the accuracy rate is somewhat lower than that of other tasks. (2) For task B-A and B-C, as shown in Figure 10e–h, the accuracy can reach 99.1% and 99.5%, respectively. Only a small number of samples are misclassified, and the accuracy is high. For task C-A, as shown in Figure 10i,j, the accuracy can reach 99.3%. For task C-B, as shown in Figure 10k,l, the analysis accuracy is 99.9%. Only one sample is misclassified, and the accuracy is very high.

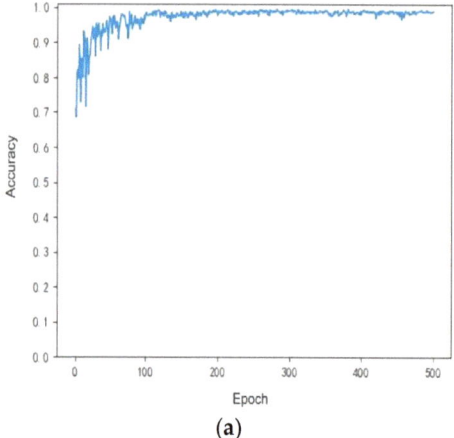

(a)

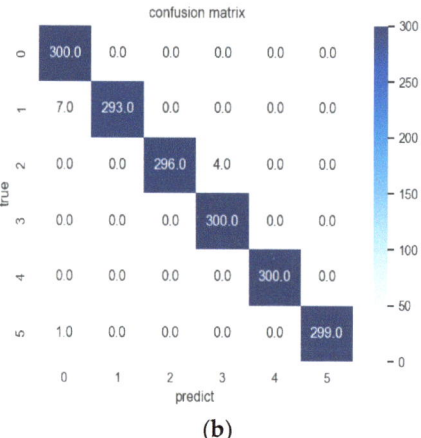

(b)

Figure 10. *Cont.*

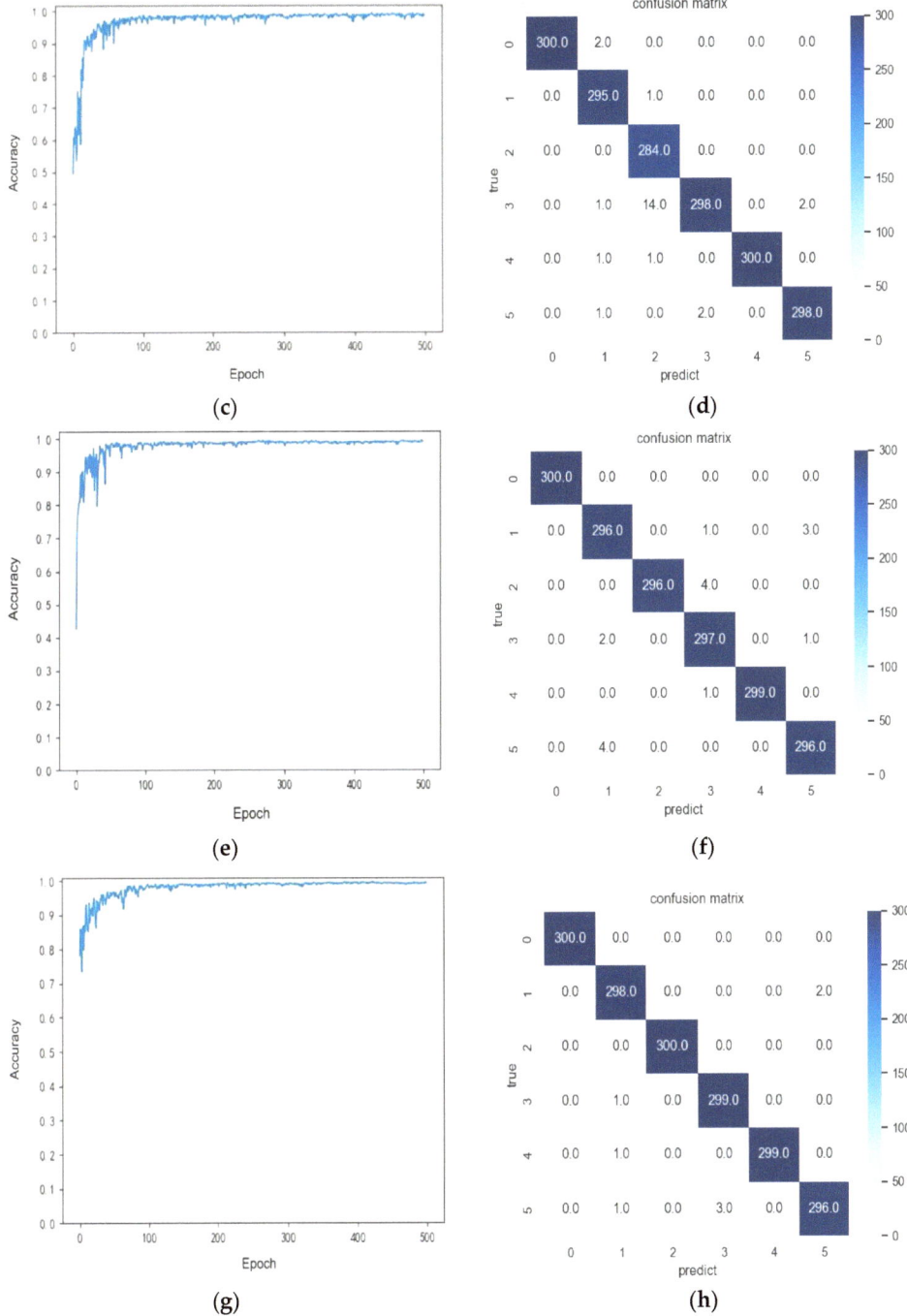

Figure 10. *Cont.*

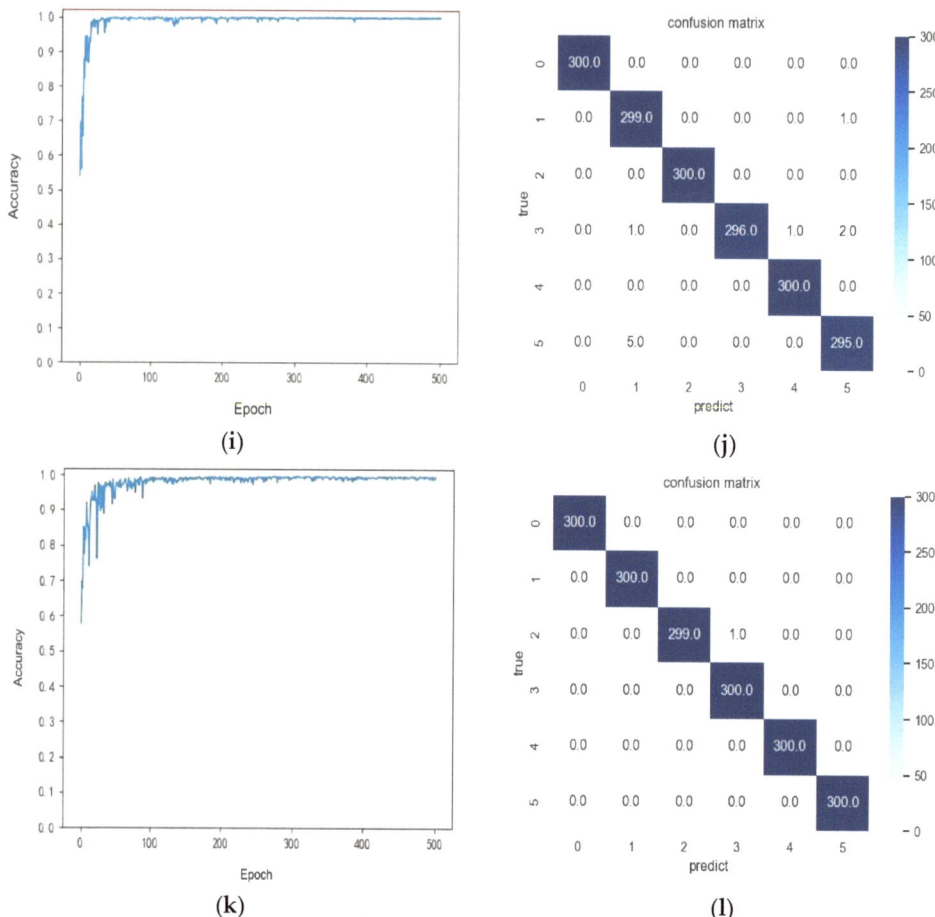

Figure 10. Accuracy curve and confusion matrix of 500 iterations under different tasks: (**a**,**b**) Task A-B accuracy curve and confusion matrix; (**c**,**d**) Task A-C accuracy curve and confusion matrix; (**e**,**f**) Task B-A accuracy curve and confusion matrix; (**g**,**h**) Task B-C accuracy curve and confusion matrix; (**i**,**j**) Task C-A accuracy curve and confusion matrix; (**k**,**l**) Task C-B accuracy curve and confusion matrix.

4.3. Computational Expense

This paper experimentally verifies the use of a notebook CPU AMD Ryzen 7 4800 H. The simulation takes 3193 s on the public data set and 1544 s on the PT500mini mechanical bearing fault simulation test bench data set. If the network structure is determined, the fixed structure is loaded onto the airborne chip. The judgment time of new samples is very short. It can meet the real-time requirements and conform to the actual project.

5. Conclusions

In this paper, a multi-scale attention mechanism domain adversarial neural network for bearing fault diagnosis (MADANN) is proposed, which includes a feature extractor, domain discriminator, feature classifier, and category domain adaptation design based on the maximum mean discrepancy. A feature extractor combining multi-scale and attention mechanism is designed to extract multi-scale and more discriminative features, and the source domain and the target domain are mapped to the feature space and the label prediction space. The maximum mean difference alignment is introduced into the label prediction space, and it is used to reduce the difference in data distribution between the source domain and the target domain in the prediction label space, as well as to improve the ability of the feature extractor to extract domain invariant features. Domain adversarial learning is introduced between the domain discriminator and feature extractor, and it is used to realize feature domain adaptation. For the variable load problem, this paper uses the open data set to verify that the accuracy of the proposed method is better than other methods. For the variable speed problem, this paper uses the data set collected from the mechanical bearing fault simulation test bed to verify that the proposed method also has high accuracy. The results of case analysis show that the method proposed in this paper can accurately diagnose faults in the case of no label in the target domain, variable load, and variable speed, and it is more suitable for engineering practice.

However, the method proposed in this paper does not consider the following situations: (1) under the actual variable working conditions of rolling bearings, the target working conditions will generate new faults that have never occurred under the source working conditions, and how to diagnose the new faults have not been considered. (2) There is a problem of data imbalance between the source domain samples and the target domain samples. Serious data imbalance will lead to a strong imbalance in the distribution of fault samples, and how to diagnose the imbalance samples is not considered. In the future, in view of the above two problems, relevant research will be carried out on how to accurately classify new faults under variable conditions and how to solve the problem of data imbalance.

Author Contributions: Conceptualization, Q.Z., N.T., X.F. and C.W.; methodology, N.T. and X.F.; validation, N.T. and X.F.; formal analysis, N.T., X.F. and C.W.; resources, Q.Z. and C.W.; data curation, N.T. and X.F.; writing—original draft preparation, N.T. and X.F.; writing—review and editing, N.T., X.F. and C.W.; project administration, Q.Z.; funding acquisition, Q.Z., C.W., H.P. and C.B. All authors have read and agreed to the published version of the manuscript.

Funding: This work was supported in part by the National Key Research and Development Program of China under Grant 2021YFB3301300; in part by the National Natural Science Foundation of China under Grant 62203213; in part by the Natural Science Foundation of Jiangsu Province under Grant BK20220332; in part by the Open Project Program of Fujian Provincial Key Laboratory of Intelligent Identification and Control of Complex Dynamic System under Grant 2022A0004.

Data Availability Statement: The data used in this study are self-text and self-collection.

Conflicts of Interest: The authors declare no conflict of interest.

References

1. Zhou, Y.; Dong, Y.; Zhou, H.; Tang, G. Deep dynamic adaptive transfer network for rolling bearing fault diagnosis with considering cross-machine instance. *IEEE Trans. Instrum. Meas.* **2021**, *70*, 3525211. [CrossRef]
2. Chen, C.; Lu, N.; Jiang, B.; Xing, Y. A data-driven approach for assessing aero-engine health status. *IFAC-Pap. Online* **2022**, *55*, 737–742. [CrossRef]
3. Wang, C.; Lu, N.; Cheng, Y.; Jiang, B. A data-driven aero-engine degradation prognostic strategy. *IEEE Trans. Cybern.* **2021**, *51*, 1531–1541. [CrossRef] [PubMed]
4. Hu, Q.; Si, X.; Qin, A.; Lv, Y.; Liu, M. Balanced adaptation regularization based transfer learning for unsupervised cross-domain fault diagnosis. *IEEE Sens. J.* **2022**, *22*, 12139–12151. [CrossRef]
5. Lei, Y.; Yang, B.; Jiang, W.; Jia, F.; Li, N.; Nandi, A.K. Applications of machine learning to machine fault diagnosis: A review and roadmap. *Mech. Syst. Signal Process.* **2020**, *138*, 106578. [CrossRef]
6. Su, K.; Liu, J.; Xiong, H. A multi-level adaptation scheme for hierarchical bearing fault diagnosis under variable working conditions. *J. Manuf. Syst.* **2022**, *64*, 251–260. [CrossRef]
7. Zhang, S.; Zhang, S.; Wang, B.; Habetler, T.G. Deep learning algorithms for bearing fault diagnostics–a comprehensive review. In Proceedings of the 2019 IEEE 12th International Symposium on Diagnostics for Electrical Machines, Power Electronics and Drives (SDEMPED), Paris, France, 27–30 August 2019.

8. Xu, J.; Tong, S.; Cong, F.; Zhang, Y. The application of time–frequency reconstruction and correlation matching for rolling bearing fault diagnosis. *ARCHIVE Proc. Inst. Mech. Eng. Part C J. Mech. Eng. Sci.* **2015**, *229*, 33291–33295. [CrossRef]
9. Zheng, J.; Cao, S.; Pan, H.; Ni, Q. Spectral envelope-based adaptive empirical Fourier decomposition method and its application to rolling bearing fault diagnosis. *ISA Trans.* **2022**, *129*, 476–492. [CrossRef]
10. Wang, Z.; Zhang, Q.; Xiong, J.; Xiao, M.; Sun, G.; He, J. Fault diagnosis of a rolling bearing using wavelet packet denoising and random forests. *IEEE Sens. J.* **2017**, *17*, 5581–5588. [CrossRef]
11. Li, H.; Liu, T.; Wu, X.; Chen, Q. An optimized VMD method and its applications in bearing fault diagnosis. *Measurement* **2020**, *166*, 108185. [CrossRef]
12. Liu, S.; Sun, Y.; Zhang, L. A novel fault diagnosis method based on noise-assisted MEMD and functional neural fuzzy network for rolling element bearings. *IEEE Access* **2018**, *6*, 27048–27068. [CrossRef]
13. Goyal, D.; Dhami, S.; Pabla, B. Non-contact fault diagnosis of bearings in machine learning environment. *IEEE Sens. J.* **2020**, *20*, 4816–4823. [CrossRef]
14. Zuo, L.; Xu, F.; Zhang, C.; Xiahou, T.; Liu, Y. A multi-layer spiking neural network-based approach to bearing fault diagnosis. *Reliab. Eng. Syst. Saf.* **2022**, *225*, 108561. [CrossRef]
15. Wang, Z.; Yao, L.; Cai, Y. Rolling bearing fault diagnosis using generalized refined composite multiscale sample entropy and optimized support vector machine. *Measurement* **2020**, *156*, 107574. [CrossRef]
16. Tan, H.; Xie, S.; Liu, R.; Ma, W. Bearing fault identification based on stacking modified composite multiscale dispersion entropy and optimised support vector machine. *Measurement* **2021**, *186*, 110180. [CrossRef]
17. Wang, Z.; Yao, L.; Chen, G.; Ding, J. Modified multiscale weighted permutation entropy and optimized support vector machine method for rolling bearing fault diagnosis with complex signals. *ISA Trans.* **2021**, *114*, 470–484. [CrossRef]
18. Lu, C.; Wang, Z.; Zhou, B. Intelligent fault diagnosis of rolling bearing using hierarchical convolutional network based health state classification. *Adv. Eng. Inform.* **2017**, *32*, 139–151. [CrossRef]
19. Gao, S.; Pei, Z.; Zhang, Y.; Li, T. Bearing fault diagnosis based on adaptive convolutional neural network with Nesterov Momentum. *IEEE Sens. J.* **2021**, *21*, 9268–9276. [CrossRef]
20. Wang, Y.; Ding, X.; Zeng, Q.; Wang, L.; Shao, Y. Intelligent rolling bearing fault diagnosis via vision ConvNet. *IEEE Sens. J.* **2021**, *21*, 6600–6609. [CrossRef]
21. Sun, J.; Wen, J.; Yuan, C.; Liu, Z.; Xiao, Q. Bearing fault diagnosis based on multiple transformation domain fusion and improved residual Dense Networks. *IEEE Sens. J.* **2022**, *22*, 1541–1551. [CrossRef]
22. Sadoughi, M.; Hu, C. Physics-Based convolutional neural network for fault diagnosis of rolling element bearings. *IEEE Sens. J.* **2019**, *19*, 4181–4192. [CrossRef]
23. Udmale, S.; Singh, S.; Singh, R.; Sangaiah, A.K. Multi-Fault bearing classification using sensors and ConvNet-Based transfer learning approach. *IEEE Sens. J.* **2020**, *20*, 1433–1444. [CrossRef]
24. Liu, D.; Cui, L.; Cheng, W.; Zhao, D.; Wen, W. Rolling bearing fault severity recognition via data mining integrated with convolutional neural network. *IEEE Sens. J.* **2022**, *22*, 5768–5777. [CrossRef]
25. Lu, S.; Qian, G.; He, Q.; Liu, F.; Liu, Y.; Wang, Q. In situ motor fault diagnosis using enhanced convolutional neural network in an embedded system. *IEEE Sens. J.* **2020**, *20*, 8287–8296. [CrossRef]
26. Li, G.; Wu, J.; Deng, C.; Chen, Z. Parallel multi-fusion convolutional neural networks based fault diagnosis of rotating machinery under noisy environments. *ISA Trans.* **2021**, *128*, 545–555. [CrossRef]
27. Wang, H.; Liu, Z.; Peng, D.; Qin, Y. Understanding and learning discriminant features based on multi-attention 1DCNN for wheelset bearing fault diagnosis. *IEEE Trans. Ind. Inform.* **2020**, *16*, 5735–5745. [CrossRef]
28. Guo, X.; Chen, L.; Shen, C. Hierarchical adaptive deep convolution neural network and its application to bearing fault diagnosis. *Measurement* **2016**, *93*, 490–502. [CrossRef]
29. Magar, R.; Ghule, L.; Li, J.; Zhao, Y.; Farimani, A.B. FaultNet: A Deep Convolutional Neural Network for bearing fault classification. *IEEE Access* **2021**, *9*, 25189–25199. [CrossRef]
30. Wang, Q.; MiChau, G.; Fink, O. Domain adaptive transfer learning for fault diagnosis. In Proceedings of the 2019 Prognostics and System Health Management Conference (PHM-Paris), Paris, France, 2–5 May 2019.
31. Li, J.; Huang, R.; He, G.; Wang, S.; Li, G.; Li, W. A deep adversarial transfer learning network for machinery emerging fault detection. *IEEE Sens. J.* **2020**, *20*, 8413–8422. [CrossRef]
32. Lu, N.; Xiao, H.; Sun, Y.; Han, M.; Wang, Y. A new method for intelligent fault diagnosis of machines based on unsupervised domain adaptation. *Neurocomputing* **2021**, *427*, 96–109. [CrossRef]
33. Wu, Y.; Zhao, R.; Ma, H.; He, Q.; Du, S.; Wu, J. Adversarial domain adaptation convolutional neural network for intelligent recognition of bearing faults. *Measurement* **2022**, *195*, 111150. [CrossRef]
34. Wu, Z.; Zhang, H.; Guo, J.; Ji, Y.; Pecht, M. Imbalanced bearing fault diagnosis under variant working conditions using cost-sensitive deep domain adaptation network. *Expert Syst. Appl.* **2022**, *193*, 116459. [CrossRef]
35. Liu, Y.; Shi, K.; Li, Z.; Ding, G.F.; Zou, Y.S. Transfer learning method for bearing fault diagnosis based on fully convolutional conditional Wasserstein adversarial Networks. *Measurement* **2021**, *180*, 109553. [CrossRef]

36. Zou, Y.; Liu, Y.; Deng, J.; Jiang, Y.; Zhang, W. A novel transfer learning method for bearing fault diagnosis under different working conditions. *Measurement* **2021**, *171*, 108767. [CrossRef]
37. Wu, Z.; Jiang, H.; Liu, S.; Yang, C. A Gaussian-guided adversarial adaptation transfer network for rolling bearing fault diagnosis. *Adv. Eng. Inform.* **2022**, *53*, 101651. [CrossRef]

Disclaimer/Publisher's Note: The statements, opinions and data contained in all publications are solely those of the individual author(s) and contributor(s) and not of MDPI and/or the editor(s). MDPI and/or the editor(s) disclaim responsibility for any injury to people or property resulting from any ideas, methods, instructions or products referred to in the content.

Article

A Decentralized LQR Output Feedback Control for Aero-Engines

Xiaoxiang Ji, Jianghong Li *, Jiao Ren and Yafeng Wu

School of Power and Energy, Northwestern Polytechnical University, Xi'an 710072, China; jixiaoxiang@mail.nwpu.edu.cn (X.J.); renjiao@mail.nwpu.edu.cn (J.R.); yfwu@nwpu.edu.cn (Y.W.)
* Correspondence: jhli@nwpu.edu.cn

Abstract: Aero-engine control systems generally adopt centralized or distributed control schemes, in which all or most of the tasks of the control system are mapped to a specific processor for processing. The performance and reliability of this processor have a significant impact on the control system. Based on the aero-engine distributed control system (DCS), we propose a decentralized controller scheme. The characteristic of this scheme is that a network composed of a group of nodes acts as the controller of the system, so that there is no core control processor in the system, and the computation is distributed throughout the entire network. An LQR output feedback control is constructed using system input and output, and the control tasks executed on each node in the decentralized controller are obtained. The constructed LQR output feedback is equivalent to the optimal LQR state feedback. The primal-dual principle is used to tune the parameters of each decentralized controller. The parameter tuning algorithm is simple to calculate, making it conducive for engineering applications. Finally, the proposed scheme was verified by simulation. The simulation results show that a high-precision feedback gain matrix can be obtained with a maximum of eight iterations. The parameter tuning algorithm proposed in this paper converges quickly during the calculation process, and the constructed output feedback scheme achieves equivalent performance to the state feedback scheme, demonstrating the effectiveness of the design scheme proposed in this paper.

Keywords: decentralized controller; aero-engine; linear quadratic regulation (LQR); Q-learning; output feedback control construction; primal-dual

Citation: Ji, X.; Li, J.; Ren, J.; Wu, Y. A Decentralized LQR Output Feedback Control for Aero-Engines. *Actuators* **2023**, *12*, 164. https://doi.org/10.3390/act12040164

Academic Editors: Ti Chen, Junjie Kang, Shidong Xu and Shuo Zhang

Received: 23 February 2023
Revised: 19 March 2023
Accepted: 28 March 2023
Published: 6 April 2023

Copyright: © 2023 by the authors. Licensee MDPI, Basel, Switzerland. This article is an open access article distributed under the terms and conditions of the Creative Commons Attribution (CC BY) license (https://creativecommons.org/licenses/by/4.0/).

1. Introduction

Early aero-engine control systems adopt a centralized control scheme, where the controller is connected to analog sensors and actuators via cables [1]. The controller executes tasks such as calculation of control law, analog-to-digital conversion (ADC), and digital-to-analog conversion (DAC), resulting in a large workload for the controller. Meanwhile, the centralized control scheme increases the difficulty and cost of upgrading and maintaining the aero-engine throughout its entire lifespan [2]. Moreover, the cables should be arranged around the rotating components. The huge quantity of cables reduces the thrust-to-weight ratio of the aero-engine. To overcome the drawbacks of the centralized control scheme, distributed control schemes for aero-engines have emerged [3].

A distributed control system (DCS) is a control architecture applied in process control or plant control [4–7]. In an aero-engine DCS, traditional analog sensors and actuators have been replaced by intelligent digital sensors and actuators. The components in the control system are connected through a universal and standardized communication interface [8,9]. Currently, aero-engine related enterprises and scientific research institutions have a basic consensus on the "functional decentralization" of the DCS, that is, some of the functions of the controller in the centralized control system, such as signal conditioning and status monitoring, are executed by intelligent sensors and actuators, while the central controller focuses on executing the core control functions of the system [10]. A simple and robust digital bus replaces the large number of cables connecting the aero-engine control components to the avionics [11,12]. Compared with centralized control, an aero-engine

DCS possesses the following advantages [13]: (1) Digital intelligent sensors and actuators reduce the workload of the central controller; (2) The digital bus reduces the weight of the aero-engine control system; (3) The modular design scheme reduces the difficulty and cost of system maintenance and upgrades.

Although the aero-engine DCS has the above advantages, it still has some inherent disadvantages due to the existence of core nodes.

1. With the improvement of aero-engine performance, the function and complexity of control tasks have greatly increased, which has increased the workload of controller. The control system needs to use high-performance, multi-core processors as its controller, which in turn puts relatively high demands on the thermal management system of the aero-engine.
2. The amount of software code in aero-engine control systems is increasing rapidly, which significantly impacts the software reliability.
3. The core control tasks of the aero-engine, such as control-law calculation, are executed in the central controller. The central controller determines the performance of the entire aero-engine control system, and its failure or damage has significant impact on the aero-engine or even the aircraft.

Decentralizing the central controller is an effective way to overcome the shortcomings of the aero-engine DCS.

Output feedback and state feedback are both utilized in designing closed-loop control systems. Generally, output feedback is easier to implement in projects, but state feedback offers better performance. However, state feedback requires full system states. In the case of aero-engines, there are numerous state variables, some of which are hard to measure. Furthermore, some state variables may possess only mathematical meaning without a physical value. Therefore, the application of the aero-engine state feedback is restricted.

Linear quadratic regulation (LQR) can derive the optimal control law for linear state feedback and facilitate the implementation of optimal closed-loop control [14]. Optimum LQR control design is achieved by solving the LQR problem [15–17]. The conventional approach to the LQR problem involves solving the Bellman equation through dynamic programming. For linear control system design, it is transformed into the solution of the algebraic Riccati equation (ARE). Both the Bellman equation and ARE are challenging to solve [18,19], and it is difficult to use a conventional microprocessor for real-time calculation [20–24].

The contributions of this paper are as follows. Based on the aero-engine DCS, we propose a decentralized controller architecture. An LQR output feedback control was constructed through system input and output, and the control tasks executed on each node in the decentralized controller are obtained. The constructed LQR output feedback is equivalent to the optimal LQR state feedback. Finally, a simple linear iterative algorithm for solving the LQR problem is introduced. It is demonstrated that the calculation of the decentralized control tasks is simple and the LQR output feedback provides equivalent performance to the LQR optimal state feedback. Meanwhile, the LQR solving algorithm is simple to calculate and fast to converge.

The remainder of this paper is organized as follows. Section 2 introduces an aero-engine linear state space model. Section 3 introduces the output feedback scheme of the decentralized control system. Section 4 introduces an output feedback scheme based on the Q-Learning LQR, equivalent to the state feedback scheme. Section 5 introduces a primal-dual principle-based parameter-tuning algorithm. Section 6 presents the simulation of the scheme proposed in this paper. Finally, Section 7 is the last section of this paper, providing the summary and conclusion.

2. Aero-Engine Model

Turbofan engines are widely used in both military and civil aviation [25,26] due to their high efficiency. However, the technical difficulty is also high [27,28], and they require advanced control systems. Here, we consider the turbofan engine as an example to study [29]. Figure 1 is the schematic diagram of a turbofan engine. We first introduced the nonlinear modeling of the turbofan engine. Then, we derived a linearized model based on the nonlinear model and obtained the structure of the linearized model. Finally, we introduced the identification of the model parameters based on the input and output data of the engine.

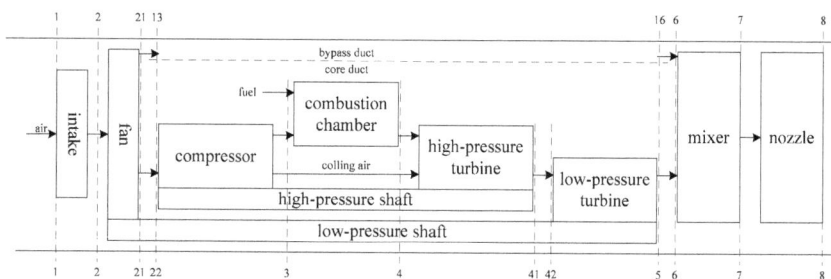

Figure 1. Schematic diagram of a turbofan engine.

2.1. Nonlinear Modeling

Due to the strong nonlinear characteristics of aero-engines, the nonlinear model of the aero-engines is briefly analyzed in this section. Based on fundamental physical principles, we model the various components of the turbofan engine. Then, using aerothermodynamic principles among components and the common working conditions of associated components, we establish joint equations. The following is a brief introduction to the physical equations involved.

(1) Intake

The intake is the first part where the air flows through the aero-engine.
The inlet temperature T_2 and the pressure P_2 of the aero-engine are:

$$\begin{cases} T_2 = T_1\left(1 + \frac{k_1-1}{2}\cdot Ma^2\right) \\ P_2 = \sigma_1 P_1\left(1 + \frac{k_1-1}{2}\cdot Ma^2\right)^{k_1/k_1-1} \end{cases} \quad (1)$$

where T_1 and P_1 are the total air temperature and the total pressure at the intake, respectively, k_1 is the air adiabatic index, σ_1 is the total pressure recovery coefficient of the intake, M_a is Mach number.

(2) Fan

The fan outlet temperature T_{21} and the pressure P_{21} of the aero-engine are:

$$\begin{cases} T_{21} = T_2\left(1 + \frac{\pi_F^{\frac{k_F-1}{k_F}}-1}{\eta_F}\right) \\ P_{21} = \pi_F P_2 \end{cases} \quad (2)$$

where k_F is the air adiabatic index, π_F is the fan pressure ratio, η_F is the efficiency of the fan.

(3) Compressor

The compressor outlet temperature T_3 and the pressure P_3 are:

$$\begin{cases} T_3 = T_{22}\left(1 + \frac{\pi_C^{\frac{k_C-1}{k_C}}-1}{\eta_C}\right) \\ P_3 = \pi_C P_{22} \end{cases} \quad (3)$$

where T_{22} and P_{22} are the temperature and pressure of the compressor inlet, respectively, k_C is the air adiabatic index, π_C is the compressor pressure ratio, η_C is the efficiency of the compressor.

(4) Combustion chamber

The outlet pressure of the combustion chamber P_4 can be expressed by the following equation:

$$P_4 = \sigma_3 P_4 \qquad (4)$$

where σ_3 represents the total pressure recovery coefficient of the combustion chamber.

(5) High-pressure turbine

The outlet temperature T_{41} and pressure P_{41} of the high-pressure turbine can be obtained as:

$$\begin{cases} T_{41} = \dfrac{q_{mg,TH} T_4 + C_{HPTCool} q_{maC,totalTcool}}{q_{mg,TH,total}} \left(1 - \left(1 - \pi_{TH}^{\frac{1-k_{TH}}{k_{TH}}}\right) \eta_{TH}\right) \\ P_{41} = \dfrac{P_4}{\pi_{TH}} \end{cases} \qquad (5)$$

where $q_{mg,TH}$ is the gas flow converted by the high-pressure turbine, T_4 is the outlet temperature of the combustion chamber, $C_{HPTCool}$ is the proportion coefficient of the high-pressure compressor bleed air used to cool the high-pressure turbine, $q_{maC,totalTcool}$ is the total flow of added air, $q_{mg,TH,total}$ is the outlet air flow of the high-pressure turbine, π_{TH} is the high-pressure turbine pressure ratio, η_{TH} is the efficiency of the high-pressure turbine, k_{TH} is the gas adiabatic index.

(6) Low-pressure turbine

The outlet temperature T_5 and pressure P_5 of the low-pressure turbine are calculated by the following equations:

$$\begin{cases} T_5 = \dfrac{q_{mg,TL} T_{42} + C_{LPTCool} q_{maC,totalTcool}}{q_{mg,TL,total}} \left(1 - \left(1 - \pi_{TL}^{\frac{1-k_{TL}}{k_{TL}}}\right) \eta_{TL}\right) \\ P_5 = \dfrac{P_{42}}{\pi_{TL}} \end{cases} \qquad (6)$$

where $q_{mg,TL}$ is the gas flow converted by the high-pressure turbine, T_{42} is the inlet temperature of the low-pressure turbine, $C_{LPTCool}$ is the proportion coefficient of the high-pressure compressor bleed air used to cool the low-pressure turbine, $q_{mg,TL,total}$ is the outlet air flow of the low-pressure turbine, π_{TL} is the low-pressure turbine pressure ratio, η_{TL} is the efficiency of the low-pressure turbine, k_{TL} is the gas adiabatic index.

(7) Bypass duct

The temperature T_6 and the pressure P_6 of the outlet of the bypass duct are:

$$\begin{cases} T_6 = T_{21} \\ P_6 = \sigma_4 P_{21} \end{cases} \qquad (7)$$

where T_{21} and P_{21} are the outlet temperature and pressure of the fan, respectively, σ_4 is the total pressure recovery coefficient of the bypass duct.

(8) Mixer

The gas flow $q_{mg,7}$ at the outlet of the mixer is the sum of the air flow $q_{ma,6}$ at the outlet of the bypass duct and the gas flow $q_{mg,5}$ at the outlet of the low-pressure turbine. Let σ_5

denotes the total pressure recovery coefficient of the mixer, then the physical parameters at the outlet of the mixer are:

$$\begin{cases} q_{mg,7} = q_{mg,5} + q_{ma,6} \\ h_7 = \dfrac{h_5 q_{mg,5} + h_6 q_{ma,6}}{q_{mg,7}} \\ P_7 = \sigma_5 \cdot \dfrac{P_5 q_{mg,5} + P_6 q_{ma,6}}{q_{mg,7}} \end{cases} \quad (8)$$

where h_5, h_6, and h_7 are the specific enthalpy of the gas at the outlet of the low-pressure turbine, air at the outlet of the bypass duct, and gas at the outlet of the mixer, respectively. From h_7, it is easy to obtain the temperature T_7 at the outlet of the mixer.

(9) Nozzle

The gas flow $q_{mg,8}$ at the nozzle outlet is:

$$q_{mg,8} = K_q \frac{P_8 A_8 q(\lambda_8)}{\sqrt{T_7}} \quad (9)$$

where K_q is the state coefficient of the nozzle, $q(\lambda_8)$ refers to the function related to the characteristics of the bypass duct and the core duct, A_8 is the sectional area of the nozzle, T_7 is the outlet temperature of the mixer.

The high-pressure turbine drives the high-pressure compressor of the turbofan engine to form the high-pressure rotor. The low-pressure turbine drives the fan to form the low-pressure rotor. After passing through the intake and fan, the gases enter the bypass duct and the core duct, respectively. The common working equations of the turbofan engine are obtained according to the flow, pressure, and power balance between the engine components.

(1) High-pressure rotor power balance

$$P_H = P_{CH} + P_{ex,H} + D_H \left(\frac{dn_H}{dt} \right) \quad (10)$$

where P_H denotes the high-pressure turbine power, P_{CH} denotes the high-pressure compressor power, $P_{ex,H}$ denotes the power lost by transmission friction force, $D_H \left(\frac{dn_H}{dt} \right)$ denotes the high-pressure rotor acceleration power. Furthermore, $D_H = (\pi/30)^2 J_H n_H$, J_H denotes the moment of inertia of the high-pressure rotor, n_H denotes the speed of the high-pressure rotor.

When the aero-engine is stable, $dn_H/dt = 0$. Ignoring the power lost by transmission friction force $P_{ex,H}$, Equation (10) can be simplified as:

$$P_H = P_{CH} \quad (11)$$

(2) Low-pressure rotor power balance

Similar to the high-pressure rotor, when the aero-engine is stable, ignoring the power lost by transmission friction force, we can obtain:

$$P_L = P_{CL} \quad (12)$$

where P_L denotes the low-pressure turbine power, P_{CL} denotes the low-pressure compressor power.

(3) Fan air flow balance

After flowing through the fan, the gas is divided into two parts, one part enters the bypass duct and the other enters the high-pressure compressor. Ignoring the gas flow loss, then:

$$q_{maF} = q_{maC} + q_{ma6} \quad (13)$$

where q_{maF} is the air flow from the fan, q_{maC} is the air flow into the high-pressure compressor, q_{ma6} is the air flow into the bypass duct.

(4) High-pressure turbine gas flow balance

$$q_{mg,4} = q_{maC} + q_{mf} \tag{14}$$

where $q_{mg,4}$ is the gas flow into the high-pressure turbine, q_{maC} is the gas flow from the high-pressure compressor, q_{mf} is the fuel flow.

(5) Low-pressure turbine gas flow balance

The low-pressure turbine inlet gas flow is equal to the high-pressure turbine outlet gas flow $q_{mg,42}$:

$$q_{mg,42} = q_{mg,4} + C_{HPTCool} q_{maC,totalTcool} \tag{15}$$

where $C_{HPTCool}$ is the proportion coefficient of the high-pressure compressor bleed air used to cool the high-pressure turbine, $q_{maC,totalTcool}$ is the total air flow into the high-pressure turbine.

(6) Nozzle gas flow balance

$$q_{mg,8} = q_{mg,5} + q_{ma,16} \tag{16}$$

where $q_{mg,8}$ is the gas flow at the inlet of the nozzle, $q_{mg,5}$ is the gas flow at the low-pressure turbine outlet, $q_{ma,16}$ is the gas flow into the bypass duct.

2.2. Linear State Space Model

We derived the component-level nonlinear model of the aero-engine, along with the common working equations. However, the current control system is primarily designed based on a linear system. Therefore, a linear model of the aero-engine is needed. In this section, the system identification method is applied to obtain the model of the aero-engine. Firstly, the structural identification of the model is carried out.

The state equation and the output equation of the aero-engine nonlinear state space model are:

$$\begin{cases} \dot{x} = f(x, u) \\ y = g(x, u) \end{cases} \tag{17}$$

where x is the state vector, y is the output vector, u is the control input, and the dimensions of x, y, u are n, m, r, respectively. It is assumed that the state equation and the output equation of the aero-engine are differentiable within the flight envelope.

Based on the principle of rotor dynamics, taking the aero-engine as a rigid body, let $\dot{w}$ denote the angular acceleration, ΔQ denote the torque difference between the two shafts, and I denote the mass moment of inertia. The acceleration of the system can be obtained from Newtonian mechanics as:

$$\dot{w} = \frac{\Delta Q}{I} \tag{18}$$

Let N denote the aero-engine shaft speed, f is the function between shaft speed and fuel quantity, the aero-engine shaft dynamic equation is:

$$\dot{n} = \frac{f(N, W_f)}{I} \tag{19}$$

Let J_H denote the moment of inertia of the high-pressure rotor, ΔM_H the residual torque of the high-pressure rotor, J_L the moment of inertia of the low-pressure rotor, ΔM_L

the residual torque of the low-pressure rotor, then, there are two rotor moment balance equations for the aero-engine:

$$\begin{cases} \left(\frac{\pi}{30}\right) \cdot J_H \cdot \frac{dN_C}{dt} = \Delta M_H \\ \left(\frac{\pi}{30}\right) \cdot J_L \cdot \frac{dN_F}{dt} = \Delta M_L \end{cases} \quad (20)$$

The residual torque of the high-pressure rotor and the residual torque of the low-pressure rotor are:

$$\begin{cases} \Delta M_H = M_{TH} - M_{CH} = \Delta M_H \cdot (N_C, N_F, W_f) \\ \Delta M_L = M_{TL} - M_{CL} = \Delta M_L \cdot (N_C, N_F, W_f) \end{cases} \quad (21)$$

Linearization is carried out at a steady-state operating point (x_0, u_0) of the aero-engine:

$$\begin{cases} \Delta M_H = \left.\frac{\partial \Delta M_H}{\partial N_C}\right|_{(x_0,u_0)} \Delta N_C + \left.\frac{\partial \Delta M_H}{\partial N_F}\right|_{(x_0,u_0)} \Delta N_F + \left.\frac{\partial \Delta M_H}{\partial W_f}\right|_{(x_0,u_0)} \Delta W_f \\ \Delta M_L = \left.\frac{\partial \Delta M_L}{\partial N_C}\right|_{(x_0,u_0)} \Delta N_C + \left.\frac{\partial \Delta M_L}{\partial N_F}\right|_{(x_0,u_0)} \Delta N_F + \left.\frac{\partial \Delta M_L}{\partial W_f}\right|_{(x_0,u_0)} \Delta W_f \end{cases} \quad (22)$$

From Equations (20) and (22), the linearized shaft dynamic equation is:

$$\begin{cases} \dot{N}_C = a_{11} \cdot \Delta N_C + a_{12} \cdot \Delta N_F + b_{11} \cdot \Delta W_f \\ \dot{N}_F = a_{21} \cdot \Delta N_C + a_{22} \cdot \Delta N_F + b_{21} \cdot \Delta W_f \end{cases} \quad (23)$$

where

$$a_{11} = \left(\frac{30}{\pi \cdot J_H}\right) \cdot \left.\frac{\partial \Delta M_H}{\partial N_C}\right|_{(x_0,u_0)}$$

$$a_{12} = \left(\frac{30}{\pi \cdot J_H}\right) \cdot \left.\frac{\partial \Delta M_H}{\partial N_F}\right|_{(x_0,u_0)}$$

$$a_{21} = \left(\frac{30}{\pi \cdot J_L}\right) \cdot \left.\frac{\partial \Delta M_L}{\partial N_C}\right|_{(x_0,u_0)}$$

$$a_{22} = \left(\frac{30}{\pi \cdot J_L}\right) \cdot \left.\frac{\partial \Delta M_L}{\partial N_F}\right|_{(x_0,u_0)}$$

$$b_{11} = \left(\frac{30}{\pi \cdot J_H}\right) \cdot \left.\frac{\partial \Delta M_H}{\partial w_f}\right|_{(x_0,u_0)}$$

$$b_{22} = \left(\frac{30}{\pi \cdot J_L}\right) \cdot \left.\frac{\partial \Delta M_L}{\partial W_f}\right|_{(x_0,u_0)}$$

Equation (23) can be written as:

$$\begin{cases} \dot{x} = Ax + Bu \\ y = Cx + Du \end{cases} \quad (24)$$

where x, y, u are the state vector, the output vector, and the control input of the aero-engine state space model, respectively. The model described by Equation (24) is a second-order system, and the structural identification of the system model is completed. The linear model of the turbofan engine is represented as a second-order system. Next, we introduce the method for identifying the parameters of the system model.

MATLAB software was used for identification. By running the open-loop control system of the engine model and providing an input at a certain steady-state point, the input and output data of the model can be saved when the system stabilizes, thus obtaining the data from the engine at that steady-state point. Then, a small step signal based on the input data is added to each variable, and the input and output data are saved when the system reaches a steady state. By subtracting the input and output data of the steady-state point from the input and output data of the system's step response, incremental data can be obtained that can be used for identification. After preprocessing the data, it is imported into the system identification toolbox, and the N4SID method is used to identify the linear model of the system.

Based on the method of system identification described above, we provide two linear models of the turbofan engine.

The first is a single-variable model for the turbofan engine. Here, taking the high-pressure rotor-speed control as an example, the high-pressure rotor speed $y = N_h$ is the controlled variable of the control system, and the main fuel flow $u = W_f$ is the control input. The system identification in this section is based on the input and output data of the system, i.e., the model of the turbofan engine is a black-box model, and the specific meaning of the state variables in the model is not limited. At the steady-state point where the relative speed is 85%, height $H = 0$, mach number $M_a = 0$, and the sample time is 20 ms, the discrete time linear state space model of the turbofan engine obtained through the system identification is:

$$\begin{cases} x[k+1] = G_S x[k] + H_S u[k] \\ y[k] = C_S x[k] + D_S u[k] \end{cases} \quad (25)$$

where $x = \begin{bmatrix} x_1 & x_2 \end{bmatrix}^T$ is the state variable, which has no physical meaning. Because this is not a physical variable, it cannot be directly measured by the sensors. Therefore, the model established by Equation (25) cannot theoretically realize state feedback control by conventional methods.

The coefficient matrix of Equation (25) is:

$$\begin{cases} G_S = \begin{bmatrix} g_{11} & g_{12} \\ g_{21} & g_{22} \end{bmatrix} = \begin{bmatrix} 0.9854 & 0.0156 \\ -1.3474 & 0.5935 \end{bmatrix} \\ H_S = \begin{bmatrix} h_{11} \\ h_{21} \end{bmatrix} = \begin{bmatrix} 0.1875 \\ 19.3024 \end{bmatrix} \\ C_S = \begin{bmatrix} 1 & 0 \end{bmatrix} \\ D_S = 0 \end{cases}$$

The second presented system is a multi-variable model for the turbofan engine. Here, the target of the control system is set to achieve a change in the thrust of the turbofan engine according to the angle of the throttle lever. As thrust cannot be measured through the sensors, parameters that are highly correlated with thrust are selected to represent the thrust characteristics. Here, according to the aero-engine principle, the controlled variable can be selected as high-pressure rotor speed n_H and turbine pressure ratio $P_i T = \pi_{TL} \pi_{TH}$, the reference input is selected as fuel flow W_f and nozzle area A_8. At the steady-state point with the power lever angle $PLA = 35°$, height $H = 0$, mach number $M_a = 0$, sample time 20 ms, the discrete time linear state space model of the turbofan engine obtained through the system identification is:

$$\begin{cases} x[k+1] = G_M x[k] + H_M u[k] \\ y[k] = C_M x[k] + D_M u[k] \end{cases} \quad (26)$$

where $u = \begin{bmatrix} W_f & A_8 \end{bmatrix}^T$ is the reference input, $y = \begin{bmatrix} n_H & P_i T \end{bmatrix}^T$ is the controlled variable, $x = \begin{bmatrix} x_1 & x_2 \end{bmatrix}^T$ is the state variable, which has no physical meaning.

The coefficient matrix of Equation (26) is:

$$\begin{cases} G_M = \begin{bmatrix} g_{11} & g_{12} \\ g_{21} & g_{22} \end{bmatrix} = \begin{bmatrix} 0.9609 & -0.0679 \\ -0.0028 & 0.9098 \end{bmatrix} \\ H_M = \begin{bmatrix} h_{11} & h_{12} \\ h_{21} & h_{22} \end{bmatrix} = \begin{bmatrix} 0.3305 & 0.1688 \\ 0.0758 & 0.2097 \end{bmatrix} \times 10^{-3} \\ C_M = \begin{bmatrix} 1 & 0 \\ 0 & 1 \end{bmatrix} \\ D_M = 0 \end{cases}$$

In this section, we report the single-variable and multi-variable linear models of the turbofan engine, respectively. Their state variables have no practical physical meaning. Because the value of the state variables cannot be measured using sensors, the state feedback scheme that relies on sensor measurements is unavailable.

3. Output Feedback of Decentralized Control System

The decentralized control system is proposed based on the development of communication technology, computer technology, data storage technology, sensor technology, and other supporting technologies. The traditional aero-engine DCS architecture is shown in Figure 2 [30]. In the DCS, the sensors and actuators are typically digital and smart, and are connected by bus to the controller. Signal conditioning, ADC, DAC, etc., are executed by smart sensors and smart actuators. Meanwhile, the controller focuses on core control tasks such as control-law calculation. With the development of the aero-engine, the tasks of the controller become more complex, and new-high performance microprocessors are needed to act as controllers. Therefore, the performance and reliability of the microprocessor acting as the controller have significant impact on the entire aero-engine control system.

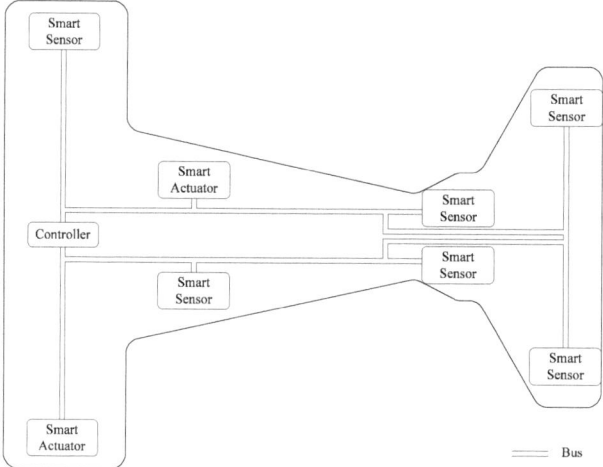

Figure 2. Overview of a traditional aero-engine DCS.

The design scheme of the decentralized controller proposed in this paper uses a group of low-price, high-reliability, but low-performance microprocessor nodes to form a decentralized network, as shown in Figure 3. Each microprocessor node is equipped with computing, memory, and communication capabilities and can transmit data to each other, as represented by the circles in Figure 3. Using this decentralized network to replace the DCS controller, as shown in Figure 2, the network acts as the controller of the control system, representing a decentralized aero-engine control scheme. Some microprocessors in the network possess the capability of bus communication, enabling the entire decentralized network to communicate with the smart sensors and actuators in the aero-engine. In this paper, we refer to this kind of control system as the software-defined control system (SDCS). In Figure 3, the green circles represent the regular nodes, and the blue circles represent the nodes with bus-communication capability.

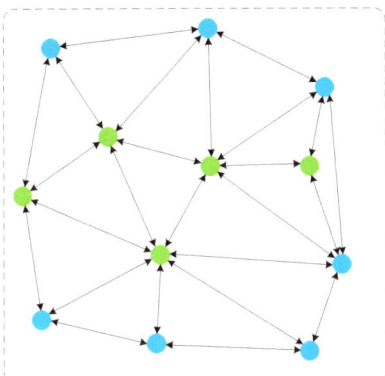

Figure 3. Overview of the decentralized controller.

The system control tasks are virtualized into multiple sub-tasks, named the virtual control tasks (VCTs). Each VCT establishes a mapping relationship with the microprocessors in the network. Since each VCT is part of the system control task, the task of each decentralized microprocessor is part of the controller task in DCS, which greatly reduces the workload of each microprocessor, and allows the use of low-price, high-reliability, low-performance microprocessors. At the same time, the amount of software code in each microprocessor is greatly reduced, which can increase the software reliability of the microprocessors.

Despite typically having inferior performance compared with state feedback, output feedback has been widely utilized due to its good economy and feasibility. Figure 4 is the structural block diagram of the output feedback system.

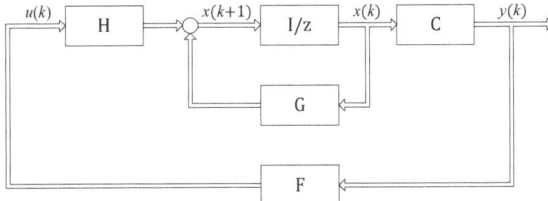

Figure 4. Output feedback control.

The dynamic equation of the feedback system is:

$$\begin{cases} x[k+1] = Gx[k] + Hu[k] \\ y[k] = Cx[k] \\ u[k] = Fy[k] \end{cases} \quad (27)$$

where $x[k] \in \mathbb{R}^n$ denotes the internal states, $u[k] \in \mathbb{R}^m$ denotes the input of the system, $y[k] \in \mathbb{R}^p$ denotes the output of the system, $G \in \mathbb{R}^{n \times n}$ is the system matrix, $H \in \mathbb{R}^{n \times m}$ is the control matrix, $C \in \mathbb{R}^{p \times n}$ is the output matrix, $F \in \mathbb{R}^{m \times p}$ is the feedback gain matrix.

In this section, $\gamma = \{v_1, v_2, \cdots, v_n\}$ denotes all of the nodes in the network, $\gamma_A = \{va_1, va_2, \cdots, va_a\}$ is the set of the nodes that send data to the actuators, $\gamma_S = \{vs_1, vs_2, \cdots, vs_s\}$ is the set of the nodes that receive sensor measurements.

As shown in Figure 5, the network acts as the system output feedback controller, i.e., the feedback F in Figure 4, sy_j denotes the sensor measuring the output, y_j is its measured value, i.e., the system output. Upon receiving sensor measurements, the network transmits the data to the system input after calculation. The system description of the nodes in the network during the running process is:

$$u_i[k] = \sum_{sy_j \in \gamma_S} f_{ij} y_j[k] \tag{28}$$

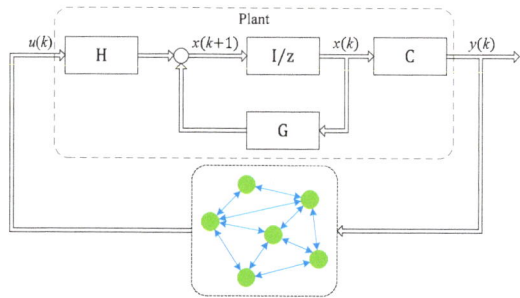

Figure 5. The networks acts as the output feedback controller.

The linear weighting process of the entire network can be expressed as:

$$u[k] = \underbrace{\begin{bmatrix} f_{11} & f_{12} & \cdots & f_{1p} \\ f_{21} & f_{22} & \cdots & f_{2p} \\ \vdots & \vdots & \ddots & \vdots \\ f_{m1} & f_{m2} & \cdots & f_{mp} \end{bmatrix}}_{F} y[k] = Fy[k] \tag{29}$$

In Figure 5, the state space of the plant is:

$$\begin{cases} x[k+1] = Gx[k] + Hu[k] \\ y[k] = Cx[k] \end{cases} \tag{30}$$

Then, the state space description of the control system in Figure 5 is:

$$\begin{cases} x[k+1] = (G + HFC)x[k] \\ y[k] = Cx[k] \end{cases} \tag{31}$$

In this section, we have briefly introduced the SDCS, and derived the model of the SDCS output feedback. In the next section, based on the LQR output feedback, we obtain the VCTs executed by the microprocessor.

4. Q-Learning Based LQR Output Feedback Control

The characteristic of the SDCS architecture is that the controller is decentralized. Therefore, it is required that the tasks executed on each node, i.e., VCTs, are also decentralized. The objective of this section is to obtain the VCTs executed on these decentralized nodes.

4.1. General LQR Problem Solving

The objective of the control system's state equation derived in Section 3 is to determine the optimal control sequence $u[k]$. When controllability and observability conditions of the system are satisfied, the following cost function can be minimized:

$$J = \sum_{i=0}^{+\infty} j(x[i], u[i]) \tag{32}$$

where j denotes the one-step cost of executing control u in the state x, and J denotes the long-term cost with an expected optimal cost as:

$$J^* = \min \sum_{i=0}^{+\infty} j(x[i], u[i]) \tag{33}$$

The optimal control sequence is:

$$u^*[k] = \arg\min \sum_{i=0}^{+\infty} j(x[i], u[i]) \tag{34}$$

For the LQR problem:

$$j(x[k], u[k]) = x^T[k] Q x[k] + u^T[k] R u[k] \tag{35}$$

where $Q \in \mathbb{R}^{p \times p}$, $R \in \mathbb{R}^{m \times m}$, and meet $Q = Q^T \geq 0$, $R = R^T > 0$.

According to the infinite time Bellman equation:

$$I(x[k]) = \min\{j(x[k], u[k]) + I(x[k+1])\} \tag{36}$$

Suppose the Bellman Equation (36) has the form of $I(x[k]) = x[k]^T P x[k]$, and satisfies $P = P^T > 0$:

$$I(x[k+1]) = (Gx[k] + Hu[k])^T P (Gx[k] + Hu[k]) \tag{37}$$

From Equations (36) and (37):

$$x^T P x = \min_r \left[x^T \left(Q + G^T P G \right) x + 2x^T G^T P H u + u^T R u + u^T H^T P H u \right] \tag{38}$$

When the system meets the controllability condition (G, H) and observability condition $(G, \sqrt{Q})$, then $\sqrt{Q}^T \sqrt{Q} = Q$, $Q = C^T \tilde{Q} C$. Taking the partial derivative of the expression inside the square brackets in Equation (38) $\partial / \partial r = 0$, the optimal state feedback of the system is:

$$u^* = -\left(R + H^T P H \right)^{-1} H^T P G x \tag{39}$$

It can be obtained from Equations (38) and (39):

$$x^T P x = x^T \left\{ Q + G^T P G - G^T P H \left(R + H^T P H \right)^{-1} H^T P G \right\} x \tag{40}$$

Then,

$$P = Q + G^T P G - G^T P H \left(R + H^T P H \right)^{-1} H^T P G \tag{41}$$

The above method is a conventional LQR problem-solving method based on the Bellman equation. Equation (41) is discrete time ARE. It can be concluded that the solved P is unique.

4.2. Construction of the Output Feedback Control

The cost function of Equation (33) is rewritten into the form of the Bellman equation,

$$I(x[k]) = j(x[k], u[k]) + I(x[k+1]) \tag{42}$$

We use the same Q function in [18]

$$Q(x[k], r[k]) = j(x[k], u[k]) + I(x[k+1]) \tag{43}$$

In the LQR problem, the Q-function above can be written as,

$$\begin{cases} Q(x[k], u[k]) &= x^T[k]Qx[k] + u^T[k]Ru[k] + x^T[k+1]Px[k+1] \\ &= x^T[k]Qx[k] + u^T[k]Ru[k] + (Gx[k] + Hu[k])^T P(Gx[k] + Hu[k]) \\ &= \begin{bmatrix} x[k] \\ u[k] \end{bmatrix}^T \begin{bmatrix} Q + G^T PG & G^T PH \\ H^T PG & R + H^T PH \end{bmatrix} \begin{bmatrix} x[k] \\ u[k] \end{bmatrix} \end{cases} \tag{44}$$

Let the I^* denotes the optimal cost function, F^* denotes the optimal feedback strategy, then the Q-function can be written as:

$$Q^*(x[k], u[k]) = j(x[k], u[k]) + I^*(x[k+1]) \tag{45}$$

The optimal strategy is:

$$F^* \tilde{x}[k] = \arg\min_r Q^*(x[k], u[k]) \tag{46}$$

The optimal LQR controller u^* is obtained by solving $(\partial/\partial r)Q^* = 0$, and the result is consistent with the result obtained based on the Bellman equation.

In order to develop a new output feedback Q-function, the internal state of the system is parameterized and constructed. Based on the state space description of Equation (30), and the recursive equation of the discrete-time equation:

$$\begin{aligned} x[k-N+1] &= Gx[k-N] + Hu[k-N] \\ x[k-N+2] &= Gx[k-N+1] + Hu[k-N+1] = G^2 x[k-N] + GHu[k-N] + Hu[k-N+1] \\ x[k-N+3] &= Gx[k-N+2] + Hu[k-N+2] = G^3 x[k-N] + G^2 Hu[k-N] + GHu[k-N+1] + Hu[k-N+2] \\ &\vdots \\ x[k-N+N] &= x[k] = G^N x[k-N] + G^{N-1} Hu[k-N] + \cdots + G^2 Hu[k-3] + GHu[k-2] + Hu[k-1] \end{aligned} \tag{47}$$

It is easy to obtain:

$$\begin{bmatrix} y[k-N] \\ y[k-N+1] \\ \vdots \\ y[k-2] \\ y[k-1] \end{bmatrix} = \begin{bmatrix} C \\ CG \\ \vdots \\ CG^{N-2} \\ CG^{N-1} \end{bmatrix} x[k-N] + \begin{bmatrix} 0 & 0 & \cdots & 0 & 0 & 0 \\ CH & 0 & \cdots & 0 & 0 & 0 \\ \vdots & \vdots & \ddots & \vdots & \vdots & \vdots \\ CG^{N-3}H & CG^{N-2}H & \cdots & CH & 0 & 0 \\ CG^{N-2}H & CG^{N-3}H & \cdots & CGH & CH & 0 \end{bmatrix} \begin{bmatrix} u[k-N] \\ u[k-N+1] \\ \vdots \\ u[k-2] \\ u[k-1] \end{bmatrix} \tag{48}$$

Let
$$V_N \triangleq \begin{bmatrix} C^T & (CG)^T & \cdots & (CG^{N-2})^T & (CG^{N-1})^T \end{bmatrix}^T$$
$$U_N = \begin{bmatrix} G^{N-1}H & G^{N-2}H & \cdots & GH & H \end{bmatrix}$$
$$T_N = \begin{bmatrix} 0 & 0 & \cdots & 0 & 0 & 0 \\ CH & 0 & \cdots & 0 & 0 & 0 \\ \vdots & \vdots & \ddots & \vdots & \vdots & \vdots \\ CG^{N-3}H & CG^{N-2}H & \cdots & CH & 0 & 0 \\ CG^{N-2}H & CG^{N-3}H & \cdots & CGH & CH & 0 \end{bmatrix}$$
$$\bar{y}[k-N, k-1] = \begin{bmatrix} y[k-N] \\ y[k-N+1] \\ \vdots \\ y[k-2] \\ y[k-1] \end{bmatrix}$$
$$\bar{u}[k-N, k-1] = \begin{bmatrix} u[k-N] \\ u[k-N+1] \\ \vdots \\ u[k-2] \\ u[k-1] \end{bmatrix}$$

Then, the Equations (47) and (48) can be written as:

$$x[k] = G^N x[k-N] + U_N \bar{u}[k-N, k-1] \tag{49}$$

$$\bar{y}[k-N, k-1] = V_N x[k-N] + T_N \bar{u}[k-N, k-1], \tag{50}$$

$\bar{y}[k-N, k-1]$ and $\bar{r}[k-N, k-1]$ are measurable system outputs and inputs, respectively. When the $[G, C]$ is observable:

$$\bar{x}[k] = \underbrace{G^N \left(V_N^T V_N \right)^{-1} V_N^T \bar{y}[k-N, k-1]}_{M_y} + \underbrace{U_N - G^N \left(V_N^T V_N \right)^{-1} V_N^T T_N \bar{u}[k-N, k-1]}_{M_u} = M_y \bar{y}[k-N, k-1]$$
$$+ M_u \bar{u}[k-N, k-1] = x[k] \tag{51}$$

The parameters in the above equation are determined by the system coefficients, which require complete system information. However, it is often difficult to obtain the complete system information. Therefore, we designed an optimal output feedback control strategy based on Q-learning, which does not require the complete system information. From the Equations (44) and (51):

$$Q(x[k], u[k]) = \begin{bmatrix} \bar{u}[k-N, k-1] \\ \bar{y}[k-N, k-1] \\ u[k] \end{bmatrix}^T \begin{bmatrix} L_{11} & L_{12} & L_{13} \\ L_{21} & L_{22} & L_{23} \\ L_{31} & L_{32} & L_{33} \end{bmatrix} \begin{bmatrix} \bar{u}[k-N, k-1] \\ \bar{y}[k-N, k-1] \\ u[k] \end{bmatrix} \tag{52}$$

Let
$$\bar{z}[k] = \begin{bmatrix} \bar{u}^T[k-N, k-1] & \bar{y}^T[k-N, k-1] & u^T[k] \end{bmatrix}^T$$
$$L = \begin{bmatrix} L_{11} & L_{12} & L_{13} \\ L_{21} & L_{22} & L_{23} \\ L_{31} & L_{32} & L_{33} \end{bmatrix}$$

The Equation (52) can be written as:

$$Q(x[k], u[k]) = \bar{z}^T[k] L \bar{z}[k] \qquad (53)$$

where

$$L_{11} = M_u^T (Q + G^T PG) M_u$$
$$L_{12} = M_u^T (Q + G^T PG) M_y$$
$$L_{13} = M_u^T G^T PH$$
$$L_{21} = M_y^T (Q + G^T PG) M_u$$
$$L_{22} = M_y^T (Q + G^T PG) M_y$$
$$L_{23} = M_y^T G^T PH$$
$$L_{31} = H^T PG M_u$$
$$L_{32} = H^T PG M_y$$
$$L_{33} = R + H^T PH$$

Since $Q = Q^T$, $R = R^T$, $P = P^T$, then, $L = L^T$

Equation (53) is a new Q-function relating to the LQR problem of system input and output data. Minimize the $Q(x[k], u[k])$ in Equation (53) by solving $\partial Q / \partial r = 0$. Then, the desired control law can be obtained as follows:

$$u^* = -(L_{33})^{-1} \left(L_{31} \bar{u}[k-N, k-1] + L_{32} \bar{y}[k-N, k-1] \right) \qquad (54)$$

From L_{31}, L_{32}, L_{33}, and Equation (54):

$$u^* = -(R + H^T PH)^{-1} \left(H^T PG M_u \bar{u}[k-N, k-1] + H^T PG M_y \bar{y}[k-N, k-1] \right) = -(R + H^T PH)^{-1} H^T PG \left(M_u \bar{u}[k-N, k-1] + M_y \bar{y}[k-N, k-1] \right) \qquad (55)$$

From Equation (51):

$$u^* = -\left(R + H^T PH \right)^{-1} H^T PG x[k] \qquad (56)$$

It can be seen that the obtained optimal output feedback controller described in this paper is equivalent to the optimal state feedback controller, without requiring all of the internal state variables. This overcomes the limitation of acquiring all internal state variables of the system in practical engineering applications of state feedback control.

4.3. Acquisition of the VCTs

The characteristic of the SDCS is that the controller is decentralized, so it is essential to obtain execution of the control tasks on each decentralized node, i.e., VCTs.

From Equation (54), the control task of the optimal output feedback LQR includes three parts: the construction of $\bar{u}$, the construction of $\bar{y}$, and the weighted sum of $\bar{u}$ and $\bar{y}$. It is easy to obtain two of the VCTs:

VCT_1: from the sampled data, according to $\bar{u}$ in Equation (49), to construct u-based $\bar{u}$;

VCT_2: from the sampled data, according to $\bar{y}$ in Equation (50), to construct y-based $\bar{y}$;

For the latter task, which is the weighted summing of $\bar{u}$ and $\bar{y}$, the number of virtualized VCTs depends on the dimensions of $\bar{u}$ and $\bar{y}$, as well as the microprocessor performance. Let VCT_3 denote the VCT set of the virtualization of the weighted summation task. Some simple examples are as follows.

Example 1: when the dimensions of $\bar{u}$ and $\bar{y}$ are low, they are virtualized into a single VCT

VCT_3: The output feedback $-(L_{33})^{-1}\left(L_{31}\bar{u} + L_{32}\bar{y}\right)$ is obtained;

Example 2: when the dimensions of $\bar{u}$ and $\bar{y}$ are high, they are virtualized into three VCTs

VCT_{31}: The output feedback $-(L_{33})^{-1}L_{31}\bar{u}$ is obtained;
VCT_{32}: The output feedback $-(L_{33})^{-1}L_{32}\bar{y}$ is obtained;
VCT_{33}: The total feedback $VCT_{31} + VCT_{32}$ is obtained.
$VCT_3 = VCT_{31} \cup VCT_{32} \cup VCT_{33}$.

In addition, virtualization can also be carried out by dividing $-(L_{33})^{-1}L_{31}$ and $-(L_{33})^{-1}L_{32}$ by rows; a detailed description is not provided here.

According to the description given above of the virtualization scheme for the LQR output feedback control task, for VCT_1 and VCT_2 we remove the earliest sampling data of u and y in each sampling period, obtain the latest u and y, then construct $\bar{u}$ and $\bar{y}$ according to Equations (54) and (55). For VCT_3, the essence of the calculation is the weighted sum. The calculated amounts of the VCTs above are very small, and most low-performance microprocessors can be used for real-time processing.

5. Parameter Tuning Based on the Primal-Dual Method

We derived the VCTs of SDCS. In the method used in the current study, the parameter tuning of VCTs requires solving the LQR problem. Usually, dynamic programming (DP) is applied to solve the Bellman equation. In recent years, with the development of convex analysis and semi-definite programming (SDP) [31], SDP-based LQR-solving methods have emerged. In addition, research on LQR problems based on reinforcement learning (RL) has received widespread attention. Most of these solve the Bellman equation based on the DP of sampled data, meanwhile making use of the monotonicity of functions or the properties of contraction mappings to ensure convergence [32].

The solution of the primal problem can be achieved by solving the dual problem, making the problem easier to solve. Usually, the primal constrained problem is transformed into a dual unconstrained problem through the Lagrange function. When the described problem meets the Karush–Kuhn–Tucker (KKT) conditions, the above scheme can be adopted. The duality of RL for LQR is discussed and studied in [33].

In this section, we assume that the system (G, H) is stable. The LQR problem described in Equation (32) can be expressed in the following form:

$$J = \sum_{i=0}^{+\infty} \begin{bmatrix} x[k] \\ u[k] \end{bmatrix}^T \underbrace{\begin{bmatrix} Q & 0 \\ 0 & R \end{bmatrix}}_{L_{QR}} \begin{bmatrix} x[k] \\ u[k] \end{bmatrix} = \sum_{i=0}^{+\infty} \begin{bmatrix} x[k] \\ u[k] \end{bmatrix}^T L_{QR} \begin{bmatrix} x[k] \\ u[k] \end{bmatrix} \tag{57}$$

Let

$$L_F = \begin{bmatrix} G & H \\ FG & FH \end{bmatrix}, \quad \Lambda = \begin{bmatrix} I_n & 0 \\ F & I_m \end{bmatrix}$$

Then,

$$\Lambda^{-1} = \begin{bmatrix} I_n & 0 \\ -F & I_m \end{bmatrix}$$

So, $\Lambda^{-1} L_F \Lambda = G + HF$, that is, $L_F \sim G + HF$.

Let $\begin{bmatrix} Q + G^T PG & G^T PH \\ H^T PG & R + H^T PH \end{bmatrix} = \tilde{P}$ (in Equation (44)). $\tilde{P}$ is represented as,

$$\tilde{P} = \begin{bmatrix} \tilde{P}_{11} & \tilde{P}_{12} \\ \tilde{P}_{12}^T & \tilde{P}_{22} \end{bmatrix} \tag{58}$$

where $\tilde{P}_{11} = Q + G^TPG$, $\tilde{P}_{12} = G^TPH$, $\tilde{P}_{22} = R + H^TPH$.

The primal problem and dual problem introduced in [33] are:
Primal problem
$$J_p = \min \text{Tr}(L_{QR}S)$$
$$s.t. L_F S L_F^T + \Gamma = S$$

where S and Γ are the symmetric positive definite matrix.

Dual problem
The Lagrange function is introduced:
$$L(\acute{P}, \grave{P}, F, S) = \text{Tr}(L_{QR}S) + \text{Tr}\left(\left(L_F S L_F^T + \Gamma - S\right)\acute{P}\right) + \text{Tr}(-S\grave{P}) \tag{59}$$

where $\acute{P}$ is an arbitrarily fixed symmetric matrix, $\grave{P}$ is an arbitrarily fixed symmetric positive semidefinite matrix cone. Then, the dual problem of the primal problem is:
$$J_d = \sup_{\acute{P}, \grave{P}} \inf_{S, F} L(\acute{P}, \grave{P}, F, S)$$

Making use of theorem 1, attribute 7, and lemma 5 of [33], it is calculated that:

1. $J_p = J_d$
2. If $(\hat{S}, \hat{F})$ is the optimal point of the primal problem, $(\acute{P}^*, \grave{P}^*)$ is the optimal point of the dual problem of the primal problem, then $(\hat{S}, \hat{F}, \acute{P}^*)$ satisfies the KKT condition of $(S, F, \acute{P})$:
$$L_F S L_F^T + \Gamma - S = 0 \tag{60}$$

$$S > 0, \tag{61}$$

$$L_F^T \acute{P} L_F - \acute{P} + L_{QR} = 0, \tag{62}$$

$$2\left(\acute{P}_{12}^T + \acute{P}_{22}F\right)[G \quad H]S[G \quad H]^T = 0, \tag{63}$$

3. Define $Y(\acute{P}) = L_F^T \acute{P} L_F + L_{QR}$, then, if $\acute{P}_1 > \acute{P}_2$, $Y(\acute{P}_1) > Y(\acute{P}_2)$, and Y is a symmetric positive semi-definite cone. A matrix norm $\|\cdot\|$ makes Y a contractive mapping. There is a unique symmetric matrix $\overline{P}$ that makes $Y(\overline{P}) = \overline{P}$, that is, Y has a unique fixed point $\overline{P}$.
4. The $\acute{P}_t$ in the following algorithm converges to the optimal solution $\acute{P}^*$ of the dual problem, that is, the optimal solution $\tilde{P}^*$ of the matrix $\tilde{P}$ required in Equation (58).

Algorithm steps:
1. Initialization of F_0, $\varepsilon > 0$, and $t = 0$.
2. Calculate $L_{F_0} = \begin{bmatrix} G & H \\ F_0 G & F_0 H \end{bmatrix}$.
3. Solve $\acute{P}_0$ from $L_{F_0}^T \acute{P}_0 L_{F_0} + L_{QR} = \acute{P}_0$.
4. Repeat.
5. $t = t + 1$.
6. Dual update: $\acute{P}_{t+1} = L_{F_t}^T \acute{P}_t L_{F_t} + L_{QR}$.
7. Primal update: $F_{t+1} = -\left(\acute{P}_{22}\right)_{t+1}^{-1}\left(\acute{P}_{12}\right)_{t+1}^T$.
8. $L_{F_{t+1}} = \begin{bmatrix} G & H \\ F_{t+1}G & F_{t+1}H \end{bmatrix}$.
9. Until $\|F_{t+1} - F_t\| \le \varepsilon$.

The optimal solution $\tilde{P}^*$ of $\tilde{P}$ in Equation (58) can be obtained from the derived $\acute{P}$. By combining Equations (51)–(53) and (58), the parameters of each VCT of SDCS can be obtained.

6. Simulation Analyses

From Section 5, we can conclude that the parameter tuning algorithm is simple to calculate, and solving the algorithm process does not involve the complex ARE or Lyapunov equation. It can be calculated with high-reliability low-performance microprocessors.

The VCTs obtained in this paper do not involve complex calculations, making them suitable for meeting the requirements of the SDCS. Additionally, the iterative process of the VCTs' parameter tuning algorithm also avoids complex calculations. The purposes of the simulation described here are: 1. to verify the convergence rate of the parameter tuning algorithm; and 2. to verify whether the output feedback control constructed in this study achieves equivalent performance to the conventional LQR state feedback control.

Based on the aero-engine model established in Section 2, the analysis in this section includes the following two simulations:

SIMULATION 1.

This simulation focuses on the single-variable aero-engine control. We used the high-pressure rotor speed-control model at the steady point of the relative speed of 83% obtained in Section 2. The eigenvalues of the system matrix of the model are 0.9231 and 0.6576, so it is asymptotically stable. Let $\tilde{Q} = 1$, $R = 0.49857$, then $Q = [1\ 0;0\ 0]$. The initial feedback gain matrix is $F = [0\ \ 0]$, and it is calculated that $L_{QR} = [1\ 0\ 0;0\ 0\ 0;0\ 0\ 0.49875]$. According to the scheme for solving ARE in Equation (41), the optimal feedback gain matrix is $F^* = [-0.9120\ \ -0.0252]$. The convergence details during the iteration are reported in Table 1.

Table 1. Convergence details of Simulation 1.

ε	Steps	F	$\|F_t - F_{t+1}\|$	$\|F_t - F^*\|$
0.1	3	$[-0.8712\ \ -0.0261]$	0.0659	0.0274
0.01	4	$[-0.9266\ \ -0.0256]$	6.0304×10^{-4}	0.0140
0.001	4	$[-0.9266\ \ -0.0256]$	6.0304×10^{-4}	0.0140
0.0007	4	$[-0.9266\ \ -0.0256]$	6.0304×10^{-4}	0.0140
0.0006	8	$[-0.9120\ \ -0.0252]$	3.5857×10^{-5}	4.6999×10^{-5}

The calculation results of $\acute{P}, L_{31}, L_{32}, L_{33}$ are shown in Equation (64), and the parameters of the three VCTs are obtained. Figure 6 shows convergence process of the parameters during the iterative process:

$$\begin{aligned} \acute{P} &= \begin{bmatrix} 3.2878 & 0.0530 & 0.9692 \\ 0.0530 & 0.0014 & 0.0267 \\ 0.9692 & 0.0267 & 1.0460 \end{bmatrix} \\ L_{31} &= [\ 0.3772\ \ 0.6980\ ] \\ L_{32} &= [\ -1.2037\ \ 2.0982\ ] \\ L_{33} &= 1.4060 \end{aligned} \tag{64}$$

From Table 1 and Figure 6, the feedback gain matrix F converges rapidly, and F_t converges to F^* within eight iterations. The simulation results show that the LQR parameter tuning algorithm proposed in this paper has the characteristics of fewer convergence steps and fast convergence speed. Due to its advantages of simple calculation and rapid convergence, microprocessors with low price, high reliability, and low performance can be utilized for real-time parameter tuning.

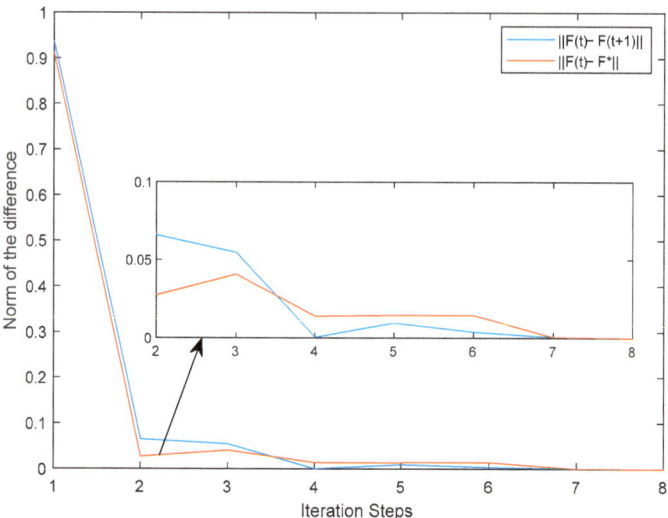

Figure 6. Convergence trajectory of $\|F_t - F_{t+1}\|$ and $\|F_t - F^*\|$ during the iterative process.

The following simulation was conducted to verify the performance of the decentralized LQR output feedback control proposed in this paper. Generally, the performance of optimal state feedback is superior to that of output feedback. Through the design described in Section 4, the LQR output feedback control proposed in this paper can theoretically achieve equivalent performance to the LQR state feedback. This scheme overcomes the shortcomings of state feedback while achieving comparable performance. This simulation compares the LQR output feedback control proposed in this paper with the conventional LQR state feedback control, to validate the aforementioned conclusions.

Figure 7 shows the input and the response curves of the high-pressure rotor speed of the turbofan engine when the output feedback scheme proposed in this paper and the conventional state feedback scheme were adopted, respectively. It can be seen from Figure 7 that the high-pressure rotor speed response curves of the turbofan engine were almost identical when the two schemes were adopted. However, in reality, the two response curves did not completely overlap. As shown in Table 1, this is because there is a certain error between the feedback gain matrices calculated by the primal-dual principle (the output feedback scheme in this paper) and the ARE (the conventional state feedback scheme), respectively. According to the simulation results presented in Figure 7, the output feedback scheme proposed in this paper can achieve equivalent performance to the conventional state feedback, which is consistent with the theoretical analysis in this paper. However, because the state variables of the high-pressure rotor speed model have no physical meaning, the conventional LQR state feedback scheme cannot be used for system design in practical engineering applications. On the other hand, the scheme proposed in this paper can be applied in actual engineering applications, because of the output feedback.

SIMULATION 2.

This simulation focused on the multiple variables of aero-engine control. We used the dual-variable control model of the operating point at a PLA angle of 35°, obtained in Section 2.

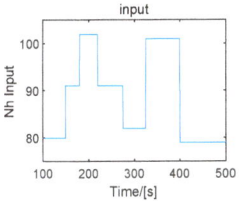
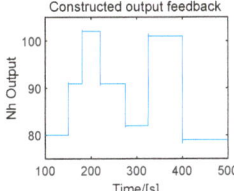
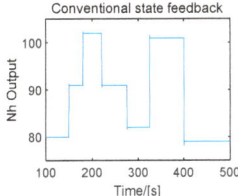

Figure 7. High-pressure rotor speed response curve.

The eigenvalues of the system matrix of the model are 0.9644 and 0.9063, so it is asymptotically stable. Let $\tilde{Q} = 1$, $R = [1\ 0; 0\ 1]$, then $Q = [1\ 0; 0\ 1]$. The initial feedback gain matrix is $F = [0\ 0; 0\ 0]$, and it is calculated that $L_{QR} = [1\ 0\ 0\ 0; 0\ 1\ 0\ 0; 0\ 0\ 1\ 0; 0\ 0\ 0\ 1]$. According to the scheme for solving ARE in Equation (41), the optimal feedback gain matrix is $F^* = [-0.0038\ 0.0017; -7.2976 \times 10^{-4}\ -0.0010]$. The convergence details obtained during the iteration are shown in Table 2.

Table 2. Convergence details of Simulation 2.

ε	Steps	F	$\|F_t - F_{t+1}\|$	$\|F_t - F^*\|$
0.1	1	$\begin{bmatrix} -0.0038 & 0.0017 \\ -7.2977 \times 10^{-4} & -0.001 \end{bmatrix}$	0.0041	0.0041
0.01	1	$\begin{bmatrix} -0.0038 & 0.0017 \\ -7.2977 \times 10^{-4} & -0.001 \end{bmatrix}$	0.0041	0.0042
0.001	2	$\begin{bmatrix} -0.0038 & 0.0017 \\ -7.2976 \times 10^{-4} & -0.001 \end{bmatrix}$	4.8053×10^{-9}	6.9260×10^{-8}
0.0001	2	$\begin{bmatrix} -0.0038 & 0.0017 \\ -7.2976 \times 10^{-4} & -0.001 \end{bmatrix}$	4.8053×10^{-9}	4.1779×10^{-5}
4.0×10^{-9}	5	$\begin{bmatrix} -0.0038 & 0.0017 \\ -7.2976 \times 10^{-4} & -0.001 \end{bmatrix}$	3.8062×10^{-9}	5.6807×10^{-8}

The calculation result of $\acute{P}$, L_{31}, L_{32}, L_{33} is shown in Equation (65), and the parameters of the three VCTs were obtained. The convergence of the parameters during the iterative process is shown in Figure 8.

$$\acute{P} = \begin{bmatrix} 13.5543 & -7.2729 & 0.0038 & 7.2976 \times 10^{-4} \\ -7.2729 & 11.3840 & -0.0017 & 0.0010 \\ 0.0038 & -0.0017 & 1.0 & 3.4001 \times 10^{-7} \\ 7.2976 \times 10^{-4} & 0.0010 & 3.4001 \times 10^{-7} & 1.0 \end{bmatrix}$$
$$L_{31} = 1.0 \times 10^{-5} \times \begin{bmatrix} 0.0545 & 0.0123 & 0.1123 & 0.0288 \\ 0.0162 & 0.0166 & 0.0317 & 0.0334 \end{bmatrix} \quad (65)$$
$$L_{32} = \begin{bmatrix} 0.0018 & -0.0009 & 0.0018 & -0.0009 \\ 0.0004 & 0.0004 & 0.0003 & 0.0004 \end{bmatrix}$$
$$L_{33} = \begin{bmatrix} 1.0 & 0.0 \\ 0.0 & 1.0 \end{bmatrix}$$

As shown in Table 2 and Figure 8, the feedback gain matrix F converged rapidly, and F_t converged to F^* within five iterations. In fact, only one or two iterations were needed to obtain a feedback gain matrix F with sufficient accuracy. The simulation results show that the LQR parameter tuning algorithm proposed in this paper has the characteristics of fewer convergence steps and fast convergence speed. Due to its advantages of simple calculation and rapid convergence, microprocessors with low price, high reliability, and low performance can be utilized for real-time parameter tuning.

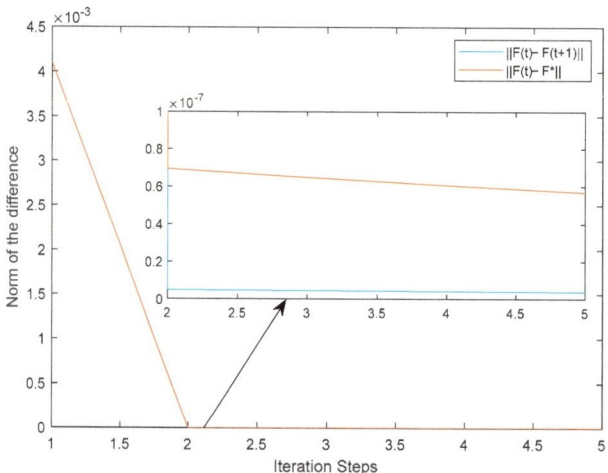

Figure 8. Convergence trajectory of $\|F_t - F_{t+1}\|$ and $\|F_t - F^*\|$ during the iterative process.

Figure 9 shows the input and the response curves of the high-pressure rotor speed of the turbofan engine when the output feedback scheme and the conventional state feedback scheme were adopted, respectively. Figure 10 shows the input and the response curves of the pressure ratio when the two feedback schemes were adopted, respectively. It can be seen from Figure 9 that the high-pressure rotor speed response curves of the turbofan engine were almost identical when the two schemes were adopted, and the same applies for the pressure ratio response curves in Figure 10. However, in both Figures 9 and 10, the system response of the two feedback schemes do not completely overlap. This is because there is a certain error between the feedback gain matrices calculated by the primal-dual principle and the ARE. According to the simulation results presented in Figures 9 and 10, the output feedback scheme proposed in this paper can achieve equivalent performance to conventional state feedback, which is consistent with the theoretical analysis in this paper. However, because the state variables of the dual-variable control model have no physical meaning, the conventional LQR state feedback scheme cannot be used for system design in practical engineering applications. On the other hand, the scheme proposed in this paper can be applied in actual engineering applications because of the output feedback.

According to the results of Simulations 1 and 2, the constructed output feedback scheme can achieve performance equivalent to that of the state feedback scheme. Furthermore, the proposed parameter tuning algorithm exhibits fast convergence speed and high accuracy. These simulation results show the effectiveness of the design scheme presented in this paper.

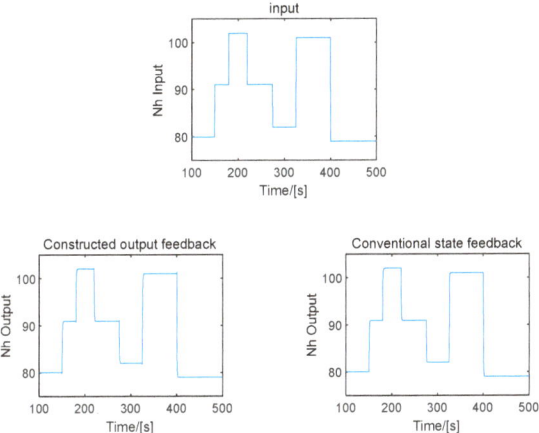

Figure 9. High-pressure rotor speed response curve.

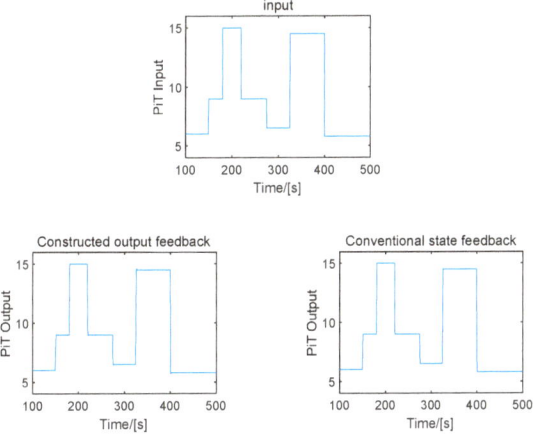

Figure 10. Pressure ratio response curve.

7. Conclusions

We propose a decentralized design scheme for controllers of aero-engines, the SDCS. In SDCS, a group of low-price, high-reliability, but low-performance microprocessors act as the controller of the aero-engine control system. An output feedback scheme based on Q-Learning LQR is constructed, and the VCTs of the SDCS are obtained. The constructed output feedback scheme is equivalent to the optimal state feedback control. A controller parameter tuning algorithm based on the primal-dual principle is introduced, which features simple calculation and can be implemented using low-cost, high-reliability, low-performance microprocessors for real-time calculation, suitable for the hardware characteristics of SDCS. Finally, we conducted simulation for verification. According to the simulation results, the controller parameter tuning algorithm proposed in this paper converges quickly and has high convergence accuracy. The single-variable model takes only a maximum of eight iterations to obtain a high-precision feedback gain matrix, and the multi-variable model takes a maximum of only five iterations. In addition, the constructed output feedback scheme achieves control performance equivalent to the conventional state feedback, which is suitable for the actual engineering application of the aero-engine control system. The simulation results demonstrate the effectiveness of the design scheme presented in this paper.

Author Contributions: Conceptualization, X.J. and J.L.; data curation, J.R.; investigation, X.J. and J.R.; methodology, X.J., J.L. and Y.W.; software, X.J. and J.R.; supervision, J.L. and Y.W.; validation, X.J. and J.R.; writing—original draft, X.J.; writing—review & editing, X.J., J.L., J.R. and Y.W. All authors have read and agreed to the published version of the manuscript.

Funding: This research received no external funding.

Data Availability Statement: The data used to support the findings of this study are included within the article.

Conflicts of Interest: The authors declare that they have no competing financial and non-financial interest.

References

1. Thompson, H.; Fleming, P. Distributed Aero-Engine Control Systems Architecture Selection Using Multi-Objective Optimisation. In Proceedings of the 5th IFAC Workshop on Algorithms & Architecture for Real Time Control (AARTC' 98), Cancun, Mexico, 15–17 April 1998.
2. Skira, C.; Agnello, M. Control Systems for the Next Century's Fighter Engines. *J. Eng. Gas Turbines Power* **1992**, *114*, 749–754. [CrossRef]
3. Culley, D.; Thomas, R.; Saus, J. Concepts for Distributed Engine Control. In Proceedings of the 43rd AIAA/ASME/SAE/ASEE Joint Propulsion Conference & Exhibit, Cincinnati, OH, USA, 8–11 July 2007.
4. Chen, T.; Shan, J. Distributed Tracking of Multiple Under-actuated Lagrangian Systems with Uncertain Parameters and Actuator Faults. In Proceedings of the 2019 American Control Conference (ACC), Philadelphia, PA, USA, 10–12 July 2019.
5. Chen, T.; Shan, J. Distributed Tracking of a Class of Underactuated Lagrangian Systems with Uncertain Parameters and Actuator Faults. *IEEE Trans. Ind. Electron.* **2020**, *67*, 4244–4253. [CrossRef]
6. Chen, T.; Shan, J. Distributed Fixed-time Control of Multi-agent Systems with Input Shaping. In Proceedings of the 2018 IEEE International Conference on Information and Automation (ICIA), Wuyishan, China, 11–13 August 2018.
7. Chen, T.; Shan, J. Distributed Adaptive Attitude Control for Multiple Underactuated Flexible Spacecraft. In Proceedings of the 2018 Annual American Control Conference (ACC), Milwaukee, WI, USA, 27–29 June 2018.
8. Thompson, H.; Benitez-Perez, H.; Lee, D.; Ramos-Hernandez, D.; Fleming, P.; Legge, C. A CANbus-Based Safety-Critical Distributed Aeroengine Control Systems Architecture Demonstrator. *Microprocess. Microsyst.* **1999**, *23*, 345–355. [CrossRef]
9. Chen, L.; Guo, Y. Design of the Distributed Control System Based on CAN Bus. *Comput. Inf. Sci.* **2011**, *4*, 83–89. [CrossRef]
10. Lv, C.; Chang, J.; Bao, W.; Yu, D. Recent Research Progress on Airbreathing Aero-engine Control Algorithm. *Propuls. Power Res.* **2022**, *11*, 1–57. [CrossRef]
11. Xu, Y.; Pan, M.; Huang, J.; Zhou, W.; Qiu, X.; Chen, Y. Estimation-Based and Dropout-Dependent Control Design for Aeroengine Distributed Control System with Packet Dropout. *Int. J. Aerosp. Eng.* **2022**, *2022*, 8658704. [CrossRef]
12. Fan, X.; Pan, M.; Huang, J. Response Time Analysis of TTCAN Bus for Aero-engine Distributed Control System. In Proceedings of the Asia-Pacific International Symposium on Aerospace Technology (APISAT 2015), Cairns, QLD, Australia, 25–27 November 2015.
13. Schley, W. Distributed Flight Control and Propulsion Control Implementation Issues and Lessons Learned. In Proceedings of the ASME 1998 International Gas Turbine and Aeroengine Congress and Exhibition, Stockholm, Sweden, 2–5 June 1998.
14. Zulkifli, N.; Ramli, M. State Feedback Controller Tuning for Liquid Slosh Suppression System Utilizing LQR-LMI Approach. In Proceedings of the 2021 IEEE International Conference on Automatic Control & Intelligent Systems (I2CACIS), Shah Alam, Malaysia, 26–26 June 2021.
15. Zhou, Y.; Xu, G.; Wei, S.; Zhang, X.; Huang, Y. Experiment Study on the Control Method of Motor-Generator Pair System. *IEEE Access* **2017**, *6*, 925–936. [CrossRef]
16. Wu, A.; Qian, Y.; Liu, W.; Sreeram, V. Linear Quadratic Regulation for Discrete-Time Antilinear Systems: An Anti-Riccati Matrix Equation Approach. *J. Frankl. Inst.* **2016**, *353*, 1041–1060. [CrossRef]
17. Bhawal, C.; Qais, I.; Pal, D. Constrained Generalized Continuous Algebraic Riccati Equation (CGCAREs) Are Generically Unsolvable. *IEEE Control Syst. Lett.* **2019**, *3*, 192–197. [CrossRef]
18. Rizvi, S.; Lin, Z. Output Feedback Optimal Tracking Control Using Reinforcement Q-Learning. In Proceedings of the 2018 Annual American Control Conference (ACC), Milwaukee, WI, USA, 27–29 June 2018.
19. Lewis, F.; Vamvoudakis, K. Reinforcement Learning for Partially Observable Dynamic Processes: Adaptive Dynamic Programming Using Measured Output Data. *IEEE Trans. Syst. Man Cybern. Part B* **2011**, *41*, 14–25. [CrossRef] [PubMed]
20. Salhi, S.; Salhi, S. LQR Control of a Grid Side Converter of a DFIG Based WECS: LMI Approach Based on Lyapunov Condition. In Proceedings of the 2019 16th International Multi-Conference on Systems, Signals & Devices (SSD), Istanbul, Turkey, 21–24 March 2019.
21. Nguyen, T.; Gajic, Z. Solving the Matrix Differential Riccati Equation: A Lyapunov Equation Approach. *IEEE Trans. Autom. Control* **2010**, *55*, 191–194.
22. Bi, H.; Chen, D. The Estimation of the Solutions Matrix of the Perturbed Discrete Time Algebraic Riccati Equation. In Proceedings of the 10th World Congress on Intelligent Control and Automation, Beijing, China, 6–8 July 2012.

23. Vargas, F.; Gonzalez, R. On the Existence of a Stabilizing Solution of Modified Algebraic Riccati Equations in Terms of Standard Algebraic Riccati Equations and Linear Matrix Inequalities. *IEEE Control Syst. Lett.* **2020**, *4*, 91–96. [CrossRef]
24. Chen, M.; Zhang, L.; Su, H.; Chen, G. Stabilizing Solution and Parameter Dependence of Modified Algebraic Riccati Equation with Application to Discrete-Time Network Synchronization. *IEEE Trans. Autom. Control* **2016**, *61*, 228–233. [CrossRef]
25. Chapman, J.; Litt, J. Control Design for an Advanced Geared Turbofan Engine. In Proceedings of the 53rd AIAA/SAE/ASEE Joint Propulsion Conference, Atlanta, GA, USA, 10–12 July 2017.
26. Mawid, M.; Park, T.; Sekar, B.; Arana, C. Application of Pulse Detonation Combustion to Turbofan Engines. *J. Eng. Gas Turbines Power* **2003**, *125*, 270–283. [CrossRef]
27. Pavlenko, D.; Dvirnyk, Y.; Przysowa, R. Advanced Materials and Technologies for Compressor Blades of Small Turbofan Engine. *Aerospace* **2021**, *8*, 1. [CrossRef]
28. Gharoun, H.; Keramati, A.; Nasiri, M. An Integrated Approach for Aircraft Turbofan Engine Fault Detection Based on Data Mining Techniques. *Expert Syst.* **2019**, *36*, e12370. [CrossRef]
29. Li, J. Turbofan Engine H_∞ Output Feedback Control and Delay Compensation Strategy. Master's Thesis, Dalian University of Technology, Dalian, China, 2021.
30. Belapurkar, R. Stability and Performance of Propulsion Control Systems with Distributed Control Architectures and Failures. Ph.D. Thesis, The Ohio State University, Columbus, OH, USA, 2012.
31. Balakrishnan, V.; Vandenberghe, L. Semidefinite Programming Duality and Linear Time-Invariant Systems. *IEEE Trans. Autom. Control* **2003**, *48*, 30–41. [CrossRef]
32. Gattami, A. Generalized Linear Quadratic Control. *IEEE Trans. Autom. Control* **2010**, *55*, 131–136. [CrossRef]
33. Lee, D.; Hu, J.H. Primal-Dual Q-Learning Framework for LQR Design. *IEEE Trans. Autom. Control* **2019**, *64*, 3756–3763. [CrossRef]

Disclaimer/Publisher's Note: The statements, opinions and data contained in all publications are solely those of the individual author(s) and contributor(s) and not of MDPI and/or the editor(s). MDPI and/or the editor(s) disclaim responsibility for any injury to people or property resulting from any ideas, methods, instructions or products referred to in the content.

Article

Trajectory Tracking and Adaptive Fuzzy Vibration Control of Multilink Space Manipulators with Experimental Validation

Chenlu Feng, Weidong Chen, Minqiang Shao * and Shihao Ni

State Key Laboratory of Mechanics and Control for Aerospace Structures, Nanjing University of Aeronautics and Astronautics, Nanjing 210016, China
* Correspondence: m.q.shao@nuaa.edu.cn

Abstract: This paper investigates the problem of modeling and controlling a space manipulator system with flexible joints and links. The dynamic model of the flexible manipulator system is derived by using the Lagrange equation and the floating frame of reference formulation, where the assumed mode method is adopted to discretize flexible links, while the flexible joints are regarded as linear torsion springs. The natural characteristics of a single flexible link manipulator, under three different boundary conditions, are compared to reveal the effect of the flexibility of joints on the manipulator system and to choose suitable assumed modes. Furthermore, singular perturbation theory is introduced to decompose the system into a slow subsystem that describes the rigid-body motion, and a fast subsystem that describes the elastic vibration. Since the system is underactuated, a compound control strategy, which consists of the underactuated computed torque controller and the adaptive fuzzy controller, is presented to improve the accuracy of the trajectory tracking of the flexible joints and to suppress the elastic vibration of the flexible links, in the meantime. Both numerical simulation and experimentation are performed to verify the effectiveness of the proposed compound controller, and a comparison with the proportional-derivative (PD) controller is provided to highlight its superiority in suppressing the residual vibration of the tip.

Keywords: flexible space manipulator; flexible joint; singular perturbation theory; underactuated computed torque controller; adaptive fuzzy controller

Citation: Feng, C.; Chen, W.; Shao, M.; Ni, S. Trajectory Tracking and Adaptive Fuzzy Vibration Control of Multilink Space Manipulators with Experimental Validation. *Actuators* **2023**, *12*, 138. https://doi.org/10.3390/act12040138

Academic Editor: André Preumont

Received: 9 February 2023
Revised: 9 March 2023
Accepted: 21 March 2023
Published: 25 March 2023

Copyright: © 2023 by the authors. Licensee MDPI, Basel, Switzerland. This article is an open access article distributed under the terms and conditions of the Creative Commons Attribution (CC BY) license (https://creativecommons.org/licenses/by/4.0/).

1. Introduction

Space manipulators play an important role in on-orbit activities, such as construction, inspection, and transportation [1,2]. With the development of space technology, lighter and larger space manipulators are increasingly applied because of their advantages of being lightweight, having low energy consumption, quick responses [3]. However, there exist obvious flexible characteristics in the space manipulator. They are mainly caused by the structural flexibility of the links and the flexibility of the joints with harmonic gear reducers. The elastic vibrations generated by these two kinds of flexibility are highly coupled, which complicates the dynamic characteristics of the space manipulator system [4] and puts forward greater requirements for controlling. In addition, the coupling between the free-floating base and the manipulator brings challenges to the control of the manipulator [5]. Structural vibrations will be obvious when the flexible manipulators perform on-orbit missions, especially for large space flexible manipulators, which may lead to a catastrophic failure. On the other hand, flexible manipulators usually have high dimensional orders, low damping ratios, and parameter uncertainties in dynamics. Therefore, it is necessary and challenging work to investigate the control strategy of the space flexible manipulator.

The focus of controlling a flexible manipulator is to track the desired trajectories and suppress the vibration of the flexible parts. The dynamics and control of the manipulators have been studied for a long time. Based on differences in modeling, control, and experimental studies, Dwivedy and Eberhard [6] summarized the original works in the field of the dynamic performance of flexible robots. In recent years, many researchers have extensively studied the problem of modeling and planning and paid more attention to the controlling of the flexible space manipulator, which has been applied to different robot platforms [7–9]. At present, advanced intelligent materials have been applied in the research of some active control methods [10–12].

A high-precision mathematical model is crucial for controller design. However, a high-precision mathematical model is often difficult to be obtained due to the uncertainty and error in the model. Yang [13] creatively introduced a set of filtered error variables and asymptotic filters as well as an auxiliary system, while two novel continuous integral robust control algorithms have been synthesized, via an improved backstepping framework, for a class of high-order systems suffering from both matched and mismatched disturbances. The validation of the proposed controller is performed on a single-link rigid manipulator and a two-link rigid manipulator, respectively. Shawky [14] used a nonlinear controller via the state-dependent Riccati equation (SDRE) to compensate for the uncertainties of the single-link flexible manipulator system with a rigid joint. The simulation results verified the effectiveness of the SDRE controller. De Luca [15] considered two model classes: robots with elastic joints and rigid links, and robots with flexible links and rigid joints. In view of the small deformation, the elasticity of the elastic joint was modeled as a linear spring. Then, model-based feedforward laws were derived for two basic motion tasks, although the generalization of the control scheme to a multilink flexible arm was a problem. The intelligent control technology, which does not depend on models, has paid more attention to the suppression of the vibration of the flexible links. Qiu et al. [16] introduced a hybrid control strategy of optimal trajectory planning and diagonal recurrent neural network (DRNN) control to suppress the vibration of a single-link flexible manipulator with a rigid joint, both during and after the point-to-point motion. Experimental results demonstrated that planning an optimal trajectory could cause fewer vibrations and that the DRNN controller was superior to the classical PD controller on vibration suppression. Malzahn et al. [17] presented a conjunction of a model-free independent joint control strategy for vibration damping with a neural-network-based payload estimation and an inverse kinematics model based on multilayer perceptron (MLP) networks for a multilink-flexible robot arm under gravity, the control architecture enabled the robot arm to catch multiple balls sequentially thrown by a human. Neural networks have been used in controlling flexible manipulators because of their strong nonlinear fitting ability, yet training requires a lot of available training data, and the training time is often long. Cao et al. [18] developed a fuzzy self-tuning proportional-integral-derivative (PID) controller and applied it to a two-link flexible manipulator featuring a piezoelectric ceramics (PZT) actuator, and the experimental results showed that the controller could effectively suppress the vibration. Owing to the existence of nonlinear factors and parameter uncertainties, model-based control methods cannot maintain the required accuracy. Wei et al. [19] designed model-free fuzzy logic control laws to suppress the vibrations of the single-link flexible piezoelectric manipulator. Experimental results showed that the adopted fuzzy control algorithms could substantially suppress the larger amplitude vibrations. Qiu et al. [20] utilized a Takagi–Sugeno model-based fuzzy neural network control (TS-FNN) scheme to suppress the residual vibrations of the two-link flexible manipulator with rigid joints. Experimental results demonstrated that the designed controller could reduce the residual vibrations quicker than the traditional linear PD controller. Tracking the desired trajectory of a flexible joint is the focus of controlling flexible manipulators. In order to achieve high-precision tracking of the revolving angles and vibration suppression of the elastic part, Zhang et al. [21] developed an adaptive iterative learning control (AILC) law for a two-link rigid–flexible coupled manipulator system with rigid joints in a three-dimensional (3D) space. The computed torque method (CTM) has been maturely applied to the tracking of the joint. Mehrzad et al. [22] designed a modified CTM to control the manipulator motion. Using numerical simulations, the performance of the proposed control system was evaluated for trajectory tracking. In addition, the assumed modes method was mostly used for modeling flexible links in the aforementioned references.

At present, most research paid more attention to rigid–flexible coupled space manipulator systems with a rigid joint. However, the flexibilities of the joints and links exist objectively only in practice. Thus, there is limited research on the types of assumed modes when using the assumed modes method to model flexible manipulators. Most controllers have high requirements for model accuracy. In practice, frictions, interstices, and impacts in gear transmission of manipulator systems are hard to model, and accurate mathematical models are hard to be obtained. In addition, according to the singular perturbation model of the space manipulator system, the fast-varying subsystem featured by the flexible vibration has the form of a linear equation. Although many controllers exist for linear systems, such as the linear quadratic regulator (LQR) controllers [23], state feedback control, etc., their performance greatly relies on the accuracy of the mathematical model of the controlled system. Therefore, it is necessary to investigate model-independent controllers. A fuzzy control does not require an accurate mathematical model and has the characteristic of good anti-interference. Moreover, an adaptive fuzzy controller (AFC) combines the advantages of traditional fuzzy control and an adaptive learning algorithm and, thus, can cope with the modeling error and the external disturbance excitation in the motion. Hence, it is a more desirable choice to design the AFC for the reduced flexible vibration of the system.

The objectives of this paper are to improve the accuracy of trajectory tracking for the flexible joints and suppress the elastic vibrations of the flexible links by using a compound control strategy that consists of the underactuated computed torque controller and the adaptive fuzzy controller. Simulations and experiments, then, confirmed the effectiveness of the proposed controller. A conclusion was drawn based on the reported results. The main contributions of this paper are as follows:

1. A mathematical model of a multilink flexible space manipulator system with flexible joints and links was established.
2. The dynamic responses and natural characteristics of the flexible manipulator under three different kinds of mode shapes are compared.
3. Based on the underactuated characteristic of the system, the underactuated CTM is designed to achieve high-precision performance on flexible joint trajectory tracking, and the non-model adaptive fuzzy controller is adopted to suppress the elastic vibrations of the flexible links.

2. Dynamics of Flexible Space Manipulator System

2.1. Mathematical Model of Flexible Space Manipulator System

A common space flexible manipulator system is shown in Figure 1. It consists of flexible links with a uniform cross-sectional area, flexible joints, and a free-floating base. A pair of PZT actuators are attached to the root of each flexible link [19] and a tip payload is attached to the distal end of the last flexible link. The entire system rotates in the horizontal plane, driven by electric motors. The kinematic and dynamic symbols of the space manipulator system used in this paper are listed in Table 1. Unless otherwise specified, all reference frame systems are inertial frames. In Figure 1, O_0 is the position of the center of mass (CM) of the free-floating base, O_i $(i = 1, \ldots, n)$ is the position of the ith joint's CM, $\mathbf{p}_i \in \mathbf{R}^2$ $(i = 1, \ldots, n)$ is the position vector of the ith joint's CM in the inertial frame Σ_d, $\mathbf{r}'_i \in \mathbf{R}^2$ $(i = 1, \ldots, n)$ is the position vector of a point P on the ith link in the frame $\hat{\Sigma}_i$, and $\mathbf{r}_i \in \mathbf{R}^2$ $(i = 1, \ldots, n)$ is the position vector of a point on the ith link in the inertial frame Σ_d, respectively.

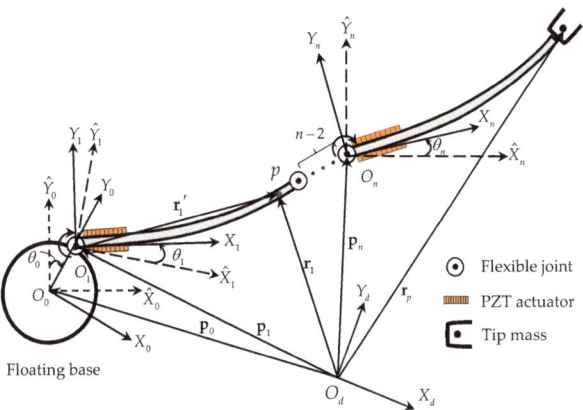

Figure 1. Schematic of a flexible manipulator system.

Table 1. Kinematic and dynamic symbols of the space manipulator system.

Symbol	Representation
$J_0, J_i (i = 1, \ldots, n)$	Moment of inertia of the base, the ith rotor
$m_0, m_i (i = 1, \ldots, n)$	mass of the base, the ith rotor
$l_0, l_1 (i = 1, \ldots, n)$	length of the base, the ith link
$\rho_i (i = 1, \ldots, n)$	The linear density of the ith link
$I_i (i = 1, \ldots, n)$	The moment of inertia of the ith link
$E_i (i = 1, \ldots, n)$	The elastic modulus of the ith link
m_p	The mass of the tip payload
J_p	Moment of inertia of tip payload
$k_i (i = 1, \ldots, n)$	The ith spring rate coefficient
$\alpha_i (i = 1, \ldots, n)$	The theoretical rotation angle of the ith rotor
$\theta_i (i = 1, \ldots, n)$	The actual rotation angle of the ith joint
$\tau_i (i = 1, \ldots, n)$	The theoretical torque of the ith joint
$\sigma_i (i = 1, \ldots, n)$	The elastic deformation of the ith joint
$\Sigma_0, \Sigma_i (i = 1, \ldots, n)$	Base, link frame system
$\hat{\Sigma}_i (i = 1, \ldots, n)$	Joint frame system
Σ_d	Inertial frame system

The flexibility of the space manipulator comes from the flexibility of the joints and the structural flexibility of the links. In practice, the joint deformations are small, and, thus, the elasticity in the joints can be modeled as a spring [15]. All electric motors were assumed as uniform rotors with their centers of mass on the rotation axes [24]. Figure 2 shows the revolute joint model established in this paper. According to geometry, one has:

$$\sigma_i = \alpha_i - \theta_i. \tag{1}$$

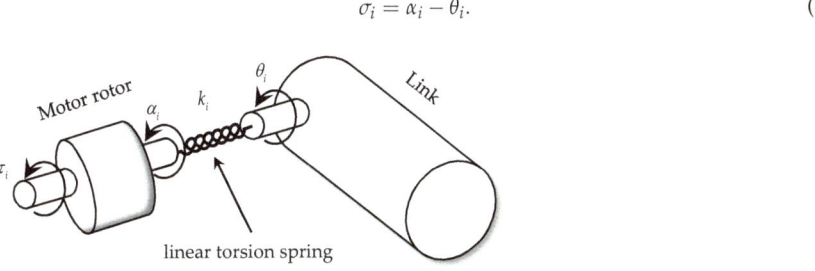

Figure 2. Schematic of the joint model.

Thus, by using the assumed mode method and the assumption of small deformations, the *ith* link's bending deformation can be expressed in terms of *m* mode shapes as:

$$u_i(x,t) = \sum_{s=1}^{m} q_i^s(t)\varphi_s^i(x), \tag{2}$$

where φ_s^i and q_i^s denote the *ith* modal shape and the corresponding generalized coordinate, respectively.

Thus, the position vector $\mathbf{r}'_i \in \mathbf{R}^2$ is:

$$\mathbf{r}'_i = \begin{bmatrix} x \\ u_i(x,t) \end{bmatrix}. \tag{3}$$

The torque of the *jth* PZT attached to the *ith* link can be expressed as $M_i^j = K_a U_j$, where K_a is a constant related to the natural characteristics of piezoelectric ceramics.

By using the floating frame of the reference formulation, the motion of the flexible manipulator system was regarded as the superposition of the large-scale rigid body motion and the deformation of flexible links. Hence, the manipulator system comprises $n+1$ bodies and n hinges. The three kinds of frames are listed in Table 1.

In order to derive the mathematical model of the manipulator system, the matrix $\mathbf{S}_i$ of converting the floating frame to the inertial frame should be given. In Figure 3, the angle transformed from frame Σ_i to frame $\hat{\Sigma}_i$ is the sum of the flexible rotation angle $\phi_{i-1}|_{x=l_{i-1}}$ of $(i-1)th$ link's end and the *ith* joint's rotation angle θ_i. Thus, the matrix of converting the floating frame to the inertial frame is:

$$\mathbf{S}_i = \mathbf{S}_{i-1}\mathbf{E}_i\mathbf{A}_i = \hat{\mathbf{S}}_{i-1}\mathbf{A}_i, \quad \hat{\mathbf{S}}_0 = \mathbf{I}_{2\times 2}, \tag{4}$$

where $\mathbf{A}_i = \begin{bmatrix} \cos(\theta_i) & -\sin(\theta_i) \\ \sin(\theta_i) & \cos(\theta_i) \end{bmatrix}$, $\mathbf{E}_i = \begin{bmatrix} \cos(\phi_{i-1}) & -\sin(\phi_{i-1}) \\ \sin(\phi_{i-1}) & \cos(\phi_{i-1}) \end{bmatrix}\bigg|_{x=l_{i-1}}$, $\mathbf{E}_1 = \mathbf{I}_{2\times 2}$.

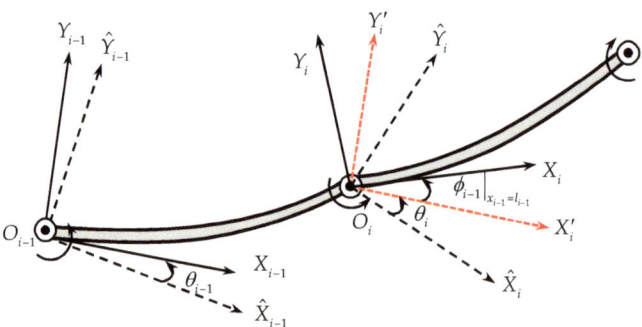

Figure 3. Schematic of two adjacent links in a manipulator system.

Based on the assumption of a small deformation, the matrix $\mathbf{E}_i$ can be simplified to:

$$\mathbf{E}_i = \begin{bmatrix} 1 & -\phi_{i-1} \\ \phi_{i-1} & 1 \end{bmatrix}\bigg|_{x=l_{i-1}}. \tag{5}$$

Therefore, the position vector $\mathbf{r}_i(x)$ in the inertial frame can be written as:

$$\mathbf{r}_i(x) = \mathbf{S}_i\mathbf{r}'_i + \mathbf{p}_i, \mathbf{p}_{i+1} = \mathbf{p}_i + \mathbf{S}_i\mathbf{r}'_i(l) = \mathbf{r}_i(l). \tag{6}$$

The velocity of $\mathbf{r}_i(x)$ on the *ith* flexible link is:

$$\dot{\mathbf{r}}_i(x) = \dot{\mathbf{S}}_i \mathbf{r}'_i + \mathbf{S}_i \dot{\mathbf{r}}'_i + \dot{\mathbf{p}}_i, \tag{7}$$

where $\dot{\mathbf{S}}_i = \dot{\hat{\mathbf{S}}}_{i-1} \mathbf{A}_i + \hat{\mathbf{S}}_{i-1} \dot{\mathbf{A}}_i$, $\dot{\hat{\mathbf{S}}}_i = \dot{\mathbf{S}}_i \mathbf{E}_{i+1} + \mathbf{S}_i \dot{\mathbf{E}}_{i+1}$, $\dot{\mathbf{A}}_i = \mathbf{B}\mathbf{A}_i \dot{\theta}_i$, $\dot{\mathbf{E}}_i = \mathbf{B}\dot{\phi}_i\big|_{x=l}$, $\mathbf{B} = \begin{bmatrix} 0 & -1 \\ 1 & 0 \end{bmatrix}$.

To derive the equation of motion for the space manipulator, which consists of lumped parameter parts (the joint rotations) and distributed parameter parts (the link deformations), by using the Lagrange formulation, the kinetic energy T and the potential energy V of the system must be computed. The kinetic energy T receives contributions from the base, links, joints, and tip payload and is given by:

$$T = T_0 + T_l + T_\alpha + T_p, \tag{8}$$

where

$$T_0 = \frac{1}{2}(m_0 \dot{\mathbf{r}}_0^T \dot{\mathbf{r}}_0 + J_0 \dot{\varphi}_0^2), \tag{9}$$

$$T_l = \frac{1}{2}\sum_{i=1}^{n} \int_0^{l_i} \rho_i \dot{\mathbf{r}}_i^T \dot{\mathbf{r}}_i dx, \tag{10}$$

$$T_\alpha = \frac{1}{2}\sum_{i=1}^{n} (J_i \dot{\alpha}_i^2 + m_i \dot{\mathbf{r}}_i^T(0)\dot{\mathbf{r}}_i(0)), \tag{11}$$

$$T_P = \frac{1}{2} m_P \dot{\mathbf{r}}_n^T(l)\dot{\mathbf{r}}_n(l) + \frac{1}{2}J_p (\dot{\varphi}_{ne})^2. \tag{12}$$

In Equation (12), $\dot{\varphi}_{ne} = \sum_{j=1}^{n} \dot{\theta}_j + \sum_{k=1}^{n} \dot{\phi}_k\big|_{x=l_k}$.

The links of the space manipulator are modeled as Euler–Bernoulli beams. Since the space manipulator system is in a weightless environment, the effect of gravity can be ignored. Therefore, the potential energy V is only contributed to by the links and joints, i.e.,

$$V = V_l + V_\alpha, \tag{13}$$

where

$$V_\alpha = \frac{1}{2}\sum_{i=1}^{n} k_i (\alpha_i - \theta_i)^2, \tag{14}$$

$$V_l = \frac{1}{2}\sum_{i=1}^{n} \int_0^{l_i} \left[EI_i \left(\frac{\partial \phi_i}{\partial x}\right)^2 \right] dx. \tag{15}$$

The generalized Lagrange Equation [14] of the second kind is:

$$\frac{d}{dt}\left(\frac{\partial T}{\partial \dot{q}_j}\right) - \frac{\partial T}{\partial q_j} + \frac{\partial V}{\partial q_j} = f_j. \tag{16}$$

The system dynamic equations can be obtained by substituting Equations (8) and (13) into Equation (16) and is given by:

$$\begin{cases} \mathbf{J}\ddot{\alpha} + \mathbf{K}_\sigma(\alpha - \theta_{sub}) = \tau_{n \times 1}, \\ \mathbf{M}(\theta, \mathbf{q}) \begin{bmatrix} \ddot{\mathbf{x}}_0 \\ \ddot{\theta} \\ \ddot{\mathbf{q}} \end{bmatrix} + \mathbf{F}(\theta, \dot{\theta}, \mathbf{q}, \dot{\mathbf{q}}) + \begin{bmatrix} \mathbf{0}_{3 \times 1} \\ -\mathbf{K}_\sigma(\alpha - \theta_{sub}) \\ \mathbf{K}_q \mathbf{q} \end{bmatrix} = \begin{bmatrix} \mathbf{0}_{(n+3) \times 1} \\ \tau'_{mn \times 1} \end{bmatrix}, \end{cases} \tag{17}$$

where $\boldsymbol{\theta} = [\theta_0, \theta_1, \cdots, \theta_n]^T$, $\boldsymbol{\alpha} = [\alpha_1, \alpha_2, \cdots, \alpha_n]^T$, $\boldsymbol{\tau}_{n\times1} = [\tau_1, \tau_2, \cdots, \tau_n]^T$ is the torque vector of the joint motor. Moreover, $\mathbf{q} = [q_1^1, q_1^2, \cdots, q_1^m, \cdots, q_i^1, q_i^2, \ldots, q_i^m, \cdots q_n^m]^T$ is a vector consisting of flexible links modal coordinates, $\boldsymbol{\theta}_{sub} = [\theta_1, \cdots, \theta_n]^T$, $\mathbf{x}_0 = [x_0, y_0]^T$ is the position vector of the free-floating base, $\mathbf{M}(\boldsymbol{\theta}, \mathbf{q})$ is the symmetric inertial matrix, $\mathbf{F}(\boldsymbol{\theta}, \dot{\boldsymbol{\theta}}, \mathbf{q}, \dot{\mathbf{q}})$ is the coupled term characterizing the interactions between centrifugal force and Coriolis force, $\mathbf{J} = diag(J_1, \ldots, J_n)$, $\mathbf{K}_\sigma = diag(k_1, \ldots, k_n)$, and $\mathbf{K}_q$ are the joint rotors mass matrix, joints stiffness matrix, and links stiffness matrix, respectively. $\mathbf{K}_q$ is expressed as:

$$\mathbf{K}_q = \begin{bmatrix} \int_0^{l_1} EI(\frac{d^2\Phi^1}{dx^2}) \times (\frac{d^2\Phi^1}{dx^2})^T dx & & \\ & \ddots & \\ & & \int_0^{l_n} EI(\frac{d^2\Phi^n}{dx^2}) \times (\frac{d^2\Phi^n}{dx^2})^T dx \end{bmatrix}, \quad (18)$$

where $\boldsymbol{\Phi}^i = [\varphi_1^i, \ldots, \varphi_m^i]^T$ is a vector consisting of the *ith* link modal shape functions and $\boldsymbol{\tau}'_{mn\times1}$ is the torque vector of PZT. The generalized force of the *jth* PZT attached to the *ith* link can be expressed as:

$$\tau_i^j = \int_{b_{ij}}^{b_{ij}+s_{ij}} \sum_{v=1}^m M_i^j \frac{d\varphi_v^i(x)}{dx} = \sum_{v=1}^m M_j^i (\frac{d\varphi_v^i(b_{ij}+s_{ij})}{dx} - \frac{d\varphi_v^i(b_{ij})}{dx}), \quad (19)$$

where b_{ij} and s_{ij} are the start- and endpoints of the *jth* PZT attached to the *ith* link, respectively.

The torque vector $\boldsymbol{\tau}'_{mn\times1}$ can be expressed as:

$$\boldsymbol{\tau}'_{mn\times1} = \mathbf{P}_{nm\times nm} \times \mathbf{M}_{nm\times1} = \begin{bmatrix} \mathbf{P}_1 & & \\ & \ddots & \\ & & \mathbf{P}_n \end{bmatrix} \times \begin{bmatrix} \mathbf{M}_1 \\ \vdots \\ \mathbf{M}_n \end{bmatrix}, \quad (20)$$

where $\mathbf{M}_i = [M_1 \cdots M_m]^T$, the matrix $\mathbf{P}_i$ can be expressed as:

$$\mathbf{P}_i = \begin{bmatrix} \frac{d\varphi_1^i(b_{i1}+s_{i1})}{dx} - \frac{d\varphi_1^i(b_{i1})}{dx} & \cdots & \frac{d\varphi_1^i(b_{im}+s_{im})}{dx} - \frac{d\varphi_1^i(b_{im})}{dx} \\ \vdots & \ddots & \vdots \\ \frac{d\varphi_m^i(b_{i1}+s_{i1})}{dx} - \frac{d\varphi_m^i(b_{i1})}{dx} & \cdots & \frac{d\varphi_m^i(b_{im}+s_{im})}{dx} - \frac{d\varphi_m^i(b_{im})}{dx} \end{bmatrix}. \quad (21)$$

Substituting $M_i^j = K_a U_j$ into Equation (20) yields:

$$\boldsymbol{\tau}'_{mn\times1} = \mathbf{P} \times \begin{bmatrix} \mathbf{M}_1 \\ \vdots \\ \mathbf{M}_n \end{bmatrix} = \mathbf{P} \times \begin{bmatrix} \mathbf{K}_1 & & \\ & \ddots & \\ & & \mathbf{K}_n \end{bmatrix} \times \begin{bmatrix} \mathbf{U}_1 \\ \vdots \\ \mathbf{U}_n \end{bmatrix} = \mathbf{K}_v \mathbf{U}, \quad (22)$$

where $\mathbf{K}_1 = \begin{bmatrix} K_a & & \\ & \ddots & \\ & & K_a \end{bmatrix}_{m\times m}$.

The first and the final equations of the dynamic model (17) are referred to as the motor and link equations, respectively.

2.2. Natural Characteristics of the Flexible Links under Different Boundary Conditions

Employing the assumed mode method, one can easily obtain the approximated link deformation by using only a finite number of modes. However, the model accuracy is not only affected by the number of modes but also by the kind of selected modes. Many researchers have investigated the errors introduced by modal truncation, and the cantilever beam mode and the simply supported beam mode are commonly used [25,26]. In fact, the boundary conditions of the flexible links of the space manipulators are different from those of cantilever beams and simply supported beams; thus, using either one of the two modes may introduce errors. Considering the actual boundary condition of the space manipulator, the natural characteristics of the flexible links, under three different boundary conditions, are compared with each other to investigate the influence of the flexibility of the joints. The three boundary conditions are:

1. Fixed-free boundary condition, which is a single cantilever beam (SCB) boundary condition.
2. Fixed-inertial load boundary condition, which is a single rigid joint and flexible link manipulator (SRF) boundary condition.
3. Elastic load-inertial load boundary condition, which is a single flexible joint and flexible link manipulator (SFF) boundary condition.

As mentioned previously, the link is modeled as the Euler–Bernoulli beam. The free vibration differential equation of the Euler–Bernoulli beam [10] is:

$$\rho \frac{\partial^2 u}{\partial t^2} + EI \frac{\partial^4 u}{\partial x^4} = 0, \tag{23}$$

where ρ is mass per unit length and EI is the bending stiffness.

The dimensionless parameters are defined as:

$$\xi = \frac{x}{l}, \Gamma_p = \frac{J_p}{\rho l^3}, M_p = \frac{m_p}{\rho l}, s = \left(\frac{\rho l^4 \omega^2}{EI}\right)^{\frac{1}{4}}. \tag{24}$$

The ratio of the stiffness of the flexible joint to the bending stiffness of the flexible link is defined as:

$$k_m = \frac{kl}{EI}. \tag{25}$$

The general solution of Equation (23) can be expressed as:

$$u(x,t) = \varphi(x)\sin(\omega t) = l\varphi(\xi)\sin(\omega t). \tag{26}$$

Substituting Equation (26) into Equation (23) yields a dimensionless expression:

$$\varphi(\xi)^{(4)} - s^4\varphi(\xi) = 0. \tag{27}$$

The general solution of Equation (27) can be expressed as:

$$\varphi(\xi) = A_1 \cdot \cos(s \cdot \xi) + A_2 \cdot \sin(s \cdot \xi) + A_3 \cdot \cosh(s \cdot \xi) + A_4 \cdot \sinh(s \cdot \xi). \tag{28}$$

The dimensionless boundary condition formulas corresponding to the three boundary conditions are:

1. SCB boundary condition:

$$\varphi(0) = 0, \left.\frac{\partial \varphi(\xi)}{\partial \xi}\right|_{\xi=0} = 0, \left.\frac{\partial^2 \varphi}{\partial \xi^2}\right|_{\xi=l} = 0, \left.\frac{\partial^3 \varphi}{\partial \xi^3}\right|_{\xi=l} = 0. \tag{29}$$

2. SRF boundary condition:

$$\varphi(0) = 0, \left.\frac{\partial \varphi(\xi)}{\partial \xi}\right|_{\xi=0} = 0, \left.\frac{\partial^2 \varphi}{\partial \xi^2}\right|_{\xi=1} - s^4 \Gamma_p \left(\left.\frac{\partial \varphi}{\partial \xi}\right|_{\xi=1}\right) = 0, \left.\frac{\partial^3 \varphi}{\partial \xi^3}\right|_{\xi=1} + s^4 M_p \frac{d^2}{dt^2}(\varphi|_{\xi=1}) = 0. \tag{30}$$

3. SFF boundary condition:

$$\begin{aligned}
&\varphi(0) = 0, \\
&\left.\frac{\partial^2 \varphi(\xi)}{\partial \xi^2}\right|_{\xi=0} - k_m \left.\frac{\partial \varphi(\xi)}{\partial \xi}\right|_{\xi=0} = 0, \\
&\left.\frac{\partial^2 \varphi}{\partial \xi^2}\right|_{\xi=1} - s^4 \Gamma_p \left(\left.\frac{\partial \varphi}{\partial \xi}\right|_{\xi=1}\right) = 0, \\
&\left.\frac{\partial^3 \varphi}{\partial \xi^3}\right|_{\xi=1} + s^4 M_p \frac{d^2}{dt^2}(\varphi|_{\xi=1}) = 0.
\end{aligned} \tag{31}$$

Substituting the general solution (28) into the aforementioned boundary condition formulas, respectively, provides the three corresponding frequency equations:

1. SCB boundary condition:

$$\cosh(s)^2 + 2\cosh(s)\cos(s) - \sinh(s)^2 + \sin(s)^2 + \cos(s)^2 = 0. \tag{32}$$

2. SRF boundary condition:

$$\begin{aligned}
&-2s\sinh(s)\left(s^2\Gamma_p - M_p\right)\cos(s) + 2s^4\Gamma_p M_p + 2 \\
&+ \left(\left(-2s^4\Gamma_p M_p + 2\right)\cos(s) - 2s\sin(s)\left(s^2\Gamma_p + M_p\right)\right)\cosh(s) = 0
\end{aligned} \tag{33}$$

3. SFF boundary condition:

$$\begin{aligned}
&2s\sinh(s)\left(s^4\Gamma_p M_p + s^2\Gamma_p k_m - M_p k_m - 1\right)\cos(s) \\
&-2s^4\Gamma_p M_p k_m + 4\sinh(s)\sin(s)s^2 M_p - 2k_m \\
&+\left[\left(\left(2\Gamma_p M_p k_m + 4\Gamma_p\right)s^4 - 2k_m\right)\cos(s) - \right. \\
&\left. 2s\sin(s)\left(s^4\Gamma_p M_p - s^2\Gamma_p k_m - M_p k_m - 1\right)\right]\cosh(s) = 0
\end{aligned} \tag{34}$$

The parameters M_p, k_m, and Γ_p directly affect the natural characteristics of the manipulator system, thus, the effect of the parameters on the natural characteristics is worth studying in detail. According to the obtained frequency Equations (32)–(34), the relationship curve between the dimensionless parameters and the dimensionless natural frequencies can be drawn.

In many missions, space manipulators only need to grasp light objects, in which case the dead-weight load ratio M_p is small. Setting $M_p = 0.1$, the 3D surfaces of the first two dimensionless frequencies s_1 and s_2 versus the stiffness ratio k_m and the moment of inertia Γ_p are shown in Figures 4 and 5, respectively. The dimensionless frequency surfaces of SRF are below the dimensionless frequency surfaces of SFF, as Γ_p increases, the first-order dimensionless frequency surface of SRF approaches the first-order dimensionless frequency surface of SFF, while the second-order dimensionless frequency surfaces are far away from each other, as shown in Figures 4 and 5, respectively. With the increase in the stiffness ratio k_m, the first two dimensionless frequency surfaces of SRF or SFF gradually move away from those of SCB, respectively, while the first two dimensionless frequency surfaces of SFF gradually approach the surfaces of SRF. It can be concluded that Γ_p and k_m have different influences on each order of the dimensionless frequency surface.

Grabbing and releasing payloads are important for space manipulators to perform on-orbit missions. Therefore, it is necessary to analyze the effect of the load ratio M_p on the natural characteristics of the flexible manipulators. When $k_m = 0.5$, the 3D surfaces of the first two dimensionless frequencies s_1 and s_2 versus M_p and Γ_p are shown in Figures 6 and 7, respectively. Following the increase of the parameter M_p, the first-order dimensionless frequency surface of SFF is far away from that of SRF, while the second-order dimensionless frequency surfaces approach each other, as shown in Figures 6 and 7, respectively. Therefore, an increase in both M_p and Γ_p can reduce the dimensionless frequencies of SRF and SFF, although the effect of M_p and Γ_p on the differences in the dimensionless frequencies of the same order of SRF and SFF is the opposite.

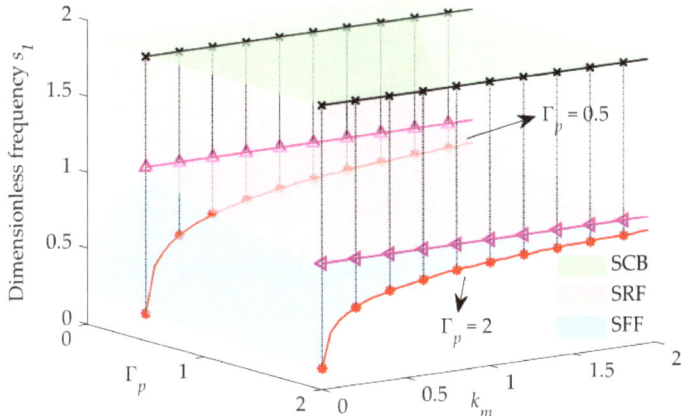

Figure 4. 3D surfaces of the first frequency versus k_m and Γ_p ($M_p = 0.1$).

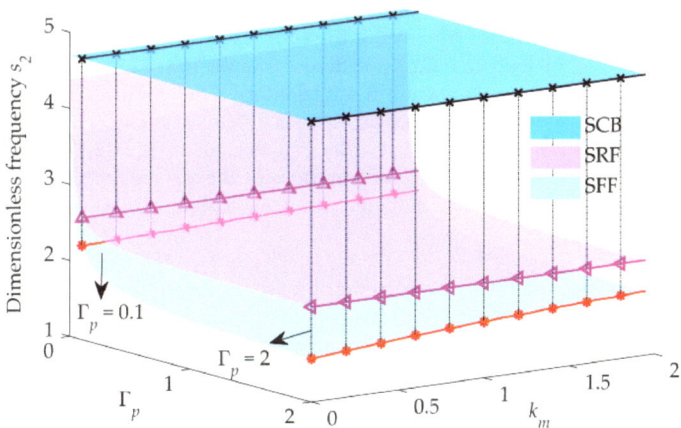

Figure 5. 3D surfaces of the second frequency versus k_m and Γ_p ($M_p = 0.1$).

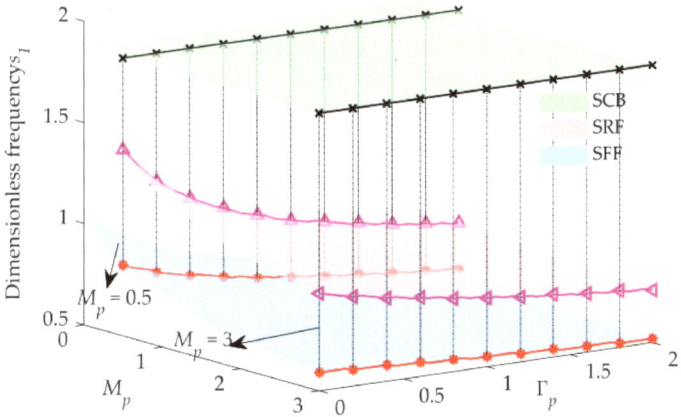

Figure 6. 3D surfaces of the first frequency versus M_p and Γ_p ($k_m = 0.5$).

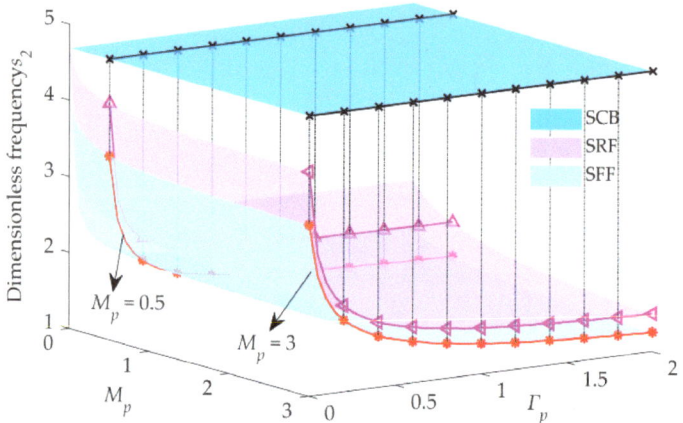

Figure 7. 3D surfaces of the second frequency versus M_p and Γ_p ($k_m = 0.5$).

Therefore, due to joint flexibility, the frequencies of SFF are lower than those of SCB and SRF. Moreover, if the stiffness of the joint is large and the moment of inertia in the tip load is small, the differences in the lower-order frequencies among SCB, SRF, and SFF are not significant. It can be inferred that in this case, whichever mode is selected as the assumed mode has no obvious difference in the model. However, in the opposite case, the differences in lower-order frequencies among SCB, SRF, and SFF are significant. Thus, it is necessary to be careful in choosing the assumed mode. Hence, the corresponding dynamic responses are studied in the next section.

2.3. Dynamic Response of a Flexible Space Manipulator System

The dynamic model of the space manipulator system has been established in Section 2.1. A simulation of the flexible two-link manipulator system was performed using MATLAB to analyze the dynamic response in this section. As mentioned in the previous section, to compare the difference in dynamic responses among the three assumed modes, the parameter k_m should be small. The parameters of the system are listed in Table 2. Both flexible joint motors are commanded to output the following sinusoidal force,

$$\tau = \begin{cases} \sin(\pi t) N, 0\text{ s} \leq t \leq 2\text{ s} \\ 0 N, 2\text{ s} < t \leq 10\text{ s} \end{cases} \quad (35)$$

Table 2. The values of system parameters.

Symbol	Value	Link1	Link2
m_0	17.23 kg	\	\
m_p	2.0 kg	\	\
J_0	0.087 kg·m²	\	\
J_p	0.005 kg·m²	\	\
E_1, E_2	\	72.0 GPa	72.0 GPa
l_0	0.12 m	\	\
ρ_1, ρ_2	\	1.620 kg/m	1.620 kg/m
J_1, J_2	\	0.005 kg·m²	0.005 kg·m²
k_1, k_2	\	500 Nm/rad	500 Nm/rad
I_1, I_2	\	4.50×10^{-8} m⁴	4.50×10^{-8} m⁴
l_1, l_2	\	2.0 m	2.0 m

The terminal deformations of the two links are shown in Figure 8. The results demonstrate that the dynamic response amplitude of SFF is the largest, although there is no significant difference in the dynamic response among the three assumed modes. The single-sided amplitude spectrum results show that the low-order natural frequencies are also close, as shown in Figure 9. Therefore, even if k_m is small, the selection of the three kinds of assumed modes has no significant influence on the dynamic response. Therefore, when the stiffness of the joint is difficult to obtain and the model accuracy requirement is not high, it is acceptable to choose the modes of SRF or SFF. However, if the stiffness of the joint can be accurately determined, the mode of SFF is more suitable because the boundary condition is more similar to the actual system. In addition, because of the coupled dynamic characteristics, the internal resonance phenomenon occurs between joint 1 and joint 2 after 2 s. Hence, higher requirements on the performance of the controller are raised.

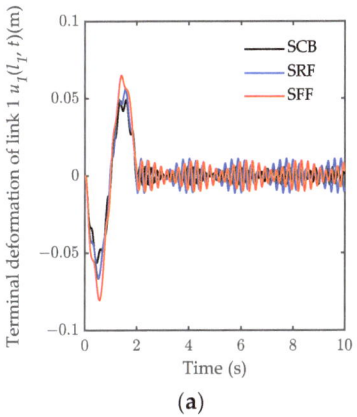 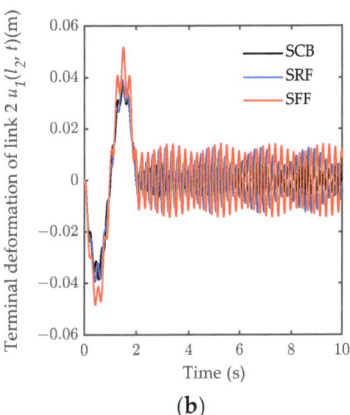

Figure 8. The terminal deformations of the two links: (**a**) the terminal deformation of link1; (**b**) the terminal deformation of link 2.

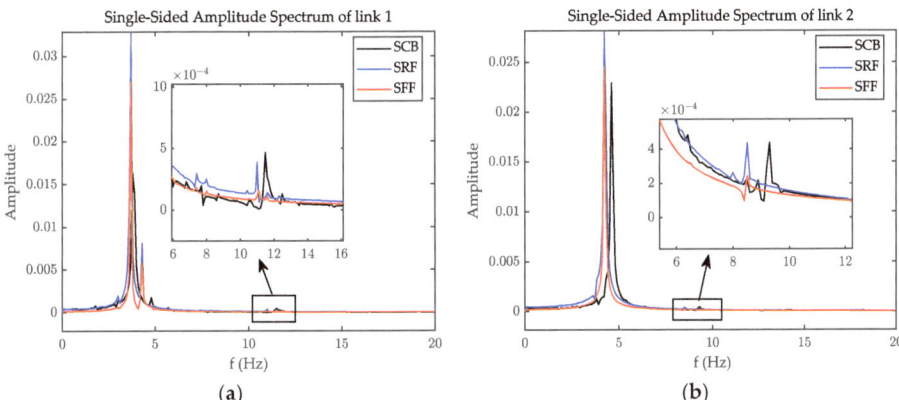

Figure 9. The single-sided amplitude spectrum: (**a**) the single-sided amplitude spectrum of link1; (**b**) the single-sided amplitude spectrum of link2.

3. Controller Design

3.1. Singular Perturbation Model of Flexible Space Manipulator System

When using the motor torque as the only input to control the output at the end of the flexible link, the system exhibits non-minimum phase characteristics [27]. Thus, it is not an easy task to suppress vibrations. In general, the rigid body motion of the system and the elastic vibration of the system occur in different timescales [28], and the frequency of rigid body motion is much less than one of the elastic vibration. Additionally, the system equation is nonlinear and coupled, thus, the calculation is hard to perform. Therefore, based on the assumption that huge differences in the frequency domain, between the rigid motion and the elastic vibration, the singular perturbation method was introduced to decompose the system into a slow subsystem, which describes the rigid-body motion, and a fast subsystem, which describes the elastic vibration, after which a composite controller can be designed [28]. The dynamic model can be transformed into a singular perturbation model as described:

Define a matrix $\mathbf{D}$

$$\mathbf{D} = \mathbf{M}^{-1} = \begin{bmatrix} \mathbf{D}_{11}(\theta, \mathbf{q}) & \mathbf{D}_{12}(\theta, \mathbf{q}) \\ \mathbf{D}_{21}(\theta, \mathbf{q}) & \mathbf{D}_{22}(\theta, \mathbf{q}) \end{bmatrix}. \tag{36}$$

From the system Equation (17), one can obtain:

$$\begin{bmatrix} \ddot{\mathbf{x}}_0 \\ \ddot{\theta} \end{bmatrix} = -\mathbf{D}_{11}\mathbf{F}_1 - \mathbf{D}_{12}\mathbf{F}_2 - \mathbf{D}_{12}\mathbf{K}_q \mathbf{q} + \mathbf{D}_{11}\begin{bmatrix} \mathbf{0}_{3\times 1} \\ \mathbf{K}_\sigma \sigma \end{bmatrix} + \mathbf{D}_{12}\tau', \tag{37}$$

$$\ddot{\mathbf{q}} = -\mathbf{D}_{21}\mathbf{F}_1 - \mathbf{D}_{22}\mathbf{F}_2 - \mathbf{D}_{22}\mathbf{K}_q \mathbf{q} + \mathbf{D}_{21}\begin{bmatrix} \mathbf{0}_{3\times 1} \\ \mathbf{K}_\sigma \sigma \end{bmatrix} + \mathbf{D}_{22}\tau', \tag{38}$$

$$\ddot{\sigma} = -\mathbf{J}^{-1}\mathbf{K}_\sigma \sigma + \mathbf{J}^{-1}\tau - \ddot{\theta}_{sub}, \tag{39}$$

where $\sigma = \alpha - \theta_{sub}$. Define a singular perturbation factor $\varepsilon^2 = 1/\min(\mathbf{K}_\sigma, \mathbf{K}_q)$, and use the factor to define:

$$\overline{\mathbf{K}}_\sigma = \varepsilon^2 \mathbf{K}_\sigma, \ \overline{\mathbf{K}}_q = \varepsilon^2 \mathbf{K}_q, \ \mathbf{z}_\sigma = \sigma \frac{1}{\varepsilon^2}, \ \mathbf{z}_q = \mathbf{q}\frac{1}{\varepsilon^2}. \tag{40}$$

From Equation (22), one obtains $\tau' = \mathbf{K}_v \mathbf{U}$. Thus, $\mathbf{K}_v = \varepsilon \overline{\mathbf{K}}_v$ due to $O(K_a) = O(\varepsilon)$. Substituting Equation (40) into Equations (37)–(39) yields:

$$\varepsilon^2 \ddot{\mathbf{z}}_\sigma = -\mathbf{J}^{-1}\overline{\mathbf{K}}_\sigma \mathbf{z}_\sigma + \mathbf{J}^{-1}\tau - \ddot{\theta}_{sub}, \tag{41}$$

$$\begin{bmatrix} \ddot{\mathbf{x}}_0 \\ \ddot{\theta} \end{bmatrix} = -\mathbf{D}_{11}(\theta, \varepsilon^2 \mathbf{z}_q)\mathbf{F}_1(\theta, \dot{\theta}, \varepsilon^2 \mathbf{z}_q, \varepsilon^2 \dot{\mathbf{z}}_q) - \mathbf{D}_{12}(\theta, \varepsilon^2 \mathbf{z}_q)\mathbf{F}_2(\theta, \dot{\theta}, \varepsilon^2 \mathbf{z}_q, \varepsilon^2 \dot{\mathbf{z}}_q) \\ -\mathbf{D}_{12}(\theta, \varepsilon^2 \mathbf{z}_q)\overline{\mathbf{K}}_q \mathbf{z}_q + \mathbf{D}_{11}(\theta, \varepsilon^2 \mathbf{z}_q)\begin{bmatrix} \mathbf{0}_{3\times 1} \\ \overline{\mathbf{K}}_\sigma \mathbf{z}_\sigma \end{bmatrix} + \varepsilon \mathbf{D}_{12}(\theta, \varepsilon^2 \mathbf{z}_q)\overline{\mathbf{K}}_v \mathbf{U} \tag{42}$$

$$\varepsilon^2 \ddot{\mathbf{z}}_q = -\mathbf{D}_{21}(\theta, \varepsilon^2 \mathbf{z}_q)\mathbf{F}(\theta, \dot{\theta}, \varepsilon^2 \mathbf{z}_q, \varepsilon^2 \dot{\mathbf{z}}_q)_1 - \mathbf{D}_{22}(\theta, \varepsilon^2 \mathbf{z}_q)\mathbf{F}_2(\theta, \dot{\theta}, \varepsilon^2 \mathbf{z}_q, \varepsilon^2 \dot{\mathbf{z}}_q) \\ -\mathbf{D}_{22}(\theta, \varepsilon^2 \mathbf{z}_q)\overline{\mathbf{K}}_q \mathbf{z}_q + \mathbf{D}_{21}(\theta, \varepsilon^2 \mathbf{z}_q)\begin{bmatrix} \mathbf{0}_{3\times 1} \\ \overline{\mathbf{K}}_\sigma \mathbf{z}_\sigma \end{bmatrix} + \varepsilon \mathbf{D}_{22}(\theta, \varepsilon^2 \mathbf{z}_q)\overline{\mathbf{K}}_v \mathbf{U} \tag{43}$$

where superscript "¯" indicates the value of the variables at $\varepsilon = 0$.

If $\varepsilon = 0$, from Equation (41), one can obtain:

$$\overline{\mathbf{z}}_\sigma = \overline{\mathbf{K}}_\sigma^{-1}(\tau_s - \mathbf{J}\ddot{\overline{\theta}}_{sub}). \tag{44}$$

In Equation (44), τ_s is the value of τ at $\varepsilon = 0$, which can also be written as $\overline{\tau}$. Substituting Equation (44) into Equation (43) provides:

$$\overline{\mathbf{z}}_q = \overline{\mathbf{K}}_q^{-1}\overline{\mathbf{D}}_{22}^{-1}(\overline{\theta}, 0)[-\overline{\mathbf{D}}_{21}(\overline{\theta}, 0)\overline{\mathbf{F}}_1(\overline{\theta}, \dot{\overline{\theta}}, 0, 0) - \overline{\mathbf{D}}_{22}(\overline{\theta}, 0)\overline{\mathbf{F}}_2(\overline{\theta}, \dot{\overline{\theta}}, 0, 0) + \overline{\mathbf{D}}_{21}(\overline{\theta}, 0)\begin{bmatrix} \mathbf{0}_{3\times 1} \\ \overline{\mathbf{K}}_\sigma \overline{\mathbf{z}}_\sigma \end{bmatrix}). \tag{45}$$

Finally, substituting Equations (44) and (45) into Equation (42) and using the inverse formula yields:

$$\begin{bmatrix} \ddot{\mathbf{x}}_0 \\ \ddot{\overline{\theta}} \end{bmatrix} = (\overline{\mathbf{D}}_{11}(\overline{\theta}) - \overline{\mathbf{D}}_{12}(\overline{\theta})\overline{\mathbf{D}}_{22}^{-1}(\overline{\theta})\overline{\mathbf{D}}_{21}(\overline{\theta}))(\begin{bmatrix} 0_{3\times 1} \\ \tau_s \end{bmatrix} - \begin{bmatrix} 0_{3\times 1} \\ \mathbf{J}\ddot{\theta}_{sub} \end{bmatrix} - \overline{\mathbf{F}}_1(\overline{\theta}, \dot{\overline{\theta}})). \qquad (46)$$

Equation (46) is the quasi-steady-state equation for the system and the slow subsystem. Defining the following boundary layer correction terms as:

$$\eta_1 = \mathbf{z}_\sigma - \overline{\mathbf{z}}_\sigma, \eta_2 = \varepsilon \dot{\mathbf{z}}_\sigma, \beta_1 = \mathbf{z}_q - \overline{\mathbf{z}}_q, \beta_2 = \varepsilon \dot{\mathbf{z}}_q. \qquad (47)$$

Substituting Equation (47) into Equations (41)–(43) gives the fast subsystem equation,

$$\frac{d\eta}{d\gamma} = \mathbf{A}_f \eta + \mathbf{B}_f \mathbf{U}_f, \qquad (48)$$

where $\eta = \begin{bmatrix} \eta_1 \\ \beta_1 \\ \eta_2 \\ \beta_2 \end{bmatrix}$, $\mathbf{U}_f = \begin{bmatrix} \tau_f \\ \tau' \end{bmatrix}$, $\mathbf{A}_f = \begin{bmatrix} 0 & 0 & \mathbf{I} \\ -\mathbf{J}^{-1}\overline{\mathbf{K}}_\sigma & 0 & 0 \\ \overline{\mathbf{D}}_{21_{12}}\overline{\mathbf{K}}_\sigma & -\overline{\mathbf{D}}_{22}\overline{\mathbf{K}}_q & 0 \end{bmatrix}$, $\mathbf{B}_f = \begin{bmatrix} 0 & 0 \\ \mathbf{J}^{-1} & 0 \\ 0 & \overline{\mathbf{D}}_{22} \end{bmatrix}$, $\overline{\mathbf{D}}_{21_{12}}$

is a submatrix of $\overline{\mathbf{D}}_{22}$, which does not contain elements in the first column of the matrix $\overline{\mathbf{D}}_{22}$.

3.2. Computed Torque Controller Designed for Joints

The system of space manipulators with a freebase is underactuated. To design the underactuated computed torque controller, the parameter δ is introduced. Based on the slow subsystem Equation (46), the control torque was designed as:

$$\begin{bmatrix} 0 \\ \tau_s \end{bmatrix} = (\overline{\mathbf{M}}_{11}(\overline{\theta}) + \begin{bmatrix} 0 \\ & \mathbf{J} \end{bmatrix})\begin{bmatrix} \delta \\ \mathbf{u}_s \end{bmatrix} + \overline{\mathbf{F}}_1(\overline{\theta}, \dot{\overline{\theta}}), \qquad (49)$$

where $\mathbf{u}_s$ is the reference input, determined by the outer loop control. Substituting Equation (49) into Equation (46) provides:

$$\begin{bmatrix} \delta \\ \mathbf{u}_s \end{bmatrix} = \begin{bmatrix} \ddot{\mathbf{x}}_0 \\ \ddot{\overline{\theta}} \end{bmatrix} = \begin{bmatrix} \ddot{\mathbf{x}}_0 \\ \ddot{\theta}_0 \\ \ddot{\theta}_{sub} \end{bmatrix}. \qquad (50)$$

In this way, the inner loop control is completed by introducing model-based torque. A PD controller was introduced, and the reference input $\mathbf{u}_s$ was obtained by:

$$\mathbf{u}_s = -\mathbf{K}_p \mathbf{e}_s - \mathbf{K}_d \dot{\mathbf{e}}_s + \ddot{\theta}_d, \qquad (51)$$

where $\mathbf{K}_p$ and $\mathbf{K}_d$ are the position feedback gain matrix and velocity feedback gain matrix, respectively, both of which are positive definite, and $\mathbf{e}_s = \theta_d - \overline{\theta}_{sub}$ is the error between desired position θ_d, and actual output joint position $\overline{\theta}_{sub}$. From Equation (44), one can easily deduce that the variable $\overline{\mathbf{z}}_\sigma$ is bounded, which indicates the elastic deformation vector $\overline{\sigma} = 0$ at $\varepsilon = 0$ (Since $\mathbf{z}_\sigma = \sigma \frac{1}{\varepsilon^2}$, and $\overline{\mathbf{z}}_\sigma$ is bounded at $\varepsilon = 0$, then, $\overline{\sigma} = 0$ at $\varepsilon = 0$ necessarily). Therefore, $\overline{\theta}_{sub} = \overline{\alpha}$, $\mathbf{e}_s = \theta_d - \overline{\alpha}$, motor position can be used for the inner loop control.

Substituting Equation (51) into Equation (50) yields:

$$\begin{cases} \delta = \begin{bmatrix} \ddot{\mathbf{x}}_0 \\ \ddot{\theta}_0 \end{bmatrix} \\ \ddot{\mathbf{e}}_s + \mathbf{K}_p \mathbf{e}_s + \mathbf{K}_v \dot{\mathbf{e}}_s = 0 \end{cases}, \qquad (52)$$

where $(\mathbf{e}_s, \dot{\mathbf{e}}_s) = (\mathbf{0}, \mathbf{0})$ is the globally asymptotically stable equilibrium point since $\mathbf{K}_p$, $\mathbf{K}_d$ are positive definite, and δ represents the perturbed acceleration of the base body, whereby the effect cannot be ignored for a small mass base.

δ can be solved from Equation (49):

$$\delta = \overline{\mathbf{M}}^{-1}{}_{11_{11}}(-\overline{\mathbf{M}}_{11_{12}}\mathbf{u}_s - \overline{\mathbf{F}}_1(\theta, \dot{\theta})), \tag{53}$$

where $\overline{\mathbf{M}}_{11_{11}} = \overline{\mathbf{M}}(1:3, 1:3)$, $\overline{\mathbf{M}}_{11_{12}} = \overline{\mathbf{M}}(1:3, 4:n+1)$.

3.3. Adaptive Fuzzy Controller Designed for Piezo Actuator

The system decomposed into a slow subsystem and a fast subsystem, as noted in Section 3.1. The fast subsystem, Equation (48), is linear. The LQR controller introduced for the linear systems by researchers [28] may be used to suppress the vibrations. However, the effectiveness of the LQR controller designed for fast subsystems relies heavily on modeling accuracy. In practice, joint friction, structural damping, etc., are difficult to be modeled. Furthermore, the modal truncation introduced by the assumed mode method reduces the accuracy of the model. Therefore, a direct adaptive fuzzy controller is presented to suppress the vibration of the flexible links.

The fuzzy system of a space manipulator system can be described as $\hat{\mathbf{F}}(\mathbf{q}|\gamma)$. A fuzzy controller is designed by using product inference engine, gauss fuzzier, and a center averaging defuzzifier. According to the controller design method, based on a traditional fuzzy system, the robust fuzzy adaptive control law is designed as:

$$\boldsymbol{\tau}'_{mn\times 1} = \hat{\mathbf{F}}(\mathbf{q}|\gamma) - \mathbf{K}_D\mathbf{s} - \mathbf{W}\mathrm{sgn}(\mathbf{s}), \tag{54}$$

where $\mathbf{s} = d\mathbf{e}_f + \Lambda \mathbf{e}_f$, and $\mathbf{K}_D$, $\mathbf{W}$, and Λ are weight matrices. The sign function in Equation (54) is designed to address the problem of external disturbances. However, the sign function may lead to a high-frequency chattering phenomenon. To avoid high-frequency chattering, the sign function can be substituted for by the saturation function. Thus, the adaptive fuzzy controller is designed as:

$$\boldsymbol{\tau}'_{mn\times 1} = \hat{\mathbf{F}}(\mathbf{q}, \dot{\mathbf{q}}|\gamma) - \mathbf{K}_D\mathbf{s} - \mathbf{W}sat(\mathbf{s}), \tag{55}$$

where $sat(\mathbf{s}) = \begin{cases} 1, \mathbf{s} > \Delta \\ \frac{1}{\Delta}\mathbf{s}, |\mathbf{s}| \leq \Delta \\ -1, \mathbf{s} < \Delta \end{cases}$, and the parameter Δ is generally set to a small value.

The adaptive law is designed as:

$$\dot{\gamma}_i = -\zeta_i^{-1}\mathbf{s}_i \xi(\dot{\mathbf{q}}), \, i = 1, 2, \cdots, n, \tag{56}$$

where ζ_i ($\zeta_i > 0$) is called the adaptive parameter.

The fuzzy system is designed as:

$$\hat{\mathbf{F}}(\dot{\mathbf{q}}|\gamma) = \begin{bmatrix} \hat{F}_1(\dot{q}_1^1) \\ \vdots \\ \hat{F}_2(\dot{q}_1^m) \\ \vdots \\ \hat{F}_{m\times n}(\dot{q}_n^m) \end{bmatrix} = \begin{bmatrix} \gamma_1^T \xi^1(\dot{q}_1^1) \\ \vdots \\ \gamma_2^T \xi^2(\dot{q}_1^m) \\ \vdots \\ \gamma_{m\times n}^T \xi^{m\times n}(\dot{q}_n^m) \end{bmatrix}. \tag{57}$$

The basic structure of the presented fuzzy controller is shown in Figure 10. The link end displacement $\mathbf{e}_f$ and its speed $d\mathbf{e}_f$ are used as control inputs, while $\mathbf{E}_f$ and $d\mathbf{E}_f$ are the fuzzy quantities corresponding to the two inputs. The control torque τ_f of the piezoelectric actuator is obtained by defuzzing the fuzzy quantity M, obtained by the fuzzy logic inference.

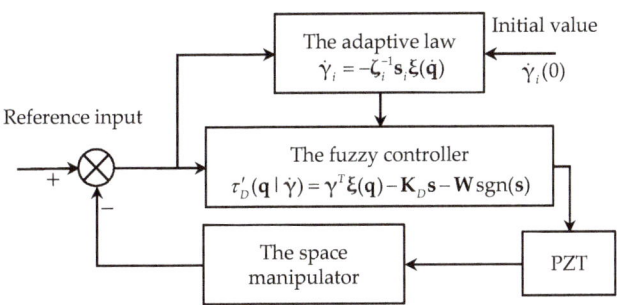

Figure 10. Schematic of adaptive fuzzy controller.

4. Numerical Simulation

The effectiveness of the presented control strategy is examined on a two-link flexible manipulator system, as shown in Figure 11. A pair of PZT actuators are attached to the root of each flexible link and a tip payload is attached to the distal end of the second flexible link. The main parameters of the manipulator system are listed in Table 3.

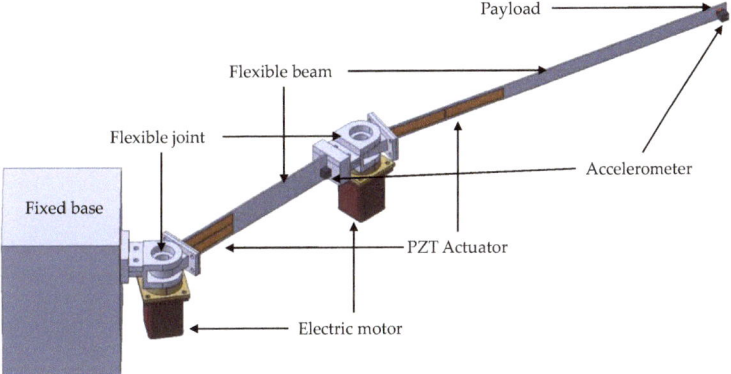

Figure 11. Two-link flexible manipulator with fixed base.

Table 3. Kinematic and dynamic parameters of the space manipulator system.

Symbol	Value	Link1	Link2
m_p	0.1 kg	\	\
J_p	1.25×10^{-6} kg·m^2	\	\
E_1, E_2	\	200.0 GPa	200.0 GPa
ρ_1, ρ_2	\	0.5688 kg/m	0.3160 kg/m
J_1, J_2	\	0.001 kg·m^2	0.001 kg·m^2
k_1, k_2	\	9.6 Nm/rad	9.6 Nm/rad
I_1, I_2	\	2.40×10^{-11} m^4	1.33×10^{-11} m^4
l_1, l_2	\	0.5 m	0.25 m

The membership functions of $\mathbf{e}_f$ and $\mathbf{de}_f$ used in the two cases are shown in Figures 12 and 13, respectively. Define nine levels of fuzzy value: PB (positive big), PM (positive middle), PS (positive small), PO (positive zero), ZO (zero), NO (negative zero), NS (negative small), NM (negative middle), and NB (negative big).

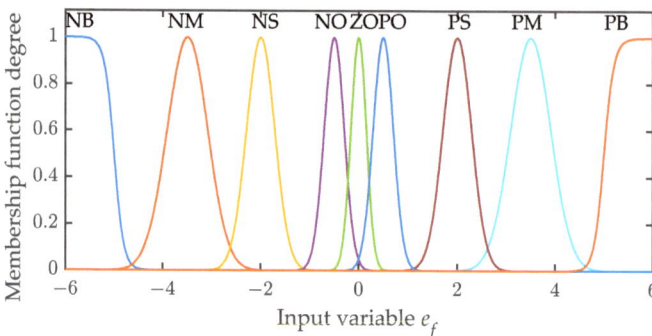

Figure 12. Membership function of input variable e_f.

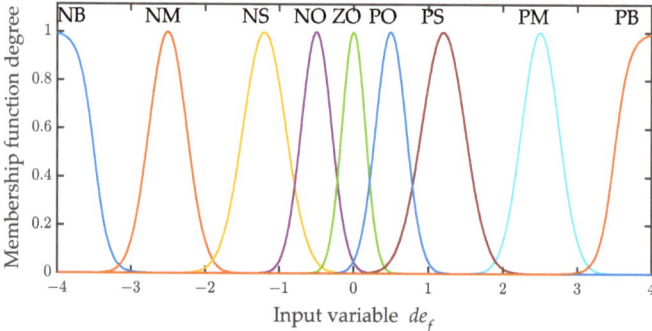

Figure 13. Membership function of input variable de_f.

In this section, the performance of the proposed control scheme is evaluated for two cases: setpoint motion control and periodic motion control. The PD and CTM controllers were firstly used to track the joint trajectories with their performance comparisons. The PD controller and AFC were, then, adopted to suppress the elastic vibration of the links and were compared with each other. Moreover, to ensure the fairness of the comparisons, the parameters $\mathbf{K}_p$ and $\mathbf{K}_d$ in the inner loop controller of CTM were set to the same as the proportional control gain and differential control gain of the PD controller, respectively, and the performance of AFC and one PD were compared when the maximum voltage of PZT was limited. A set of gains for the PD controller for better control performance was tuned by numerical simulation.

4.1. Case 1: Setpoint Motion Control

The manipulator is commanded to track a desired cycloid trajectory to a desired position. The desired position of both links was set to $\pi/2$ rad, and the required time for the two links to reach the desired position was set to 5.0 s. The desired cycloid trajectory is $\theta_d = \pi \times [t/5 - \sin(2\pi t/5)/2\pi]/2$. The tip vibration caused by the motor stopping is called residual vibration. In order to demonstrate the control effect on the residual vibration, the simulation time was set to 10 s.

The comparison of tracking the control performance between the proposed computed torque method and the PD controller is shown in Figure 14. Figure 14b,e show the superior performance of CTM in tracking the desired trajectory, especially at the inflection point of the trajectory. Moreover, the output torque of CTM was smoother than that of PD, as evidenced in Figure 14c,f. The simulation result of acceleration response is shown in Figure 15. The amplitude of the residual vibration with AFC costs only 0.1 s to attenuate to 10% of its maximum amplitude, the PD controller costs 0.2 s, and uncontrol costs 0.18. The

AFC provides better performance than PD in rapidly suppressing the residual vibration. Additionally, compared to the uncontrol, the RMS of the acceleration response with AFC attenuated by 95%, whereas one with the PD controller only attenuated by 23%. The residual vibration attenuates under the uncontrol condition was due to the joint control torques, which kept the joint positions stable.

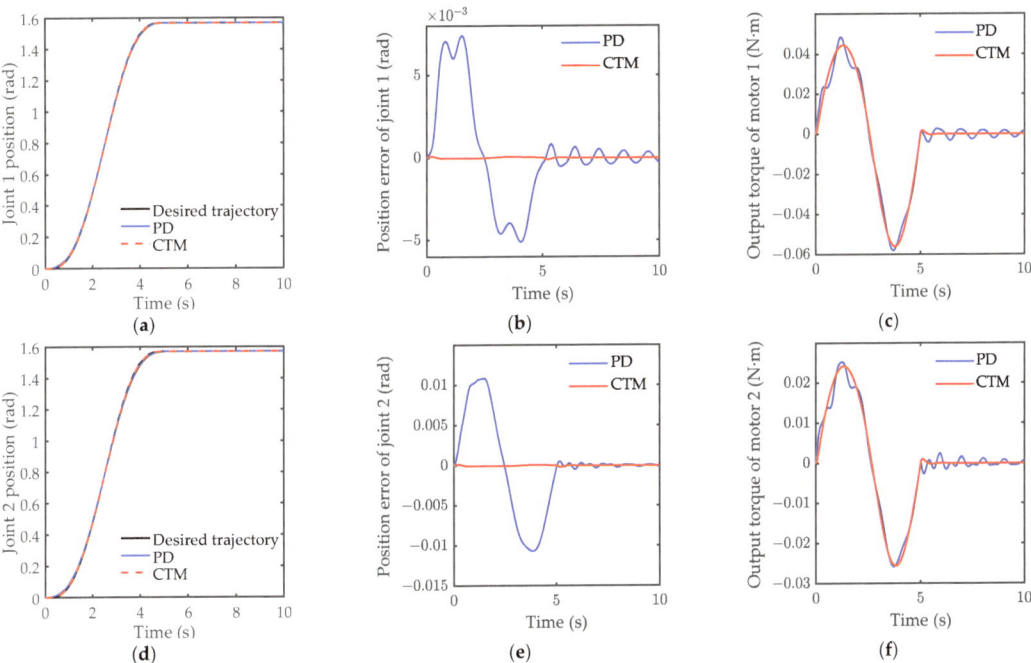

Figure 14. Simulation result of motion case 1: (**a**) joint 1 position; (**b**) position error of joint 1; (**c**) output torque of motor 1; (**d**) joint 2 position; (**e**) position error of joint 2; (**f**) output torque of motor 2.

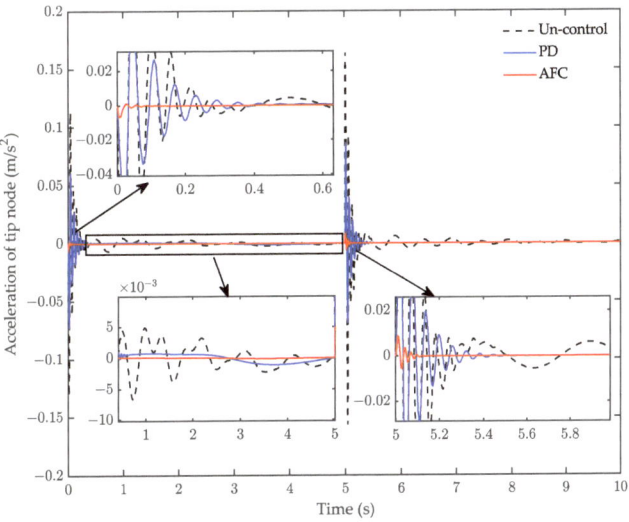

Figure 15. Simulation result of acceleration response for case 1.

4.2. Case 2: Periodic Motion Control

In case 2, a desired period sinusoidal trajectory $\theta_d(t) = A_m \sin(2\pi f_1 t + b) + A_m/2$ is selected to test the effectiveness of the proposed controller in a periodic motion, where $A_m = \pi/4$, $f_1 = 0.1$, and $b = 0$. The motion time was 30 s and the simulation time was 35 s. A comparison of tracking is shown in Figure 16. Here, CTM is seen to still provide better performance than PD in tracking the periodic motion. The trajectory of PD obviously buffets near the desired trajectory. It is worth noting that the tracking error of CTM visibly increases at 30 s, yet rapidly reduces after 30 s. Figure 16c,f show the torques of the two joints, respectively. It can be seen that the torques of the CTM are obviously smaller than for PD. Figure 17 shows a comparison of the control effect on the tip vibrational acceleration. Compared to the uncontrol, the RMS of the acceleration response with AFC attenuated by 95%, and one with the PD controller only attenuated by 62%. Obviously, AFC maintains a good performance in suppressing the vibration. In addition, in the two cases, the frequency of the elastic vibration was greater than 70 Hz, and one for the rigid motion was less than 3 Hz, which indicates that the elastic vibration and the rigid body motion occur in different timescales, hence, the introduction of the singular perturbation theory is suitable. The elastic deformation is very small in numerical simulation, which demonstrates that the small deformation assumption of the flexible joint is suitable.

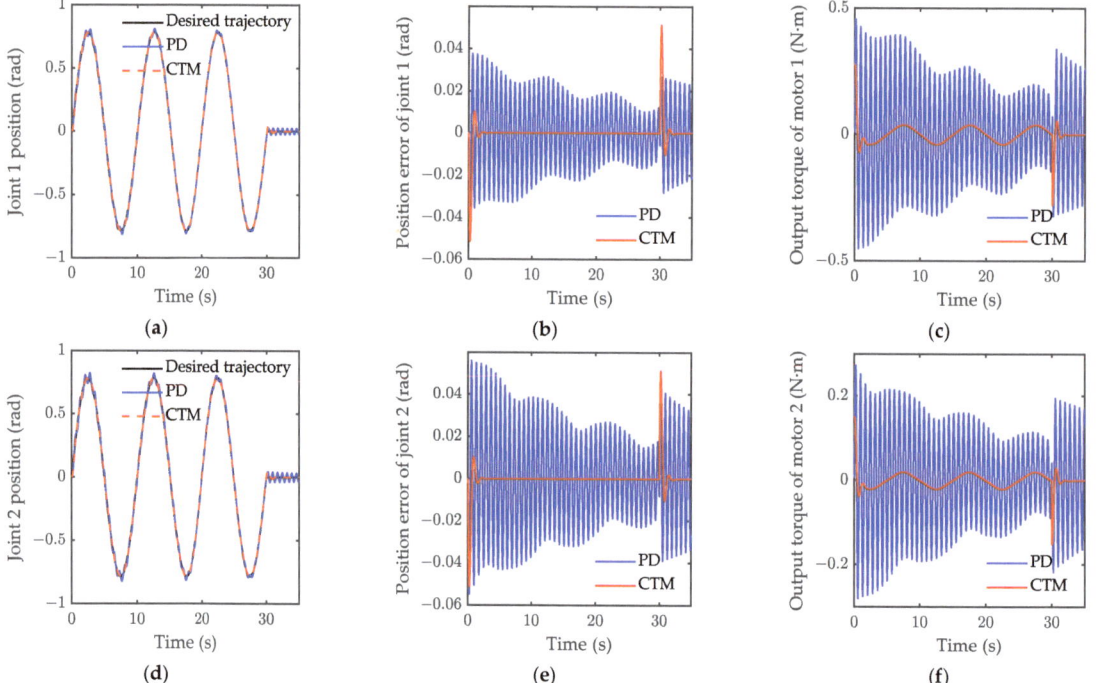

Figure 16. Simulation result of motion for case 2: (**a**) joint 1 position; (**b**) position error of joint 1; (**c**) output torque of motor 1; (**d**) joint 2 position; (**e**) position error of joint 2; (**f**) output torque of motor 2.

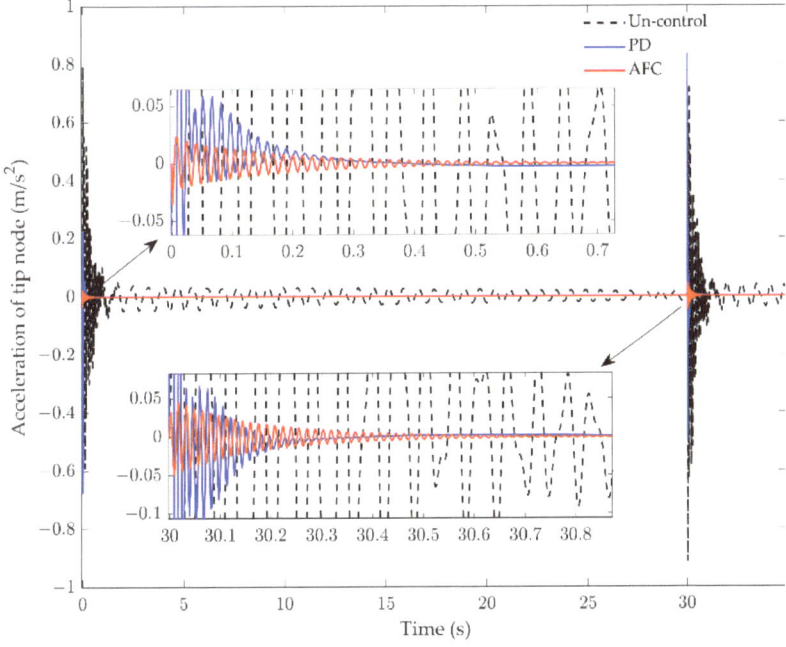

Figure 17. Simulation result of acceleration response for case 2.

5. Experimental Results

Experiments were performed to verify the simulations in case 1 and case 2. The experimental setup is schematically shown in Figure 18. The flexible links are clamped at the shafts of the motors through couplings, and the diagonal rope is supported on each joint to counter the gravity on the flexible link. Two AC rotary servomotors (designed by YASKAWA and SGM7J-01AFC6S) were used to drive the flexible links, and a built-in 24-bit absolute encoder installed within the motor was applied to calculate the rotating angle. Each flexible link had a PZT actuator (MFC, M-8514-P1) attached to the root, and the PZT actuators were used to suppress the elastic vibration on the flexible links. Each flexible link had an accelerometer (PCB 333B32) attached at the end to collect the vibration acceleration signal. The velocity signal and displacement signal were obtained using the first and second integration of the acceleration signal, respectively. In practice, the zero drift and high-frequency noise of the sensor cannot be avoided. Hence, a low-pass digital filter was designed to address the issue. A small-time delay could be introduced by the digital filter. However, the time delay was accepted in the experiment. Furthermore, the ambient laboratory temperature was maintained at a constant value to reduce the drift caused by the temperature changes. The main parameters of the experimental model are shown in Table 3. A motion controller (GALIL DMC1846) interfaced with a high-performance PC was used to snatch data and process data. The control voltage signals for the motor and the PZT actuators were sent to the servo driver and HVPZT amplifier using the GALIL motion controller, respectively. To remove the effect of the residual modes and high-frequency noise, a low-pass digital filter was designed for the vibrational acceleration signal.

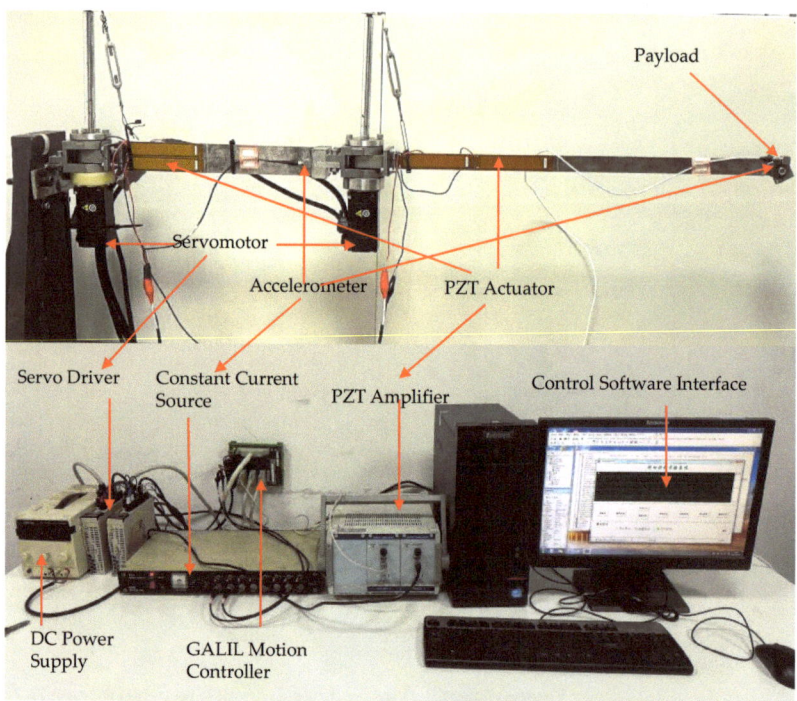

Figure 18. Scheme of the two-link flexible manipulator experimental setup.

The experimental results of the setpoint motion control are shown in Figures 19 and 20. Figure 19 shows that the CTM performs better than PD in the experiment. Figure 20 indicates that both PD and AFC can suppress residual vibration, although AFC had a better performance than the PD controller in suppressing the residual vibrations. The root mean square (RMS) and the amplitude of the whole course corresponding responses are listed in Table 4. Compared to the uncontrol, the RMS of the acceleration response with AFC was attenuated by 78.3%, while one with the PD controller attenuated by only 44.9%. This indicates that AFC provides better performance in suppressing the vibrations. Figure 19a,b demonstrate that serious buffet occurs under PD control, and the position error of PD is larger than for CTM. The position error of joint 2 is larger than for joint 1 under PD control during the motion, as shown in Figure 19c,d. The experimental performance of CTM was worse than the simulated performance. The position error was more significant during the first 3 s. CTM cannot perform high-precision trajectory tracking in the experiment. A possible reason is that the mathematical model does not consider the influence of nonlinear factors, such as friction and clearance, in the actual system.

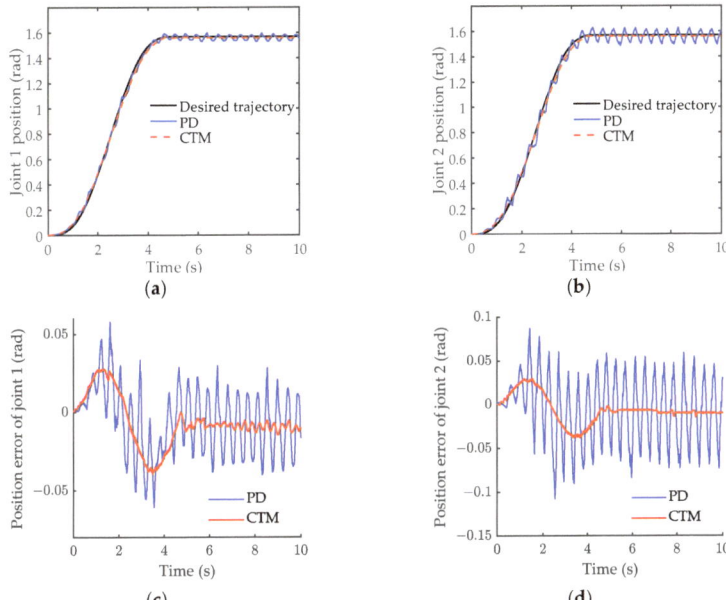

Figure 19. Experiment result of joint 1 position for case 1: (**a**) joint 1 position; (**b**) joint 2 position; (**c**) position error of joint 1; (**d**) position error of joint 2.

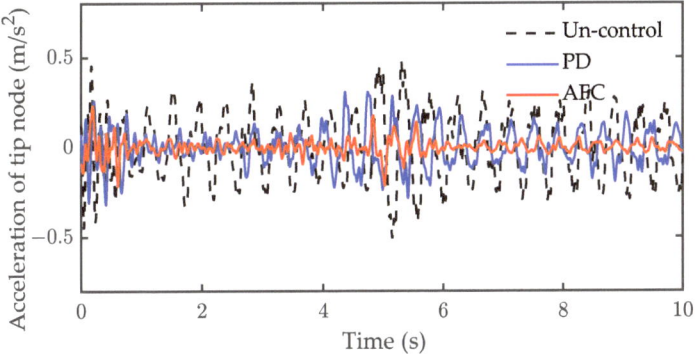

Figure 20. Experiment result of acceleration response for case 1.

Table 4. Vibration suppression effect of the experimental system for case 1.

Controller	Uncontrol	PD	AFC
RMS (m/s^2)	0.1847	0.1017	0.04
Amplitude (m/s^2)	0.5048	0.3186	0.2309

Experimental results under periodic motion control are shown in Figures 21–24, respectively. The results of the tracking shown in Figure 21 demonstrate that the joint positions are tracked well once the CTM is employed. Figure 22 shows the acceleration response of the tip node when the adaptive parameter is $\zeta_1 = \zeta_2 = \zeta = 2000$. The RMS and the amplitude of the whole course response are listed in Table 5. Both AFC and PD can suppress the vibration of the tip node. Compared with uncontrol, the RMS of AFC was attenuated by 62.5% and one for the PD controller was attenuated by 48.9%. In this case, the control performance of AFC was slightly better than of PD. The value of the adaptive parameter ζ

is important for the control performance of AFC. The experimental results for the further adjustment of the parameters ζ are shown in Figures 23 and 24. Figures 23 and 24 show the control performance of AFC when $\zeta = 5000$ and $\zeta = 20{,}000$, respectively. When $\zeta = 5000$, the vibration is rapidly suppressed by employing AFC, the RMS of AFC was attenuated by 81.3% compared with the uncontrol. However, increasing the adaptive parameter to $\zeta = 20{,}000$ does not necessarily improve the performance of the AFC, the RMS of AFC was only attenuated by 57.1% compared to the uncontrol. When the value of the adaptive parameter ζ was small, the convergence of the AFC algorithm was slow and, thus, the AFC performance was poor. Overall, the acceleration can be suppressed quicker as the adaptive parameter increases. However, when the value of the adaptive parameter ζ was too large, the gain of the AFC was large, resulting in instability and causing additional vibrations. In light of these results on simulation, $4000 < \zeta < 6000$ is suitable for this experimental system.

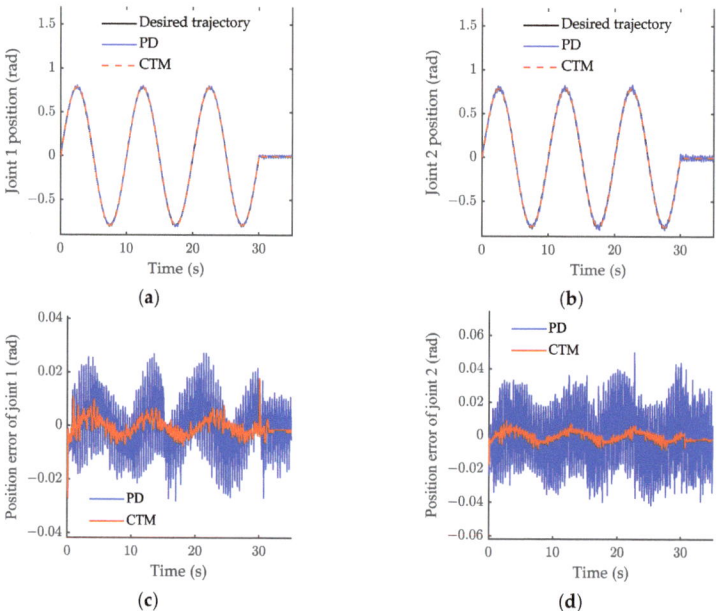

Figure 21. Experiment result of joint 1 position for case 2: (**a**) joint 1 position; (**b**) joint 2 position; (**c**) position error of joint 1; (**d**) position error of joint 2.

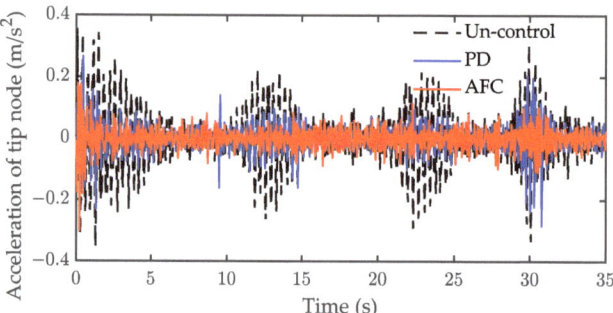

Figure 22. Experiment result of acceleration response for case 2 ($\zeta = 2000$).

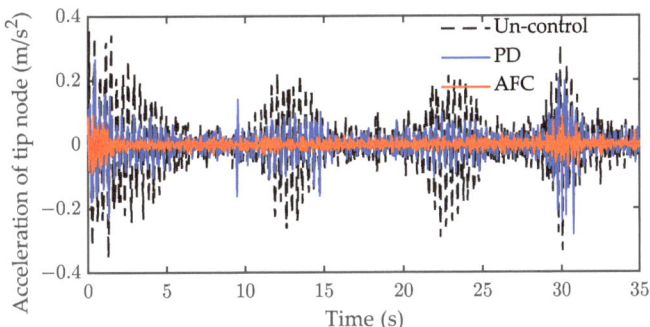

Figure 23. Experiment result of acceleration response for case 2 ($\zeta = 5000$).

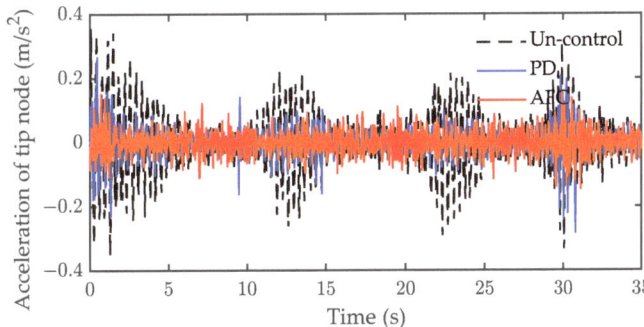

Figure 24. Experiment result of acceleration response for case 2 ($\zeta = 20{,}000$).

Table 5. Vibration suppression effect of the experimental system for case 2 ($\zeta = 2000$).

Controller	Uncontrol	PD	AFC
RMS (m/s^2)	0.0808	0.0413	0.0303
Amplitude (m/s^2)	0.3500	0.2819	0.2997

6. Conclusions

In this paper, a dynamic mathematical model for a flexible space manipulator was derived. To choose a suitable mode, the natural characteristics of the link under three different boundary conditions were compared to each other, and the dynamic response results of the simulation showed that the internal resonance behavior existed in the system. A computed torque controller was designed to track the angle of the joints. Furthermore, a model-independent adaptive fuzzy controller was proposed to suppress the elastic vibration. The simulations of the two cases were performed in a two-link flexible manipulator system, and the results show that the proposed control strategy had a good performance in rapidly tracking trajectory and effectively suppressing the flexible vibration. Experiments were also performed to verify the proposed control strategy and simulation results. Compared with the uncontrol, the RMS of the vibration response with AFC was attenuated by 78.3% and 81.3% in the two cases, respectively. In addition, to improve the accuracy of the mathematic model of the space manipulator system, the damping of the joint and link should be considered in future modeling studies, and the effect of the internal resonance phenomenon on the stability of the control system should be studied in further detail in the future.

Author Contributions: Conceptualization, C.F.; methodology, C.F.; validation, C.F. and M.S.; resources, S.N.; writing—original draft, C.F.; writing—review and editing, W.C.; supervision, W.C. and M.S.; project administration, W.C.; funding acquisition, M.S. All authors have read and agreed to the published version of the manuscript.

Funding: This work was supported by the National Natural Science Foundation of China under Grant No. 12102174, and the Research Fund of State Key Laboratory of Mechanics and Control for Aerospace Structures (Nanjing University of Aeronautics and Astronautics) (Grant No. MCMS-I-0122K01).

Data Availability Statement: Not applicable.

Acknowledgments: This work was supported by the National Natural Science Foundation of China under Grant No. 12102174, and the Research Fund of State Key Laboratory of Mechanics and Control for Aerospace Structures (Nanjing University of Aeronautics and Astronautics) (Grant No. MCMS-I-0122K01). The authors gratefully acknowledge these agencies for their support.

Conflicts of Interest: The authors declare no conflict of interest.

References

1. Sachdev, S.; Harvey, W.; Gibbs, G.; Marcotte, B.; Buckley, N.; Braithwaite, T.; Rey, D. Canada and the International Space Station Program: Overview and Status since IAC 2005. In Proceedings of the 57th International Astronautical Congress, Valencia, Spain, 2–6 October 2006.
2. Friend, R.B. Orbital Express program summary and mission overview. In Proceedings of the SPIE Defense and Security Symposium, Orlando, FL, USA, 15 April 2008.
3. Alandoli, E.A.; Lee, T.S. A Critical Review of Control Techniques for Flexible and Rigid Link Manipulators. *Robotica* **2020**, *38*, 2239–2265. [CrossRef]
4. Subedi, D.; Tyapin, I.; Hovland, G. Review on Modeling and Control of Flexible Link Manipulators. *Model. Identif. Control* **2020**, *41*, 141–163. [CrossRef]
5. Yu, X.; Chen, L. Dynamic modeling and control of a free-flying space robot with flexible-link and flexible-joints. In Proceedings of the IEEE International Conference on Robotics & Automation, Hong Kong, China, 31 May–7 June 2014.
6. Dwivedy, S.K.; Eberhard, P. Dynamic analysis of flexible manipulators, a literature review. *Mech. Mach. Theory* **2006**, *41*, 749–777. [CrossRef]
7. Wei, J.; Cao, D.; Wang, L.; Huang, H.; Huang, W. Dynamic modeling and simulation for flexible spacecraft with flexible jointed solar panels. *Int. J. Mech. Sci.* **2017**, *130*, 558–570. [CrossRef]
8. Kiang, C.T.; Spowage, A.; Yoong, C.K. Review of control and sensor system of flexible manipulator. *J. Intell. Robot. Syst.* **2015**, *77*, 187–213. [CrossRef]
9. He, F.; Huang, Q. Time-Optimal Trajectory Planning of 6-DOF Manipulator Based on Fuzzy Control. *Actuators* **2022**, *11*, 332. [CrossRef]
10. Sabatini, M.; Gasbarri, P.; Monti, R.; Palmerini, G.B. Vibration control of a flexible space manipulator during on orbit operations. *Acta Astronaut.* **2012**, *73*, 109–121. [CrossRef]
11. Zhang, Q.; Li, C.; Zhang, J.; Zhang, J. Smooth adaptive sliding mode vibration control of a flexible parallel manipulator with multiple smart linkages in modal space. *J. Sound Vib.* **2017**, *411*, 1–19. [CrossRef]
12. Zhang, W. The impulse spectrum method for vibration suppression of a flexible multilink robot arm. *J. Vib. Control.* **2018**, *24*, 3865–3881. [CrossRef]
13. Yang, G. Asymptotic tracking with novel integral robust schemes for mismatched uncertain nonlinear systems. *Int. J. Robust Nonlinear Control.* **2023**, *33*, 1988–2002. [CrossRef]
14. Shawky, A.; Ordys, A.; Grimble, M.J. End-point control of a flexible-link manipulator using H∞ nonlinear control via a state-dependent Riccati equation. In Proceedings of the International Conference on Control Applications, Glasgow, UK, 18–20 September 2002.
15. De Luca, A. Feedforward/Feedback Laws for the Control of Flexible Robots. In Proceedings of the 2000 ICRA, San Francisco, CA, USA, 24–28 April 2000.
16. Qiu, Z.; Zhang, W. Trajectory planning and diagonal recurrent neural network vibration control of a flexible manipulator using structural light sensor. *Mech. Syst. Signal Process.* **2019**, *132*, 563–594. [CrossRef]
17. Malzahn, J.; Phung, A.S.; Bertram, T. A Multi-Link-Flexible Robot Arm Catching Thrown Balls. In Proceedings of the 7th German Conference on Robotics, Munich, Germany, 21–22 May 2012.
18. Qingsong, C.; Ailan, Y. Optimal actuator placement for vibration control of two-link piezoelectric flexible manipulator. In Proceedings of the International Conference on Mechanic Automation and Control Engineering, Wuhan, China, 3 August 2010.
19. Wei, J.; Qiu, Z.; Han, J.; Wang, Y. Experimental Comparison Research on Active Vibration Control for Flexible Piezoelectric Manipulator Using Fuzzy Controller. *J. Intell. Robot. Syst.* **2010**, *59*, 31–56. [CrossRef]

20. Qiu, Z.; Li, C.; Zhang, X. Experimental study on active vibration control for a kind of two-link flexible manipulator. *Mech. Syst. Signal Process.* **2019**, *118*, 623–644. [CrossRef]
21. Zhang, J.; Dai, X.; Huang, Q.; Wu, Q. AILC for Rigid-Flexible Coupled Manipulator System in Three-Dimensional Space with Time-Varying Disturbances and Input Constraints. *Actuators* **2022**, *11*, 268. [CrossRef]
22. Soltani, M.; Keshmiri, M.; Misra, A.K. Dynamic analysis and trajectory tracking of a tethered space robot. *Acta Astronaut.* **2016**, *128*, 335–342. [CrossRef]
23. Shao, M.; Huang, Y.; Silberschmidt, V.V. Intelligent Manipulator with Flexible Link and Joint: Modeling and Vibration Control. *Shock Vib.* **2020**, *2020*, 4671358. [CrossRef]
24. Spong, M.W. Review article: Modeling and control of elastic joint robots. *Math. Comput. Model.* **1989**, *12*, 912. [CrossRef]
25. Gao, H.; He, W.; Zhou, C.; Sun, C. Neural Network Control of a Two-Link Flexible Robotic Manipulator Using Assumed Mode Method. *IEEE Trans. Ind. Inform.* **2019**, *15*, 755–765. [CrossRef]
26. Li, Y.; Tang, B.; Shi, Z.; Lu, Y. Experimental study for trajectory tracking of a two-link flexible manipulator. *Int. J. Syst. Sci.* **2000**, *31*, 3–9. [CrossRef]
27. Vakil, M.; Fotouhi, R.; Nikiforuk, P.N. On the zeros of the transfer function of a single flexible link manipulators and their non-minimum phase behaviour. *Proc. Inst. Mech. Eng. Part C* **2010**, *224*, 2083–2096. [CrossRef]
28. Shao, Z.; Zhang, X. Intelligent control of flexible-joint manipulator based on singular perturbation. In Proceedings of the International Conference on Automation and Logistics, Hong Kong, China, 16–20 August 2010.

Disclaimer/Publisher's Note: The statements, opinions and data contained in all publications are solely those of the individual author(s) and contributor(s) and not of MDPI and/or the editor(s). MDPI and/or the editor(s) disclaim responsibility for any injury to people or property resulting from any ideas, methods, instructions or products referred to in the content.

Article

Model-Based Systems Engineering Approach for the First-Stage Separation System of Launch Vehicle

Wenfeng Zhang *, Zhendong Liu, Xiong Liu, Yili Jin, Qixiao Wang and Rong Hong

Shanghai Aerospace Systems Engineering Institute, Shanghai 201109, China
* Correspondence: zhangwenfeng4312@163.com

Abstract: This paper proposes a model-based systems engineering (MBSE) methodology to design a first-stage separation system for a launch vehicle. It focuses on the whole process of system modeling, such as modeling the requirements analysis, logical architecture design, physical architecture design, and system verification and validation. Finally, the component requirements are obtained as the baseline for the component design. Requirements analysis is carried out by identifying stakeholders with the cycle modeling for this system and the use of case modeling to ensure that the requirements are comprehensive and correct. Additionally, the standard system requirements are obtained and baselined. Based on system requirements, the trade-off analysis of hierarchical functional architecture and key indicators was mainly carried out to design the logical architecture. Once the logical architecture was decided, the logical architecture was allocated to the physical architecture to be implemented. Several physical architectures are analyzed hierarchically to seek the optimal architectures. Then, other CAE analysis tools were integrated to verify the physical architecture design. All these processes are modeled and integrated as the authority system model, which benefits the system engineer for managing the requirement changes easier and rapidly provides multi-views for different roles.

Keywords: model-based systems engineering; first-stage separation system; logical architecture; physical architecture; trade-off

Citation: Zhang, W.; Liu, Z.; Liu, X.; Jin, Y.; Wang, Q.; Hong, R. Model-Based Systems Engineering Approach for the First-Stage Separation System of Launch Vehicle. *Actuators* **2022**, *11*, 366. https://doi.org/10.3390/act11120366

Academic Editor: Ronald M. Barrett

Received: 21 October 2022
Accepted: 1 December 2022
Published: 7 December 2022

Publisher's Note: MDPI stays neutral with regard to jurisdictional claims in published maps and institutional affiliations.

Copyright: © 2022 by the authors. Licensee MDPI, Basel, Switzerland. This article is an open access article distributed under the terms and conditions of the Creative Commons Attribution (CC BY) license (https://creativecommons.org/licenses/by/4.0/).

1. Introduction

Launch vehicles play a fundamental and critical role for human beings to explore space, and their technical capabilities determine the depth and extent of a country's space exploration activities. In recent years, new technologies such as full three-dimensional computer-aided design (CAD) and product data management (PDM) platforms have been widely used in the development of next-generation launch vehicles [1], which have improved the product development efficiency of enterprises. In the field of system design, however, the document-based systems engineering method is still too dominant to be adopted, which restricts the complex system design capabilities. Therefore, it is urgent to explore and apply a new model-based system design paradigm to improve the leap-forward development and manage innovative and complex launch vehicle system designs [2].

As a promising technology to manage system design better, MBSE is a hotspot technology for the complex system development paradigm [3]. MBSE takes system modeling as its core and supports the processes of system requirements capture, function analysis, architecture design, comprehensive performance simulation, etc. Moreover, the system model is integrated with the other models developed in the life cycle and works as the authority model, which is an important way to decrease misunderstanding and improve the design consistency for complex systems [4]. Based on lots of industry practice, several MBSE methodologies were proposed, such as MagicGrid, Arcadia, and Harmony SE. These methodologies are a little abstract at the top level, which is a little difficult to guide in engineering projects [5]. David Kalsow and etc. developed the CubeSat project as a reference model using the MBSE approach from 2012 to 2018 [6–10]. These papers introduced an MBSE roadmap for a satellite project [7]. A mission-specific satellite model was established [8], while a correspondent validation strategy was developed [9]. All the information is a little fuzzy to be projected to the true engineering system design for the whole project. Moreover, logical architecture and physical architecture are not clearly stated in the CubeSat reference model [11]. This paper aims to develop the first-stage separation system of launch vehicles using the MBSE approach from the viewpoint of a system engineer. Additionally, it clearly introduces the full process, including requirement analysis, logical architecture design, physical architecture design, the validation and verification of these designed architectures, and finally, the component requirements are obtained. Moreover, while the whole authority model is built, there are lots of scenarios to use this system model in, such as requirement change analysis and providing different views for different roles and applications.

2. Research Progress

The concept of model-based design, model-based development, and model-based definition started with CAD and CAE in the 1960s. With the development of the CAD model as the baseline for system engineering, the PDM platform was developed in the 1980s, and PDM was integrated with CAD to improve the configuration management in collaboration with the model-based design. However, the activity of the system design is still document based, which is not formalized to support future intelligent design. In 2007, the model-based system design concept was explored, applied, and developed. A new paradigm named MBSE has been developed to cope with the increasing complexity of space missions and the development of complex space products.

At present, there are three leading ecosystems of MBSE [12]. One ecosystem is the OMG ecosystem. This ecosystem is led by the INCOSE association, supported by the Dassault core platform, and is applied to leading enterprises such as NASA, Boeing, Lockheed Martin, etc., which develop lots of achievements such as standards, tools, and lots of applications [13]. Standards include SysML language, SysPhs, methodology, etc. Tools include Magicdraw, Rhaposody, EA, etc. [14]. Another ecosystem is the France ecosystem. This ecosystem is driven by the PolarSys organization, relying on the application practices of leading enterprises, such as Thales and Siemens, and is supported by the core platform of Siemens [15]. A series of standard achievements, such as Capella, have been formed, and the Thales ecology has been continuously expanded [16]. The third ecosystem is the ISO ecosystem. This ecosystem is driven by the ISO organization and relies on the application practice of some leading enterprises, such as NASA and General Motors, with OPM as the core tool support [17,18]. A series of standard achievements, such as the OPL language, are formed to continuously expand the OPM ecosystem [19].

The most important feature of the ecosystem is the trinity system, which combines academic research in institutions such as INCOSE, software tool development in tool suppliers such as Dassault, and engineering applications in industries such as NASA and Dassault Aviation. The three-in-one system enhances each other and grows together to become the industry benchmark in their respective fields [20].

The exploration of the MBSE development paradigm in China is still at a preliminary stage. Along with the successful practice of MBSE in other countries, the exploration of the MBSE development paradigm in China has also entered a prosperous state, among which the Aviation Industry Corporation of China (AVIC, Beijing, China), Tsinghua University(Beijing, China), Beihang University(Beijing, China), China Aerospace Science and Industry Corporation Limited (CASIC, Beijing, China) and China Aerospace Science and Technology Corporation (CASC, Beijing, China)have carried out a lot of practical exploration work, forming a series of theoretical methods, software tools, practical cases, and other achievements [21].

3. The First-Stage Separation System Design Based on MBSE

3.1. The First-Stage Separation System

The separation system is an important sub-system of the launch vehicle. Its main function is to separate the parts of the rocket that have completed their scheduled work during the flight. By reducing the mass that is useless in further flight, the mass characteristics of the launch vehicle can be improved, and its carrying capacity can be increased. The first-stage separation system is mainly composed of three parts according to the functions of stage separation, namely, the connecting and unlocking device, the impulse separation device, and the detonation device. As a part of the launch vehicle, the separation system is related to every system on the launch vehicle, such as the general design, electrical system, power system, and structural system. Therefore, the design of the separation system has a wide representation in the launch vehicle design process.

3.2. Modeling Framework of the Separation System Adopting MBSE Approach

The MBSE modeling approach is a collection of related processes, methods, and tools that support the systems engineering regulations in a model-driven environment. It is the top-level guidance for system design. The magic grid methodology is typically adopted and applied to NASA's spacecraft and Dassault Aviation's aircraft. However, the public magic grid methodology is abstracted and described from the modeling approach viewpoint. This paper tailors the magic grid methodology and proposes the model framework and modeling process for the first-stage separation system design viewpoint, as shown in Figure 1.

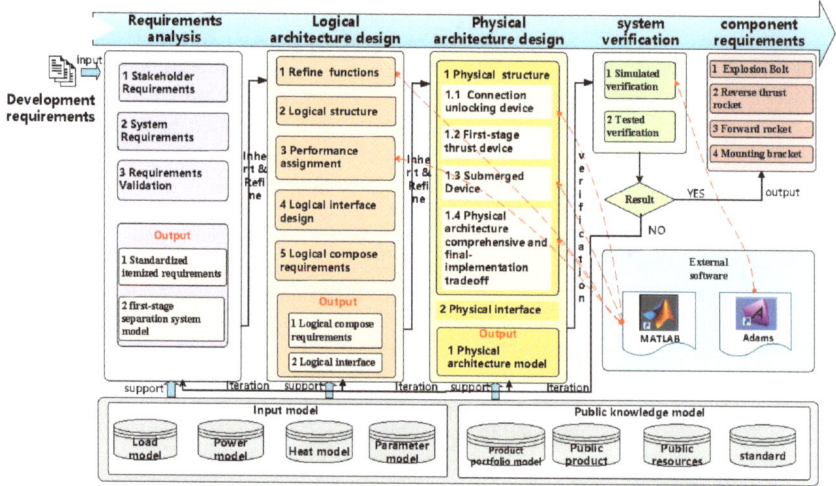

Figure 1. Model framework of the separation system.

3.3. Requirements Analysis of the First-Stage Separation System

Compared with traditional requirements analysis, requirements analysis by the MBSE approach formalizes the requirements expression, provides the stakeholder and full cycle viewpoint to check the requirements, builds the operational scenario to capture requirements, and flows down from the stakeholders' requirements to system requirements and then to component requirements by quantitative analysis, architecture trade-off, and concept design. All these processes are standardized and formalized through SysML language.

The process of requirements analysis and modeling based on MBSE is as follows: first, we should capture requirements in three ways. One way is to identify the stakeholders to check who proposed the requirements, as shown in Figure 2. The second way is to define the life cycle process to check when the requirements were proposed, as shown in Figure 3. The third way is to design the scenario to check where the requirement was proposed, as shown in Figure 4. The stakeholders' model, lifecycle process model, and use case model are baselined for requirement analysis. as shown in Figures 2–4. Secondly, once the requirements are fully captured, the requirements should be itemized and standardized, following the industry standard specification. Thirdly, any functional requirements should be checked, and the performance of the function should be evaluated. Through the functional requirements comparison matrix with performance requirements, there are some performance requirements figured out. For example, tall functions in the first-stage separation system need to work safely under complex force constraints such as the first-stage engine thrust, the second-stage engine thrust, the electrical connector pull-out force, and the trachea pull-out force, etc. In order to validate and verify the requirements of safety separation, the minimum separation gap should be quantitively defined according to the experience or design mechanism. Through the continuous iteration of the above process, 58 requirements for the first-stage separation system were finally formed and can be used as the design basis for the separation system, as shown in Figure 5.

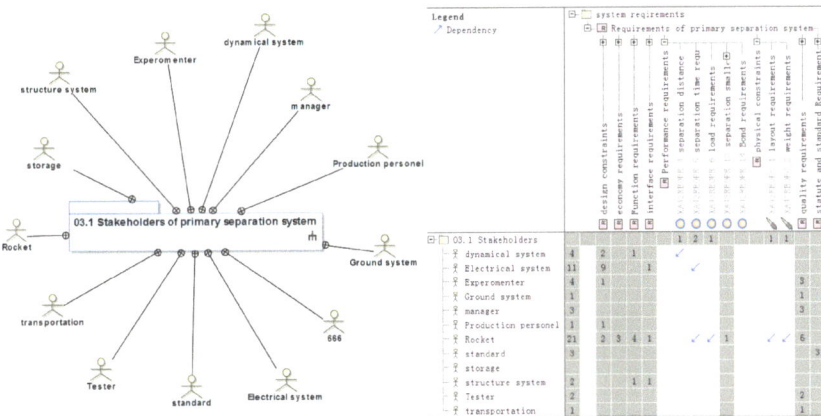

Figure 2. Who proposes requirements.

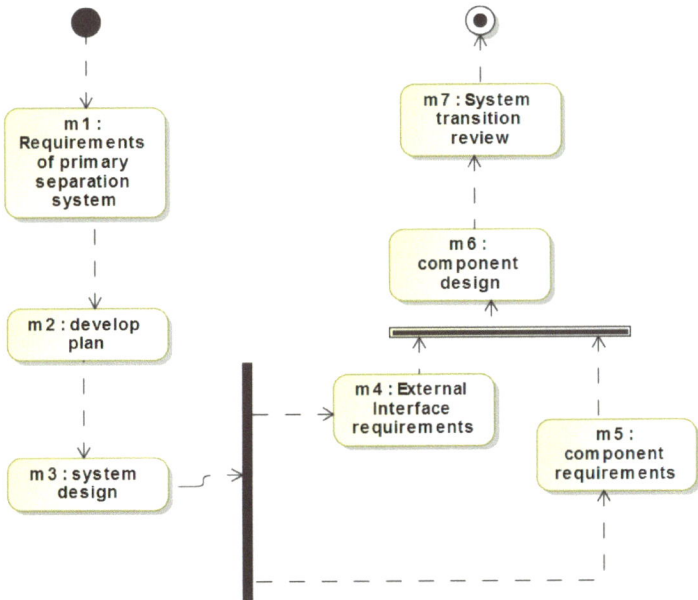

Figure 3. When requirements are proposed.

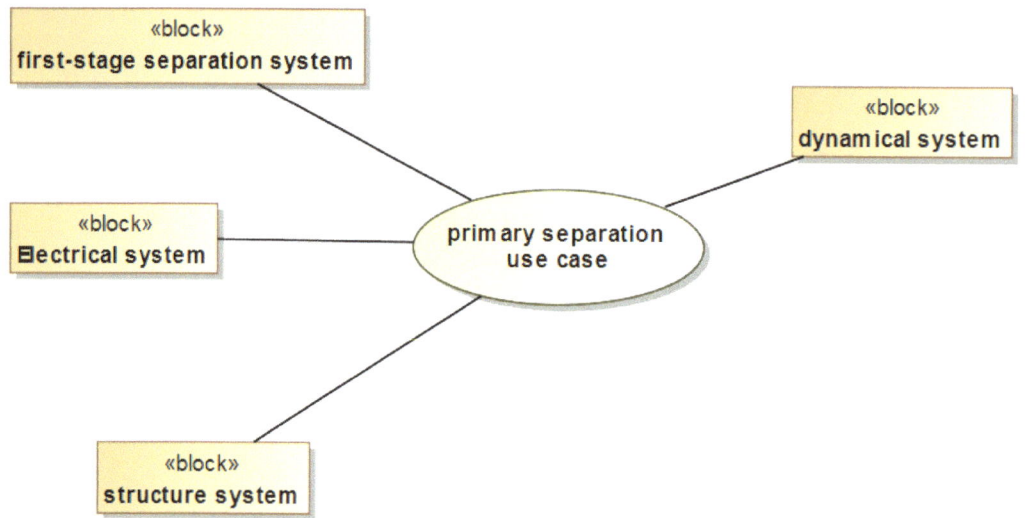

Figure 4. Where requirements are proposed.

#	Name
1	⊟ R Requirements of primary separation system
2	⊞ R quality requirements
17	⊞ R interface requirements
21	⊞ R statute and standard Requirements
25	⊞ R Function requirements
30	⊟ R Performance requirements
31	○ XA43REOPR_5 separation time requirements
32	○ XA43REOPR_6 load requirements
33	○ XA43REOPR_4 separation distance requirements
34	⊞ ○ XA43REOPR_13 separation smallest distance requirements
38	○ XA43REOPR_14 Bond requirements
39	⊞ R design constraints
52	⊞ R economy requirements
56	⊟ R physical constraints
57	✎ XA43REOPC_1 layout requirements
58	✎ XA43REOPC_2 weight requirements

Figure 5. Requirements for the first-stage separation system.

3.4. Logic Architecture Design of the Separation System for Key Parameters

The main purpose of the logic architecture design is to obtain the best system architecture among lots of alternatives. The traditional logical architecture design using the document-based systems engineering paradigm usually selects the empirical alternative and calculates the total impulse and number of rockets in multiple groups through simulation verification. The MBSE approach emphasizes the forward design process. The process is to analyze and refine the functions, perform the comprehensive clustering of functions to obtain components, and obtain the optimal composition of each component through index allocation. Then, the interaction and combination of functions are synthesized, and the interface design is performed. Finally, the optimal alternatives are obtained. For this system, the modeling process is as follows:

Firstly, the top-level functionality is obtained based on the system requirements, and the functional architecture is defined. When analyzing the operational scenario for the top-level functionality, there are two alternative logical architectures. One is separation once the other is separation twice. Based on the effective criteria, these two alternatives are traded off quantitatively, and then the proposal of separation once is better than the other, which is shown in Figures 6 and 7.

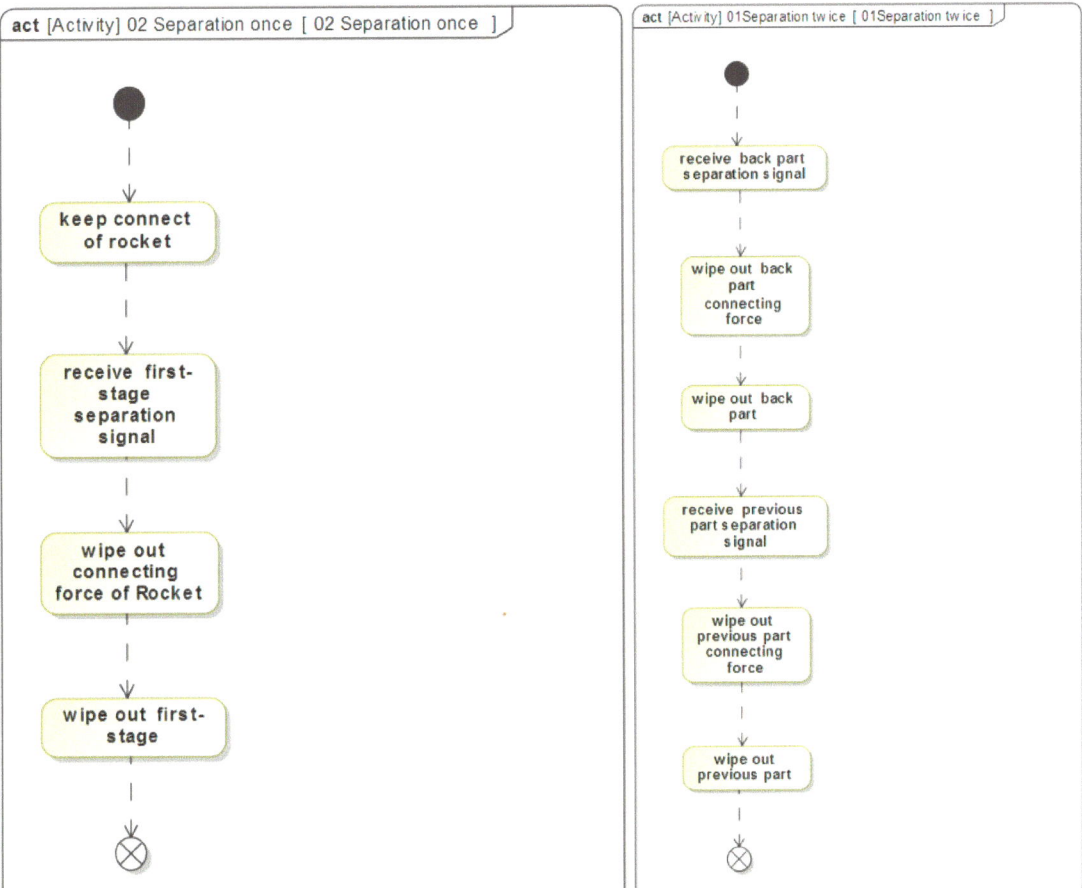

Figure 6. Two alternatives of separation once and separation twice.

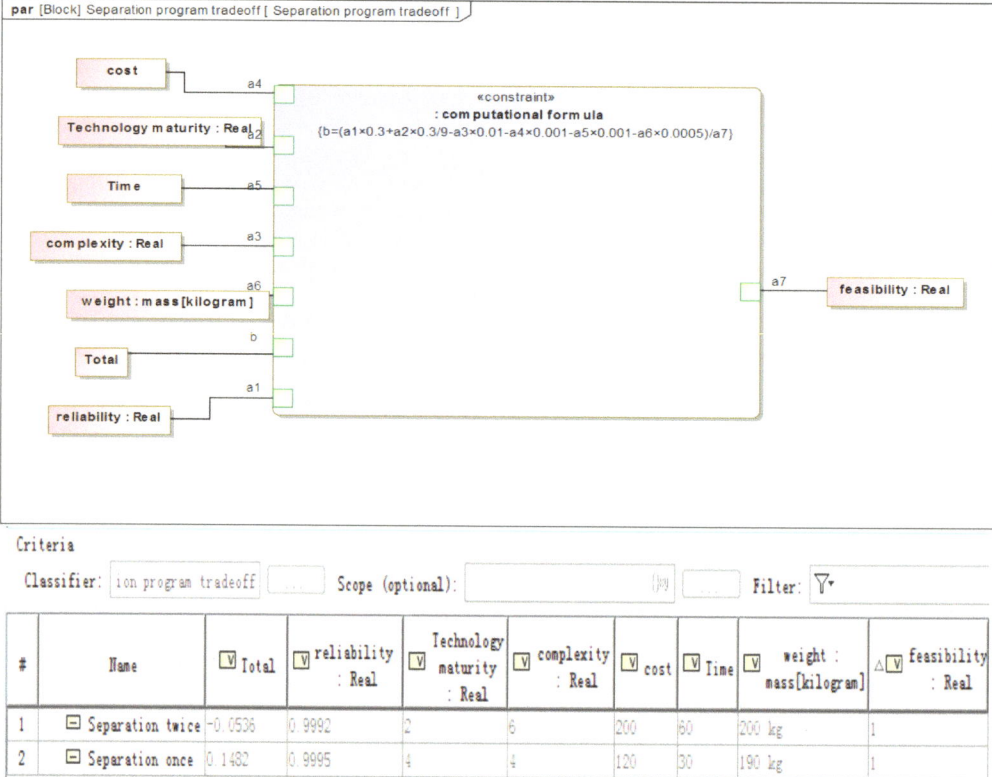

Figure 7. The tradeoffs result in functional architecture.

To design the first-stage separation system of the launch vehicle, we should identify the measure of effectiveness (MOE) from the system requirements. MOE includes the reliability, weight, cost, and safety separation. etc. Then, MOE should be transformed into the measure of performance (MOP), which should be baselined for the index allocation. For example, the safety separation should be refined by the separation time, separation gap, and so on. Finally, all the MOP should be allocated to technical performance measures, which could be allocated to logical components. The index flows down from MOE to MOP and from MOP to TPM is complicated because of the complicated system, as shown in Figure 8. Anyway, there are three core parameters, such as the connection forces to combine the first and second stages of the rocket, the thrust force to throw away the first sub-stage, and the push force to ensure that the secondary propellant sinks to the bottom. There are still several alternatives for the parameter flowing down from the top level to the low level. The system engineer would build the evaluation criteria and then select the best alternative through trade-off analysis. Finally, the core parameters are calculated, as shown in Figure 9. Additionally, the three core parameter values are as follows:

1. The connection device needs to provide a connection force greater than 940,286 N.
2. The first thrust device needs to provide a total axial impulse greater than 168,000 N/s.
3. The device, to ensure propellant sinking to the bottom, should provide an axial thrust force greater than 293 N.

By integrating specific functions and assigning them to the logical entities and logical interfaces, the logical architecture of the first-stage separation system can be built, which is shown in Figure 10, and all of the indexes have been allocated to each logical component.

As a result, the interfaces and parameters of the logical architecture will be determined. Hence, the logical architecture is shown in Figure 10.

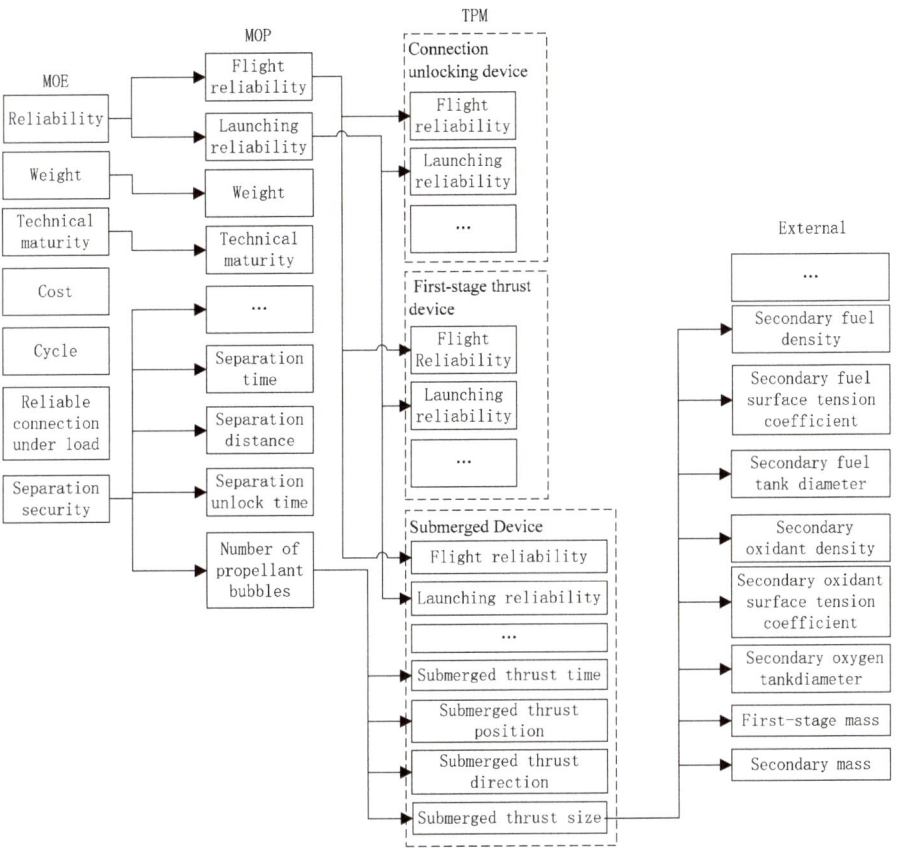

Figure 8. Flow down from MOEs to MOP to TPMs.

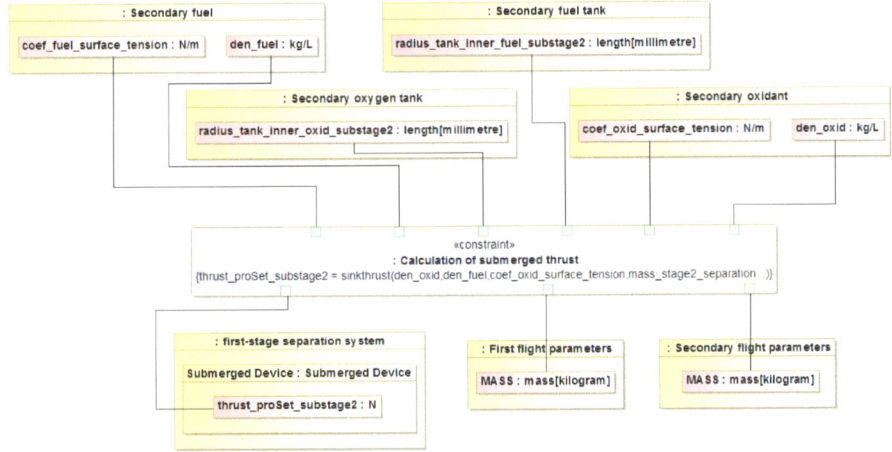

Figure 9. Calculation of submerged thrust.

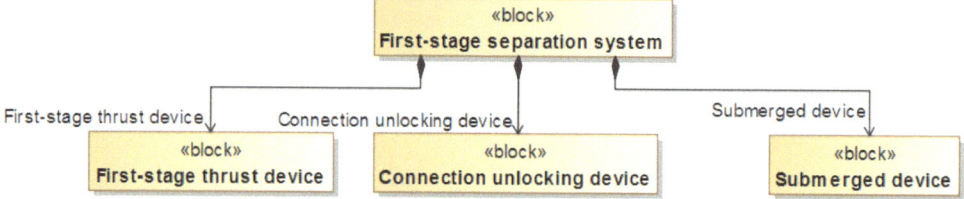

Figure 10. Logical compose of first-stage separation system.

3.5. Physical Architecture Design of the First-Stage Separation System Based on Model Selection Analysis

The physical architecture design of the first-stage separation system is a set of product units that achieve the required functionality and performance of the system within the performance constraints specified by the logical architecture. The physical architecture design is, therefore, the process of implementing a solution based on the logical architecture design.

Compared with the traditional separation system design, the physical architecture design of the first-stage separation system based on MBSE has a similar work content. It mainly completes the trade-off analysis of the physical implementation options, as well as the product selection analysis, and performs the comprehensive analysis of the physical composition to form the final physical architecture design results. Finally, the final selection results are taken as the basis for forming the final physical architecture design result and creating the interface model, and organizing and summarizing the interface information.

The specific modeling process is as follows: create a physical architecture model according to the logical architecture and establish a model of the inheritance relationship from the physical architecture to the logical architecture. First, a trade-off analysis of the implementation of the separation forces is carried out. By the quantitative trade-off analysis of the pneumatic connection unlocking device, explosive bolts, and linear connection unlocking device, the explosive bolts are chosen as the final solution. Then, the trade-off analysis of the types of explosive bolts is carried out, including the BLS-300C24-1 explosive bolt and BLS-300C24-2 explosive bolt. After quantitative trade-off analysis, BLS-300C24-1 is selected. After a series of analyses, the first-stage separation system physical architecture scheme is determined, as shown in Figure 11. Finally, the physical architecture design is completed by defining the physical interface model.

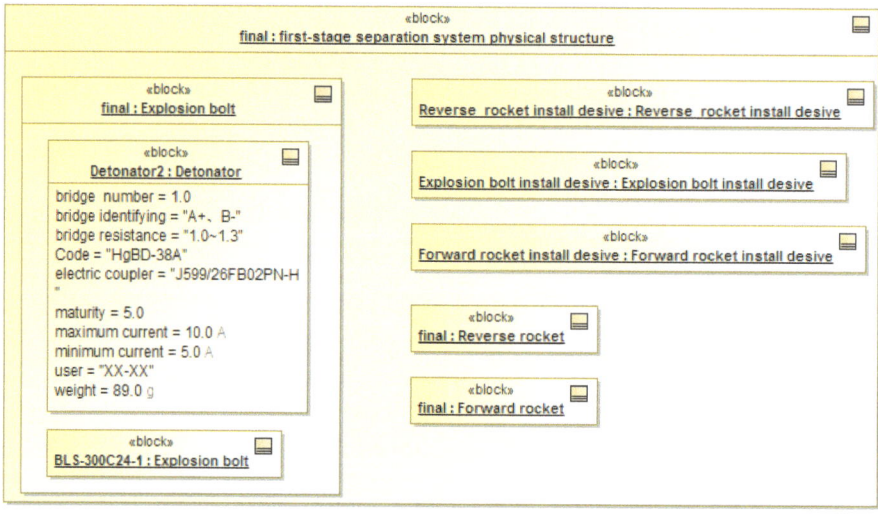

Figure 11. Physical architecture of first-stage separation system.

3.6. Validation and Verification of the First-Stage Separation System Based on the Object-Oriented Method

Compared with the traditional design method, this model-based first-stage separation system verification adopts "object-oriented concepts" that take the physical architecture as the test object, and validates objects from the functional verification, performance index verification, and interface verification.

In the functional verification process, the interaction between the components and the external system in the main functional scenario of the system is established using the SysML sequence diagram to complete the traditional time sequence design process. The mapping between the functions and requirements is described by establishing a relationship matrix (RM) to check that each function meets the requirements.

The verification of the performance indicators is the core element of the validation of the separation system. The key performance parameters for the separation process, including the separation distance, separation gap, and separation speed, need to be analyzed by ADAMS for rigid body dynamics. In this case, the standard data files and interface models are used to integrate the system model with the CAE simulation model.

3.7. Component Requirements for the First-Stage Separation System

After the architecture of the separation system is verified, the functions, indicators, interface models, and itemized requirements of each component can be generated from the system model. The component model and itemized requirements are baselined, and then the component designer takes it as the design input to move to the next step.

4. The Application Advantages of Using System Model in Engineering Activity

4.1. Requirement Changes Analysis Using the System Model

With t, the integrated system model, we can build the requirement change impact analysis process to support the system design, as shown in Figure 12. This integrated system model is split into two models. One model is the master model, and the other is the branch model. The branch model supports the requirement change analysis. Once the requirement is changed, the change impact domain analysis map is figured out through the meta-chain in the branch model. It is easier and quicker to locate the chain between the changed requirements and the related model elements, such as the architecture design model, verification model, and so on. Then, the related models are redesigned. Moreover, the verification model is simulated again, as shown in Figure 13. Compared with the master model and branch model, the change report is generated for the system engineer. Based on these two results analyses, the changed requirements and new system architecture are decided to be acceptable or not. If the change is accepted, the branch model will be merged with the master model, as shown in Figures 12 and 13.

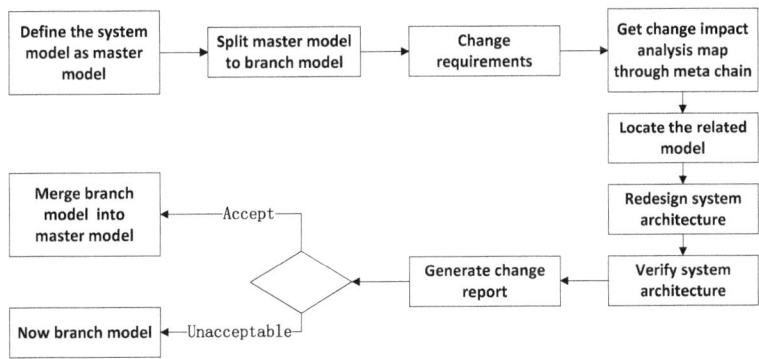

Figure 12. Requirement change impact analysis process.

#	Name	:first-stage separation system.separation time : time[second]	:first-stage separation system.separation velocity : m/s	:first-stage separation system.separation distance : length[metre]	:first-stage separation system.time of come away : time[second]	:first-stage separation system.First-stage thrust device.total impulse : N·s
1	provide separation force 11	4.5 s	4.9951 m/s	3.9169 m	1.2149 s	152000 N·s
2	provide separation force-Select	4.5 s	5.5769 m/s	4.8479 m	1.1299 s	168000 N·s
3	provide separation force1	4.5 s	4.6576 m/s	5.2765 m	1.3051 s	168000 N·s

Figure 13. The change in key parameters.

4.2. Multi-Views Based on Authority Model

The integrated system model is a full-feature model which includes the system design process and system definition. Different stakeholders from different viewpoints can obtain various design view models from this system model. In the design process of the separation system, the interface control document, component requirements, flight procedures, system layout, etc., can be generated from the system model automatically, as shown in Figures 14–17.

#	Part A	Port A	Port A Features	Item Flow
1	P Forward rocket install de...	Bracket : install interface		
2	P Forward rocket : Forward rocket	inout Installation interface...	F inout force : force V direction V size	
3	P Explosion bolt : Explosion bolt	inout HgBD-38A : Detonator...	F in ignition : signal F inout install : force V bridge number : Real V bridge identifying : String V bridge resistance : String V Code : String V electric coupler : String V maturity : Real V maximum current : electric curr V minimum current : electric curr V user : String V weight : mass[gram]	signal
4	P Explosion bolt : Explosion bolt	inout Explosion bolt inter...	F inout force : force V direction V size	

Figure 14. Interface control documents.

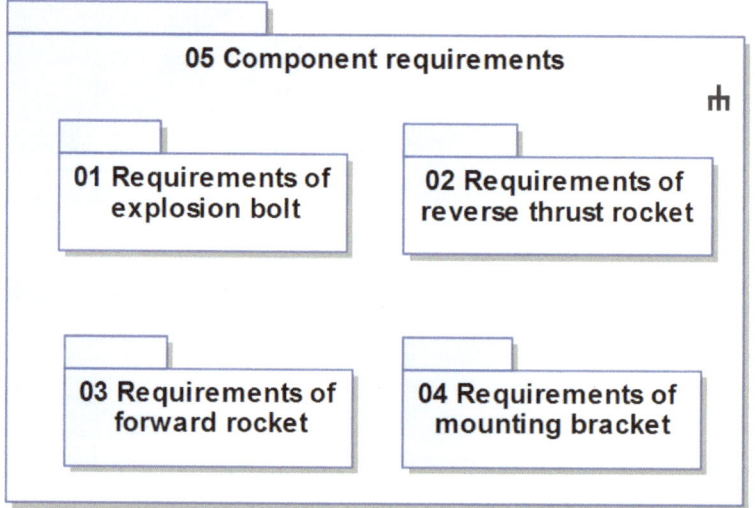

Figure 15. Component requirements.

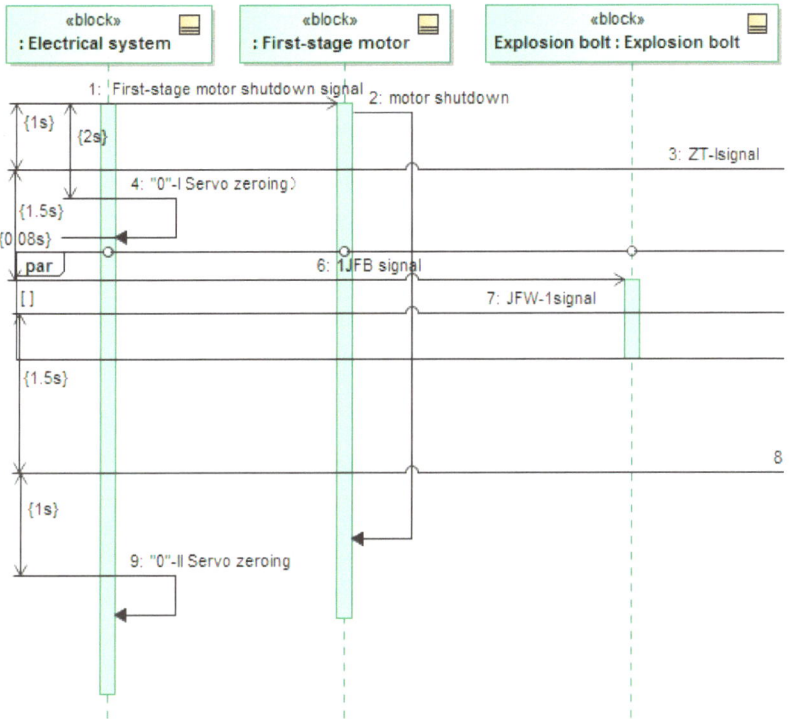

Figure 16. Flight procedures.

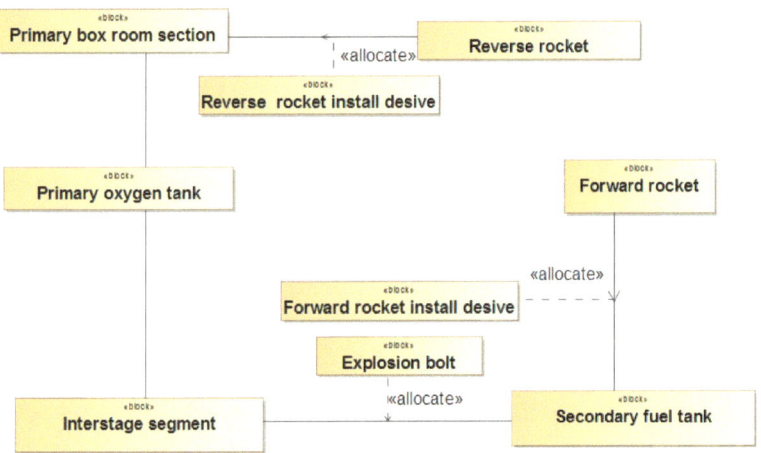

Figure 17. System layout.

5. Conclusions

In this paper, the MBSE approach is applied to design the first-stage separation system. This paper tailored the magic grid methodology and proposed the modeling framework, which could be suitable for similar dynamic and transient systems, such as the rocket system. A set of modeling for the requirements of the flow down is demonstrated; the logical architecture and physical architecture are designed and follows the framework. Moreover, the system design is verified and validated by integrating the Adams simulation model, which makes the system model as the authority model. In addition, using this system model in a requirement change analysis scenario shows that the MBSE approach has obvious advantages for system engineering. Of course, it is more rapid to provide multi-view information for other users, which is better than a document-based approach.

Author Contributions: W.Z. and Z.L. completed preliminary research; Z.L. conceived and wrote the paper; X.L., Y.J., Q.W. and R.H. supervised the overall work and reviewed it. All authors have read and agreed to the published version of the manuscript.

Funding: This work was funded by the National Key R&D Program of China under Grant 2020YFB1708100 (system modeling theory for complex product development integrated the process of design, manufacture and service), and the "14th Five-Year Plan" Major program of Advance research on Civil Aerospace Technology under Grant D020101 (design and verification technology of aerospace transportation system under MBSE paradigm).

Institutional Review Board Statement: Not applicable.

Informed Consent Statement: Not applicable.

Data Availability Statement: Not applicable.

Conflicts of Interest: The authors declare no conflict of interest.

References

1. Redmon, J.; Shirley, M.; Kinard, P. A Large-Scale Design Integration Approach Developed in Conjunction with the Ares Launch Vehicle Program. In Proceedings of the 50th AIAA Aerospace Sciences Meeting including the New Horizons Forum and Aerospace Exposition, Nashville, TN, USA, 9–12 January 2012; p. 881.
2. Henderson, K.; Salado, A. Value and benefits of model-based systems engineering (MBSE): Evidence from the literature. *Syst. Eng.* **2021**, *24*, 51–66. [CrossRef]
3. Wade, J.; Verma, D.; McDermott, T.; Boehm, B. The SERC 5-year technical plan: Designing the future of Systems Engineering Research. In Proceedings of the Ninth International Conference on Complex Systems Design & Management, Paris, France, 18–19 December 2018; Volume 1, p. 241.
4. Mann, C.J.H. A practical guide to SysML: The systems modeling language. *Kybernetes* **2009**, *38*, 989–994. [CrossRef]
5. Huldt, T.; Stenius, I. State-of-practice survey of model-based systems engineering. *Syst. Eng.* **2019**, *22*, 134–145. [CrossRef]
6. Kaslow, D.; Ayres, B.; Cahill, P.T.; Hart, L. A model-based systems engineering approach for technical measurement with application to a CubeSat. In Proceedings of the 2018 IEEE Aerospace Conference, Big Sky, MT, USA, 3–10 March 2018; pp. 1–10.
7. Kaslow, D.; Anderson, L.; Asundi, S.; Ayres, B.; Iwata, C.; Shiotani, B.; Thompson, R. Developing a cubesat model-based system engineering (mbse) reference model-interim status. In Proceedings of the 2015 IEEE Aerospace Conference, Big Sky, MT, USA, 7–14 March 2015; pp. 1–16.
8. Kaslow, D.; Ayres, B.; Cahill, P.T.; Hart, L.; Yntema, R. Developing a CubeSat Model-Based Systems Engineering (MBSE) Reference Model—Interim Status #3. In Proceedings of the IEEE Aerospace Conference, Big Sky, MT, USA, 4–11 March 2017.
9. Kaslow, D.; Ayres, B.; Cahill, P.T.; Hart, L.; Levi, A.G.; Croney, C. Developing an mbse cubesat reference model–interim status# 4. In Proceedings of the 2018 AIAA SPACE and Astronautics Forum and Exposition, Orlando, FL, USA, 17–19 September 2018; p. 5328.
10. Kaslow, D.; Madni, A.M. *Validation and Verification of MBSE-Compliant CubeSat Reference Model*; Disciplinary Convergence in Systems Engineering Research; Springer: Cham, Switzerland, 2018; pp. 381–393.
11. Kaslow, D.; Cahill, P.T.; Ayres, B. Development and application of the CubeSat system reference model. In Proceedings of the 2020 IEEE Aerospace Conference, Big Sky, MT, USA, 7–14 March 2020; pp. 1–15.
12. De Saqui-Sannes, P.; Vingerhoeds, R.A.; Garion, C.; Thirioux, X. A taxonomy of MBSE approaches by languages, tools and methods. *IEEE Access* **2022**, *10*, 120936–120950. [CrossRef]
13. Holladay, J.B.; Knizhnik, J.; Weiland, K.J.; Stein, A.; Sanders, T.; Schwindt, P. MBSE Infusion and Modernization Initiative (MIAMI):"Hot" benefits for real NASA applications. In Proceedings of the 2019 IEEE Aerospace Conference, Big Sky, MT, USA, 2–9 March 2019; pp. 1–14.
14. Estefan, J.A. Survey of model-based systems engineering (MBSE) methodologies. *Incose MBSE Focus Group* **2007**, *25*, 1–12.
15. Pagnanelli, C.A.G.; Carson, R.S.; Palmer, J.R.; Crow, M.E.; Sheeley, B.J. 4.5. 3 Model-Based Systems Engineering in an Integrated Environment. In Proceedings of the INCOSE International Symposium, Rome, Italy, 9–12 July 2012; Volume 22, pp. 633–649.
16. Roques, P. MBSE with the ARCADIA Method and the Capella Tool. In Proceedings of the 8th European Congress on Embedded Real Time Software and Systems (ERTS 2016), Toulouse, France, 27–29 January 2016.
17. Forte, S.; Göbel, J.C.; Dickopf, T. System of systems lifecycle engineering approach integrating smart product and service ecosystems. *Proc. Des. Soc.* **2021**, *1*, 2911–2920. [CrossRef]
18. Parrott, E.; Trase, K.; Green, R.; Varga, D.; Powell, J. *NASA GRC MBSE Implementation Status*; GSFC MBSE Workshop: Washington, DC, USA, 17 February 2016; No. GRC-E-DAA-TN29928.

19. Hause, M.C.; Day, R.L. Frenemies: Opm and SysML together in an MBSE model. In Proceedings of the INCOSE International Symposium, Orlando, FL, USA, 20–25 July 2019; Volume 29, pp. 691–706.
20. Schindel, W.D. Realizing the Promise of Digital Engineering: Planning, Implementing, and Evolving the Ecosystem. In Proceedings of the INCOSE International Symposium, Detroit, MI, USA, 25–30 June 2022; Volume 32, pp. 1114–1130.
21. Chang, S.; Wang, Y. Civil aircraft IVHM system analysis using model based system engineering. In Proceedings of the 2017 Second International Conference on Reliability Systems Engineering (ICRSE), Beijing, China, 10–12 July 2017; pp. 1–5.

MDPI AG
Grosspeteranlage 5
4052 Basel
Switzerland
Tel.: +41 61 683 77 34

Actuators Editorial Office
E-mail: actuators@mdpi.com
www.mdpi.com/journal/actuators

Disclaimer/Publisher's Note: The title and front matter of this reprint are at the discretion of the Guest Editors. The publisher is not responsible for their content or any associated concerns. The statements, opinions and data contained in all individual articles are solely those of the individual Editors and contributors and not of MDPI. MDPI disclaims responsibility for any injury to people or property resulting from any ideas, methods, instructions or products referred to in the content.

www.ingramcontent.com/pod-product-compliance
Lightning Source LLC
LaVergne TN
LVHW072320090526
838202LV00019B/2321